MODERN PHYSICS

(For the Students of B.E., B.Tech., B.Sc. & A.M.I.E.)

B. L. THERAJA

S. CHAND PUBLISHING

S Chand And Company Limited

(ISO 9001 Certified Company)

S Chand And Company Limited
(ISO 9001 Certified Company)
Head Office: D-92, Sector–2, Noida – 201301, U.P. (India), Ph. 91-120-4682700
Registered Office: A-27, 2nd Floor, Mohan Co-operative Industrial Estate, New Delhi – 110 044, Phone: 011-49731800
www.**schandpublishing.com**; e-mail: **info@schandpublishing.com**

Marketing Offices:

Chennai : Ph: 23632120; chennai@schandpublishing.com
Guwahati : Ph: 2738811, 2735640; guwahati@schandpublishing.com
Hyderabad : Ph: 40186018; hyderabad@schandpublishing.com
Jalandhar : Ph: 4645630; jalandhar@schandpublishing.com
Kolkata : Ph: 23357458, 23353914; kolkata@schandpublishing.com
Lucknow : Ph: 4003633; lucknow@schandpublishing.com
Mumbai : Ph: 25000297; mumbai@schandpublishing.com
Patna : Ph: 2260011; patna@schandpublishing.com

First Edition 1995
Subsequent Editions and Reprints 1976, 78, 79, 80, 81, 82, 83, 84, 85, 86, 87, 89, 90, 91, 92, 93, 93, 95, 96, 98, 99
Thoroughly Revised and Enlarged Edition 2008
Reprints 2009, 2012, 2013, 2014, 2015, 2016, 2018 (Twice), 2019, 2021, 2022

Reprint 2024

ISBN: 978-81-219-0163-5 **Product Code:** H3MOP68PHYS10ENAP08O

PRINTED IN INDIA

By Vikas Publishing House Private Limited, Plot 20/4, Site-IV, Industrial Area Sahibabad, Ghaziabad – 201 010 and Published by S Chand And Company Limited, A-27, 2nd Floor, Mohan Co-operative Industrial Estate, New Delhi – 110 044.

Dedicated to the loving
memory of my brother-in-law
Shri Hans Bagai
who led a life of truth and service

Preface to the Sixteenth Edition

The author feels great pleasure in presenting the updated sixteenth edition of this popular book which has been in the service of its esteemed readers since 1975. Opportunity has been taken to include solutions of A.M.I.E. papers. Some of the latest questions from the B.E., B.Sc. (Engg.) and B.Sc. (General) examinations of various Indian Universities have also been added. Special feature of the book is that all the diagrams are redrawn & made by computer. The size of the book is also changed as per the present trend of various popular textbooks. I must thank my publishers, particularly Mrs. Nirmala Gupta, Chairperson to present the book in this nicely form.

I am also grateful to Sri Rupesh Gupta, M.Sc. (Physics), Subject Editor for his Editorial Contribution in the book.

In the end, I would like to record my gratitude to all my readers–both here and abroad–for their kind and consistent patronage and would request them to send in their valuable suggestions for the improvement of the book.

B. L. Theraja

Techouse
8/80 Punjabi Bagh
New Delhi-110026

Preface to the First Edition

The book is based primarily on the syllabus for the recently-introduced paper of Modern Physics under the new scheme of A.M.I.E. examination applicable with effect from May 1974. This fascinating subject deals with the invisible silent world of atoms, atomic nuclei and elementary particles where customary physical laws almost cease to operate. In this eerie world, particles lose their dimensions and acquire the properties of waves and again waves begin to behave as particles ! Electrons and other fundamental particles can pass through impenetrable barriers or they vanish altogether leaving only photons in their place ! Consequently, full understanding of the laws of this world of the ultrasmall requires the creation of bizarre conceptions beyond the scope of pictorial imagery.

Keenly conscious of the conceptual difficulties involved in fully understanding (particularly by engineering students who are trained to handle matter in bulk) the ultimate truth about atomic structure, crystal lattices and their description by Miller indices, mysteries of nuclear disintegration and nuclear fission and fusion processes etc., author has put in lot of labour to make the description of such intricate atomic phenomena as lucid and limpid as possible.

The book contains a large number of both solved and unsolved problems. Although many of these problems amount to little more than numerical substitution, author's experience has shown that they give the student lot of self-confidence and provide him useful ideas about the order of the magnitude of the atomic entities.

It gives me great pleasure to thank my publishers for expeditious printing of the book through their press manager Sh. Ravi Gupta, B.E. (Mech). I am also indbeted to Sh. R.C. Kumar, MBA, for the highly aesthetic and imaginative cover design proposed by him.

B.L. Theraja

Glen Gowan
Mussoorie (U.P.)

CONTENTS

Important Derived Units in SI System of Units

The *SI* system of units is an extension of rationalized *MKSA* (metre-kilogram-second-ampere) system which has four basic or fundamental units. The *SI* system has six basic units : **metre, kilogram, second, ampere, kelvin** and **candela**. All other units can be expressed in terms of these units and hence are known as derived units. Some of the important derived units required for the purpose of studying this book are as under :

1. **Mass.** It is the quantity of matter contained in a body.

   Its absolute unit is **kilogram (kg)**

   The bigger units commonly used are quintal and tonne. 1 quintal = 100 kg and 1 tonne = 1000 kg = 10 quintal

2. **Force.** Its unit is **newton (N).**

   It may be derived from the relation $F = ma$

   If $m = 1$ kg, $a = 1$ m/s^2, then $F = 1$ N

   Hence, one newton may be defined as that force which can give an acceleration of 1 m/s^2 to a body of mass 1 kg.

3. **Weight.** It is the force with which a body is pulled by earth downwards. Its units are the same as those of force given above.

4. **Work.** Its unit is **joule.** It can be defined from the equation Work = $F \times S$

   If $F = 1$ N, $S = 1$ m, work = 1 joule (J). Hence one joule of work is defined as the work done by a force of 1 N in moving a body through a distance of 1 m along its direction of application.

5. **Special unit of work.** In Modern Physics, a special unit of work is commonly employed. It is called **electron volts (eV).**

   $$1 \text{ eV} = 1.6 \times 10^{-19} \text{ J}$$

   $$1 \text{ MeV} = 1.6 \times 10^{-13} \text{ J}$$

   It will be seen that electron-volt is extremely small.

6. **Power.** Its unit is **watt** (W). One watt represents work rate of 1 joule/second.

   $$1 \text{ W} = 1 \text{ J/s}$$

   (*i*) Power in watts = F × v = force × velocity where F is in N and v in m/s.

   Also power in watts = Tω

   where T is in N—m and ω in radian/second.

7. **Heat Energy.** It is measured in **kilo-calorie** (kcal). One kilo-calorie is the heat energy required to heat 1 kg of water through 1 °K (or 1 °C).

ELECTRIC AND MAGNETIC FIELDS 1

Coulomb's Law of Electric Force — Electric Field — Intensity of Electric Field — Potential at a Point — Field Around Conductors — Electron-volt — Total Energy of a Revolving Charge — Laws of Magnetic Force — Magnetic Field Strength or Magnetising Force — Magnetic Flux Density — Force on a Current-carrying Conductor — Motion of a Charged Particle in Transverse Magnetic Field — Highlights.

1.1. Coulomb's Law of Electric Force

According to this law, the force between two *point* charges

(*i*) is directly proportional to the product of their strengths

(*ii*) varies inversely as the square of the distance between them

(*iii*) is inversely proportional to the absolute permittivity (ε) of the surrounding medium.

Mathematically speaking,

$$F \propto \frac{Q_1 Q_2}{\varepsilon d^2}$$

or

$$F = k.\frac{Q_1 Q_2}{\varepsilon d^2}$$

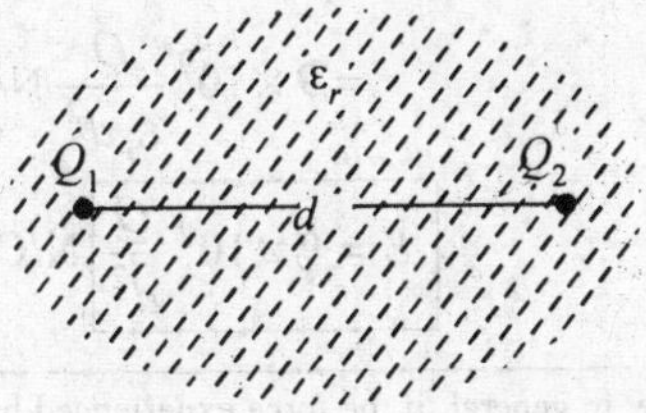

Fig. 1.1

where k is the constant of proportionality whose value depends on the system of units employed. In the M.K.S.A. as well as SI systems, $k = 1/4\pi$.

$$\therefore \quad F = \frac{Q_1 Q_2}{4\pi\varepsilon d^2} = \frac{Q_1 Q_2}{4\pi\varepsilon_0\varepsilon_r d^2}$$

Force is in newtons (N) provided Q_1, Q_2 are in coulombs (C)

d is in metres (m)

ε is in farad/metre (F/m)

$$\text{Now} \quad \frac{1}{4\pi\varepsilon_0} = \frac{1}{4\pi \times 8{\cdot}854 \times 10^{-12}} = 9 \times 10^9$$

$$\therefore \quad F = 9 \times 10^9 \frac{Q_1 Q_2}{\varepsilon_r d^2} \quad \text{N – in medium}$$

$$\text{or} \quad \boxed{F = 9 \times 10^9 \frac{Q_1 Q_2}{d^2}} \quad \text{N – in vacuum}$$

> **Note.** The charge carried by an electron is negative and has a numerical value of $1{\cdot}602 \times 10^{-19}$ C. Obviously, one coulomb equals the charge carried by $1.602 \times 10^{-19} = 6{\cdot}24 \times 10^{18}$ electrons. If a body has a negative charge of one coulomb, it means that it has an *excess* of $6{\cdot}24 \times 10^{18}$ electrons. Similarly, a body having a positive charge of one coulomb has a *deficiency* of $6{\cdot}24 \times 10^{18}$ electrons.

1.2. Electric Field

In the space around every charged body, there is an invisible something which affects other charged bodies. We call this something as *electric field.* It is usually represented by lines of force.

1.3. Intensity of Electric Field (E)

It may be defined in the following two ways :

(*i*) *The intensity of electric field at any point around a charged body is numerically equal to the force experienced by a unit positive charge placed at that point.*

Suppose it is required to find the value of E at point A (Fig. 1.2). For that purpose, imagine a positive charge of one coulomb placed at A. Whatever the force experienced by this charge, it represents the electric intensity at that point. The value of this force, as found from Coulomb's law, is

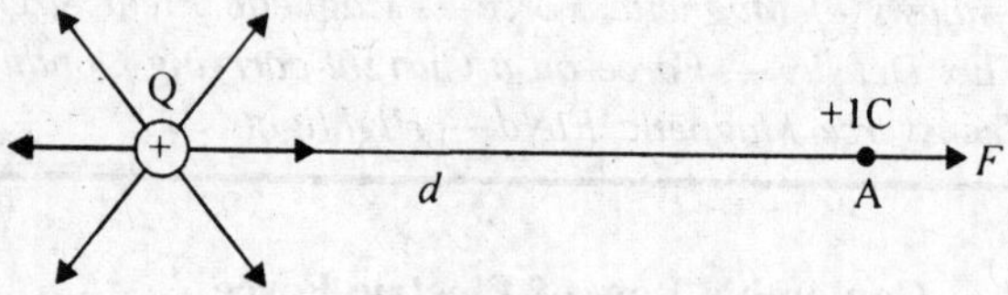

Fig. 1.2

$$F = \frac{Q \times 1}{4\pi\varepsilon d^2} = \frac{Q}{4\pi\varepsilon d^2}\ \text{N}$$

$$\therefore \quad E = \frac{Q}{4\pi\varepsilon_0\varepsilon_r d^2}\ \text{N/C}$$

$$= 9 \times 10^9 \frac{Q}{\varepsilon_r d^2}\ \text{N/C} \qquad \text{– in medium}$$

$$\therefore \quad \boxed{E = 9 \times 10^9 \frac{Q}{d^2}\ \text{N/C}} \qquad \text{– in vacuum}$$

> **Note.** In general, if the force experienced by a charge of Q coulomb when placed in an electric field is F newton, then electric intensity (which is force per unit charge) is given by
> $$E = F/Q \text{ or } F = QE$$

(*ii*) *Electric intensity* (E) *at any point around a charge is given by the potential gradient at that point.*

Suppose there are two very close points having a separation of dx and potential difference of dV, then potential gradient is dV/dx.

By definition,

$$\boxed{E = \frac{dV}{dx}} \text{ volt/metre}$$

In Fig. 1.3 are shown two plates M and N having a potential difference of 12 V. Obviously, this p.d. of 12 V has *uniformly* dropped over a distance of 0.6 m.

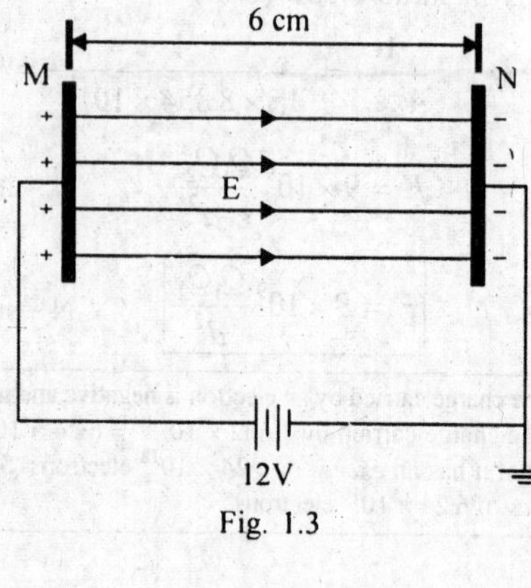

Fig. 1.3

$$\therefore \quad \text{potential gradient} = \frac{dV}{dx}$$

$$= \frac{12}{0{\cdot}06} = 200 \text{ V/m}$$

$$\therefore \quad E = 200 \text{ V/m}$$

In general if V volts is the p.d. between two such plates which are separated by a distance of d metres, then

$$\boxed{E = \frac{V}{d}} \text{ volt/metre}$$

1.4. Potential at a Point

The potential at any point around a charged body is *numerically equal to the work done in shifting a positive charge of one coulomb from infinity to that point.*

As shown in Fig. 1.4, suppose it is required to find the potential of point A distant d metres from a point charge of Q coulomb.

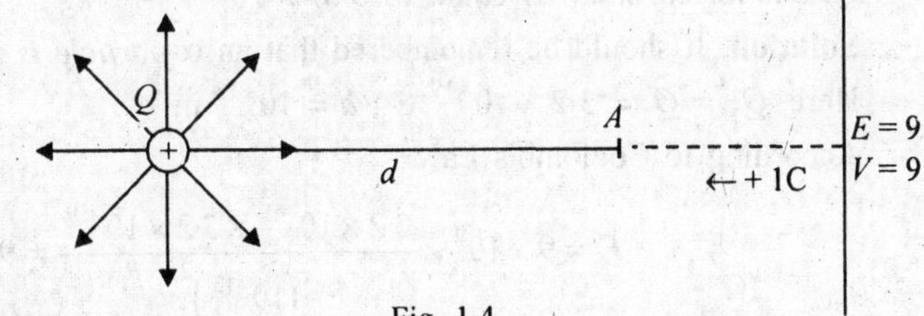

Fig. 1.4

For this purpose, we will have to find the amount of work which would be needed to shift a positive charge of one coulomb from infinity to point A against the ever-increasing repulsive force due to Q. It can be easily shown that work done is

$$W = \frac{Q}{4\pi\varepsilon d} \text{ joules}$$

By definition, this represents the potential of point A.

$$\therefore \quad V = \frac{Q}{4\pi\varepsilon_0\varepsilon_r d} \text{ J/C or volt}$$

$$\therefore \quad V = 9 \times 10^9 \frac{Q}{\varepsilon_r d} \text{ volt} \qquad \text{– in medium}$$

$$\boxed{V = 9 \times 10^9 \frac{Q}{d}} \text{ volt} \qquad \text{– in vacuum}$$

Note. If a charge of Q is shifted against a potential difference of V, then work done or energy spent $W = QV$ joules.

1.5. Field Around Conductors

The following points about the electric fields in and around conductors are worth noting :

(*i*) The field is zero inside a conductor.

(*ii*) The field at the surface of a conductor is everywhere normal to the surface.

(*iii*) The charge on a conductor resides entirely on its outer surface.

(*iv*) The potential anywhere inside a charged sphere is the same as at its surface.

(*v*) The field and potential *outside* a charged conducting sphere are the same as if the charge on it were concentrated at its centre.

1.6. Electron-volt (eV)

It is the unit of energy commonly used in Atomic Physics. It is defined as *the energy gained by an electron when it is accelerated through a potential difference of one volt.*

$$1 \text{ eV} = \text{electron charge in coulomb} \times \text{volt}$$
$$= 1{\cdot}6 \times 10^{-19} \times 1 = 1{\cdot}6 \times 10^{-19} \text{ joule}$$

Bigger units employed are :

1 mega-electron-volt (MeV) = 10^4 eV

∴ 1 MeV = $1{\cdot}6 \times 10^{-13}$ J

1 giga-electron-volt (GeV) = 10^9 eV

∴ 1 GeV = $1{\cdot}6 \times 10^{-10}$ J

Example 1.1. *Calculate the electrostatic force of repulsion between two α-particles when at a distance of 10^{-13} m from each other. Charge of an α-particle is $3{\cdot}2 \times 10^{-19}$ C. If mass of each α-particle is $6{\cdot}68 \times 10^{-27}$ kg, compare this force with the gravitational force between them. Take the gravitational constant as equal to $6{\cdot}57 \times 10^{-11}$ N-m²/kg².*

Solution. It should be remembered that an *α-particle is the nucleus of a helium atom.*

Here, $Q_1 = Q_2 = 3{\cdot}2 \times 10^{-19}$ C ; $d = 10^{-13}$ m

According to Coulomb's Law

$$F = 9 \times 10^9 \times \frac{3{\cdot}2 \times 10^{-19} \times 3{\cdot}2 \times 10^{-19}}{\left(10^{-13}\right)^2} = 9{\cdot}2 \times 10^{-2}\,\text{N}$$

The force of gravitational attraction between the two particles is given by

$$F = G.\frac{m_1 m_2}{d^2} = 6.67 \times 10^{-11} \times \frac{(6.68 \times 10^{-27})^2}{(10^{-13})^2}$$

$$= 2{\cdot}97 \times 10^{-37}\,\text{N}$$

Obviously, this force is negligible as compared to the electrostatic force between the two particles.

Example 1.2. *Calculate the distance of separation between two electrons (in vacuum) for which the electric force between them is equal to the gravitational force on one of them at the earth's surface.*

Given : mass of electron = $9{\cdot}1 \times 10^{-31}$ kg and charge = $1{\cdot}6 \times 10^{-19}$ C.

Solution. Gravitational force on one electron

$= mg$ newton $= 9{\cdot}1 \times 10^{-31} \times 9{\cdot}81$ N

Electrostatic force between the two electrons

$$= 9 \times 10^9 \frac{Q^2}{d^2} = \frac{9 \times 10^9 \times (1{\cdot}6 \times 10^{-19})^2}{d^2}\,\text{N}$$

Equating the two forces, we get

$9 \times 10^9 \times 2{\cdot}56 \times 10^{-38}/d^2 = 9{\cdot}1 \times 10^{-31} \times 9{\cdot}81$; $\quad$ **d = 5·08 m**

Example 1.3. *Two flat metal plates 2 cm apart have potential difference of 400 V between them. A proton with a positive charge of $1{\cdot}6 \times 10^{-19}$ C is situated between the plates. Find (i) the electric intensity between the plates and (ii) the force on the proton.*

Solution. (*i*) $E = \dfrac{V}{d} = \dfrac{400}{0{\cdot}02}$ = **20,000 V/m**

(*ii*) $F = QE = 1{\cdot}6 \times 10^{-19} \times 20{,}000 =$ **$3{\cdot}2 \times 10^{-15}$ N**

Example 1.4. *Two brass plates are arranged horizontally, one 2 cm above the other and the lower plate is earthed. The plates are charged to a p.d. of 6000 V. A drop of oil with an electric charge of $-1{\cdot}6 \times 10^{-19}$ C is in equilibrium between the plates so that it neither rises nor falls. What is the mass of the oil drop ?*

Solution. The oil drop as located in the electric field between the two plates is shown in Fig. 1.5. The two forces acting on this negatively-charged drop are

(*i*) electric force F acting upwards *i.e.* towards the positive plate, and

(*ii*) gravitational force 'mg' acting downwards.

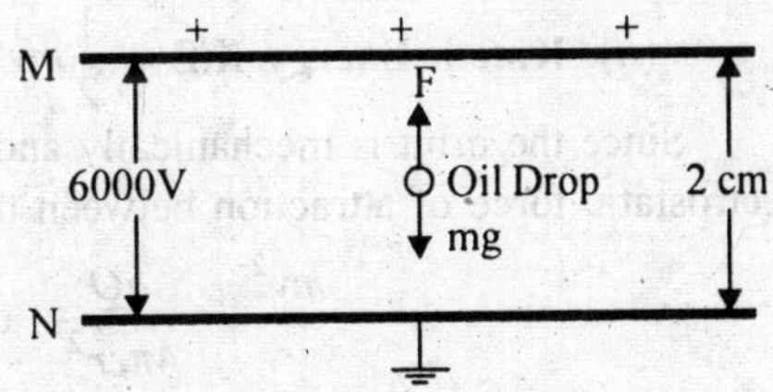

Fig. 1.5

The drop would be in equilibrium when the two forces are equal and opposite.

$$E = \frac{V}{d} = \frac{6000}{0{\cdot}02} = 3 \times 10^5 \text{ V/m}$$

$$F = Q.E. = 1{\cdot}6 \times 10^{-19} \times 3 \times 10^5 = 4{\cdot}8 \times 10^{-14} \text{ N}$$

Wt. of the drop = $mg = m \times 9.81$ N

Equating the two forces, we have

$m \times 9{\cdot}81 = 4{\cdot}8 \times 10^{-14}$; $m = \mathbf{4{\cdot}89 \times 10^{-15}}$ **kg**

Example 1.5. *The potential difference between the cathode and anode of a radio valve is 150 V. Calculate the velocity with which an electron, starting from cathode with negligible velocity, strikes the anode. Mass of an electron = $9{\cdot}1 \times 10^{-31}$ kg and charge = $1{\cdot}6 \times 10^{-19}$ C.*

Solution. The energy gained by the electron during its speeding up from cathode to anode is

$$\text{Energy} = QV = 1{\cdot}6 \times 10^{-19} \times 150 \text{ J}$$

If v is the striking velocity. then kinetic energy of the electron is $\frac{1}{2}.mv^2$

$$\therefore \frac{1}{2} \times 9{\cdot}1 \times 10^{-31} \times v^2 = 1{\cdot}6 \times 10^{-19} \times 150 \quad \therefore \mathbf{v = 7{\cdot}3 \times 10^6 \text{ m/s}}$$

Example 1.6. *Calculate the energy in (i) eV, (ii) MeV and (iii) joules gained by an α-particle when it is accelerated through a p.d. of 5 MV.*

Solution. It should be kept in mind that an α particle has a positive charge twice in magnitude as compared to an electronic charge.

(*i*) Energy = $2e \times V = 2 \times (5 \times 10^6) = \mathbf{10^7}$ **eV**

(*ii*) Energy = $10^7/10^6$ = **10 MeV**

(*iii*) Energy = $10 \times 1{\cdot}6 \times 10^{-13} = \mathbf{1{\cdot}6 \times 10^{-12}}$ **J**

1.7. Total Energy of a Revolving Charge

Consider the case of a small particle of mass m and charge $-q$ circling around a fixed charge of $+$Q in an orbit of radius r. The *total* energy of this revolving charge consists of two parts :

(*i*) **Potential Energy.** It is the amount of work done in bringing this $-q$ charge from infinity to where it is *i.e.* at a distance of r from the fixed charge $+Q$. This energy would be taken negative because $-q$ is brought not against repulsion but by the attraction of the charge $+Q$.

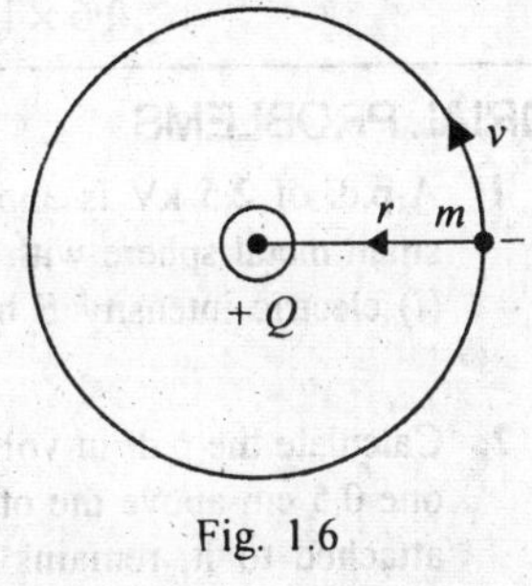

Fig. 1.6

∴ P.E. = potential × charge – Art. 1.4

$$= \frac{Q}{4\pi\varepsilon r} \times -q = -\frac{qQ}{4\pi\varepsilon r}$$

(*ii*) **Kinetic Energy.** K.E. = $\frac{1}{2}mv^2$. Its value can be found from the following considerations.

Since the orbit is mechanically and electrically stable, the centripetal force must equal the electrostatic force of attraction between the two charges.

$$\therefore \quad \frac{mv^2}{r} = \frac{qQ}{4\pi\varepsilon r^2} \quad \text{or} \quad mv^2 = \frac{qQ}{4\pi\varepsilon r}$$

$$\therefore \quad K.E. = \frac{1}{2} \times \frac{qQ}{4\pi\varepsilon r} = \frac{qQ}{8\pi\varepsilon r}$$

Hence, total energy of the revolving charge is

$$\therefore \quad E_t = \text{K.E.+P.E.}$$

$$\therefore \quad = \frac{qQ}{8\pi\varepsilon r} = \frac{qQ}{4\pi\varepsilon r}$$

$$\text{or} \quad E_t = -\frac{qQ}{8\pi\varepsilon r} = -\frac{qQ}{8\pi\varepsilon_0\varepsilon_r r}$$

$$\therefore \quad \boxed{E_t = -\frac{qQ}{8\pi\varepsilon_0 r}} \quad \text{– in vacuum}$$

The negative sign indicates that the charge $-q$ is *bound* to the nucleus and also that this much energy would be required to remove the charge $-q$ away from the fixed charge $+Q$.

Note. Incidentally, it may be noted that, numerically speaking, potential energy is twice the kinetic energy.

Example 1.7. *Calculate the total energy of an electron which revolves around hydrogen nucleus in an orbit of radius $0{\cdot}528 \times 10^{-10}$ m. Express the answer in joules and electron-volts.*

Solution. $E_t = \frac{qQ}{8\pi\varepsilon_0 r}$

Here, electron charge

$q = -1{\cdot}6 \times 10^{-19}$ C (Fig. 1.7)

Charge on hydrogen nucleus,

$Q = +1.6 \times 10^{-19}$ C

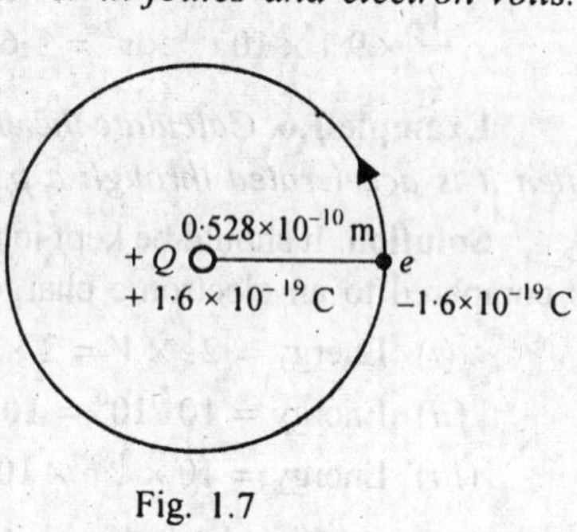

Fig. 1.7

$$\therefore \quad E_t = -\frac{(1.6 \times 10^{-19})^2}{8\pi \times 8.854 \times 10^{-12} \times 0{\cdot}528 \times 10^{-10}} \text{ J}$$

$$= -21{\cdot}8 \times 10^{-19} \text{ J}$$

Now $\quad 1 \text{ eV} = 1{\cdot}6 \times 10^{-19}$ J

$$\therefore \quad E_t = \frac{-21{\cdot}8 \times 10^{-19}}{1{\cdot}6 \times 10^{-19}} = -13.6 \text{ eV}$$

TUTORIAL PROBLEMS

1. A p.d. of 2·5 kV is applied between two parallel metal plates 5 cm apart and placed in air. A small metal sphere with a charge of 3×10^{-9} C is located midway between the plates. Calculate, (*i*) electric intensity E between the plates and (*ii*) the force experienced by the metal sphere.
[(*i*) 5×10^4 V/m; (*ii*) $1{\cdot}5 \times 10^{-4}$ N]

2. Calculate the p.d. in volts necessary to be maintained between two horizontal conducting plates, one 0·5 cm above the other, so that a small oil drop of mass $1{\cdot}31 \times 10^{-13}$ kg with two electrons attached to it, remains in equilibrium between them. Which plate would be at the positive potential ?
[20.1 kV; top]

3. A charged oil drop of mass of 10^{-13} kg is observed to remain stationary in the space between two horizontal charged plates; the electrical force being sufficient to counteract its weight. Find the charge on the drop if this occurs when the p.d. between the plates is 3 kV and their distance apart 1 cm. [$3{\cdot}3 \times 10^{-19}$ C]

4. In Millikan's oil drop experiment, an oil drop was prevented from falling or rising by applying a p.d. of 5750 V between parallel horizontal plates 1·5 cm apart. Assuming that the drops each carried one electron charge, what was the radius of the drop ? Density of oil = 0·92 g/cm³. [$1{\cdot}18 \times 10^{-4}$ cm]

1.8. Laws of Magnetic Force

The magnetic force between two isolated magnetic poles (Fig. 1.8) varies

(*i*) directly as the product of their pole strengths

(*ii*) inversely as the square of the distance between them

(*iii*) inversely as the absolute permeability of the surrounding medium.

Mathematically speaking,

$$F \propto \frac{m_1 m_2}{\mu d^2}$$

or $$F = k.\frac{m_1 m_2}{\mu d^2}$$

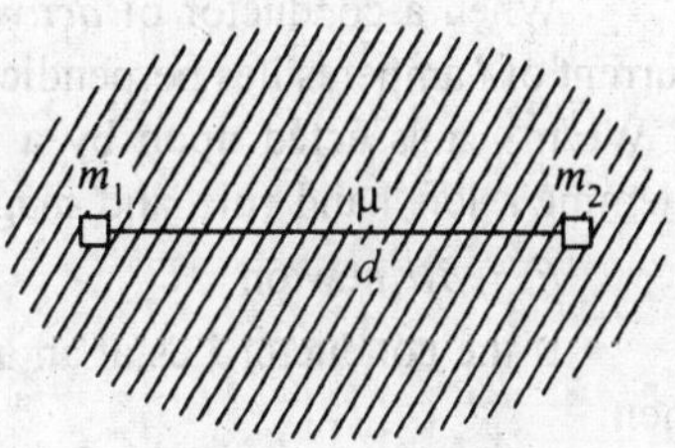

Fig. 1.8

where k is the constant of proportionality whose value depends on the system of units employed. In the MKSA as well as SI system, the value of $k = 1/4\pi$.

$$\therefore \quad \boxed{F \frac{m_1 m_2}{4\pi\mu_0\mu_r d^2}} \text{ newtons}$$

Force F is in newtons (N) provided

m_1, m_2 are in weber (Wb)

d is in metre (m)

and μ is in henry/metre (H/m)

1.9. Magnetic Field Strength or Magnetising Force (H)

The magnetising force at any point around a magnetic pole may be defined as *equal to the force experienced by a North pole of one weber placed at that point.*

For example, if the force experienced by a unit *N*-pole is 20 N, then field strength of magnetising force at that point is 20 N/Wb or 20 AT/m.

In Fig. 1.9, value of H at point A can be found by using Law of Magnetic Force.

$$F = \frac{m \times 1}{4\pi\mu d^2}$$

$$= \frac{m}{4\pi\mu_0\mu_r d^2} \text{ N}$$

Fig. 1.9

$$\therefore \quad H = \frac{m}{4\pi\mu_0\mu_r d^2} \text{ N/C or AT/m} \quad \text{– in medium}$$

$$\boxed{H = \frac{m}{4\pi\mu_0 d^2}} \text{ N/C or AT/m} \quad \text{– in vacuum}$$

In general, if at any point within a magnetic field, the force experienced by a pole of m Wb is F newton, then $H = F/m$. Conversely, if a pole of strength m Wb is placed in a field of strength H, force experienced by it is $F = mH$.

1.10 Magnetic Flux Density (B)

It is given by the perpendicular flux incident per unit area. As shown in Fig. 1.10, if a flux of Φ weber falls perpendicularly on an area of A m², then

$$B = \frac{\Phi}{A} \text{ Wb/m}^2$$

It may be noted that

$$\boxed{B = \mu H = \mu_0 \mu_r H}$$

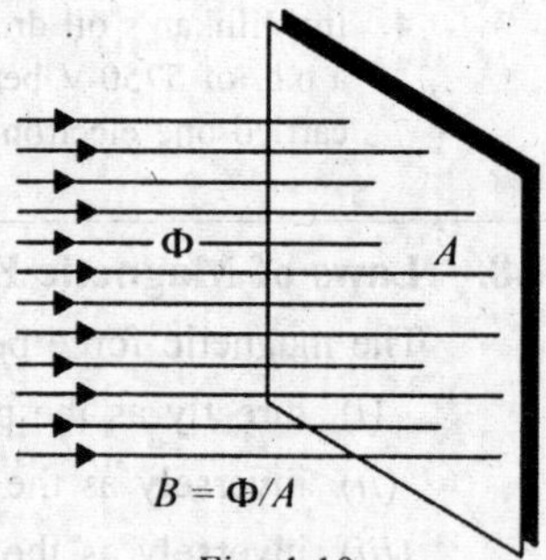

Fig. 1.10

1.11. Force on a Current-Carrying Conductor

When a conductor of *active* length l metres and carrying a current of I amperes lies perpendicular to a magnetic field of flux density B Wb/m², it is acted upon by a force F whose direction is given by Fleming's left hand rule and magnitude by

$$F = BIl \text{ newton}$$

If the conductor lies at an angle θ with the direction of the field, then

$$F = BIl \sin\theta \text{ newton}$$

In Fig. 1.11, a conductor A lies in a magnetic field and carries current 'outwards' perpendicular to the plane of the paper. The direction of the magnetic force as found by using Fleming's Left hand rule is as shown.

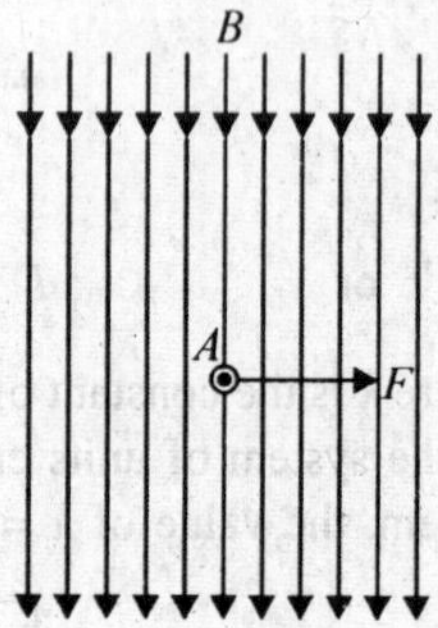

Fig. 1.11

1.12. Force on a Moving Charge in a Magnetic Field

An electrically-charged body in an electric field is acted upon by a force whether it is at rest or moving. However, in a magnetic field, charged body experiences the force *only when it is moving*.

When a charge of $-Q$ coulomb moves with a velocity of v m/s perpendicular to a magnetic field of flux density B Wb/m², then force acting on it is

$$\boxed{F = BQv} \text{ newton}$$

Its direction is also given by Fleming's Left hand rule.*

If the direction of motion of the charged particle makes an angle θ with the direction of B, then

$$F = BQv \sin\theta \text{ newton}$$

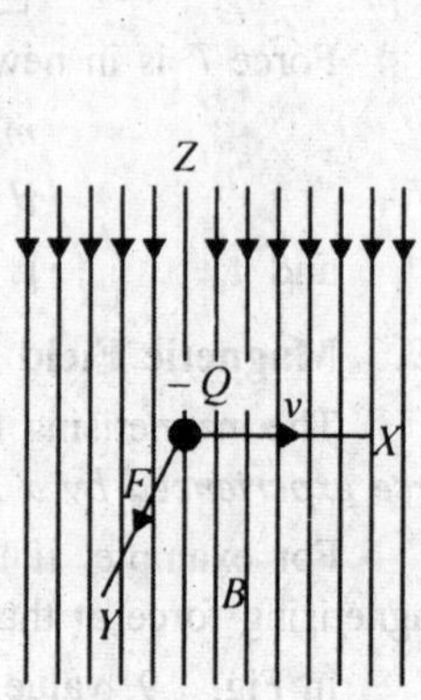

Fig. 1.12

It should be noted that the directions of B, v and F are mutually perpendicular to each other (Fig. 1.12).

1.13. Motion of a Charged Particle in Transverse Magnetic Field

In Fig. 1.13 (*a*) is shown a particle of mass m and charge Q moving with a constant velocity of v at right angles to a uniform magnetic field of flux density B. The particle experiences a force

*However, while applying this rule, middle finger must point towards the flow of positive charge. For example, in the case of electron motion middle finger must point in a direction opposite to the direction of motion of the electron.

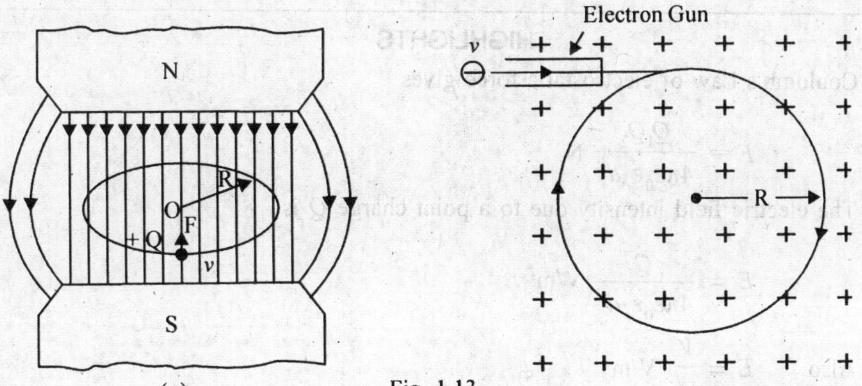

(*a*) Fig. 1.13 (*b*)

$F = BQv$ newtons which is at right angles to both B and v. Since this force is constant in magnitude and always perpendicular to the direction of motion of the particle, it imparts a constant radial acceleration ($= v^2/R$) to the particle. Consequently, the particle describes a circular path whose radius may be found by equating the magnetic force to the centripetal force.

$$\therefore \qquad \frac{mv^2}{R} = BQv$$

$$\therefore \qquad R = \frac{m}{Q}\cdot\frac{v}{B}$$

The time-period T, *i.e.* time required to complete one revolution, is given by

$$\therefore \qquad T = \frac{2\pi R}{v}$$

Eliminating v, we have

$$\boxed{T = \frac{2\pi m}{BQ} = \frac{2\pi}{B}\cdot\frac{m}{Q}}$$

The fact that T is independent of v and R is made use of in the working of a cyclotron (Art. 16.4). Fig. 1.13 (*b*) shows the circular path executed by an electron released by an electron gun with an initial velocity of v. The field is perpendicular to the plane of the paper and is directed downwards (as shown by crosses).

Example 1.8. *An a-particle with a charge of $3{\cdot}2 \times 10^{-19}$ C and mass of $6{\cdot}68 \times 10^{-27}$ kg moves through a transverse magnetic field of flux density $1{\cdot}5$ Wb/m^2. If the velocity of the α-particle is 5×10^6 m/s, find (i) force on the particle and (ii) the radius of its circular path.*

Solution. (*i*) $F = BQv = 1{\cdot}5 \times 3{\cdot}2 \times 10^{-19} \times 5 \times 10^6$

$= \mathbf{2{\cdot}4 \times 10^{-12}\ N}$

(*ii*) $R = \dfrac{m}{Q}\cdot\dfrac{v}{B} = \dfrac{6{\cdot}68 \times 10^{-27}}{3{\cdot}2 \times 10^{-19}} \times \dfrac{5 \times 10^{o}}{1{\cdot}5}$

$= \mathbf{0.07m = 7cm.}$

HIGHLIGHTS

1. Coulomb's Law of electrostatic force gives

$$F = \frac{Q_1Q_2}{4\pi\varepsilon_0\varepsilon_r d^2}\ \text{N}$$

2. The electric field intensity due to a point charge Q is

$$E = \frac{Q}{4\pi\varepsilon_0\varepsilon_r d^2}\ \text{V/m}$$

Also $$E = \frac{V}{d}\ \text{V/m}$$

3. Force on a charge Q placed in electric field of intensity E is given by

$$F = QE \text{ newton}$$

4. Electron-volt (eV) is the unit of energy.

$$1 \text{ eV} = 1{\cdot}6 \times 10^{-19} \text{ J}$$

5. Magnetic force between two magnetic poles is given by

$$F = \frac{m_1 m_2}{4\pi\mu_0\mu_r d^2}\ \text{N}$$

6. The force experienced by a pole of m Wb when placed in a magnetic field of strength H N/Wb is

$$\text{F} = mH \text{ newton}$$
$$= mB/\mu \text{ newton}$$

7. Force on a current-carrying conductor lying in a magnetic field is $F = BIl$ newton.
8. Force on a charged particle moving in a transverse magnetic field is

$$F = BQv \text{ newton}$$

9. Radius and time-period of such a circling particle are given by

$$R = \frac{m}{Q}\cdot\frac{v}{B}\ \text{metre}$$

$$T = \frac{2\pi m}{BQ}\ \text{second}$$

THE ELECTRON

2

Discovery of Electron — Motion of Electron in Electric Fields — Motion of Electron in a Magnetic Field — Electron Motion in Crossed Electric and Magnetic Fields — Thomson's Method for Measuring e/m for Electron — Mass of an Electron — Millikan's Oil Drop Experiment for Electron Charge — Positive Rays — Measurement of q/m for Positive Rays — Isotopes — Isotopic Masses and Abundances — Highlights.

2.1. Discovery of Electron

With the development of vacuum pumps and availability of fairly high voltages in the middle of 19th century, physicists began to experiment on the conduction of electricity by gases using arrangement like the one shown in Fig. 2.1. It is found that gases at ordinary pressure are good insulators and need very high voltages to produce measurable conduction. If the pressure is reduced to about 10^{-3} of the normal atmospheric pressure (*i.e.* to 100 N/m^2) gases are found to conduct sufficiently well and produce spectacular and complex displays as in neon lamps. However, when pressure is further reduced to about 10^{-5} atmosphere (or 1 N/m^2) gases again become poor conductors and the visual patterns in the gases fade away. However, at low pressures of about 10^{-6} atmosphere (*i.e.* 0·1 N/m^2), a new phenomenon is observed. The glass tube starts glowing with an eerie greenish light mainly over the region opposite to cathode C. It was soon shown that fluorescence of glass was due to its bombardment by invisible emanations from the cathode called *cathode rays*. These cathode rays were later discovered to be tiny negatively-charged particles which we now call *electrons*.

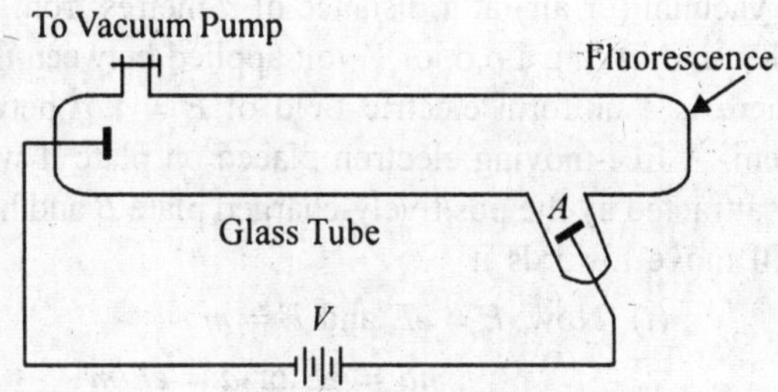

Fig. 2.1

The early experiments revealed the following properties of the cathode rays :

(*i*) They excite fluorescence in many crystalline materials such as Willemite and zinc sulphide and also in glass. The minerals which show fluorescence are called *phosphors*.

(*ii*) They travel in straight lines.

This can be conveniently shown by Crookes' cross experiment.

(*iii*) They produce heat in any material on which they impinge.

(*iv*) They can pass through thin metal foils.

This fact was demonstrated by Heinrich Hertz and his assistant P. Lenard (1862-1947) who passed cathode rays through thin aluminium foils.

(*v*) They possess momentum which means that they have mass and velocity.

This fact was demonstrated in 1870 by Crookes.

(*vi*) They are negatively charged.

This was first discovered in 1895 by French physicist J.B. Perrin (1870-1942).

(*vii*) They are deflected by electric and magnetic fields.

Their deflection by magnetic field was first demonstrated in 1869 by the German chemist J.W. Hittorf (1824-1914). Later on, it was found that electric field also produces deflection.

Some of the above properties are consistent with the wave nature of the rays but properties like deflection by magnetic and electric fields and carrying of negative charge can be consistent only when we interpret cathode rays as negatively-charged particles *i.e.* electrons. This fact was firmly established by J.J. Thomson (1897) who used the property of deflection by electric and magnetic fields to measure the e/m value for the electrons (Art. 2.5).

2.2. Motion of Electron in Electric Fields

Following two cases would be considered :

(*a*) When motion is along the field. Fig. 2.2 shows two large plane parallel plates A and B situated in vacuum (or air) at a distance of d metres from each other and having a p.d. of V volt applied between them. There is a uniform electric field of $E = V/d$ between them. A free-moving electron placed on plate A would be attracted by the positively-charged plate B and hence will move towards it.

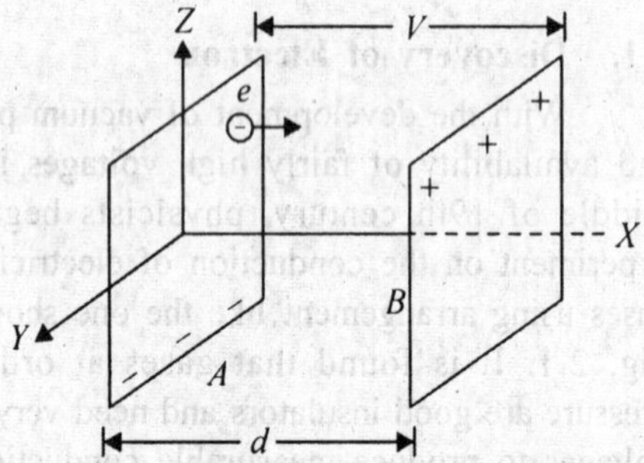

Fig. 2.2

(*i*) Now, $F = eE$ and $F = ma$

$$\therefore \quad ma = eE \text{ or } a = eE/m$$

$$\therefore \quad a = \frac{e}{m}.E = \frac{e}{m}\cdot\frac{V}{d}.$$

where m is the mass of the electron,

e its charge and a its acceleration along X-axis.

(*ii*) Velocity of the electron at any time t is given by

$$v = u + at = \text{o} + at$$

$$\therefore \quad v = \frac{e}{m}.\frac{V}{d}.t \qquad ...(i)$$

(*iii*) Distance travelled by the electron during that time is

$$x = \frac{1}{2}at^3 = \frac{1}{2}.\frac{e}{m}.\frac{V}{d}.t^2 \quad \left(S = ut + \frac{1}{2}at^2\right) \qquad ...(ii)$$

(*iv*) Eliminating t from Eqn. (*i*), we get

$$v = \sqrt{2\cdot\frac{e}{m}\cdot\frac{V}{d}\cdot x}$$

(*v*) The impact velocity v_4 can be found by putting $x = d$.

$$\therefore \quad v_4 = \sqrt{2.\frac{e}{m},\frac{V}{d}.d} = \sqrt{2.\frac{e}{m}.V}$$

(*vi*) Time of transit from A to B as given by Eqn. (*ii*) is

$$t^2 = \frac{2mxd}{eV} = \frac{2md^2}{eV}$$

$$\therefore \qquad t = \sqrt{\frac{2md^2}{eV}} = \sqrt{\frac{4d^2}{2eV/m}} = \frac{2d}{v_i}$$

(*b*) When motion is perpendicular to the field. Fig. 2.3 shows an electron with initial *X*-velocity of u_x entering an electric field (between plates *M* and *N*) which is at right angles to the initial direction of travel of the electron. While passing through the field, the electron will experience an acceleration along *Y*-axis. Hence, it will be deflected vertically and will finally strike the screen at point *S*. It is obvious that the axial velocity u_x of the electron remains unchanged throughout because there is no force and hence no acceleration along *X*-axis.

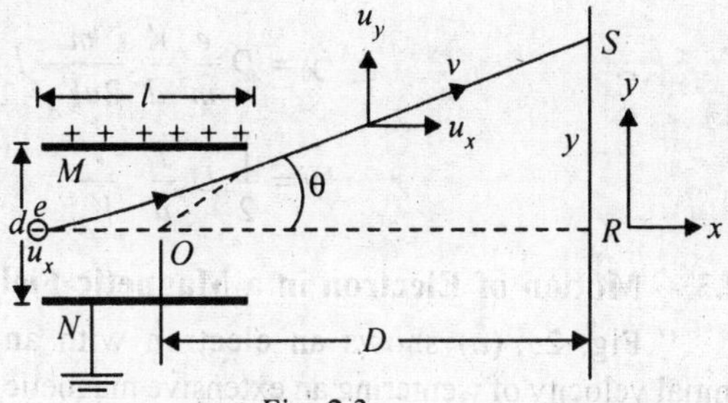

Fig. 2.3

If *V* is the p.d. between the two plates, *d* their separation, then vertical acceleration a_v is given by

$$F = m.a_y = eE = eV/d$$

$$\therefore \qquad a_y = \frac{e}{m}.\frac{V}{d}$$

If *l* is the length of the plates, then the time for which electron remains in the region between the plates is $t = l/u_x$.

The final velocity v_y of the electron (while leaving the electric field) is given by

$$v_y = 0 + a_y t$$

$$= \frac{e}{m}.\frac{V}{d}.t \qquad (\because\ v = u + at)$$

$$\text{or} \qquad v_y = \frac{e}{m}.\frac{V}{d}.\frac{l}{u_x}$$

This vertical velocity remains constant after the electron moves out of the field. From then onwards, the electron travels in a straight line because the space is field-free. Its velocity is

$$v = \sqrt{u_x^{\,2} + v_y^{\,2}}$$

The straight line when produced backwards will meet the *X*-axis at point *O* which is the mid-point of the plates. The line makes an angle of $\theta = \tan^{-1}(v_y/u_x)$ with the *X*-axis. Although, in actuality, the electron path between the plates is parabolic, there is not much loss of accuracy in assuming that the beam suddenly deflects by an angle θ at the centre *O* of the plates. The vertical deflection *y* is given by

$$y = D \tan\theta = Dv_y/u_x$$

Substituting the value of v_y, we have

$$y = D.\frac{e}{m}.\frac{V}{d}.\frac{l}{u_x^2}$$

If, before entering the field, the electron had been accelerated by a potential V_A, then

$$\frac{1}{2}mu_x^2 = eV_A$$

or $$u_x^2 = 2e\,V_A/m$$

$\therefore$ $$y = D.\frac{e}{m}.\frac{V}{d}.\frac{m}{2eV_A}.l$$

$$y = \frac{1}{2}.D.\frac{l}{d}.\frac{V}{V_A}$$

2.3. Motion of Electron in a Magnetic Field

Fig. 2.4 (*a*) shows an electron with an initial velocity of *v* entering an extensive magnetic field perpendicular to its direction of motion. The electron is subjected to a radial magnetic force (Art. 1.12) equal to *Bev*. Under the influence of this force, the electron moves in a circular orbit. The centripetal force ($= mv^2/R$) required for orbital motion is supplied by the magnetic force.

$\therefore$ $$mv^2/R = Bev$$

or $$R = \frac{m}{e}.\frac{v}{B}$$

Time period of orbital motion is

$$T = \frac{2\pi R}{v} = \frac{2\pi}{B}.\frac{m}{e}$$

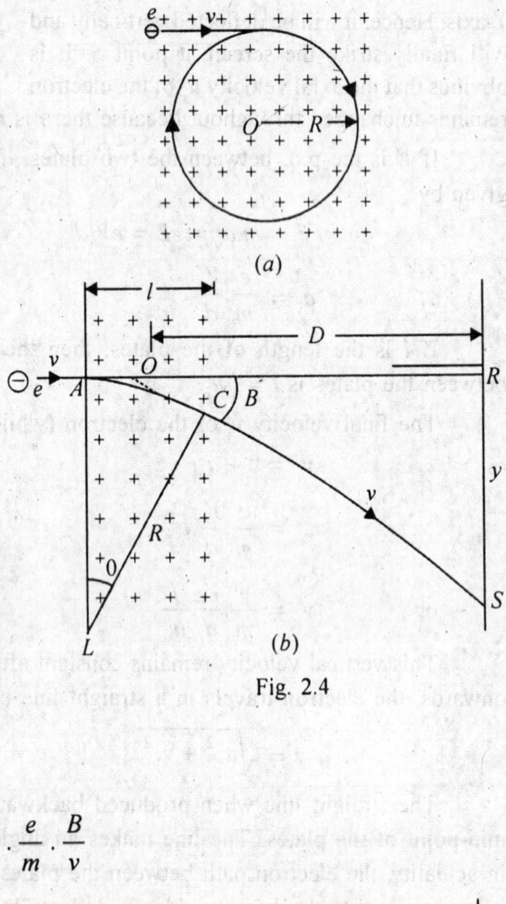

Fig. 2.4

Fig. 2.4 (*b*) shows the case when the electron passes through a magnetic field, which is confined over a small region. After describing a circular arc *AC* of radius *R*, the electron continues moving in a straight line with the same velocity *v* and strikes the screen at point *S*. The vertical deflection is given by $y = D \tan\theta$.

$\therefore$ $$y = D.\frac{AC}{R} \cong D.\frac{l}{R}$$

Now $$R = \frac{m}{e}.\frac{v}{B} \quad \therefore\ y = Dl.\frac{e}{m}.\frac{B}{v}$$

2.4. Electron Motion in Crossed Electric and Magnetic Fields

Fig. 2.5 shows an electric field existing simultaneously at right angles to a magnetic field. By using Fleming's left-hand rule, it can be seen that an electron would be deflected downwards by the magnetic field. However, it would be attracted upwards by the electric field between the two plates.

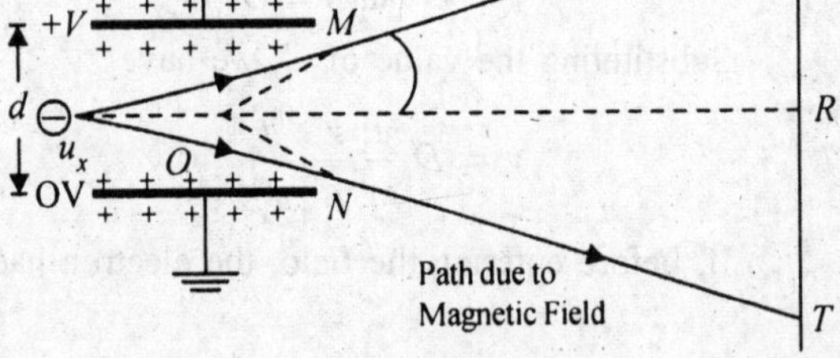

Fig. 2.5

Upward electric force = *eE*

Downward magnetic force = Bev

If the two forces are equal, then net vertical deflection suffered by the electron would be zero so that it will go straight and strike the screen at point R.

In that case, $eE = Bev$

$\therefore$ $v = E/B$

Incidentally, the above relation suggests a practical method of finding electron velocity.*

2.5. Thomson's Method for Measuring e/m for Electrons

By using the apparatus shown in Fig. 2.6, J.J. Thomson succeeded in 1897 in determining the velocity of electrons in cathode rays and also their e/m ratio.

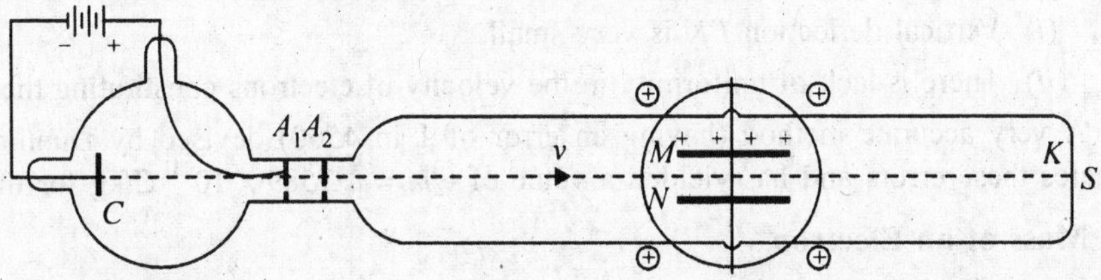

Fig. 2.6

The apparatus consists of an evacuated discharge bulb made of glass in which cathode rays are produced from cathode C. The electrons are made to pass through fine slits in anodes A_1 and A_2. In this way, they are converted into a narrow well-defined beam.

Next, the electron beam passes between cross-fields (*i.e.* electric and magnetic fields at right angles to each other) before striking the fluorescent screen S at point K. The experiment consists of two parts :

(*i*) First, the two fields are so adjusted as to make the electric and magnetic forces acting on an electron equal and opposite. In that case, the electrons go undeviated and strike the screen at point K. Since the two forces are equal

$\therefore$ $eE = Bev$

or $$v = \frac{E}{B} = \frac{V}{Bd} \quad ...(i)$$

where V is the p.d. between two plates and B the flux density of the magnetic field.

Next, electric field is switched off leaving behind only the magnetic field which deflects the electrons downwards (Fig. 2.7).

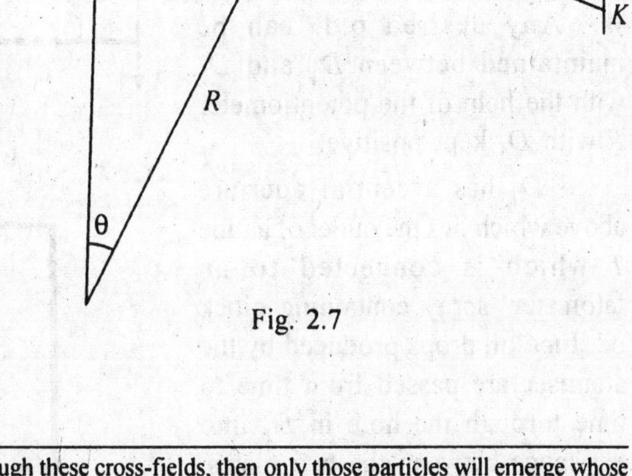

Fig. 2.7

The motion of electrons in a magnetic field is governed by the relation

$$mv^2/R = Bev$$

or $$v = \frac{BRe}{m} \quad ...(ii)$$

*If a beam of particles with a range of velocities is passed through these cross-fields, then only those particles will emerge whose velocity is $v = E/B$ whereas others would be so deflected as to hit the capacitor plates and thereby get absorbed. Such an arrangement then acts as a velocity selector.

From Eqns. (*i*) and (*ii*), we get

$$BR\frac{e}{m}=\frac{E}{B}$$

$$\therefore \quad \frac{e}{m}=\frac{E}{B^2R}=\frac{V}{dB^2R}$$

The value of R may be found with the help of Fig. 2.7 from where it is seen that

$$\tan\theta = PK/OP = LH/R$$

$$\therefore \quad R = LH \times OP/PK$$

Knowing V, d and B, value of e/m can be calculated.

Thomson's method has two main sources of error :

(*i*) Vertical deflection PK is very small.

(*ii*) There is lack of uniformity in the velocity of electrons constituting the cathode rays.

A very accurate method (having an error of 1 in 4500) devised by Dunnington in 1933 eliminates these errors and has yielded a value of $e/m = 1{\cdot}7589 \times 10^{11}$ C/kg for the electron.

2.6. Mass of an Electron

It is difficult to measure the mass of an electron directly. However, its value can be found indirectly by using the relation

$$\text{mass} = \frac{e}{e/m} = m$$

If we could measure electronic charge e, then knowing e/m from Thomson's experiment, m could be found. The value of e can be measured with the help of Millikan's oil drop experiment described below.

2.7. Millikan's Oil Drop Experiment for Electron Charge

It is an accurate method for measuring electronic charge and was devised by the American physicist R.A. Millikan (1868-1953).

As shown in Fig. 2.8, the apparatus consists of two flat and plane metal discs D_1 and D_2 arranged parallel to each other inside an earthed metallic box having three windows W_1, W_2 and W_3. Any desired p.d. can be maintained between D_1 and D_2 with the help of the potentiometer R with D_1 kept positive.

D_1 has a central aperture above which lies the outlet of a tube T which is connected to an 'atomiser' spray containing clock oil. Fine oil drops produced by the atomiser are passed from time to time through the hole in D_1 into the space between the two plates. These tiny drops are illuminated

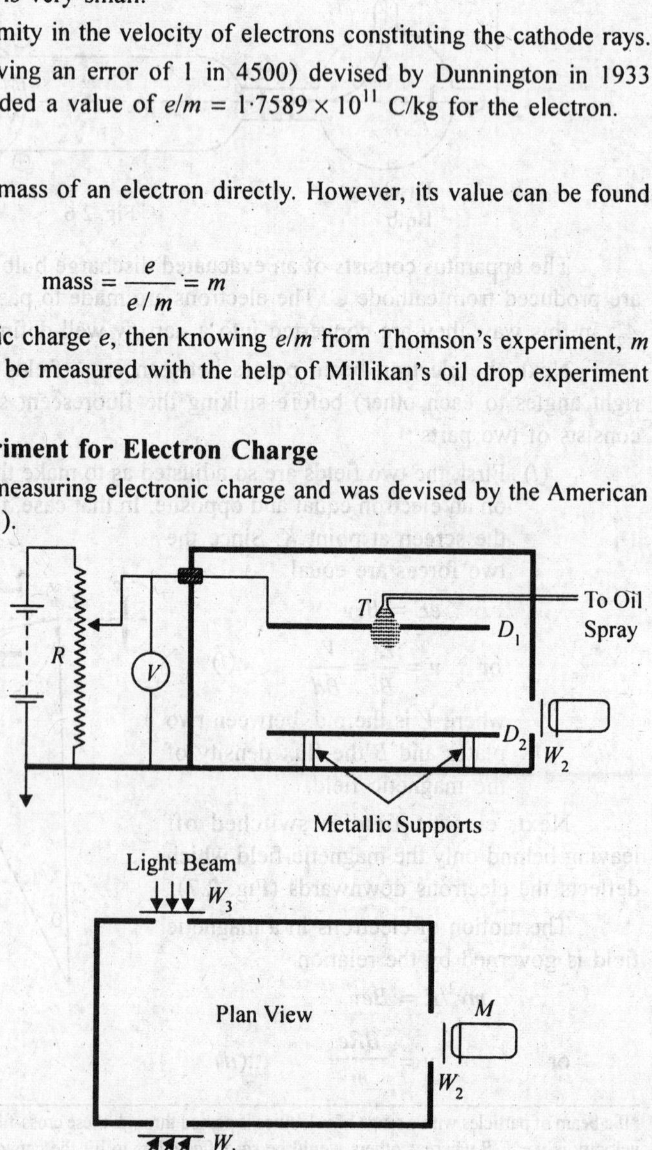

Fig. 2.8

by an intense narrow beam of light and viewed by a microscope kept outside the glass window W_2. Usually, these drops get charged by friction or by picking up stray electrons although X-rays can be used to produce further ionization if necessary. The microscope eye-piece is provided with a vertical divided scale (graticule) so that speed of fall or rise of charged particles may be measured. The experiment consists of two parts :

(*i*) In the first part of the experiment, the electric field between D_1 and D_2 is reduced to zero *i.e.* they are joined electrically and the ionizing radiation is not used. The drop falls under the influence of the gravitational field until it reaches constant terminal velocity when viscous force balances the gravitational force.

According to Stokes' law, if a spherical body of radius a falls under gravity in a fluid having a coefficient of viscosity η, it attains terminal velocity v when the retarding viscous force F becomes

$$F = 6\pi\eta av \qquad ...(i)$$

The force acts upwards *i.e.* opposite to the direction of fall of the drop. It is balanced by the gravitational force acting downwards which is

$$= \frac{4}{3}\pi a^3(\rho - \sigma)g \qquad ...(ii)$$

where ρ is the density of the drop and σ is the density of the medium in which drop is falling. It should be noted that $= \frac{4}{3}\pi a^3\rho$ is the mass of the drop and $= \frac{4}{3}\pi a^3\sigma$ is the buoyancy correction.

$$\therefore \qquad \frac{4}{3}\pi a^3(\rho - \sigma)g = 6\pi\eta av$$

$$\therefore \qquad a^2 = \frac{9\eta v}{2g(\rho - \sigma)} \qquad ...(iii)$$

Hence, if v is measured and η is known, value of a can be calculated.

(*ii*) In the second part of the experiment, an electric field E is established between D_1 and D_2 by applying a p.d. of V volts between them. Now, an additional upward electric force $= eE$ acts on the charged oil drop. If the terminal velocity of the drop is reduced to u, the equation of motion of the drop becomes

$$\frac{4}{3}\pi a^3(\rho - \sigma)g - eE = 6\pi\eta au$$

The value of E can be so adjusted as to make $u = 0$ *i.e.* upward electric force can be balanced by the downward gravitational force. In that case, the drop will remain stationary between D_1 and D_2 in the field of view of the microscope.

$$eE = \frac{4}{3}\pi^3 a(\rho - \sigma)g \qquad ...(iv)$$

Since a is known from first part of the experiment, e can be calculated from the above equation.

It is worth pointing out that *the same drop must be observed in both parts of the experiment.* It is no use observing two different drops because they may not have the same radius.

It is preferable first to make observations on the charged drop when E is on and is adjusted to keep the drop stationary and then remove E and observe the fall of the same drop under gravity.

The present-day accepted value for $e = 1{\cdot}602 \times 10^{-19}$ C.

The mass of the electron is given by

$$m = e \div e/m$$
$$= 1{\cdot}602 \times 10^{-19} \div 1{\cdot}759 \times 10^{11}$$
$$= 9{\cdot}108 \times 10^{-31} \text{ kg.}$$

Example 2.1. *How are cathode rays produced and what are their properties ?*

An electron starts from rest and moves freely in an electric field of intensity 2400 V/m. Determine (a) force on the electron, (b) its acceleration, (c) kinetic energy acquired and the velocity attained if the electron moves through a p.d. of 90 V. Electron has a charge of $1{\cdot}6 \times 10^{-19}$ C and mass $9{\cdot}12 \times 10^{-31}$ kg.

Solution. (*a*) $F = eE = 1{\cdot}6 \times 10^{-19} \times 2400 = \mathbf{3.84 \times 10^{-16}\ N}$

(*b*) $a = F/m = 3.84 \times 10^{-16}/9{\cdot}12 \times 10^{-31} = \mathbf{4{\cdot}22 \times 10^{14}\ m/s^2}$

(*c*) $\text{K.E.} = \frac{1}{2}mv^2 = eV = 1{\cdot}6 \times 10^{-19} \times 90 = \mathbf{1{\cdot}44 \times 10^{-17}\ J}$

$$\text{Also } v = \sqrt{\frac{2eV}{m}} = \sqrt{\frac{2 \times 1{\cdot}44 \times 10^{-17}}{9{\cdot}12 \times 10^{-31}}} = \mathbf{5{\cdot}63 \times 10^6\ m/s}$$

Example 2.2. *An electron, accelerated from rest, through a potential difference of 900 V enters a uniform magnetic field of 0·01 Wb/m² whose direction is perpendicular to the initial direction of the electron. Determine (i) the linear velocity of the electron and (ii) the radius of the circular path followed by the electron in the magnetic field. Take $e/m = 1{\cdot}76 \times 10^{11}$ C/kg for an electron.*

Solution. (*i*)
$$\frac{1}{2}mv^2 = eV$$
$$v = \sqrt{\frac{2eV}{m}} = \sqrt{2 \times 1{\cdot}76 \times 10^{11} \times 900} = \mathbf{1{\cdot}78 \times 10^7\ m/s}$$

(*ii*)
$$R = \frac{m}{e} \cdot \frac{v}{B} = \frac{1{\cdot}78 \times 10^7}{1{\cdot}76 \times 10^{11} \times 0{\cdot}01} = \mathbf{0{\cdot}01\ m}$$

Example 2.3. *Give the essential theory of a method for determining the e/m ratio of an electron.*

Two large parallel metal plates are 6 mm apart and have a potential drop across them of 900 V. If singly-charged positive particles enter the space between the plates and in a plane parallel to each plate, show that it is possible to arrange a uniform magnetic field so that the particles are not deviated as they pass between the plates. If magnetic flux density required for this condition to be attained is 0·5 Wb/m², find the velocity of the particles. When the magnetic field alone is applied, the particles traverse a circular track of radius 10·6 cm. Determine the mass of a particle, the charge on each being $1{\cdot}6 \times 10^{-19}$ C.

Solution. For the first part, please refer to Art. 2.6.

For no deflection condition in cross-fields,

$$v = E/B = V/Bd = 900/6 \times 10^{-3} \times 0{\cdot}5$$
$$= 3 \times 10^5 \text{ m/s}$$

And
$$R = \frac{m}{Q} \cdot \frac{v}{B}$$

$$\therefore \quad 0{\cdot}106 = \frac{m}{1{\cdot}6 \times 10^{-19}} \times \frac{3 \times 10^5}{0{\cdot}5}$$

$$\therefore \quad m = \mathbf{2{\cdot}82 \times 10^{-26}\ kg}$$

Example 2.4. *A charged oil drop is prevented from falling under gravity by a vertical electric field between two horizontal metal plates charged to a p.d. of 6920 V, the distance between the plates being 1·3 cm. When the field is cut off, the drop falls in air with a uniform velocity of $1{\cdot}9 \times 10^{-4}$ m/s. Calculate (a) the radius of the drop and (b) charge on the drop.*

Density of oil = $0{\cdot}9 \times 10^3$ kg/m³; coefficient of viscosity of air = $1{\cdot}81 \times 10^{-5}$ N-s/m². The density of air may be neglected in comparison with that of oil.

Solution. $$a^2 = \frac{9\eta v}{2g(\rho-\sigma)} \simeq \frac{9\eta v}{2g\rho}$$

Here, $\eta = 1{\cdot}81 \times 10^{-5}$ N-s/m²; $v = 1{\cdot}9 \times 10^{-4}$ m/s

$g = 9{\cdot}81$ m/s² $\quad \rho = 0{\cdot}9 \times 10^3$ kg/m³

$$\therefore \quad a^2 = \frac{9 \times 1{\cdot}81 \times 10^{-5} \times 1{\cdot}9 \times 10^{-4}}{2 \times 9{\cdot}81 \times 0{\cdot}9 \times 10^3}$$

$$\therefore \quad a = 1{\cdot}75 \times 10^{-6} \text{ m} = \mathbf{1{\cdot}75\ \mu m}$$

Now, $QE = (4/3)\pi a^3 (\rho - \sigma) g$ or $QE \simeq (4/3)\pi a^3 \rho g$

Now, $E = 6920/1{\cdot}2 \times 10^{-2}$ V/m

$$\therefore \quad Q \times \frac{6920}{1{\cdot}3 \times 10^{-2}} = \frac{4}{3}\pi(1{\cdot}75 \times 10^{-6})^3 \times 0{\cdot}9 \times 10^3 \times 9{\cdot}81$$

$$\therefore \quad Q = \mathbf{4.042 \times 10^{-19}\ C}$$

Obviously, the drop has three electrons attached to it.

2.8. Positive Rays

In 1886, German physicist E. Goldstein (1850-1931) set up a discharge tube whose cathode was perforated (Fig. 2.9). The tube was exhausted to a pressure of about 1 N/m² (*i.e.* 10^{-5} atmospheric pressure.) When a large p.d. was applied across the electrodes to cause a discharge through the residual gas, luminous streamers were found to appear in the space *behind* the cathode. These streamers were called *positive rays* because they seemed to originate from the anode and travel towards the cathode. The positive rays were found to cause fluorescence, affect photographic plates and showed deflection in magnetic and electric fields.

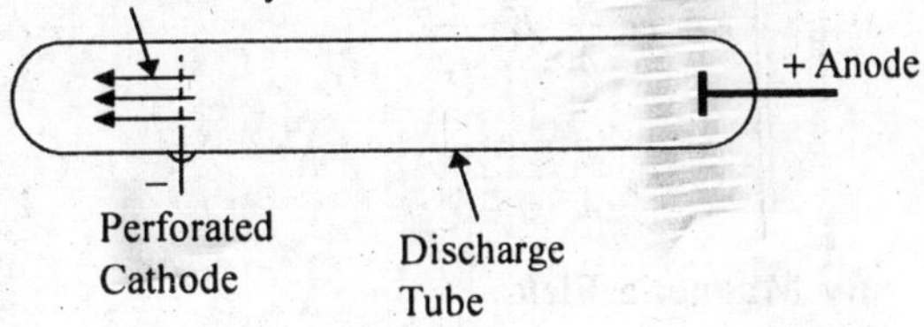

Fig. 2.9

These positive rays were, later on, found to consist of positive ions which are formed by ionization in the gas and are then drawn to the cathode by the electric field. Their q/m value was found by J.J. Thomson with the help of a mass spectrograph known after his name.

2.9. Measurement of q/m for Positive Rays

In 1911, J.J. Thomson successfully analysed positive rays by deflecting them in combined electric and magnetic fields acting in *the same direction*.

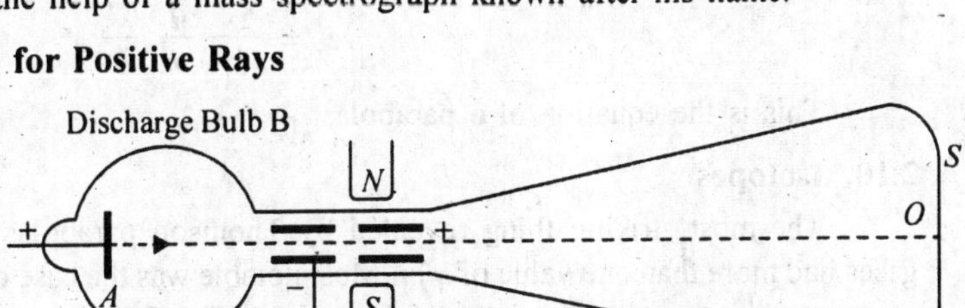

Fig. 2.10

One form of the apparatus used by him is shown in Fig. 2.10.

The main discharge took place in the bulb *B* through the low-pressure gas between the anode *A* and cathode *C*. Most of the ions produced at *A* hit the cathode and heat it. Hence, provision is made to keep the cathode cooled. Some of the ions pass through a canal in the cathode and emerge in the form of narrow beam of ions whose velocities depend on their charge and mass and on the discharge potential.

This ion beam is subjected to simultaneous magnetic and electric deflections with the help of two fields acting in *the same direction i.e.* downwards as shown in Fig. 2.11. Finally, the ions fall on a fluorescent screen or a photographic plate *S*.

Thomson found that the positive ions formed a number of parabolas on the photographic plate. Ions with the same value of q/m strike the plate at a series of points along a parabola, each point corresponding to a different value of u_x while for a different value of q/m, a different parabola is formed.

Formation of Thomson Parabolas

As seen from Fig. 2.11, electric field gives rise to a deflection along *Y*-axis whereas magnetic field produces deflection in positive ions along *Z*-axis. If a positive ion of q/m value enters the two fields with an initial axial velocity of u_x, the deflections of the ion on the screen are as under :

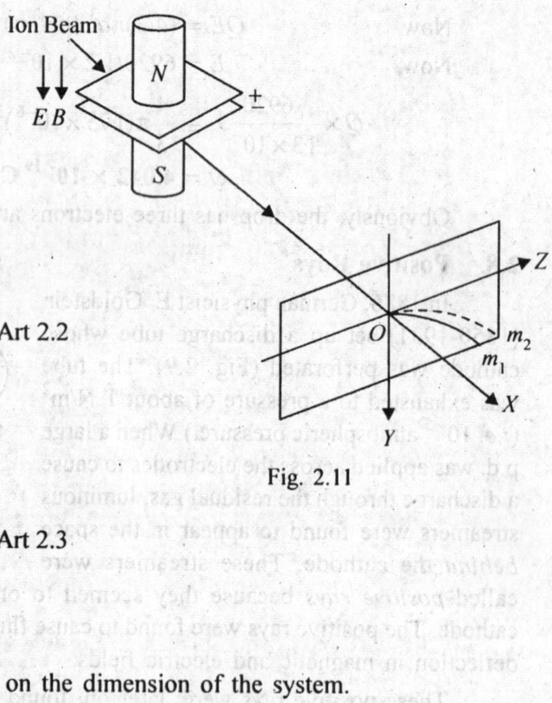

Fig. 2.11

By Electric Field

$$y = D \cdot \frac{q}{m} \cdot E \cdot \frac{l}{u_x^2} \qquad \text{—Art 2.2}$$

$$= k_1 \cdot \frac{q}{m} \cdot \frac{E}{u_x^2}$$

By Magnetic Field

$$z = D \cdot l \cdot \frac{q}{m} \cdot \frac{B}{u_x} \qquad \text{—Art 2.3}$$

$$= k_2 \cdot \frac{q}{m} \cdot \frac{B}{u_x}$$

where k_1 and k_2 are the constants depending on the dimension of the system.

Eliminating u_x from the above-given two equations, we get

$$z^2 = \frac{k_2^2}{k_1} \cdot \frac{q}{m} \cdot \frac{B^2}{E} \cdot y$$

This is the equation of a parabola.

2.10. Isotopes

The most striking thing revealed by Thomson parabolas was that some chemically pure gases had more than one value of q/m. Most notable was the case of neon which gave two parabolas corresponding to atomic weights of 20 and 22 (Fig. 2·12).

Neon was known to have an atomic weight of 20·2 and was the lightest element whose atomic weight differed significantly from a whole number. No element of atomic weight 22 was known. Since the next element sodium has an atomic weight of 23, efforts to explain the expected value of q/m corresponding to atomic weight of 22 failed at first. Finally, it was proved that neon existed in two forms which could not be distinguished chemically but which had two atomic weights of 20 and 22.

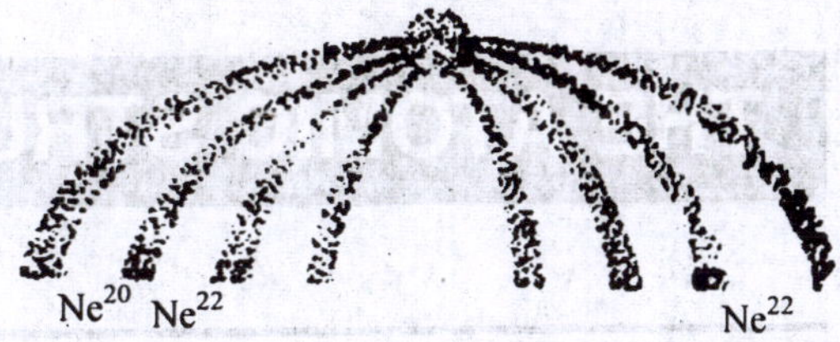

Fig. 2.12

Two atoms having same atomic number (*i.e.* Z value) but different atomic weights are known as isotopes of that element.

The two neon isotopes are written as $_{10}Ne^{20}$ and $_{10}Ne^{22}$. Their *relative* abundance is 9 : 1 respectively.* The average atomic weight of a mixture of these isotopes is

$$= \frac{(9\times 20)+(1\times 22)}{(9+1)} = 20.2$$

This value agrees with known value.

2.11. Isotopic Masses and Abundances

Although Thomson analyzer (Art. 2.9) was adequate for a general survey of masses and velocities, it could not yield accurate values of isotopic masses and their abundance. To fulfil this need, English physicist F.W. Aston (1877-1945) devised a *mass spectrograph* with which outstanding experiments on the determination of atomic weights were undertaken. It was followed by Dempster mass spectrograph which has recently been superseded by *Bainbridge magnetic deflection mass spectrograph.*

Of the 106 elements, some 284 stable** isotopic forms exist giving an average of about 3 per element. Some elements like tin have 10 isotopes extending from $_{50}Sn^{112}$ to $_{50}Sn^{124}$ whereas others like gold exist in one stable form of $_{79}Au^{197}$ only.

HIGHLIGHTS

1. When a particle of mass m and charge q passes through an electric field with an initial axial velocity u_x, it suffers vertical deflection given by

$$y = D\cdot\frac{q}{m}\cdot E\cdot\frac{l}{u_x^2}$$

$$= D\cdot\frac{q}{m}\cdot\frac{V}{d}\cdot\frac{l}{u_x^2}$$

2. Similarly, deflection produced by a magnetic field is

$$y = D\cdot l\cdot\frac{q}{m}\cdot\frac{B}{v}$$

3. The condition for no-deflection when a particle passes through crossed electric and magnetic field is, $y = \dfrac{E}{B}$

4. Cathode rays consist of a stream of electrons whereas positive rays are made up of positive ions produced in a discharge bulb.

5. Thomson's parabolas led to the important discovery of isotopes.

---

*Recent discoveries have shown that there are three isotopes of neon of atomic weights 20, 21 and 22 having relative abundance of 90·92, 0·257 and 8·82 per cent respectively.

**The number of both stable and unstable isotopes exceeds 1000.

THE ATOMIC STRUCTURE 3

Atomic Model — Thomson's Plum Pudding Model — Rutherford's Experiment on scattering of α-particles — Rutherford's Nuclear Atomic Model — Distance of Closest Approach — Major Deficiencies in Rutherford's Nuclear Model — Bohr's Atomic Model — Distance of Closest Approach — Major Deficiencies in Rutherford's Nuclear Model — Bohr's Atomic Model — Calculations Concerning Bohr's Atomic Model — Normal, Excited and Ionized Atom — Bohr Stoner Scheme of Building-up of Atoms — Electron Energy Levels in Hydrogen Atom — Bohr's Model Applied to Hydrogen Spectra — Spectral Series of Hydrogen Atom — Deficiencies in Bohr's Model — Sommerfeld's Relativistic Model — Orbital Angular Momentum of Different Orbits — Electronic Configurations of Atoms — Number of Electrons in a Shell — Electron Spin and Spin Quantum Number — Total Angular Quantum Number — Magnetic Quantum Number (m) — Four Quantum Numbers of an Electron — Values of n, l, j, Quantum Numbers — Values of n, l, j, m, Quantum Numbers — Orbital Magnetic Moment of the Electron — Spin Magnetic Moment of an Electron — Four Quantum Numbers Neglecting Spin-orbit Interaction — Different Shells, Sub-shells and Sub-Sub-shells — Pauli's Exclusion Principle — Diagrammatic Representation of Pauli's Exclusion Principle — Highlights — Can You Answer This ?

3.1. Atomic Model

By now, it is well-known that matter, electricity and radiation etc. are all atomic in character. Although, no one has so far seen individual atoms, there is no doubt that they really exist. In 1895, it was discovered by J. Perrin in Paris that the cathode rays (Art. 2.1) consist of negatively-charged particles called electrons. In 1897, J.J. Thomson measured the *e/m* ratio for an electron whereas its charge was measured by Millikan in 1906 by his famous oil-drop experiment. Mass of the electron was found by dividing charge '*e*' by the ratio *e/m*. Discovery of positive rays (Art. 2.8) during the latter part of 19th century indicated that a normal atom consisted of both negative and positive charges. But how these charges are distributed in an atom was not known at that time. Continuous efforts have been made since then to study the physical structure of an atom such as its extranuclear electronic structure chiefly with the help of spectral properties of atoms. To account for the spectroscopic data obtained experimentally over the years, several theories regarding atomic structure have been proposed from time to time which are called the *atomic models*. Various atomic models proposed by scientists over the last few decades are :

(*i*) Thomson's Plum Pudding model (*ii*) Rutherford's Nuclear model (*iii*) Bohr's model (*iv*) Sommerfeld's Relativistic model (*v*) Vector model and finally (*vi*) Wave-mechanical model.

These different models have been suggested one after the other in an effort to get a satisfactory interpretation of the experimental data which, it is hoped, will ultimately lead to a perfect and complete understanding of the physical structure of an atom.

3.2. Thomson's Plum Pudding Model

According to this model, the atom is regarded as a heavy sphere of positive charge seasoned with enough electron 'plums' to make it electrically neutral. Thomson visualized the positive charge of an atom being spread out uniformly throughout a sphere of about 10^{-10} metre radius with electrons as smaller particles distributed in circular shells as shown in Fig. 3.1. Whereas the net force exerted by the positively-charged sphere on each electron is towards the centre of the sphere, the different electrons experience mutual repulsion and get arranged in the form of circular shells.

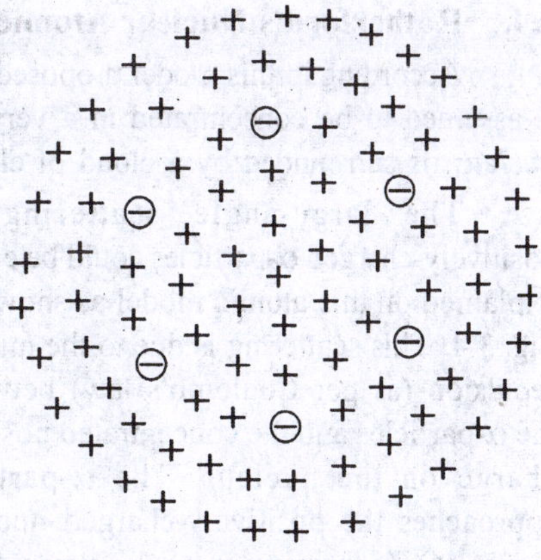

Fig. 3.1

This atomic model was given up after some time because it could not provide any satisfactory mechanism for explaining the large deflection suffered by α-particles in Rutherford's experiment briefly described below.

3.3. Rutherford's Experiment on α-particle Scattering

As shown in Fig. 3.2, high-speed α-particles (*i.e.* helium nuclei each with a charge of $+2e$) from some radioactive material like radium or radon, confined to a narrow beam by a hole in a lead block, were made to strike a very thin gold foil G. While most of the α-particles went straight through the foil as if there were nothing there (and produced scintillations on a fluorescent screen), some of them 'collided' with the atoms of the foil and were scattered around at various angles — a few being turned back towards the source itself !

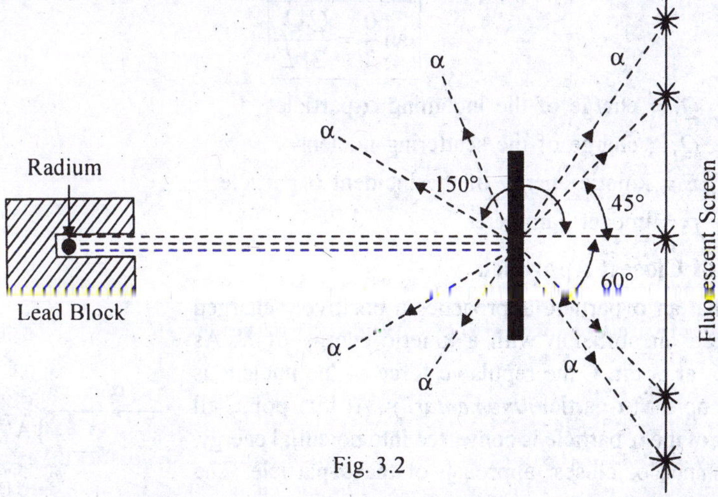

Fig. 3.2

Although some small-angle scattering could be expected from Thomson's model, large-angle scattering was absolutely not expected at all. Further detailed experiments in Rutherford's laboratory by Geiger and Marsden showed conclusively that large-angle scattering can be expected only if one assumes that a massive positive point charge exists at the centre of each gold atom as shown in Fig. 3.3.

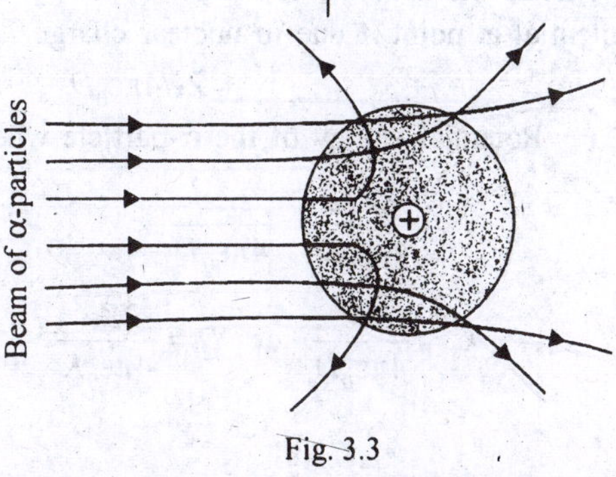

Fig. 3.3

3.4. Rutherford's Nuclear Atomic Model

According to this model proposed by Rutherford in 1911, the positive massive part of an atom is assumed to be concentrated in a very small volume at its centre. This central core, now called *nucleus*, is surrounded by a cloud of electrons which makes the entire atom electrically neutral.

The large-angle scattering of positively-charged α-particles could be easily explained on this atomic model as shown in Fig. 3.4. This scattering is due to the mutual repulsion (as per Coulomb's law) between the α-particles and the concentrated positive charge on the nucleus. The α-particle approaches the positively-charged nucleus along AO. If there were no repulsion from the nucleus, it would have passed at a distance of p from it. However, due to coulombic force of repulsion, the α-particle follows a hyperbola with nucleus as its focus.

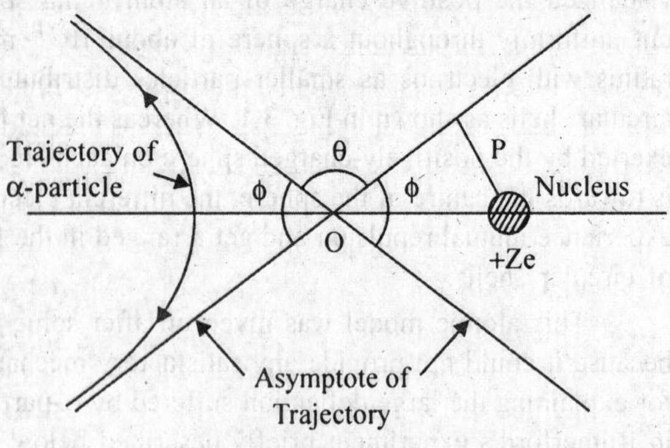

Fig. 3.4

The lines AO and $A'O$ are the asymptotes of the hyperbola and present approximately the initial and final directions of the α-particle when it has passed out of the effective range of the nuclear electric field.

As seen, the α-particle is deflected through an angle θ. The perpendicular distance from the nucleus to the line AO is called the *impact parameter* and is denoted by p. The Rutherford's scattering formula is

$$\boxed{\tan\frac{\theta}{2} = \frac{Q_1 Q_2}{2pE}}$$

where Q_1 = charge of the incoming α-particle

Q_2 = charge of the scattering nucleus

E = kinetic energy of the incident α-particle

p = impact parameter

3.5. Distance of Closest Approach

Suppose that an α-particle approaches a positively-charged nucleus for a head on collision with a kinetic energy of K. As shown in Fig. 3.5, at point A, the repulsive force of the nucleus is so strong as to stop the α-particle *momentarily*. At this point, all the kinetic energy of the α particle is converted into potential energy. Let D be the distance of closest approach of the α-particle. The potential at point A due to nuclear charge Ze is

$$= Ze/4\pi\varepsilon_0 D$$

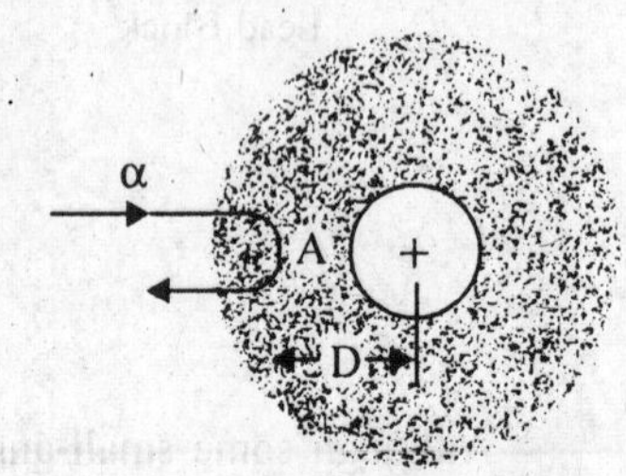

Fig. 3.5

Potential energy of the α-particle when at point A is

$$= \frac{Ze.2e}{4\pi\varepsilon_0 D} = \frac{2Ze^2}{4\pi\varepsilon_0 D} \qquad \text{(Art. 1.7)}$$

$$\therefore \quad k = \frac{2Ze^2}{4\pi\varepsilon_0 D} \quad \text{or} \quad D = \frac{2Ze^2}{4\pi\varepsilon_0 K}$$

Since α-particles are generally obtained from natural radioactive substances, their kinetic energy K is known. Hence, value of D can be found easily from the above equation.

Example 3.1. *In Geiger-Marsden experiment on α-particle scattering from gold foil, the kinetic energy of α-particles and was 7.68 MeV. Calculate the distance of closest approach of α-particle if atomic number of gold is 79.*

Solution. $$D = \frac{2Ze^2}{4\pi\varepsilon_0 K}$$

Here, $Z = 79$, $e = 1.6 \times 10^{-19}\ C$; $\varepsilon_0 = 8.854 \times 10^{-12}$ F/m

$$K = 7.68 \text{ MeV} = 7.68 \times 1.6 \times 10^{-13} \text{ J}$$

$$D = \frac{2 \times 79 \times (1{\cdot}6 \times 10^{-19})^2}{4\pi \times 8{\cdot}854 \times 10^{-12} \times 7{\cdot}68 \times 1{\cdot}6 \times 10^{-13}}$$

$$= \mathbf{2.96 \times 10^{-14}\ m}$$

3.6. Major Deficiencies in Rutherford's Nuclear Model

It was found later on that Rutherford's model had two serious drawbacks concerning

(*i*) distribution of electrons outside the nucleus and

(*ii*) the stability of the atom as a whole.

It can be shown that electrostatic forces between the positive nucleus and the static negative extra-nuclear electrons are not enough to produce equilibrium in such a *nuclear* atom. For example, consider the case of an atom* having two electrons and a nucleus with a charge of $+2e$. Suppose the electrons are symmetrically placed at a distance of r from the nucleus and are stationary. The force of attraction between the nucleus and *each* of the electrons [Fig. 3.6 (*a*)] is $F = e.2e/4\pi\varepsilon r^2 = 2e^2/4\pi\varepsilon r^2$ while the force of repulsion between the two electrons is $= e^2/4\pi\varepsilon \times 4r^2 = e^2/16\pi\varepsilon r^2$. Since the force of attraction is eight times the force of repulsion, the electrons will fall into the nucleus thereby destroying the stable structure of the atom.

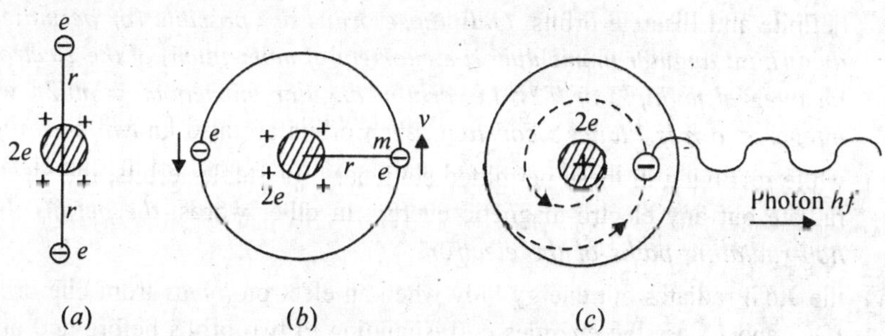

Fig. 3.6

To overcome this difficulty, Rutherford suggested that stability can be achieved (as in a solar system) by assuming that electrons, instead of being static, revolve round the nucleus with such a speed that the centrifugal force balances the attractive force exerted by the nucleus on the electrons. As seen from Fig. 3.6 (*b*), condition for stability is achieved when

$$\frac{mv^2}{r} = \frac{2e.e}{4\pi\varepsilon_0 r^2} \quad \text{or} \quad mv^2 r = \frac{2e^2}{4\pi\varepsilon_0}$$

*Of course, if we consider hydroen atom having one electron, it will simply fall into the nucleus due to its attraction. There is no other force to prevent this collapse.

In general, if Z is the atomic number, then nuclear charge is Ze, so that the above relation becomes

$$\frac{mv^2}{r} = \frac{Ze.e}{4\pi\varepsilon_0 r^2} \quad \text{or} \quad mv^2 r = \frac{Ze}{4\pi\varepsilon_0}$$

Incidentally, it may be noted that according to the above relation, it is possible to have an infinite number of orbits in which electrons can rotate.

But this assumption of revolving electrons led to serious difficulty from the point of view of the *electromagnetic theory* according to which an accelerated *charge* must continuously emit electromagnetic radiation or energy. Since an electron revolving in a circular orbit has centripetal acceleration ($= v^2/r$), it must radiate energy as per the laws of *classical electrodynamics*.

Due to this continuous loss of energy, the electrons will gradually approach the nucleus by a spiral path and finally fall into it as shown in Fig. 3.6 (*c*). Hence, it is seen that the orbital motion of the electron destroys the very purpose for which it was postulated *i.e.* the stability of the atom. Obviously, either Rutherford's nuclear atomic model with revolving electrons is defective or the classical electromagnetic theory falls in this particular case. This dilemma was solved in 1913 by Niels Bohr who proposed an improved version of Rutherford's atomic model.

3.7. Bohr's Atomic Model

This model (first proposed for hydrogen atom but later applied to other atoms as well) retains the two essential features of Rutherford's planetary model *i.e.*

(*i*) the atom has a massive positively-charged nucleus and

(*ii*) the electrons revolve round their nucleus in circular orbits—the centrifugal force being balanced, as before, by the electrostatic pull between the nucleus and electrons.

However, he extended this model further by utilizing Planck's Quantum Theory. He made the following three assumptions :

(*iii*) an electron cannot revolve round the nucleus in any arbitrary orbit but in just certain definite and discrete orbits. *Only those orbits are possible* (*or permitted*) *for which the orbital angular momentum* (*i.e. moment of momentum*) *of the electron is equal to an integral multiple of* $h/2\pi$ *i.e. orbital angular momentum* = $nh/2\pi$ *where* n *is an integer and* h *is Planck's constant*. Such orbits are also known as *stationary* orbits.

(*iv*) while revolving in these permitted stationary (or stable) orbits, the electron does not radiate out any electro-magnetic energy. In other words, *the permissible orbits are non-radiating paths of the electron.*

(*v*) the atom radiates out energy only when an electron *jumps* from one orbit to another. If E_2 and E_1 are the energies corresponding to two orbits before and after the jump, the frequency of the emitted photon is given by the relation

$$\boxed{E_2 - E_1 = hf \text{ or } \Delta E = hf}$$

where f is the frequency of the emitted radiations.

Explanation. If I is the moment of inertia of an electron and ω its angular velocity, then as per assumption (*iii*) given above

$$\omega I = nh/2\pi \qquad \text{or} \qquad (mr^2)\,\omega = nh/2\pi$$

$$\text{or} \qquad (mr^2)v/r = n{\cdot}h/2\pi \qquad \text{or} \qquad mvr = n{\cdot}h/2\pi.$$

Alternatively, since the momentum of the revolving electron is mv, its moment about the nucleus is $= mvr$

Hence, $mvr = n \cdot h/2\pi$ ...(*i*)

where $n = 1, 2, 3$ for the first, second and third orbits respectively. It is called the *principal quantum number* and because it can take whole number values only, it *fixes the sizes of the allowed orbits* (also called Bohr's circular orbits).

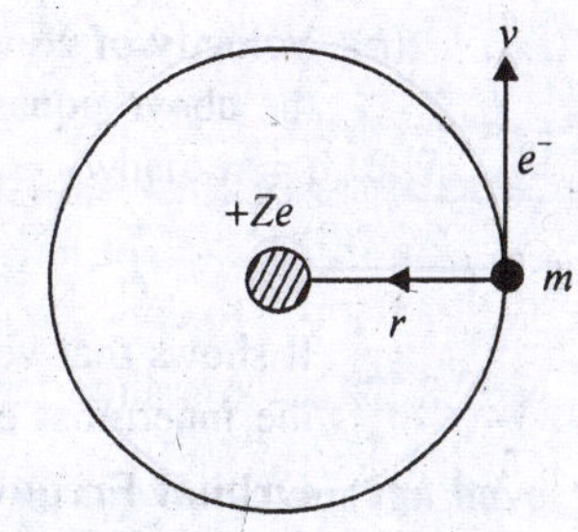

Fig. 3·7

Let the different permitted orbits have energies of E_1, E_2, E_3 etc. as shown in Fig. 3.8 (*a*). The electron can be raised from $n = 1$ orbit to any other higher orbit if it is given proper amount of energy. When it drops back to $n = 1$ orbit after a short interval of time, it gives out the energy difference ΔE in the form of a radiation as shown in Fig. 3.8 (*b*). The relation between the energy released and frequency of the emitted radiation is

$$E_2 - E_1 = hf \quad \text{or} \quad \Delta E = hf$$

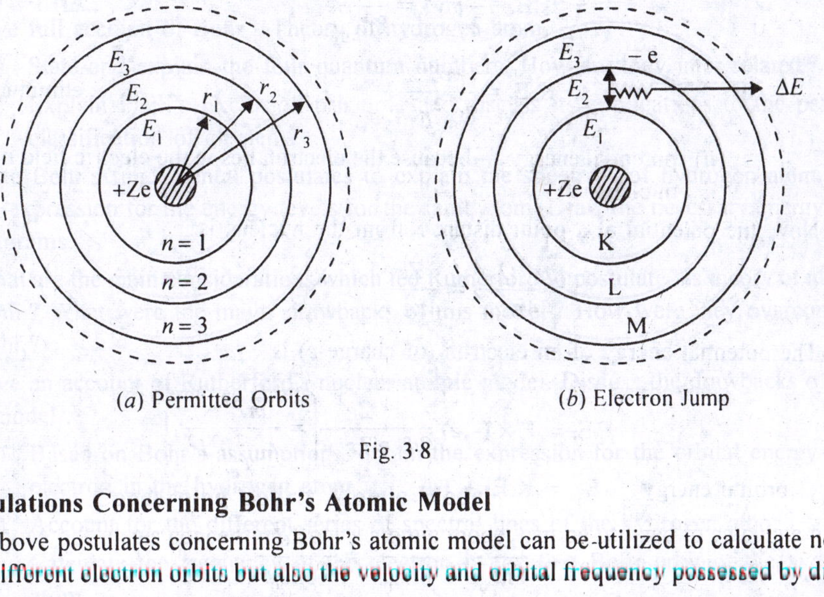

Fig. 3·8

3.8. Calculations Concerning Bohr's Atomic Model

The above postulates concerning Bohr's atomic model can be utilized to calculate not only the radii of different electron orbits but also the velocity and orbital frequency possessed by different electrons.

The following two equations may be used to find the radii of different Bohr's circular orbits.

$$mv^2/r = Ze^2/4\pi\epsilon_0 \qquad \text{—Art. 3.6}$$

$$mvr = nh/2\pi \qquad \text{— given above}$$

(*a*) **Radii of Orbits.** Eliminating v from the above two equations, we get

$$r = \frac{\epsilon_0 n^2 h^2}{\pi m Z e^2} \quad \text{or} \quad r_n = \frac{\epsilon_0 n^2 h^2}{\pi m Z e^2}$$

It is seen that the radii of the permitted orbits vary as the square of the principal quantum number n.

Also, $r_n = n^2 r_1$

For hydrogen atom, $r_1 = 0.53 \times 10^{-10}$ m, $r_2 = 2^2 \times 0.53 \times 10^{-10}$ m $= 2.12 \times 10^{-10}$ m. All values between r_1 and r_2 are forbidden.

(*b*) **Velocity of Revolving Electrons.** The velocity of a revolving electron as found from the above equations is

$$v = \frac{Ze^2}{2\varepsilon_0 nh} = 9\times 10^9 \cdot \frac{2\pi Z e^2}{nh}$$

It shows that velocity is inversely proportional to n *i.e.* $v \alpha 1/n$. Hence, the electron in the innermost orbit has the highest velocity (nearly 1/137 of the velocity of light).

(*c*) **Orbital Frequency.** The orbital rotational frequency of an electron is

$$f = \frac{v}{2\pi r} = \frac{mZ^2e^4}{4\varepsilon_0{}^2 n^3 h^3}$$

It is seen that $f \propto 1/n^3$.

(*d*) **Electron Energy.** The orbital energy of a revolving electron is of two types :

(*i*) kinetic energy due to the motion of the electron. Its value is

$$\text{K.E.} = \frac{1}{2}mv^2 = \frac{Ze^2}{8\pi\varepsilon_0 r} \qquad \text{...Art.1·7}$$

or

$$\text{K.E.} = \frac{mZ^2e^4}{8\varepsilon_0{}^2 n^2 h^2} \qquad \text{...eliminating } r$$

(*ii*) potential energy — because the electron lies in the electric field of the positive nucleus.

Now, the potential at a point distant r from the nucleus is

$$V = \frac{Q}{4\pi\varepsilon_0 r} = \frac{Ze}{4\pi\varepsilon_0 r}$$

The potential energy of an electron (of charge-e) is

$$\text{P.E.} = V\times(-e) = -\frac{Ze^2}{4\pi\varepsilon_0 r}\left(= \frac{mZ^2e^4}{4\varepsilon_0{}^2 n^2 h^2}\right)$$

∴ orbital energy E_n = K.E. + P.E.

$$= \frac{Ze^2}{8\pi\varepsilon_0 r} - \frac{Ze^2}{4\pi\varepsilon_0 r} = -\frac{Ze^2}{8\pi\varepsilon_0 r}$$

If r is eliminated from the above equation, then orbital energy*

$$\boxed{E_n = -\frac{mZ^3e^4}{8\varepsilon_0{}^2 n^2 h^2}}$$

where n = 1, 2, 3..... etc. for the energy states that it is possible for the electron to have. It will be seen that $E_n = E_1/n^2$.

Note. Numerically, potential energy is twice the kinetic energy.

Example 3.2. *Calculate for hydrogen atom* (*i*) *velocity of an electron in ground state* (*ii*) *radius of Bohr's orbit in ground state* (*iii*) *time taken by the electron to traverse the Bohr's first orbit* (*iv*) *Rydberg's constant.*

---

*The negative sign only indicates that this much energy is required to remove the electron.

Solution. (*i*) As seen from Art. 3.8

$$v = 9\times10^9 \frac{2\pi Ze^2}{nh} = 9\times10^9 \frac{2\pi e^2}{nh} \qquad (\because\ Z = 1)$$

$$= 9 \times 10^9 \times 2\pi \times (1.6 \times 10^{-19})^2/1 \times 6.625 \times 10^{-34}$$

$$= \mathbf{21.85 \times 10^5\ m/s}$$

(*ii*)
$$r = \frac{\varepsilon_0 n^2 h^2}{\pi mz\, e^2} = \frac{\varepsilon_0 h^2}{\pi m\, e^2} \qquad (\because\ n = 1 \text{ and } z = 1)$$

$$= \frac{8{\cdot}854\times10^{-12}\times\left(6{\cdot}625\times10^{-34}\right)^2}{\pi\times9{\cdot}1\times10^{-31}\times\left(1{\cdot}6\times10^{-19}\right)^2} = \mathbf{0{\cdot}53\times10^{-10}\ m}$$

(*iii*)
$$t = \frac{2\pi r}{v} = \frac{2\pi\times0{\cdot}53\times10^{-10}}{21{\cdot}85\times10^6} = \mathbf{1{\cdot}5\mu s}$$

(*iv*)
$$R = \frac{me^4}{8\varepsilon_0{}^2 h^3 c}$$

$$= \frac{9{\cdot}1\times10^{-31}\times(1{\cdot}6\times10^{-19})^4}{8\times\left(8{\cdot}854\times10^{-12}\right)^2\times(6{\cdot}625\times10^{-34})^3\times3\times10^8}$$

$$= \mathbf{10.97\times10^6\ m^{-1}}$$

Example 3.3. *Energy in a Bohr orbit is given to be equal to* $-B/n^2$ *where* $B = 2.179 \times 10^{-18}$ *joule.*

Calculate the frequency of radiation when the electron jumps from the third orbit to the second. Take $h = 6.6 \times 10^{-34}$ *J.s.*

Solution.
$$E_3 = -\frac{B}{n^2} = -\frac{2{\cdot}179\times10^{-18}}{3^2} = -2{\cdot}42\times10^{-17}\ \text{J}$$

$$E_2 = -\frac{2{\cdot}179\times10^{-18}}{2^2} = -5{\cdot}45\times10^{-17}\ \text{J}$$

Now
$$\Delta E = hf \quad \text{or} \quad f = \Delta E/h$$

$\therefore$
$$f = \frac{E_3 - E_2}{h} = \frac{[-2{\cdot}42-(-5{\cdot}45)]\times10^{-17}}{6{\cdot}6\times10^{-34}}$$

$$= \mathbf{4.6\times10^{16}\ Hz.}$$

Example 3.4. *Calculate the frequency of the electron in the Bohr's first orbit in a hydrogen atom.*

Solution. As seen from Art. 3.8, the orbital frequency of an electron is given by

$$f = \frac{mZ^2e^4}{4\varepsilon_0{}^2 n^3 h^3}\ \text{Hz}$$

Now, in hydrogen atom, $Z = 1$. Also, we are given $n = 1$. Hence

$$f = \frac{me^4}{4\varepsilon_0{}^2 h^3}$$

$$= \frac{9{\cdot}1\times10^{-31}\times(1{\cdot}6\times10^{-19})^4}{4\times(8{\cdot}854\times10^{-12})^2\times(6{\cdot}625\times10^{-34})^3}$$

$$= \mathbf{6.54\times10^{15}\ Hz.}$$

Example 3.5. *At what speed must the electron revolve round the nucleus of a hydrogen atom in order that it may not be pulled into the nucleus by electrostatic attraction ? Also calculate its orbital energy and frequency.*

Solution. It should be remembered that hydrogen atom consists of a nucleus having one proton (of charge + e) and one electron revolving round it. Also, for hydrogen atom $Z = 1$ and $n = 1$. Now

$$v = \frac{Ze^2}{2\varepsilon_0 nh}$$

$$= \frac{(1{\cdot}6\times10^{-19})^2}{2\times8{\cdot}854\times10^{-12}\times6{\cdot}62\times10^{-34}} = \mathbf{2.25\times10^6\ m/s}$$

Also,
$$E_n = -\frac{mZ^2e^4}{8\varepsilon_0{}^2n^2h^2}$$

$$= -\frac{9{\cdot}1\times10^{-31}\times(1{\cdot}6\times10^{-19})^4}{8\times(8{\cdot}854\times10^{-12})^2\times(6{\cdot}62\times10^{-34})^2}$$

$$= \mathbf{-.21.7\times10^{-19}\ J}$$

$$f = \frac{mZ^2e^4}{4\varepsilon_0{}^2n^3h^3}$$

$$= \frac{9{\cdot}1\times10^{-31}\times(1{\cdot}6\times10^{-19})^4}{4(8{\cdot}854\times10^{-12})^3\times(6{\cdot}62\times10^{-34})^3} = \mathbf{6.7\times10^{15}\ Hz = 6{,}700}\ \text{THz}$$

Example 3.6. *Show that for large values of principal quantum number, the frequencies of an electron rotating in adjacent energy levels of hydrogen atom and the radiated frequencies for a transition between these levels all approach the same value.*

Solution. As seen from Art. 3.8, orbital frequency of a rotating electron is

$$f = \frac{mZ^2e^4}{4\varepsilon_0{}^2n^3h^3} \qquad ...(i)$$

Consider two adjacent orbits having principal quantum number n and $(n + 1)$ where n is very large as compared to unity.

$$E_n = -\frac{mZ^2e^4}{8\varepsilon_0{}^2n^2h^2} \text{ and } E_{n+1} = -\frac{mZ^2e^4}{8\varepsilon_0{}^2(n+1)^2h^2}$$

$$\therefore \quad E_{n+1} - E_n = \frac{mZ^2e^4}{8\varepsilon_0{}^2h^2}\left[\frac{1}{n^2} - \frac{1}{(n+1)^2}\right]$$

$$= \frac{mZ^2e^4}{8\varepsilon_0{}^2h^2}\left[\frac{(n+1)^2 - n^2}{n^2(n+1)^2}\right]$$

$$= \frac{mZ^2e^4}{8\varepsilon_0{}^2h^2}\left[\frac{(2n+1)}{n^2(n+1)^2}\right]$$

Neglecting 1 as compared to n and $2n$, we get

$$E_{n+1} - E_n \cong \frac{mZ^2e^4}{8\varepsilon_0{}^2h^2}\left(\frac{2n}{n^1}\right) \cong \frac{mZ^2e^4}{4\varepsilon_0{}^2h^2n^3}$$

If 'f' is the radiated frequency for transition between the two levels $(n + 1)$ and n, then

$$hf = \frac{mZ^2e^4}{4\varepsilon_0^{\,2}h^2n^3} \text{ or } f = \frac{mZ^2e^4}{4\varepsilon_0^{\,2}n^3h^3} \quad ...(ii)$$

As seen, Eqns. (*i*) and (*ii*) are equal proving that frequency of revolution of an electron in its orbit is equal to the frequency of the radiation emitted when the electron jumps from the state $n_2 = (n + 1)$ to $n_1 = n$.

3.9. Normal, Excited and Ionized Atom

Consider the case of the simplest atom *i.e.* hydrogen atom. When its only electron is in its innermost orbit ($n = 1$), then the atom is said to be in its normal (or unexcited) state. Generally, it is this condition in which most of the free hydrogen atoms in a gas are found to exist at normal room temperature and pressure. However, if spark is passed through hydrogen gas contained in a vessel, then high-speed electrons produced by the spark collide with hydrogen atom and may either completely remove the $n = 1$ electron from hydrogen atom or raise it to higher permitted orbits having $n = 2, 3, 4$ etc.

When the electron is *completely* removed from the atom, the atom is said to be *ionized.* If, however, the electron is forced into an outer or higher *n*-value orbit, then the atom is said to be excited (or in an excited state). The atom does not remain in the excited state longer than 10^{-8} second because the electron under the attractive force of the nucleus jumps to the lower permitted orbit. In so doing, the electron loses the energy it had earlier gained during collision. However, the electron may return by several jumps, thereby emitting many different radiations of different frequencies.

3.10. Bohr-Stoner Scheme of Building-up of Atoms

The orbital model of hydrogen was extended by Bohr and Stoner to include all of the chemical elements. As shown in Fig. 3.9, each atom consists of a positively-charged nucleus with a certain number of electrons revolving round it in discrete and definite orbits.

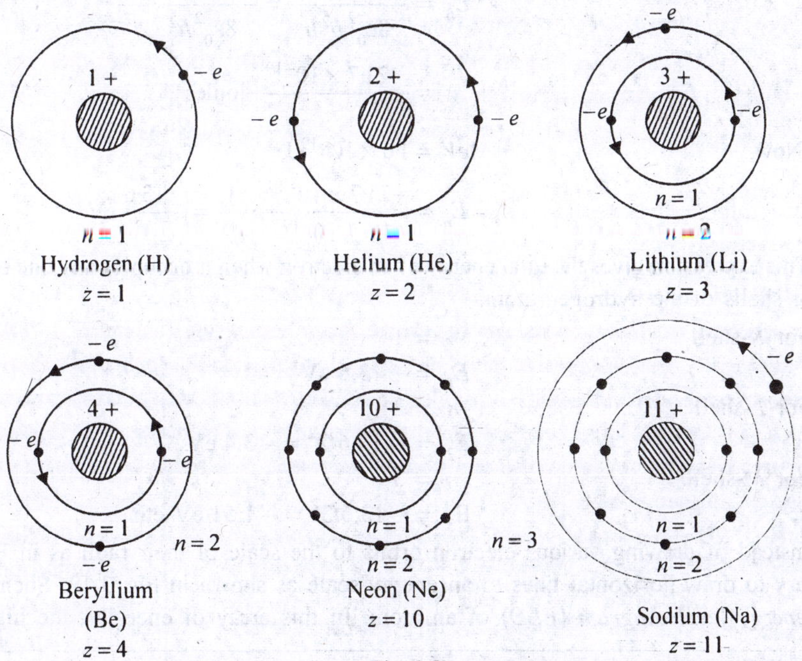

Fig. 3·9

The positive charge carried by the nucleus is numerically equal to the atomic number (Z) and it determines the number of planetary electrons an atom can have. A helium atom, with atomic number $Z = 2$, has two positive charges on the nucleus and two electrons outside in the $n = 1$ orbit. A beryllium atom (atomic number $Z = 4$) has 4 positive charges on the nucleus and four electrons outside, two revolving in $n = 1$ orbit and the other two in the $n = 2$ orbit. Similarly, sodium atom has 11 positive charges on the nucleus and 11 electrons outside – 2 in $n = 1$ orbit, 8 in $n = 2$ orbit and 1 in $n = 3$ orbit.

Bohr's orbits to which electrons are confined are called electron *shells*. As we go from one element to another in the atomic table, starting with hydrogen, electrons are added one after the other, filling one shell and then another. A shell is said to be *filled* or *complete*, when it contains $2n^2$ electrons where n is the shell or orbit number. For example, first shell is complete when it has $2 \times 1^2 = 2$ electrons, second, when it has $2 \times 2^2 = 8$ electrons and third, when it has $2 \times 3^3 = 18$ electrons etc. as shown below :

Principal quantum number	$n = 1$	$n = 2$	$n = 3$	$n = 4$
Number of electrons	2	8	18	32

However, the maximum number of electrons a shell of any n-value can have is 32. Among heavier elements there are several departures from the above rule. It may be noted that $n = 1$ orbit is known as K-shell, $n = 2$ orbit as L-shell, $n = 3$ orbit as M-shell and so on.

3.11. Electron Energy Levels in Hydrogen Atom

As seen from Art. 3.8, orbital energy of an electron revolving in nth orbit or shell* is

$$E_n = -\frac{mZ^2e^4}{8\varepsilon_0^{\ 2}n^2h^2}$$

In the case of hydrogen atom, $Z = 1$, hence

$$E_n = -\frac{me^4}{8\varepsilon_0^{\ 2}n^2h^2} = -\frac{me^4}{8\varepsilon_0^{\ 2}h^2}\cdot\frac{1}{n^2}$$

$$= -\frac{21{\cdot}7 \times 10^{-19}}{n^2} \text{ joules}$$

Now, $\quad 1eV = 1{\cdot}6 \times 10^{-19}\,\text{J}$

$$\therefore \quad E_n = \frac{-21{\cdot}7 \times 10^{-19}}{1{\cdot}6 \times 10^{-19}}\cdot\frac{1}{n^2} = -\frac{13{\cdot}6}{n^2}\text{eV}$$

This expression gives the total energy of an electron when it occupies any one of the different orbits or shells of the hydrogen atom.

For K-shell $\quad n = 1$

$E_1 = -13.6$ eV

For L-shell $\quad n = 2$

$\therefore \quad E_2 = -13.6/2^2 = -3.4$ eV

For M-shell, $\quad n = 3$

$\therefore \quad E_3 = -13.6/3^2 = -1.51$ eV etc.

Instead of drawing various electron orbits to the scale of their radii as in Fig. 3.10, it is customary to draw horizontal lines to an energy scale as shown in Fig. 3.12. Such a diagram is called *energy level diagram* (*ELD*) of an atom. In this array of energies, the higher (*i.e.* less

*The term *shell* is used to relate a group of electrons having same quantum number n. But this term is somewhat misleading because electrons are not confined to a volume that is a shell in a strictly classical sense.

negative) energies are at the bottom. The various electron jumps between allowed orbits now become vertical arrows between the energy levels (Fig. 3.12). Greater the length of the arrow, greater is the energy hf of the radiated photon.

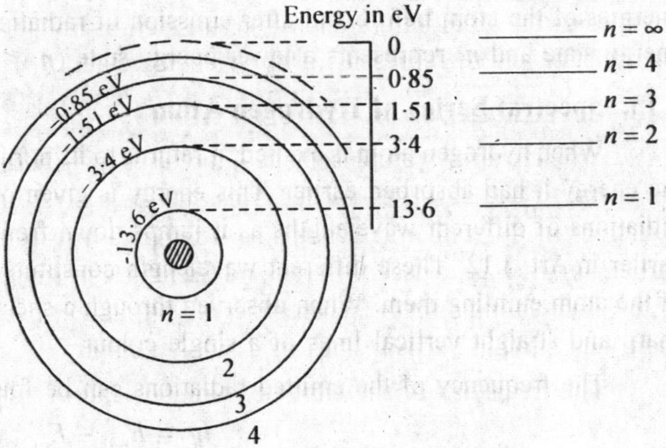

Fig. 3.10

3.12. Bohr's Model Applied to Hydrogen Spectra

Bohr's orbital model of hydrogen atom is very successful in accounting for the spectral series emitted by an excited hydrogen atom. In a normal unexcited hydrogen atom, the electron is in its lowest energy state at the bottom *i.e.* in the innermost orbit with $n = 1$ (K-shell). It is also called the *ground state*. When in the ground state, the electron is stable and revolves without emitting or absorbing any energy.

The atom becomes *excited* when the electron absorbs energy in some way or the other. There are many ways in which this can be brought about :

(*i*) an accelerated particle may be utilized to hit the planetary electron and hence impart some energy to it.

(*ii*) hydrogen may be heated so that electron absorbs thermal energy.

(*iii*) hydrogen may be illuminated, in which case, the electron absorbs energy from the photons.

In each case, due to the absorption of energy, the electron is raised to one of the higher permitted energy states or levels corresponding to different values of n.

(*a*) Suppose, the electron absorbs 13.6 eV. Then it is raised to $n = \infty$ level *i.e.* the electron leaves the atom and drifts about freely with only-its thermal kinetic energy. Since the electron is completely removed from the atom, the atom is said to be *ionized*. The energy 13.6 eV is called the ionization energy of hydrogen atom. Obviously, *ionization potential* of hydrogen atom is 13.6 V.

(*b*) Now, suppose that the electron absorbs an energy of 20 eV. This energy is not only sufficient to remove the electron completely from the atom (13.6 eV) but imparts an additional kinetic energy to the electron equal to (20 – 13.6) = 6.4 eV.

(*c*) Suppose, the electron absorbs an energy of 10.2 eV. In this case, it will be raised or excited from its ground state (n = 1) to the next higher permitted state (n = 2). It is so because, as seen from Fig. 3.10, the energy difference between these two states is (13.6 – 3.4) = 10.2 eV. This energy is called *excitation* energy.

(*d*) Similarly, for being excited from n = 1 state to n = 4 state, the electron must absorb an amount of energy = (13.6 – 0.85) = 12.75 eV.

After excitation, the electron returns to its ground state in about 10^{-8} second. If it is excited, say, to n = 4 state, then it may jump from n = 4 state to n = 1 state in one step. Alternatively, it may go 4 → 2 → 1 or 4 → 3 → 1 or 4 → 3 → 2 → 1. In each step of return journey, the electron radiates the energy difference of the two states (or levels) in the form of radiations whose frequency is given by $E_{n2} - E_{n1} = hf$ or whose wavelength is $E_{n2} - E_{n1} = hc/\lambda$ where E_{n2} and E_{n1} are the energies of the atom before and after emission of radiations. It means that n_2 represents a higher energy state and n_1 represents a lower energy state ($n_1 < n_2$).

3.13. Spectral Series of Hydrogen Atom

When hydrogen atom is excited, it returns to its normal unexcited or *ground* state by emitting the energy it had absorbed earlier. This energy is given out by the raised electron in the form of radiations of different wavelengths as it jumps down from a higher to a lower orbit as discussed earlier in Art. 3.12. These different wavelength constitute spectral series which are characteristic of the atom emitting them. When observed through a spectroscope, these radiations are imaged as sharp and straight vertical lines of a single colour.

The frequency of the emitted radiations can be found from the following relation

$$hf = E_{n2} - E_{n1}$$

Now,
$$E_{n2} = -\frac{mZ^2e^4}{8\varepsilon_0^{\,2}h^2n_2^{\,2}} = -\frac{me^4}{8\varepsilon_0^{\,2}h^2}\cdot\frac{1}{n_2^{\,2}} \qquad ...(\because Z = 1)$$

$$E_{n1} = -\frac{me^4}{8\varepsilon_0^{\,2}h^2}\cdot\frac{1}{n_1^{\,2}}$$

$\therefore$
$$hf = \frac{me^4}{8\varepsilon_0^{\,2}h^2}\left(\frac{1}{n_1^{\,2}} - \frac{1}{n_2^{\,2}}\right)$$

or
$$= \frac{me^4}{8\varepsilon_0^{\,2}h^3}\left(\frac{1}{n_1^{\,2}} - \frac{1}{n_2^{\,2}}\right)$$

Now, $c = f\lambda$ where c is velocity of light.

Hence, $f = c/\lambda$. Replacing f, we get

$$\frac{c}{\lambda} = \frac{me^4}{8\varepsilon_0^{\,2}h^3}\left(\frac{1}{n_1^{\,2}} - \frac{1}{n_2^{\,2}}\right)$$

or
$$\frac{1}{\lambda} = \frac{me^4}{8\varepsilon_0^{\,2}h^3c}\left(\frac{1}{n_1^{\,2}} - \frac{1}{n_2^{\,2}}\right)$$

Let
$$R = \frac{me^4}{8\varepsilon_0^{\,2}h^3c} = 10{\cdot}97 \times 10^6\ \text{m}^{-1}$$

= Rydberg's constant

$\therefore$
$$\frac{1}{\lambda} = R\left(\frac{1}{n_1^{\,2}} - \frac{1}{n_2^{\,2}}\right)$$

Now, wave-number (ν) is given by the reciprocal of wavelength *i.e.* $\nu = 1/\lambda$

$\therefore$
$$\nu = R\left(\frac{1}{n_1^{\,2}} - \frac{1}{n_2^{\,2}}\right) \qquad \text{– for hydrogen}$$

This is a general expression for the wave-number of radiation emitted by an electron as it jumps from higher orbit n_2 to lower orbit n_1. The different wavelengths which form different spectral series of hydrogen atom can be found by giving suitable values to n_1 and n_2. The different series observed in the hydrogen spectrum are as under :

(*a*) Lyman Series

It consists of all those wavelengths which are emitted when an electron jumps from different higher orbit to the final orbit with $n = 1$. In other words, all those electronic jumps which end at *K*-orbit give rise to Lyman series.

Here, $n_1 = 1$ and $n_2 = 2, 3, 4, \ldots\ldots, \infty$

The wavelength of the different members of this series may be found from the following relation :

$$v = \frac{1}{\lambda} = R\left(\frac{1}{1^2} - \frac{1}{n_2^2}\right) = R\left(1 - \frac{1}{n_2^2}\right)$$

(*i*) **First member :** $n_2 = 2$

$$\therefore \quad \frac{1}{\lambda} = R\left(1 - \frac{1}{2^2}\right) = \frac{3R}{4}$$

Or
$$\lambda = \frac{4}{3R} = \frac{4}{3 \times 10{\cdot}97 \times 10^6}$$
$$= 1216 \times 10^{-10}\ \text{m} = 1216\ \text{Å}$$

(*ii*) **Second member :** $n_2 = 3$

$$\frac{1}{\lambda} = R\left(1 - \frac{1}{3^2}\right) = \frac{8R}{9}$$

$$\therefore \quad \lambda = \frac{9}{8R} = 1026\ \text{Å and so on}$$

(*iii*) **Limiting member :** $n_2 = \infty$

The limiting member of the series *i.e.* the shortest wavelength of this series is given by putting $n_2 = \infty$ in the above relation.

$$\therefore \quad \frac{1}{\lambda} = R\left(1 - \frac{1}{\infty^2}\right) = R$$

Or
$$\lambda = \frac{1}{R} = \frac{1}{10{\cdot}97 \times 10^4} = 912\ \text{Å}$$

This series of wavelengths lies in the *ultraviolet* region of the spectrum. The jumps corresponding to this series are shown in Fig. 3.11.

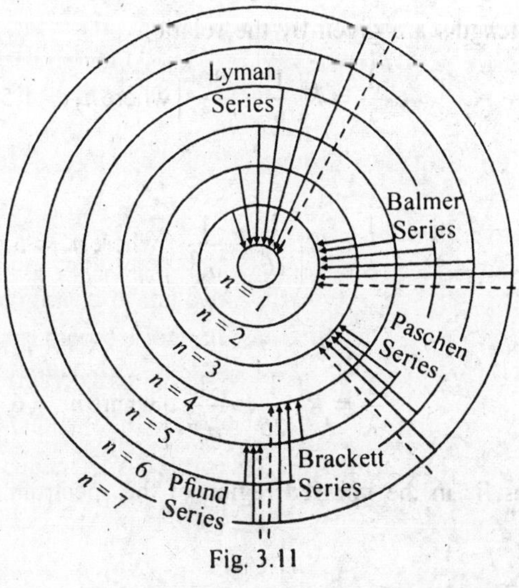

Fig. 3.11

(b) Balmer Series

It consists of all those wavelengths which are emitted when different electron jumps end at orbit (or level) with $n = 2$. Hence, difference members of this series may be found by putting $n_1 = 2$ and $n_2 = 3, 4, 5, \ldots\ldots, \infty$

$$\therefore \qquad \nu = \frac{1}{\lambda} = R\left(\frac{1}{2^2} - \frac{1}{n_2^{\,2}}\right)$$

$$\text{Or} \qquad \frac{1}{\lambda} = R = \left(\frac{1}{4} - \frac{1}{n_2^{\,2}}\right)$$

The wavelengths of different members are as under ·

(*i*) **First member :** $n_2 = 3$

$$\therefore \qquad \frac{1}{\lambda} = R\left(\frac{1}{4} - \frac{1}{3^2}\right) = \frac{5R}{36}$$

$$\text{Or} \qquad \lambda = \frac{36}{5 \times 10{\cdot}97 \times 10^6}$$

$$= 6563 \times 10^{-10}\ m = 6563\ \text{Å}$$

(*ii*) **Second member :** $n_2 = 4$

$$\frac{1}{\lambda} = R\left(\frac{1}{4} - \frac{1}{4^2}\right) = \frac{3R}{16}$$

$$\therefore \qquad \lambda = \frac{16}{3R} = 4861\ \text{Å and so on}$$

(*iii*) **Limiting member :** $n_2 = \infty$

$$\therefore \qquad \frac{1}{\lambda} = R\left(\frac{1}{4} - \frac{1}{\infty^2}\right) = \frac{R}{4}$$

$$\text{Here,} \qquad \lambda = \frac{4}{R} = 3646\ \text{Å}$$

This series lies in the visible and near ultraviolet region of solar spectrum.

(c) Paschen Series

The different wavelengths are given by the relation

$$\frac{1}{\lambda} = R\left(\frac{1}{9} - \frac{1}{n_2^{\,2}}\right) \text{ where } n_2 = 4, 5, 6 \ldots \infty$$

(d) Brackett Series

$$\frac{1}{\lambda} = R\left(\frac{1}{16} - \frac{1}{n_2^{\,2}}\right) \text{ where } n_2 = 5, 6, 7 \ldots \infty$$

(e) Pfund Series

$$\frac{1}{\lambda} = R\left(\frac{1}{25} - \frac{1}{n_2^{\,2}}\right) \text{ where } n_2 = 6, 7, 8 \ldots \infty$$

The last three series lie in the infrared region of the spectrum.

Combined Formula

The composite formula from which all series can be found is

$$\boxed{\frac{1}{\lambda} = R\left(\frac{1}{N^2} - \frac{1}{n^2}\right)}$$

where N = 1, 2, 3, ..., ∞ for Lyman, Balmer and Paschen series etc.

n = any integer greater than N

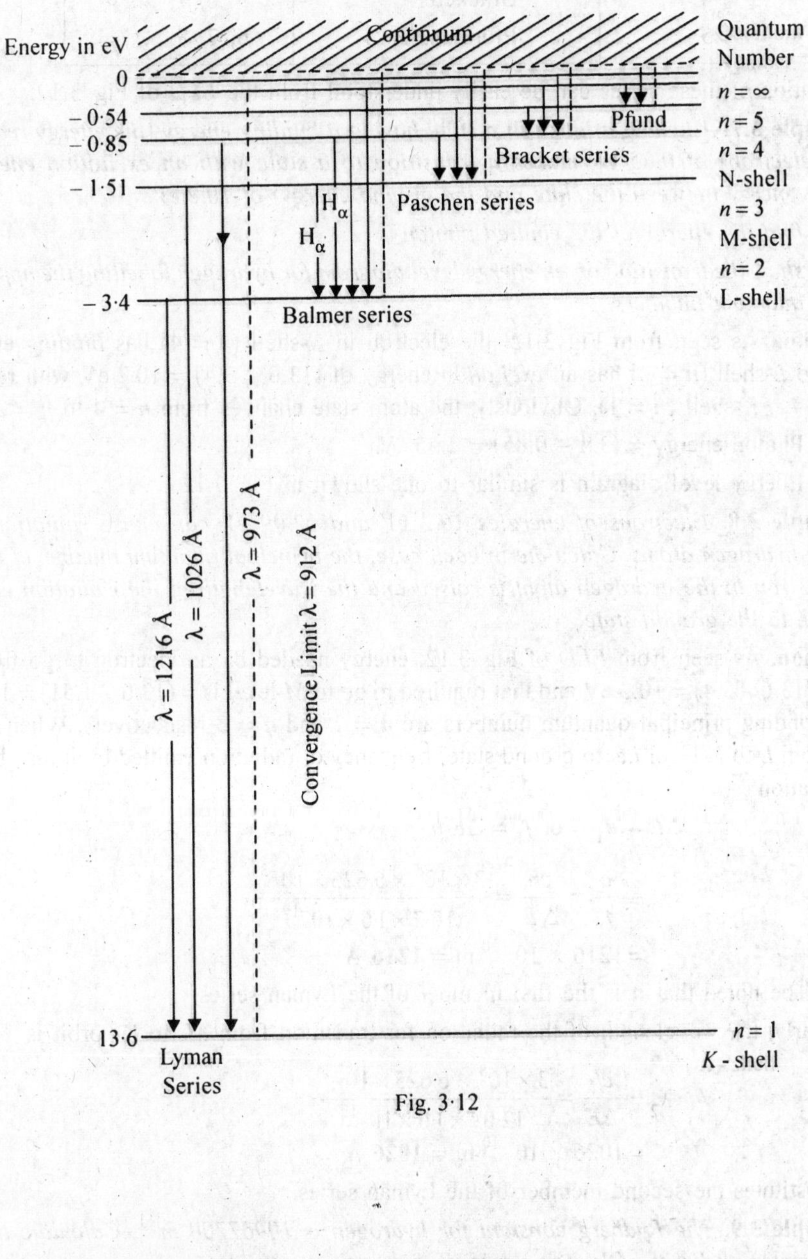

Fig. 3·12

The above results are summarised below.

Hydrogen Series

Value of n_1	Name of Series	Value of n_2
1	Lyman	2, 3, 4, ..., ∞
2	Balmer	3, 4, 5, ..., ∞
3	Paschen	4, 5, 6, ..., ∞
4	Brackett	5, 6, 7, ..., ∞
5	Pfund	6, 7, 8, ..., ∞

Formation of these series can be easily understood from the *ELD* of Fig. 3.12.

Example 3.7. *A hydrogen atom in a state having a binding energy (the energy required to remove an electron) of 0.85 eV makes a transition to a state with an excitation energy (the difference in energy between the state and the ground energy) of 10.2 eV,*

(*i*) *find the energy of the emitted photon*

(*ii*) *show the transition on an energy level diagram for hydrogen labelling the appropriate quantum numbers.*

Solution. As seen from Fig. 3.12, the electron in *N*-shell ($n = 4$) has *binding* energy of 0.85 eV. Also *L*-shell ($n = 2$) has an *excitation* energy of (13.6 – 3.4) = 10.2 eV with respect to ground state *i.e.* *K*-shell ($n = 1$). Obviously, the atom state changes from $n = 4$ to $n = 2$.

(*i*) Photon energy = (3.4 – 0.85) = 2.55 eV.

(*ii*) Energy level diagram is similar to one shown in Fig. 3.12.

Example 3.8. *Electrons of energies 10.2 eV and 12.09 eV can cause radiations to be emitted from hydrogen atoms. Calculate in each case, the principal quantum number of the orbit to which electron in the hydrogen atom is raised and the wavelength of the radiation emitted if it drops back to the ground state.*

Solution. As seen from *ELD* of Fig. 3.12, energy needed by an electron to go from *K*-to *L*-level is = (13.6 – 3.4) = 10.2 eV and that required to go to *M*-level is = (13.6 – 1.51) = 12.09 eV. The corresponding principal quantum numbers are $\boldsymbol{n} = \mathbf{2}$ and $\boldsymbol{n} = \mathbf{3}$ respectively. When electron falls back from *L*-to *K*-level *i.e.* to ground state, frequency of radiation emitted by it may be found from the relation

$$\Delta E = hf_1 = \text{or } f_1 = \Delta E/h$$

$$\lambda_1 = \frac{c}{f_1} = \frac{ch}{\Delta E} = \frac{3\times10^8\times6{\cdot}625\times10^{-34}}{10{\cdot}2\times1{\cdot}6\times10^{-19}}$$

$$= 1216\times10^{-10}\text{ m} = \mathbf{1216\ \text{Å}}$$

It will be noted that it is the first member of the Lyman series.

Similarly, the wavelength of the radiation for transition from *M*- to K- orbit is

$$\lambda_2 = \frac{ch}{\Delta E} = \frac{3\times10^8\times6{\cdot}625\times10^{-34}}{12{\cdot}09\times1{\cdot}6\times10^{-19}}$$

$$= 1026\times10^{-10}\text{ m} = \mathbf{1026\ \text{Å}}$$

It constitutes the second member of the Lyman series.

Example 3.9. *The Rydberg constant for hydrogen is 10967700* m^{-1}. *Calculate the short and long wavelength limits of Lyman series of hydrogen spectrum.*

Solution. The first member of the Lyman series with $n_2 = 1$ has the longest wavelength and the limiting member with $n_2 = \infty$ has the shortest wavelength.

First Member
$$\frac{1}{\lambda} = R\left(1 - \frac{1}{n_2^{\,2}}\right) = R\left(1 - \frac{1}{2^2}\right) = \frac{3R}{4}$$

$$\therefore \quad \lambda = \frac{4}{3R} = \frac{4}{3 \times 10967700} = 1216 \times 10^{-10} = \mathbf{1216\ Å}$$

Limiting Member
$$\frac{1}{\lambda} = R\left(1 - \frac{1}{\infty^2}\right) = R$$

$$\therefore \quad \lambda = \frac{1}{R} = \frac{1}{10967700} = \mathbf{912\ Å}$$

3.14. Deficiencies in Bohr's Theory

Bohr's theory, based on circular electron orbits, was successful in predicting accurately the positions of the spectral lines of the neutral hydrogen atom and singly-ionized helium atom in terms of but one quantum number called *principal quantum number n*. Soon after, however, it was revealed by refined spectroscopic analysis that the individual spectral lines were not really single but had a *fine structure i.e.* they consisted of several very fine lines lying close together. For example, H_α line in the Balmer series of hydrogen atom (Fig. 3.12) was found to consist of 5 component lines. This fine structure of spectral lines could not be explained on Bohr's theory according to which there is only one orbit (or energy level) for each principal quantum number. The existence of fine structure, on the other hand, suggests that for a given principal quantum number *n*, there are several orbits (or energy levels) of slightly different energies.

In brief, certain obvious deficiencies in Bohr's theory may be summarised as under :

(*i*) the planetary model introduces only one quantum number *n* whereas experimental evidence concerning fine structure of spectral lines suggests additional quantum numbers.

(*ii*) it applies to one-electron atoms and is not easily extended to describe more complicated or many-electron atoms.

(*iii*) it cannot be used to make any calculations about transitions from one level to another such as the rate at which they occur or the selection rules which apply to them.

(*iv*) it cannot be used for the quantitative explanation of chemical bonding. Thus, the calculation of bond strength (bond-breaking energy) in the simplest system like ionized hydrogen, molecule, according to this theory gives a negative value suggesting that such an ion cannot exist. Actually, however, it does exist and bond strength in this case is + 61 kcal/mole.

3.15. Sommerfeld's Relativistic Model

Sommerfeld succeeded partially in explaining fine structure by extending Bohr's theory in two ways :

(*i*) he allowed the possibility of elliptical orbits for the electrons in addition to Bohr's circular orbits.

(*ii*) he took note of the relativistic variation of electron mass with velocity which was different at different parts of the elliptical orbit.

Now, when elliptical orbits are permitted, one has to deal with two variable quantities. In polar coordinates, these are (*a*) the varying distance of the electron from the nucleus *i.e.* r and

(*b*) the varying angular position of the electron with respect to the nucleus *i.e.* the azimuthal angle ϕ.

To deal with these two variables, we need two quantum numbers :

(*i*) one is the original quantum number n of Bohr's theory which has been retained and associated with the mean separation of the electron from the nucleus. It also determines the energy of the electrons.

(*ii*) the other is a new quantum number called *orbital* (*or azimuthal*) *quantum number* l which has been introduced to characterise the angular momentum of an orbit *i.e.* it determines the orbital angular momentum of the electron. Its value varies from zero to $(n - 1)$ in steps of unity.

This orbital quantum number is useful in finding the possible elliptical orbits. Possible ellipses are such that

$$\frac{b}{a} = \frac{l+1}{n}$$

where a and b are semi-major and semi-minor axes respectively of the ellipse.

According to Sommerfeld's model, for any principal quantum number n, there are n possible orbits of varying eccentricities called *sub-orbits* or *sub-shells* (and having different values of l). These possible sub-orbits possess slightly different energies because of relativistic variation of the electron mass*. For example, for $n = 1$, number of possible sub-orbits is 1, for $n = 2$, the number of possible sub-orbits is 2 and for $n = 3$, number of possible sub-orbits is 3 and so on. These additional possible sub-orbits or sub-shells for a given principal quantum number are allotted different l-values. Different sub-orbits have been shown circular for the sake of simplicity although in practice, many are elliptical (Fig. 3.13).

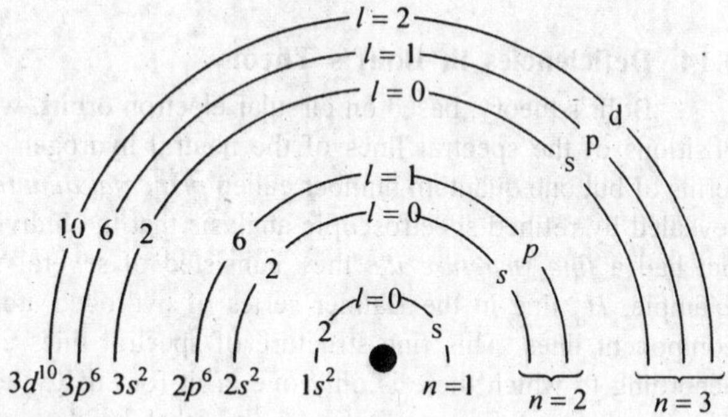

Fig. 3.13

For $n = 1$, only one sub-orbit or sub-shell is possible having $l = 0$. For $n = 2$, two sub-orbits or sub-shells are possible having l-values of 0 and 1. For $n = 3$, three sub-orbits or sub-shells are possible with $l = 0$, 1 and 2 respectively (Fig. 3.13) and so on. Obviously, the orbital quantum number l can have only integral values lying between 0 and $(n - 1)$. In other words, $0 \le l \le (n - 1)$.

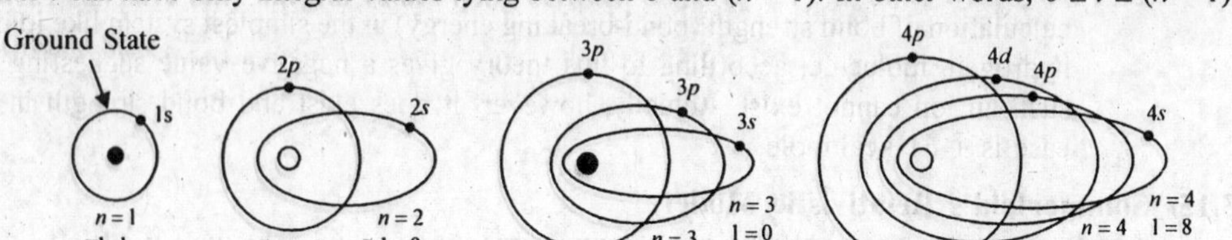

Fig. 3.14

Fig. 3.14 shows the possible orbits for values of n from 1 to 4. For example, for $n = 4$, four sub-shells are possible

*The velocity of an electron in an elliptical orbit varies (in a circle it is constant) being greater when the electron is near the nucleus and smaller when it is relatively away. Hence, according to special theory of relativity, the mass of the electron varies in the orbit.

(*i*) $n = 4, l = 0$ (*ii*) $n = 4, l = 1$ (*iii*) $n = 4, l = 2$ (*iv*) $n = 4, l = 3$.

For $l = 3$, $\frac{b}{a} = \frac{3+1}{4} = 1$ or $b = a$

Hence, $n = 4, l = 3$ sub-shell is circular as shown.

3.16. Orbital Angular Momentum of Different Orbits

The orbital angular momentum p_l (also called orbital mechanical moment) associated with different orbits is

$$p_l = l.\frac{h}{2\pi} \qquad \text{where } l = 0, 1, 2, ..., \text{etc.*}$$

It is represented vectorially as shown in Fig. 3.15 by a straight line which is parallel to the axis of rotation and whose positive sense is given by the direction of advance of a *right-hand* screw rotated with the motion.

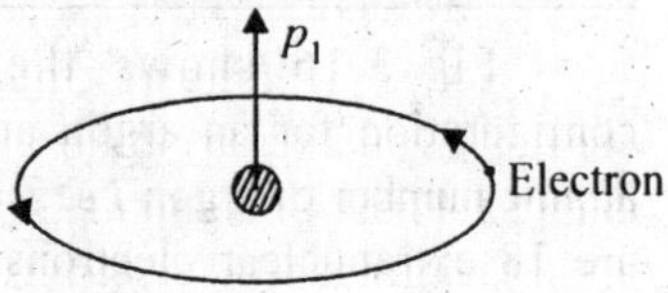

Fig. 3·15

It may, however, be mentioned here that quantum mechanical considerations indicate that $p_l \neq l\ h/2\pi$ as stated above but instead $p_l = l^{**}h/2\pi$ where $l^{**} = [l(l+1)]^{1/2}$. Accordingly. Mechanical moment of an *s*-electron $=$

$$[0(0+1)]^{1/2}.\frac{h}{2\pi} = 0$$

$$\text{mechanical moment of a } p\text{-electron} = [1(1+1)]^{1/2}.\frac{h}{2\pi} = \sqrt{2}\cdot\frac{h}{2\pi}$$

$$\text{mechanical moment of a } d\text{-electron} = [2(2+1)]^{1/2}.\frac{h}{2\pi} = \sqrt{6}\cdot\frac{h}{2\pi}$$

It is common practice to assign letters to *l*-values as given below :

orbital quantum number *l* :	0	1	2	3	4	5
electron designation :	*s*	*p*	*d*	*f*	*g*	*h*

Hence, electron in the $l = 0, 1, 2, 3, 4,...$ states are said to be in the *s, p, d, f, g*,...(continuing alphabetically) states.**

An electron with $n = 3$ and $l = 0$ is known as 3*s* electron, the number giving the *n*-value and letter the *l*-value. Similarly, an electron with $n = 2$ and $l = 1$ is known as 2*p* electron. All the electrons belonging to the same sub-shell are *equivalent* electrons and it is usual to denote the number of such electrons in an atom by an index. For example, two is electrons are written as $1s^2$, six 2*p* electrons as $2p^6$ and so on.

The different sub-shells are tabulated below :

Table 3.1

Shells	*n* \ *l*	Sub-shells 0	1	2	3	4
K	1	1*s*				
L	2	2*s*	2*p*			
M	3	3*s*	3*p*	3*d*		
N	4	4*s*	4*p*	4*d*	4*f*	
O	5	5*s*	5*p*	5*d*	5*f*	5*g*

*According to Bohr's theory. $p_l = n..h/\pi$ (Art. 3-7).

**These words stand for Sharp, Principal, Diffuse and Fundamental etc. after the nomenclature used in spectroscopy.

The maximum number of electrons allowed in any sub-shell is = 2 (2l + 1). The different sub-shells and the number of possible electrons for values of n up to 4 are tabulated below.

Table 3.2

Sub-shell	*K*	*L*		*M*			*N*			
Principal quantum number, n	1	2		3			4			
Orbital quantum number, l	0	0	1	0	1	2	0	1	2	3
Possible electrons	1s	2s	2p	3s	3p	3d	4s	4p	4d	4f
Maximum No. of electrons 2 (2l + 1)	2	2	6	2	6	10	2	6	10	14
Total = 2n^2	2	8		18			32			

Fig. 3.16 shows the electron configuration for an argon atom. Since atomic number of argon *i.e.* Z = 18, there are 18 extranuclear electrons revolving round its nucleus. The distribution of these electrons in various sub-shells is as shown in Fig. 3.17.

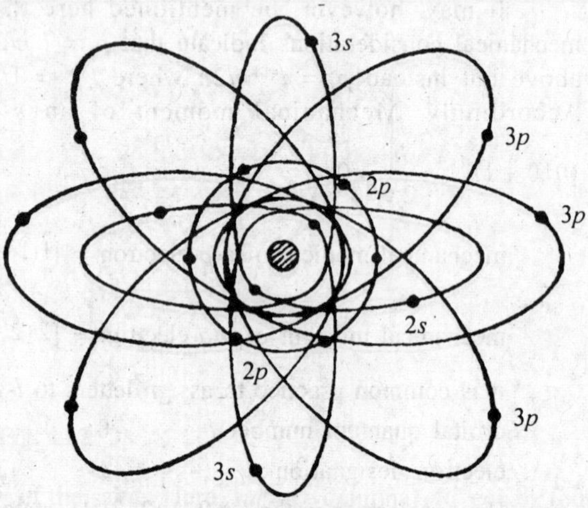

$1s^2 2s^2 2p^6 3s^2 3p^6$

Fig. 3·16

When orbital quantum number l is taken into account, the energy level diagram becomes something like that shown in Fig. 3.17. For the sake of simplicity, all the s, p, d...levels have been shown together closely. In practice, they are not so. In fact, 3d level lies much above 3s and 3p levels. Actually, 3d level lies above 4s and 4p levels. Same is the case with 4d level and so on.

Talking in terms of shells, it is seen that K-shell (n = 1) has one sub-shell with l = 0, p_l = 0. The L-shell (n = 2) has two sub-shells :

1st sub-shell :

$l = 0, p_l = 0$

2nd sub-shell :

$l = 1, p_l = \sqrt{2} \cdot h/2\pi$

Similarly, M-shell (n = 3) has following three sub-shells :

1st sub-shell :

$l = 0, p_l = 0$

2nd sub-shell : $l = 1,$ $p_l = \sqrt{2} \cdot h/2\pi$

3rd sub-shell : $l = 2,$ $p_l = \sqrt{6} h/2\pi$

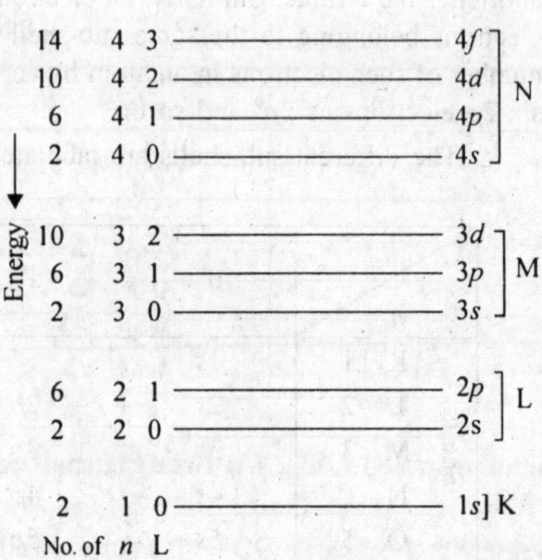

Fig. 3·17

Same is the case with other shells of higher n-values. Suitable number of electrons are added to these sub-shells to build up the periodic table in the Bohr-Stoner scheme (Art. 3.10).

3.17. Electron Configuration of Atoms

By electron configuration of an atom is meant the distribution of its electrons in various sub-shells around the nucleus.

Following three rules govern the electronic distribution :

(*a*) *total number of electrons with the same principal quantum number n is* $2n^2$.

(*b*) *in the nth shell, there are n sub-shells having different values of orbital quantum number l such as 0, 1, 2...(n – 1).*

(*c*) *each sub-shell can have a maximum of 2 (2l + 1)* electrons.

Consider the following elements :

(*i*) **Helium.** It has $Z = 2$ and so has two planetary electrons. These fill up the K-shell ($n = 1$) and both have $l = 0$. The electron configuration of helium can be written as $\boxed{1s^2}$

(*ii*) **Lithium.** It has $Z = 3$, hence its atom has 3 electrons. Two electrons fill up the K-shell having $n = 1$ and $l = 0$. The third electron occupies the $l = 0$ sub-shell of second shell *i.e.* it is $2s$ electron. Hence, electron configuration of lithium is $\boxed{1s^2}\ 2s$.

(*iii*) **Beryllium.** Since its $Z = 4$, it has four electrons. Two are $1s$ electrons and the other two are $2s$ electrons. Its electronic configuration is $\boxed{1s^2}\ 2s^2$.

(*iv*) **Neon.** Its $Z = 10$. Hence, its electronic configuration is $\boxed{1s^2 2s^2 2p^6}$

(*v*) **Sodium.** Its $Z = 11$, hence its electronic configuration is $\boxed{1s^2 2s^2 2p^6}\ 3s$

(*vi*) **Argon.** ($Z = 18$). Its electronic configuration is $\boxed{1s^2 2s^2 2p^6}\ 3s^2 3p^6$

(*vii*) **Copper** ($Z = 29$). Of the 29 electrons, two occupy the first K-shell, 8 occupy two sub-shells of the L-shell, 18 occupy the M-shell distributed as 2, 6 and 10 in the three sub-shells and the remaining one electron occupies the first sub-shell of the N-shell. Hence, electron configuration of Cu becomes $\boxed{1s^2 2s^2 2p^6 3s^2 3p^6 3d^{10}}\ 4s$.

3.18. Number of Electrons in a Shell

Electrons which have the same principal quantum number n are said to belong to the same *shell.* These shells are also designated by letters. For example, $n = 1$ shell is called K shell, $n = 2$ shell is called L-shell and $n = 3$ shell is called M-shell and so on.

For a given value of n, electrons having the same orbital quantum number l are said to belong to the same *sub-shell*. These sub-shells are also denoted by letters like s, p, d, f etc. The maximum number of electrons in a *sub-shell* is given by $2(2l + 1)$ because there are $(2l + 1)$ values of m_l and two values of m_s (Art. 3.27).

Hence, maximum number of electrons in a *shell* is given by

$$\sum_{l=0}^{l=n-1} 2(2l+1) = (2\times 1) + (2\times 3) + (2\times 5) + \ldots + 2[2(n-1)+1]$$

$$= 2[1 + 3 + 5 + \ldots + \{2(n-1)+1\}]$$

This represents an arithmetical progression with a common difference of 2, initial term of 1 and the final term $[2(n-1)+1]$. The sum of such a progression is

$$2\left[\frac{n}{2}\{2 + 2(n-1)\}\right] = 2n^2$$

When a shell contains its maximum possible number of electrons, it is said to be *filled* or *closed.* The same terms are applied to sub-shell as well.

Table No. 3.3 gives the electronic configurations of first 30 elements.

Table No. 3.3

		K-shell	*L-shell*		*M-shell*			*N-shell*
Z	*Element*	$(n=1)$	$(n=2)$		$(n=3)$			$(n=4)$
		$l=0$	$l=0;\ l=1$		$l=0;\ l=1;\ l=2$			$l=0$
		1*s*	2*s*	2*p*	3*s*	3*p*	3*d*	4*s*
1	H	1						
2	He	2						
3	Li	2	1					
4	Be	2	2					
5	B	2	2	1				
6	C	2	2	2				
7	N	2	2	3				
8	O	2	2	4				
9	F	2	2	5				
10	Ne	2	2	6				
11	Na	2	2	6	1			
12	Mg	2	2	6	2			
13	Al	2	2	6	2	1		
14	Si	2	2	6	2	2		
15	P	2	2	6	2	3		
16	S	2	2	6	2	4		
17	Cl	2	2	6	2	5		
18	A	2	2	6	2	6		
19	K	2	2	6	2	6		1
20	Ca	2	2	6	2	6		2
21	Sc	2	2	6	2	6	1	2
22	Ti	2	2	6	2	6	2	2
23	V	2	2	6	2	6	3	2
24	Cr	2	2	6	2	6	5	1
25	Mn	2	2	6	2	6	5	2
26	Fe	2	2	6	2	6	6	2
27	Co	2	2	6	2	6	7	2
28	Ni	2	2	6	2	6	8	2
29	Cu	2	2	6	2	6	10	1
30	Zn	2	2	6	2	6	10	2

3.19. Electron Spin and Spin Quantum Number

For explaining the doublets or double-line structure in the spectra of alkali metals*, Goudsmit and Uhlenbeck postulated in 1925 that an electron *spins* around its own mechanical axis as it moves in an orbit around the nucleus (Fig. 3.18). For this purpose, they introduced another quantum number called electron *spin quantum number* whose only possible value is $\frac{1}{2}$. The spin angular momentum is given by

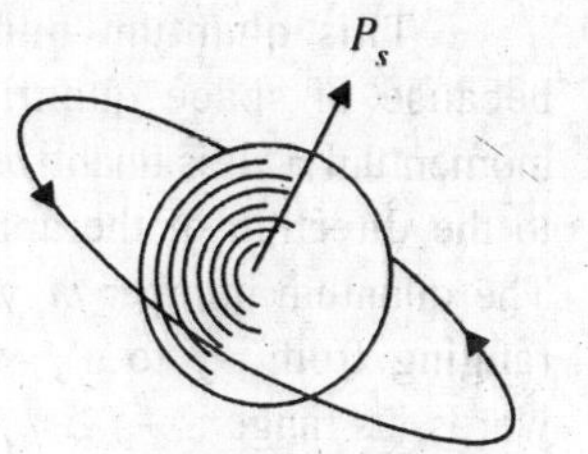

Electron Spin

Fig. 3·18

$$p_s = 8 \cdot \frac{h}{2\pi} \quad \text{where } s = \frac{1}{2}$$

Strictly speaking, $$p_s = s^* \frac{h}{2\pi}$$

where $$s^* = [s(s+1)]^{\frac{1}{2}}$$

3.20. Total Angular Quantum Number (*j*)

It has been found that coupling occurs between the spin and orbital angular momenta of an electron which has been described by another quantum number *j* called the *total angular quantum number*. The values of *j* are

$$j = l + s \text{ and } j = l - s$$

with the restriction that *j* must be positive.

The positive sign is taken when *s* is parallel to *l* and minus sign when it is antiparallel as shown in Fig. 3.19 (*a*) and (*b*).

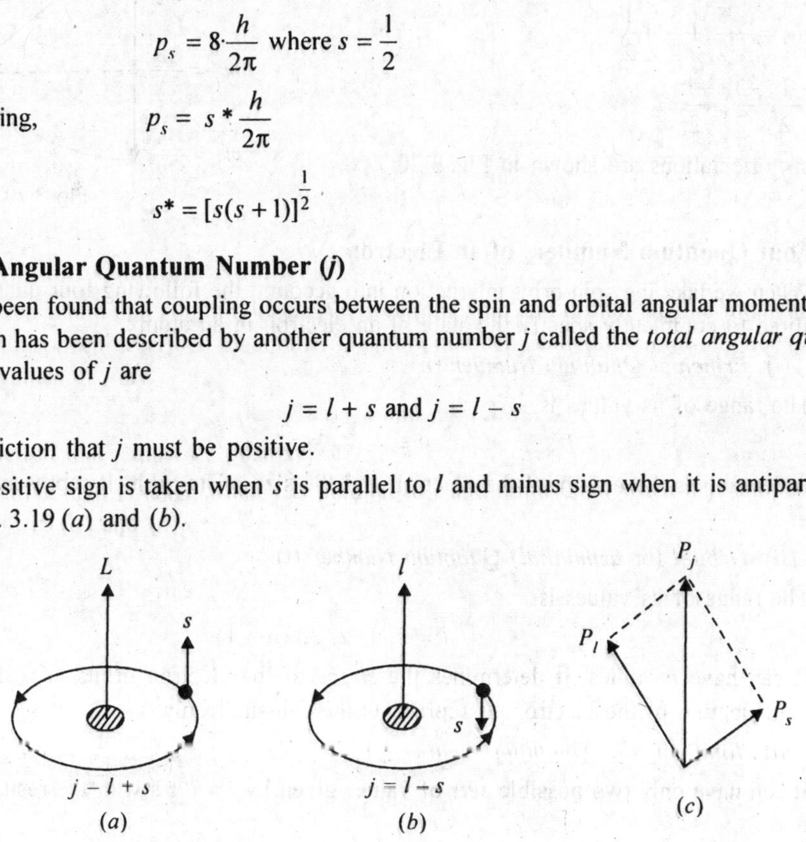

Fig. 3·19

The total angular momentum is given by

$$p_j = J^* \frac{h}{2\pi} \quad \text{where } j^* = [j(j+1)]^{\frac{1}{2}}$$

The spin and angular momenta couple vectorially as shown in Fig. 3.19 (*c*).

*This doublet structure is different from *fine* structure which consists of closely-spaced fine lines (not widely-separated ones).

3.21. Magnetic Quantum Number (m_j)

This quantum number comes into existence because of space quantization of the total angular momentum p_j. It is quantized component of j with respect to the direction of the applied external magnetic field. The quantum number m_j can have only $(2j + 1)$ values ranging from $-j$ to $+j$ with zero *excluded.* In other words, its range is $+j \geq m_j \geq -j$. For example, if $j = 5/2$, then allowed orientations of j – vector along vector B are six in number and are specified by

$$m_j = +\frac{5}{2}, +\frac{3}{2}, +\frac{1}{2}$$

$$\frac{-1}{2}, \frac{-3}{2}, \frac{-5}{2}$$

These six orientations are shown in Fig. 3.20.

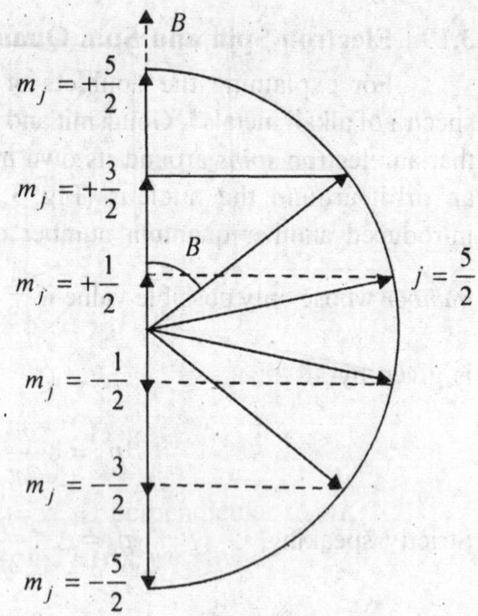

Fig. 3·20

3.22. Four Quantum Numbers of an Electron

When we take the spin-orbit interaction into account, the following four quantum numbers are required to completely specify the state of an electron in an atom :

(*i*) *Principal Quantum Number* (n)

The range of its values is

$$n = 1, 2, 3, ..., \infty$$

It determines the *energy* of the electron and the size of its orbit. It represents the shell as a whole.

(*ii*) *Orbital* (or *azimuthal*) *Quantum Number* (l).

The range of its values is

$$l = 0, 1, 2, 3...(n - 1)$$

It can have n values. It determines the *shape* of the electron orbits and also the orbital angular momentum of the electron. It represents the sub-shell only.

(*iii*) *Total Angular Quantum Number* (j)

It can have only two possible *sets* of values given by $j = l \pm s$ with the restriction that j is positive.

It determines the *total* angular momentum of the electron *i.e.* the vector sum of orbital angular momentum and spin angular momentum.

(*iv*) *Magnetic Quantum Number* (m_j)

It results from space quantization of total angular momentum of an electron. It, in fact, represents the *numerical* value of this total angular momentum vector j on the direction of the applied external field.

This quantum number can have $(2j + 1)$ values ranging from $-j$ to $+j$ *with zero excluded.* In other words, its range is $+j \geq m_j \geq -j$.

or $$m_j = \pm\frac{1}{2}, \pm\frac{3}{2}, \pm\frac{5}{2}, ... \pm j$$

It should be noted that values of m_j depend on j which itself depends on l which, in turn, is dependent on n. Hence, it will be seen that ultimately it is n which heads the hierarchy of quantum numbers.

3.23. Values of *n*, *l*, *j* Quantum Numbers

The different possible values of quantum numbers n, l and j for different shells are as under :

(*i*) $n = 1, l = 0, s = \frac{1}{2}$

$j = l + s = 0 + 1/2 = 1/2$

Hence, only one (n, l, j) state is possible *i.e.* (1, 0, 1/2)

Note. The other possible value of $j = l - s = 0 - 1/2 = -1/2$ is not allowed because it is negative.

(*ii*) $n = 2, l = 0$ and $1, s = 1/2$

(*a*) $j = l + s = 0 + 1/2 = 1/2$

(*b*) $j = l + s = 1 + 1/2 = 3/2$

$j = l - s = 1 - 1/2 = 1/2$

Hence, following *three* states are possible (Fig. 3.21)

(2, 0, $\frac{1}{2}$), (2, 1, $\frac{1}{2}$), (2, 1, 3/2)

(*iii*) $n = 3, l = 0, 1, 2, s = 1/2$.

(*a*) $j = 0 + 1/2 = 1/2$

(*b*) $j = 1 + 1/2 = 3/2$

$j = 1 - 1/2 = 1/2$

(*c*) $j = 2 + 1/2 = 5/2$

$j = 2 - 1/2 = 3/2$

Hence, following *five* states are possible.

(3, 0, 1/2), (3, 1, 1/2), (3, 1, 3/2) (3, 2, 3/2) and (3, 2, 5/2)

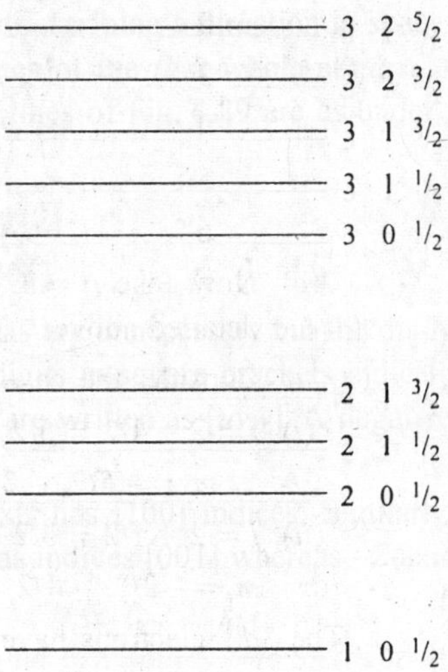

Fig. 3·21

3.24. Values of *n*, *l*, *j*, *m* Quantum Numbers

The values of different quantum numbers for electrons occupying different shells and sub-shells etc. are as under :

(*a*) Take $n = 1$. Then $l = 0$.

(*i*) $j = l + s = 0 + 1/2 = 1/2$

(*ii*) $m_j = 1/2, -1/2$

Hence, the two $1s^2$ electrons have the following four quantum numbers each.

	n	l	j	m_j
1st electron :	1	0	$\frac{1}{2}$	$+\frac{1}{2}$
2nd electron :	1	0	$\frac{1}{2}$	$-\frac{1}{2}$

(*b*) Take $n = 2$. Here, $l = 0, 1$ and $s = 1/2$

(*i*) First, take $l = 0$ $j = 0 + 1/2 = 1/2$

$m_j = + 1/2$ and $- 1/2$

Hence, $2s^2$ electrons have the following four quantum numbers each.

$$\left(2, 0, \frac{1}{2}, \frac{1}{2}\right) \text{ and } \left(2, 0, \frac{1}{2}, -\frac{1}{2}\right)$$

(*ii*) Now, take $l = 1$, $j = 1 + 1/2 = 3/2$

$m_j = - 3/2, -1/2, 1/2$ and $3/2$

When $j = 1 - 1/2 = 1/2$

$m_j = - 1/2$ and $1/2$

Hence, the $2p^6$ electrons have the following values of their four quantum numbers.

Sl. No. of Electron	n	l	j	m_j
1	2	1	3/2	– 3/2
2	2	1	3/2	– 1/2
3	2	1	3/2	+ 1/2
4	2	1	3/2	+ 3/2
5	2	1	1/2	– 1/2
6	2	1	1/2	+ 1/2

(*c*) Now, take $n = 3$. Here, $l = 0, 1, 2$. The procedure for $n = 3$ and $l = 0$ and 1 is as detailed above.

Consider the case when $n = 3$ and $l = 2$

(*i*) $j = 2 + 1/2 = 5/2$

$\therefore m_j = - 5/2, - 3/2, - 1/2, + 1/2, + 3/2$ and $+ 5/2$

(*ii*) $j = 2 - 1/2 = 3/2$

$m_j = - 3/2, - 1/2, + 1/2$ and $+ 3/2$

The $3d^{10}$ electrons have following values of their four quantum numbers.

Sl. No. of Electron	n	l	j	m_j
1	3	2	5/2	– 5/2
2	3	2	5/2	– 3/2
3	3	2	5/2	– 1/2
4	3	2	5/2	+ 1/2
5	3	2	5/2	+ 3/2
6	3	2	5/2	+ 5/2
7	3	2	3/2	– 3/2
8	3	2	3/2	– 1/2
9	3	2	3/2	+ 1/2
10	3	2	3/2	+ 3/2

It will be noted that no two electrons possess all the four quantum numbers alike. The maximum number of common quantum numbers which two electrons can have is only three but never four. This statement constitutes what is generally known as Pauli's Exclusion Principal (Art. 3.29).

3.25. Orbital Magnetic Moment of an Electron

There are two magnetic moments associated with the electron, one due to its orbital motion and the other due to spin. Since electron is negatively charged, its orbital motion is like the flow of current though not confined in a wire. Hence, a magnetic field as shown in Fig. 3.22 is set up. The direction of this field which is similar to the field around a bar magnet is as shown. The magnetic dipole moment μ_l is given by

$$\frac{\mu_l}{p_l} = \frac{e}{2m}$$

The left-hand term m_l/p_1 is called the gyromagnetic ratio (G). It is the ratio between the magnetic moment and mechanical moment.

$$\therefore \qquad \mu_l = p_l \cdot \frac{e}{2m} = l * \frac{h}{2\pi} \cdot \frac{e}{2m}$$

or $$\mu_l = l * \mu_B$$

where μ_B is a unit of magnetic moment and is called Bohr magneton.

$$\mu_B = \frac{h}{2\pi} \cdot \frac{e}{2m} = 9{\cdot}27 \times 10^{-24} \text{ A} - m^2$$

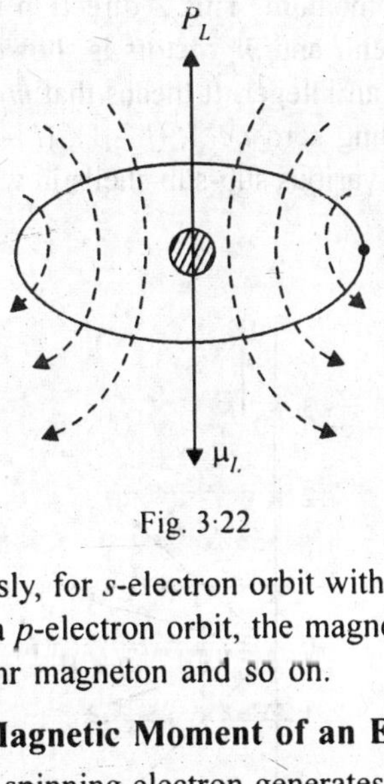

Fig. 3·22

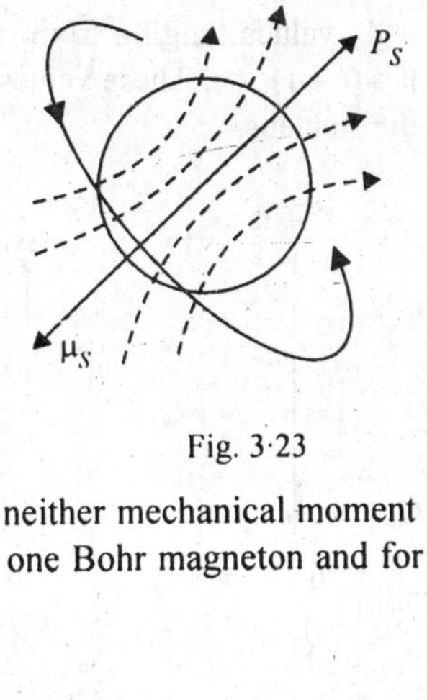

Fig. 3·23

Obviously, for *s*-electron orbit with $l = 0$, there is neither mechanical moment nor magnetic moment. For a *p*-electron orbit, the magnetic moment is one Bohr magneton and for a *d*-electron, orbit, two Bohr magneton and so on.

3.26. Spin Magnetic Moment of an Electron

Now, a spinning electron generates a magnetic field like that shown in Fig. 3.23. This field is also like that due to a bar magnet. The magnetic moment μ_s of a spinning electron is given by

$$\frac{\mu_s}{p_s} = 2 \cdot \frac{e'}{2m}$$

or $$\mu_s = p_s . 2 \cdot \frac{e}{2m} = \frac{1}{2} \cdot \frac{h}{2\pi} \cdot 2 \cdot \frac{e}{2m}$$

$$= \frac{h}{2\pi} \cdot \frac{e}{2m} = \mu_B$$

Hence, a spinning electron has a magnetic moment of one Bohr magneton.

The doubling of energy levels in atoms having one valence electron is due to the interaction between the magnetic field of the electron orbit and magnetic field of electron spin.

3.27. Four Quantum Numbers Neglecting Spin-Orbit Interaction

If we neglect the spin-orbit interaction (unlike what we did in Art. 3.22), then the state of an electron in an atom can be completely specified by the following four quantum numbers :

(*i*) *principal quantum number* (n)

(*ii*) *orbital quantum number* (l)

(*iii*) *orbital* (*or azimuthal*) *magnetic quantum number* (m_l)

(*iv*) *magnetic spin quantum number* (m_s)

The first two quantum numbers have already been discussed above but the other two which represent the projections of orbital angular momentum and spin angular momentum respectively on the applied magnetic field are described below.

(*a*) Orbital Magnetic Quantum Number (m_l).

This number arises out of the fact that an electron in an orbit constitutes a rotating charge and hence an electric current which has is associated magnetic field and magnetic moment.

It is related to the *orientation of elliptical electron orbits in space* when the atom is placed in a magnetic field. There are restrictions on the orientation of the electron orbits because of which they are said to be *space quantized.* Whereas l determines the orbital angular momentum, m_l represents the Z component of the orbital angular momentum. This Z direction is taken along the direction of the applied field. The relationship between l and m_l vectors is shown in Fig. 3.24. Permitted values of θ are those for which $m_l = l \cos \theta$ is an integer. It means that m_l can have any one of the $(2l + 1)$ values ranging from $-l$ to $+l$ including zero *i.e.* l, $(l-1)$, $(l-2)$...1, 0, -1, -2, ... $-(l-2)$, $-(l-1)$, $-l$. These values correspond to various sub-sub-shells in which electrons rotate around the nucleus.

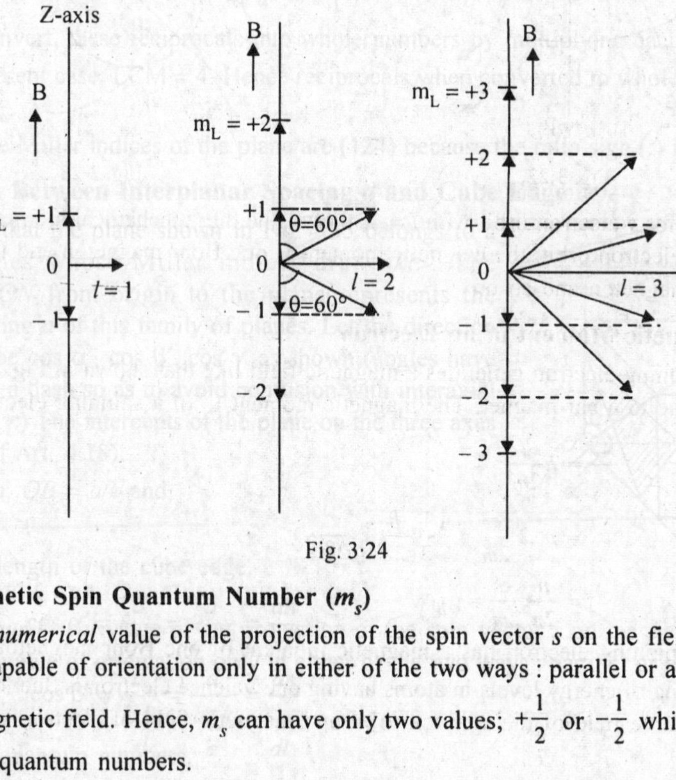

Fig. 3·24

(*b*) Magnetic Spin Quantum Number (m_s)

It is the *numerical* value of the projection of the spin vector s on the field direction. This spin vector is capable of orientation only in either of the two ways : parallel or anti-parallel to the surrounding magnetic field. Hence, m_s can have only two values; $+\frac{1}{2}$ or $-\frac{1}{2}$ which differ by unity as do all other quantum numbers.

Following points regarding the above four quantum numbers are worth noting :

(1) The principal quantum number (n) gives the position of an electron in the shell.

(2) The orbital quantum number l gives its position in the sub-shell.

(3) The orbital magnetic quantum number m_l gives its position in the sub-subshell.

(4) The magnetic spin quantum number m_3 gives direction of spin or self-rotation i.e. whether clockwise or anti-clockwise.

3.28. Different Shells, Sub-shells and Sub-subshells

The number of sub-shells and sub-subshells corresponding to various shells upto $n = 3$ are given in Table No. 3.4. Since the value of $m_3 = \pm 1/2$ for all sub-subshells, it has not been shown in the table.

Table 3.4

n shell	l *subshell*	m_l *sub-subshell*
1	0	0
2	0	0
	1	− 1 0 + 1
3	0	0
	1	− 1 0 + 1
	2	− 2 − 1 0 + 1 + 2

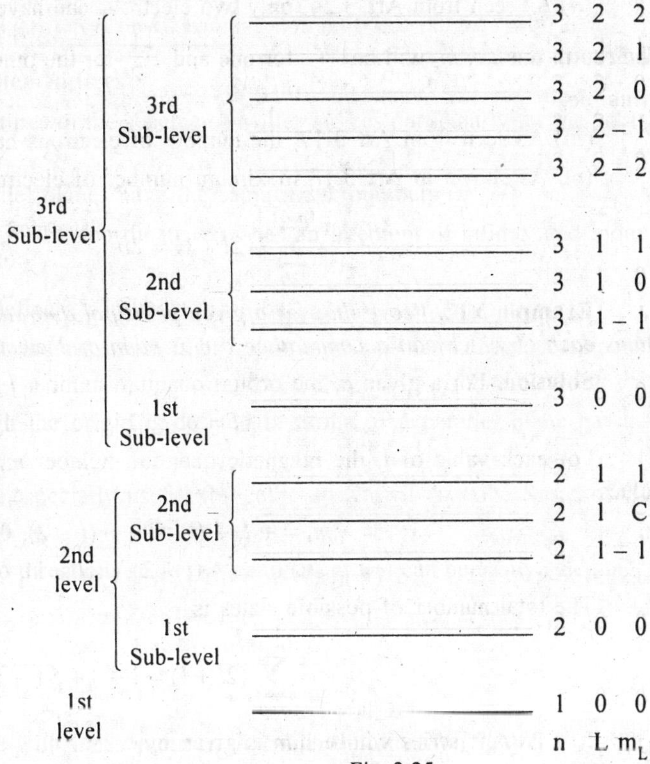

Fig. 3·25

The various levels, sub-levels and sub-sublevels have been shown in Fig. 3.25. Again, values of $m_2 = \pm 1/2$ have not been indicated in the figures because they are the same for all sub-sublevels.

Example 3.10. *How are electrons distributed in the various sub-shells for n = 3. Give the quantum numbers for the electrons in the first sub-shell.*

Solution. The shell having principal quantum number $n = 3$ has *three* sub-shells having values of $l = 0, 1, 2$.

The quantum numbers of various sub-shells are as under :

(*i*) $n = 3$, $l = 0$. The two electrons occupying it are known as 3*s* electrons.

(*ii*) $n = 3$, $l = 1$. The 6 electrons occupying it are known as 3*p* electrons.

(*iii*) $n = 3$, $l = 2$. The 10 electrons occupying it are known as 3*d* electrons.

The shapes of these sub-shells are shown in Fig. 3.14 (*c*).

The four quantum numbers of the $3s^2$ electrons are as under :

	n	l	m_l	m_s
1st electron	3	0	0	+1/2
2nd electron	3	0	0	–1/2

Example 3.11. *What is the maximum number of the electrons in an atom that can share the following quantum numbers* :

(*i*) n, l, m_l, m_s (*ii*) n, l, m_l (*iii*) l (*iv*) n ?

Solution. (*i*) As required by Pauli's Exclusion Principle (Art. 3.29) there is only one electron and none other in the entire atom which has identical values of these four quantum numbers.

(*ii*) As seen from Art. 3.24, only two electrons can have three common quantum numbers. The fourth one *i.e.* m_s will be $+\frac{1}{2}$ for one and $-\frac{1}{2}$ for the other as required by Pauli's Exclusion Principle.

(*iii*) As seen from Art. 3.17, the number of electrons having same l value is $2(2l + 1)$.

(*iv*) As shown in Art. 3.18 maximum number of electrons having same n-value is

$$\sum_{l=0}^{(n-1)} 2(2l+1) = 2n^2$$

Example 3.12. *Prove that for a given principal quantum number n, there are n^2 possible states each of which can accommodate a maximum of 2 electrons.*

Solution. For a given n, the orbital quantum number l has the following n values :

$$l = 0, 1, 2, \ldots, (n-1)$$

For each value of l, the magnetic quantum number m_l possesses the following $(2l + 1)$ values :

$$m_l = +l, +(l-1), +(l-2), 0, -1, -2 \ldots -(l-2), -(l-1), -l$$

The total number of possible states is

$$= \sum_{l=0}^{(n-1)} (2l+1) = 1 + 3 + 5 \ldots + (2n-1)$$

It is an A.P. series whose sum is given by

$$S = n\left(\frac{a+l}{2}\right)$$

$$= n\left[\frac{1+(2n-1)}{2}\right] = n^2$$

∴ Total number of states (or sub-sublevels in an *ELD*) is n^2.

Example 3.13. *Find the number of electrons in atoms which have the following levels filled in the ground state* :

(*a*) *K- and L-shells, the 3s sub-shell and one half of the 3-p sub-shell.*

(*b*) *the K-, L- and M-shells and the 4s, 4p and 4d sub-shells.*

What atoms are these ?

Solution. Table No. 3.3 may please be referred to for finding the number of electrons.

(*a*) No. of electrons in *K*-shell = 2

No. of electrons in *L*-shell = 8

No. of electrons in 3*s* sub-shell = 2

Half the No. of electrons in 3-*p* sub-shell = 3

Total number = 15

This also represents the atomic number of the element. As seen from Table No. 3.3, the element with $Z = 15$ is **phosphorus.**

(*b*) No. of electrons in *K*-shell = 2

No. of electrons in *L*-shell = 8

No. of electrons in *M*-shell = 18

No. of electrons in 4*s*-subshell = 2

No. of electrons in 4*p*-subshell = 6

No. of electrons in 4*d*-shell = 10

Total number of electrons = 46

Hence, the atomic number of the element is $Z = 46$. With reference to Table No. 3.3, the element is **Palladium.**

3.29. Pauli's Exclusion Principle

It states that in *one atom, no two electrons can have the same set of values for the four quantum numbers n, l, m_1 and m_2.* Consequently, it restricts the number of electrons an atom can have.

It has been stated in Art. 3.27 that four quantum numbers *n l*, m_l and m_s (or *n*, *l*, *j* and m_j) are required to specify the state of an electron in an atom. They give its position in the principal energy level (*n*), the sub-level (*l*), the sub-sublevel (m_l) and the direction of spin (m_s). According to this principal, *no two electrons in one atom can have all the four quantum numbers alike i.e. no two electrons can be described by an identical set of quantum numbers.* They may have three numbers alike but at least one must be different. This rule may, in some respects, be compared to the classical mechanics which states that no two objects can occupy the same space at the same time.

Consider the case of helium* which has two electrons in its atom. These electrons occupy *K*-shell ($n = 1$) and are designated as $1s^2$ electrons. Their four quantum numbers are as follows :

	n	l	m_l	m_s
1st electron	1	0	0	$+\frac{1}{2}$
2nd electron	1	0	0	$-\frac{1}{2}$

As seen, the two electrons have different sets of four numbers $(1, 0, 0, +\frac{1}{2})$ and $(1, 0, 0, -\frac{1}{2})$ as required by Pauli's Exclusion Principle.

Pauli's principle affords a very rational theoretical justification for the electronic shell structure of the atoms deduced experimentally.

(*a*) ***K*-shell, $n = 1$**

*Hydrogen atom does not demonstrate this rule for the simple reason that it has one electron only.

As discussed above, it can have only two electrons which can differ in respect of one quantum number only *i.e.* m_s.

(*b*) ***L*-shell, $n = 2$**

It has two sub-shells with $l = 0, 1$.

(*i*) For the first sub-shell, $n = 2$, $l = 0$, $m_l = 0$ and $m_s = +\frac{1}{2}$ and $-\frac{1}{2}$. Hence, this sub-shell can accommodate two electrons designated as $2s^2$.

(*ii*) For the second sub-shell, $n = 2$ and $l = 1$ so that $m_l = -1, 0, +1$ each of which can be associated with two values of $m_s = \pm\frac{1}{2}$. Hence, in all, there are $2(2l + 1) = 2(2 \times 1 + 1)$ $= 6$ electrons occupying three sub-subshells and are designated as $2p^6$ electrons.

None of these electrons will have all the four quantum numbers identical as shown in Table No. 3.5.

Table No. 3.5

Sl. No. of Electron	*n*	*l*	m_l	m_s
1	2	1	0	$+\frac{1}{2}$
2	2	1	0	$-\frac{1}{2}$
3	2	1	+1	$+\frac{1}{2}$
4	2	1	+1	$-\frac{1}{2}$
5	2	1	− 1	$+\frac{1}{2}$
6	2	1	− 1	$-\frac{1}{2}$

(*c*) ***M*-shell, $n = 3$**

It has three sub-shells having $l = 0, 1, 2$. The first and second sub-shells are completed by 2 and 6 electrons as explained above. The third sub-shell has $2(2 \times 2 + 1) = 10$ electrons which are further distributed in 5 sub-subshells.

Hence, total number of electrons required to fill *M*-shell is $= 2 + 6 + 10 = 18$.

Pauli's Exclusion Principle has been very helpful in our understanding not only of atomic states but also of chemical activity and the nature of chemical bonding, the periodic table of elements, the structure and composition of nucleus and conduction of electricity in solids.

3.30. Diagrammatic Representation of Pauli's Principle

The combination of quantum numbers of electrons allowed by Pauli's Exclusion Principle for *L*-shell ($n = 2$) are depicted diagrammatically below in Fig. 3.26. Instead of using the associated orbital and spin momentum vectors, the quantum numbers themselves have been shown by vectors for the sake of simplicity.

3.31. Quantum Numbers and Their Significance

The four quantum numbers which completely describe the state of an individual electron in an atom along with their significance are given below :

(*a*) First is **principal quantum number (*n*)** which varies from unity to infinity in steps of unity.

It arises from quantization of energy *i.e.* it controls electron energy and also determines the size of the electron orbit.

(*b*) Second is **orbital quantum number (*l*)** which takes integral values ranging from 0 to $(n - 1)$ in steps of unity.

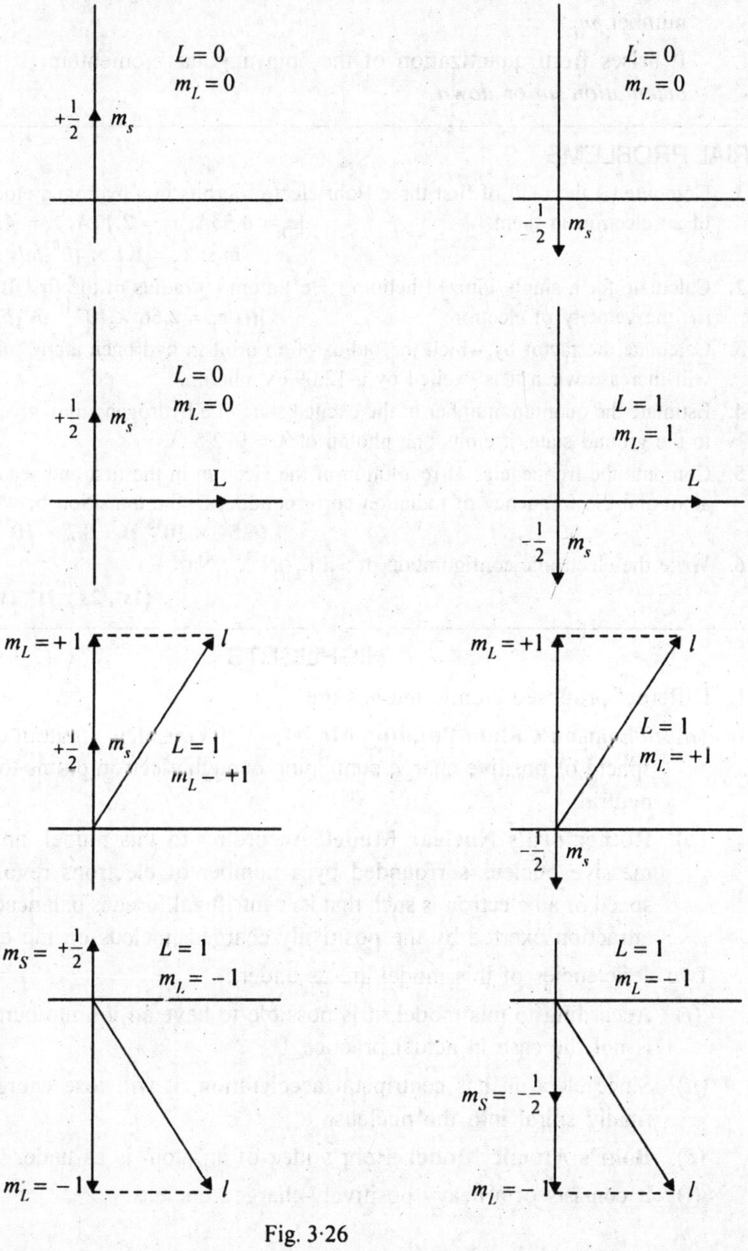

Fig. 3·26

It arises from the quantization of the orbital angular momentum *i.e.* it determines the angular momentum of the electron about the nucleus and it also determines the *shape* or eccentricities of the electron orbits.

(*c*) Third is **orbital magnetic quantum number** (m_l) which takes integral values from $-l$ to $+l$ in steps of unity.

It arises from quantization of the directions of the orbital angular momentum *i.e.* it defines the angular momentum parallel to an applied magnetic field. It determines the *orientation* of the electron orbit in space.

(*d*) Fourth is **magnetic spin quantum number** (m_s) which takes values of $\pm 1/2$, the sign depending on whether it is parallel or antiparallel to the orbital magnetic quantum number m_l.

It arises from quantization of the spin angular momentum. It determines the *spin orientation up* or *down*.

TUTORIAL PROBLEMS

1. Compute (*i*) the radii of first three Bohr electron orbits in a hydrogen atom and (*ii*) the velocity of an electron in them. **[r_1 = 0.53Å, r_2 = 2.12Å, r_3 = 4.77Å; $v_1 = 2.19 \times 10^6$ m/s; $v_2 = 1.1 \times 10^6$ m/s : $v_3 = 7.3 \times 10^5$ m/s].**

2. Calculate for a singly ionized helium ($_2He^4$) atom (*i*) radius of the first Bohr electron orbit and (*ii*) the velocity of electron. **[(*i*) $r_1 = 2.66 \times 10^{-11}$ m (*ii*) $v = 4.37 \times 10^6$ m/s]**

3. Calculate the factor by which the radius of an orbit in hydrogen atom, initially in ground state, will increase when it is excited by a 12.09 eV photon. **(9)**

4. Estimate the quantum number of the excited state of a hydrogen atom, given that in the transition to the ground state, it emits one photon of λ = 972.5 Å. **(n = 4)**

5. Compute the frequencies of revolution of the electron in the first and second orbits of hydrogen atom and the frequency of radiation corresponding to the transition between the two orbits. **(6.54×10^{15} Hz; 8.2×10^{14} Hz, 2.47×10^{15} Hz)**

6. Write the electronic configuration of : $_3Li^7$, $_7N^{14}$, $_{11}Na^{23}$ **[$1s^2$, $2s$: $1s^2\ 2s^2\ 2p^3$; $1s^2\ 2s^2\ 2p^6\ 3s$]**

HIGHLIGHTS

1. Different proposed atomic models are :

(*a*) **Thomson's Plum Pudding Model** which considers an atom to consist of a heavy sphere of positive charge containing enough electron plums to make it electrically neutral.

(*b*) **Rutherford's Nuclear Model.** According to this model, an atom consists of a massive nucleus surrounded by a number of electrons revolving in orbits. The speed of an electron is such that its centrifugal force is balanced by the electrostatic attraction exerted by the positively-charged nucleus on the electron.

Two deficiencies of this model are as under :

(*i*) According to this model, it is possible to have an infinite number of orbits which is not the case in actual practice.

(*ii*) Since electron has centripetal acceleration, it will lose energy continuously and finally spiral into the nucleus.

(*c*) **Bohr's Atomic Model**. Bohr's idea of an atom is as under :

(*i*) It consists of a heavy positively-charged nucleus.

(*ii*) The electrons move only in permitted circular orbits. Permitted orbits are those for which the orbital angular momentum of the electron equals an integral multiple of ($h/2\pi$).

(*iii*) While moving in permitted orbits, electron does not radiate out any energy. It radiates out energy only when it jumps from one orbit to another.

The above conditions can be put in the following mathematical equations :

$$mvr = n \cdot \frac{h}{2\pi} \qquad ...(i)$$

Since coulombic force balances the centrifugal force on the electron

$$\therefore \quad \frac{mv^2}{r} = \frac{Ze^2}{4\pi\varepsilon_0 r^2}$$

or

$$mv^2 r = \frac{Ze^2}{4\pi\varepsilon_0} \qquad ...(ii)$$

The radius of the non-radiating orbit is

$$r_n = \frac{\varepsilon_0 n^2 h^2}{\pi m Z e^2}$$

The velocity of the electron is given by

$$v = \frac{Ze^2}{2\varepsilon_0 nh}$$

Orbital frequency of the electron is

$$f = \frac{v}{2\pi r} = \frac{mZ^2 e^4}{4\varepsilon_0^2 n^3 h^3}$$

The total energy (*i.e.* sum of the kinetic and potential energies) of an electron is

$$E_n = -\frac{mZ^2 e^4}{8\varepsilon_0^2 n^2 h^2}$$

Bohr's theory has the following deficiencies :

(*i*) It introduces only one quantum number n whereas experimental evidence concerning fine structure of spectral lines suggests additional quantum numbers.

(*ii*) it applies to one-electron atom *i.e.* hydrogen atom and is not easily extended to describe more complicated atoms.

(*iii*) it says nothing about any rules or restrictions regarding electron transitions from one level to another.

(*d*) Sommerfeld's Relativistic Model

It is an extension of Bohr's model in two respects :

(*i*) firstly, it allows the possibility of elliptical orbits for the electrons in addition to the Bohr's circular orbits.

(*ii*) it takes into account the relativistic variation of electron mass which is different at different parts of the elliptical orbit. Introduction of elliptical paths leads to two variable quantities—one is the distance r from the nucleus and the other is the varying angular position of the electron from the nucleus *i.e.* azimuthal angle ϕ.

The two quantum numbers required to deal with these two variables are :

(*i*) principal quantum number n which is associated with the separation of the electron from the nucleus. It can have values of 1, 2, 3, ..., ∞.

(*ii*) orbital (or azimutha) quantum number l which determines the shape of the electron orbits. Its value varies from 0 to $(n - 1)$ in steps of unity.

The eccentricity of the ellipse can be found from the relation

$$\frac{b}{a} = \frac{l+1}{n}$$

2. Four Quantum Numbers of an Electron.

Two cases may be considered :

(*a*) **When spin-orbit interaction is considered.** In this case, the four quantum numbers are as follows :

(*i*) principal quantum number n which varies from 1 to ∞ in steps of unity.

(*ii*) orbital quantum number l which takes integer values from 0 to $(n - 1)$ in steps of unity.

(*iii*) total angular quantum number j which has only two possible *sets* of values $l \pm s$ with the restriction that j is positive.

(*iv*) magnetic quantum number (m_j) whose value ranges from $-j$ to $+j$ with zero excluded. Put in another way, it means

$$m_l = \pm\frac{1}{2}, \pm\frac{3}{2}, \pm\frac{5}{2} \ldots \pm j$$

(*b*) **When spin-orbit interaction is not considered.** In this case, the four quantum numbers are as under :

(*i*) principal quantum number n – same as in (*a*) above.

(*ii*) orbital quantum number l – same as in (*a*) above.

(*iii*) orbital magnetic quantum number (m_l) which takes integer values from $-l$ to $+l$ in steps of unity including zero.

(*iv*) magnetic spin quantum number (m_s) which takes values of $\pm\frac{1}{2}$ only.

Thus, the four quantum numbers, n, l, m_l and m_s (or n, l, j, m_l) can be regarded as *labels* which define completely the state of an electron in an atom and determine its energy and angular momentum. During the transition of an electron from one state to another, some or all of these quantum numbers and the associated properties change. However, intrinsic properties of the electron such as charge, rest mass and spin remain unchanged.

3. Different series constituting hydrogen spectrum are

(*i*) **Lyman Series**

$$\frac{1}{\lambda} = R\left(1 - \frac{1}{n^2}\right) \qquad \text{where } n = 2, 3, 4, \ldots\infty$$

(*ii*) **Balmer Series**

$$\frac{1}{\lambda} = R\left(\frac{1}{4} - \frac{1}{n^2}\right) \qquad \text{where } n = 3, 4, 5, \ldots\infty$$

(*iii*) **Paschen Series**

$$\frac{1}{\lambda} = R\left(\frac{1}{9} - \frac{1}{n^2}\right) \qquad \text{where } n = 4, 5, 6, \ldots\infty$$

(*iv*) **Brackett Series**

$$\frac{1}{\lambda} = R\left(\frac{1}{16} - \frac{1}{n^2}\right) \qquad \text{where } n = 5, 6, 7, \ldots\infty$$

(*v*) **Pfund Series**

$$\frac{1}{\lambda} = R\left(\frac{1}{25} - \frac{1}{n^2}\right) \qquad \text{where } n = 6, 7, 8, \ldots\infty$$

4. According to Pauli's Exclusion Principle, no two electrons in any one atom can have identical sets of quantum numbers n, l, m_l and m_s. In other words, no two electrons in an atom can have all the four quantum numbers alike.

Can you Answer This ?

1. (*a*) Give full account of Bohr's theory of hydrogen atom.

(*b*) Explain Pauli's exclusion principle and discuss its application to the periodic classification of elements.

2. Explain Pauli's principle

3. Give full account of Bohr's Theory of hydrogen atom.

4. (*a*) State and explain the four quantum numbers. How are they inter related ?

(*b*) Explain Pauli's Exclusion principle and discuss its applications to the periodic classification of elements.

5. State Bohr's fundamental postulates to explain the spectrum of hydrogen atom. Find the expression for the energy levels for the same atom. Draw the necessary energy level diagrams.

6. What are the main considerations which led Rutherford to postulate his theory of nuclear atom ? What were the major drawbacks of this model ? How were they overcome by Bohr ?

7. Give an account of Rutherford's nuclear atomic model. Discuss the drawbacks of such a model.

8. (*a*) Based on Bohr's assumptions, obtain the expression for the orbital energy of an electron in the hydrogen atom.

(*b*) Account for the different series of spectral lines of the hydrogen atom.

(*c*) Calculate the frequency of the electron in the first Bohr orbit of the hydrogen atom.

9. Write short note on electronic configuration in elements.

10. (*a*) Obtain a general expression for (*i*) the radii of Bohr's orbits and (*ii*) the orbital frequencies of the electrons occupying them stating the assumptions made.

(*b*) Calculate for hydrogen atom (*i*) velocity of the electron in the ground state (*ii*) radius of Bohr's orbit in the ground state and (*iii*) time taken for the electron to traverse the first Bohr's orbit.

11. (*a*) Using classical expression for the kinetic and potential energy of electron-proton system and the Bohr quantum condition, derive an expression for the energy levels of the hydrogen atoms.

(*b*) Estimate the energy difference between the second and fourth energy levels.

12. (*a*) State Pauli's Exclusion principle. How does this principle account for the electronic structure of an atom ? Discuss its utility in classification of elements.

(*b*) Prove that the maximum number of electrons in an atom that have the same principle quantum number is $2n^2$.

13. (*a*) What was the main result of Rutherford's alpha scattering experiment ? Briefly explain how these results ruled out Thomson's model of the atom and supported Rutherford's model.

(*b*) State the main postulates of Bohr's model of the hydrogen atom.

14. Write short note on electronic configuration in atoms.

15. Show in a diagram the transitions possible under Bohr model when hydrogen atoms in an excited state with $n = 4$ return to the ground state. Calculate the shortest and longest wavelengths of the resulting spectral lines.

Write short note on Pauli's Exclusion principle.

CRYSTALLOGRAPHY 4

What is Crystallography ? — Space Lattice — Crystal Lattice and Unit Cell — Lattice Parameters of Unit Cell — The Seven Crystal Systems — Bravais' Lattices — Symmetry Elements of a Crystalline Solid — Space Lattices of Cubic System — Calculations of Unit Cell — Crystal Structures of Commercially Important Materials — Coordination Number — Atomic Radius — Number of Atoms per Unit Cell — Density of Packing — Important Plane Systems in a Cubic Crystal — Directions, Planes and Miller Indices — Unit Face and Axial Units — Miller Indices — Miller Indices of Cubic Crystal Planes — Procedure for Finding Miller Indices — Relation Between Interplanar Spacing d and Cube Edge a — Important Features of Miller Indices — Crystal Plane and its Normal — Highlights — Can You Answer This ?

4.1. What is Crystallography ?

The study of the geometric form and other physical properties of crystalline solids by using X-rays, electron beams and neutron beams etc. constitutes the science of crystallography.

Crystalline solids are those in which the atoms or molecules are arranged in a very regular and orderly fashion in a three-dimensional pattern. Each atom or molecule is fixed at a definite *point in space at a definite distance* from and in a definite angular orientation to all others surrounding it. This internal spatial symmetry of atomic or molecular orientation is an essential features of crystalline state. The *angular arrangement of the space positions of the atoms in a crystal is called space lattice or lattice array.*

4.2. Space Lattice

A space lattice may be defined as *an infinite array of points in three dimensions in which every point has surroundings or environments identical to that of every other point in the array.*

As an example, let us consider a finite two-dimensional squares array of points shown in Fig. 4.1. If we locate ourself at any lattice point and look around in the plane of the paper, the scenery is the same irrespective of where we are. Consider the immediate surroundings of a point like *A*. If we look due North or due East from this point, we see another point at a distance of 1 unit; along North-East, we see the nearest point at a distance of $\sqrt{2}$ units and along North-North-East, the nearest point is at a distance of $\sqrt{5}$ units. The environments or surrounding scenery would remain unchanged if we shift our position to another point like *B* in the array.

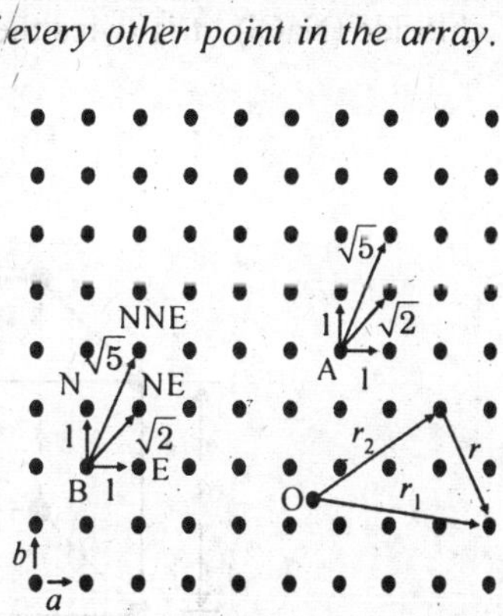

Fig. 4.1

Since this would be true for every point in the array, this array when extended infinitely satisfies the definition given above and can be called a two-dimensional square lattice.

The space lattice may be defined in a more rigorous mathematical way as under :

Choose any arbitrary point O as origin and let vectors r_1 and r_2 join any two lattice points to this origin. The difference T of the vectors should satisfy the following relation

$$T = n_1 a + n_2 b$$

where a and b are the fundamental translation vectors characteristic of the array and n_1 and n_2 are integers. When the above relation is true for all values of r_1 and r_2, the array of points is a two-dimensional lattice.

Similarly, in a 3-dimensional lattice, $T = n_1 a + n_2 b + n_3 c$.

4.3. Crystal Lattice and Unit Cell

The entire lattice structure of a crystal is found to consist of identical blocks or unit cells. The unit cell is the smallest *block or geometric figure* from which the crystal is built up by repetition in three-dimensions. The unit cell is a parallelepiped by moving which in the direction of each of the coordinate axis x, y and z, arranged parallel to edges of the figure, the crystal lattice can be constructed. As seen, this process resembles brick laying. It is also known as fundamental *elementary pattern* because it is repeated again and again to form the lattice structure of a crystal.

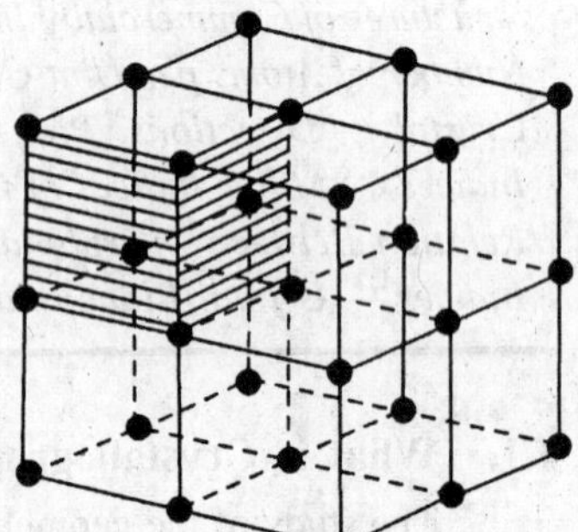

Fig. 4.2

In general, a unit cell may be defined as that *volume of a solid* from which the entire crystal can be constructed by translational repetition in three-dimensions (Fig. 4.2). In fact, unit cell is the smallest part of a crystal having all the structural properties of the given lattice.

If each atom in a lattice is replaced by a point, then each such point is called a *lattice point* and the arrangement of the points is referred to as the (three-dimensional) lattice array. The length of the side of a unit cell is the distance between atoms of the *same* kind. In the case of pure metals whose crystals have simple cubic structure [Fig. 4.3 (*a*)], it is equal to the basic distance 'a' only. However, in the case of crystals of chemical compounds like the cubic crystal of sodium chloride (NaCl), it is twice the distance 'd' between two adjacent atoms as shown in Fig. 4.3 (*b*).

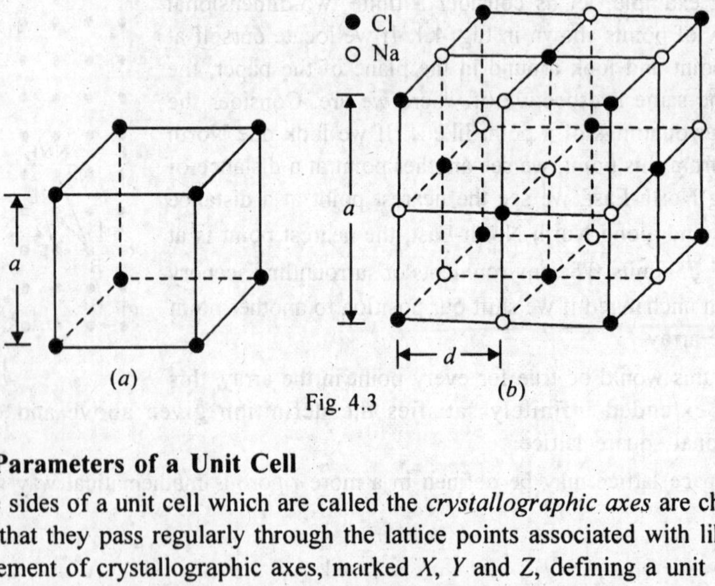

Fig. 4.3

4.4. Lattice Parameters of a Unit Cell

The three sides of a unit cell which are called the *crystallographic axes* are chosen to have directions such that they pass regularly through the lattice points associated with like-atoms. An arbitrary arrangement of crystallographic axes, marked X, Y and Z, defining a unit cell is shown in Fig. 4.4. The angles between the three axes are called *interfacial angles*. The intercepts *a*, *b* and *c* define the dimensions of a unit cell and are known as its *primitives* or *characteristic intercepts* on the axes. The primitives and interfacial angles constitute the *lattice parameters of the unit cell*. These are also called the geometrical constants of a given crystal substance.

If we know the linear dimensions *a*, *b* and *c* and also the interfacial angles α, β and γ, we know the form and actual size of the unit cell. But if we do not know actual values of primitives but only their ratio and the values of interaxial angles, then we can only determine the form of the unit cell and not its actual size.

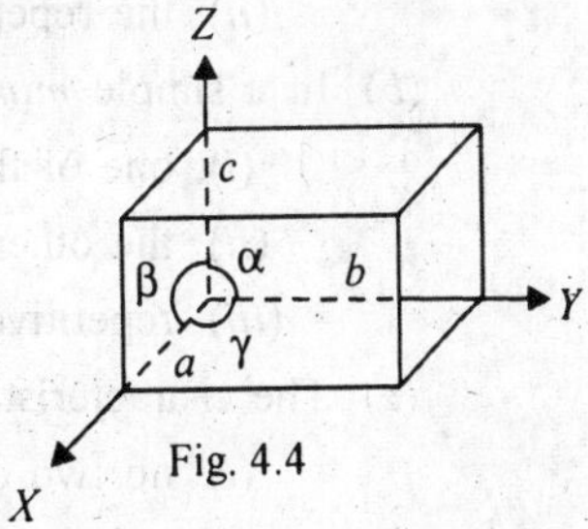

Fig. 4.4

In a space lattice, all lattice points may be included in a set of parallel and equally-spaced planes known as *lattice planes*. These planes give the orientations of the possible faces of a crystal of the substance in question. In theory, there are 320 such lattices or space groups divided into 32 crystal *classes*. These can be arranged in 7 crystal *systems* which are characterised by three symmetry elements : (*i*) the plane of symmetry (*ii*) the axes of symmetry and (*iii*) the centre of symmetry.

4.5. The Seven Crystal Systems

As stated above, there are seven crystal systems corresponding to seven distinct types of unit cells (or subunits). These unit cells are all parallelepipeds whose shapes are determined by the lengths of the three sides *a*, *b*, *c* and the values of the three interfacial angles α, β, and γ.

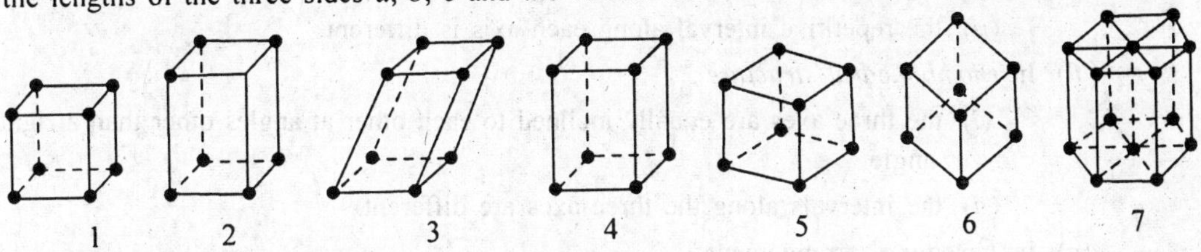

Fig. 4.5 (*a*)

The names of the seven crystal systems are :

(1) Cubic (2) Monoclinic (3) Triclinic (4) Tetragonal (5) Orthorhombic (6) Rhombohedral (Trigonal) and (7) Hexagonal.

The shapes of their unit cells are shown in Fig. 4.5 (*a*) and their lattice parameters are summarised in Table No. 4.1.

Table No. 4.1
The Crystal Systems

S. No.	*Name of system*	*Relation between primitives*	*Interfacial angles*	*Example*
1.	Cubic	$a = b = c$	$\alpha = \beta = \gamma = 90°$	NaCl, CaF_2, $NaClO_2$
2.	Monoclinic	$a \neq b \neq c$	$\alpha = \beta = 90° \neq \gamma$	Na_2SO_4, $FeSO_4$
3.	Triclinic	$a \neq b \neq c$	$\alpha \neq \beta \neq \gamma \neq 90°$	$CuSO_4$, $K_2Cr_2O_7$
4.	Tetragonal	$a = b \neq c$	$\alpha = \beta = \gamma = 90°$	$NiSO_4$, SnO_2
5.	Orthorhombic	$a \neq b \neq c$	$\alpha = \beta = \gamma = 90°$	KNO_3, $BaSO_4$, $MgSO_4$
6.	Rhombohedral (Trigonal)	$a = b = c$	$\alpha = \beta = \gamma \neq 90°$	$CaSO_4$
7.	Hexagonal	$a = b \neq c$	$\alpha = \beta = 90°$ $\gamma = 120°$	SiO_2, AgI

Following points are worth noting regarding the different structures given above.

(*a*) In simple *cubic* structure :

(*i*) the three crystal axes are at right angles

(*ii*) the repetitive intervals along the three axes are equal.

(*b*) In a simple *monoclinic* arrangement :

(*i*) one of the axes is at right angles to the other two but

(*ii*) the other two aces are not at right angles to each other and

(*iii*) repetitive interval along each axis is different.

(*c*) The characteristic features of *triclinic* structure is a complete lack of symmetry *i.e.*

(*i*) no two of the three axes are at right angles to each other

(*ii*) repetitive interval along each axis is different.

(*d*) A simple *tetragonal* structure has the following features :

(*i*) the three crystal axes are at right angles to each other

(*ii*) the repetitive intervals along two of the axes are equal

(*iii*) the interval in the third direction is different from that in the other two.

(*e*) *Orthorhombic* structure has the following features :

(*i*) the three crystal axes are at right angles

(*ii*) the repetitive interval along each axis is different.

(*f*) In *rhombohedral structure* :

(*i*) the three axes are equally inclined to each other at angles other than a right angle

(*ii*) the intervals along the three axes are different.

(*g*) In *hexagonal* arrangement :

(*i*) there are three axes in one plane 120° from each other and a fourth axis normal to this plane

(*ii*) the repetition intervals along the three coplanar axes are equal but the interval along the fourth axis is unique.

It will be seen that cubic system possesses the greatest symmetry whereas triclinic system has the least.

4.6. Bravais' Lattices

According to Bravais who introduced the space lattice idea as early as 1880, there are 14 *possible types of space lattices* in these *seven systems of crystals* : one triclinic, two monoclinic, four orthorhombic, two tetragonal, one hexagonal, one rhombohedral and three cubic as illustrated in Fig. 4.5 (*b*). In fact, it has been proved mathematically that there are only fourteen *independent* ways of arranging points in three-dimensional space such that each arrangement conforms to the definition of a space lattice.

The cubic system which is the simplest and also most common will be considered in detail.

4.7. Symmetry Elements of a Crystalline Solid

The main symmetry elements of a crystalline solid are : (*i*) axes of symmetry (*ii*) planes of symmetry and (*iii*) centre of symmetry.

Consider a simple geometrical solid body such as a cube. Since it has a certain regularity in the arrangement of its faces, it is called a symmetrical solid. In fact, a description of symmetries present in a solid is a *way of describing this regularity of arrangement of its faces.*

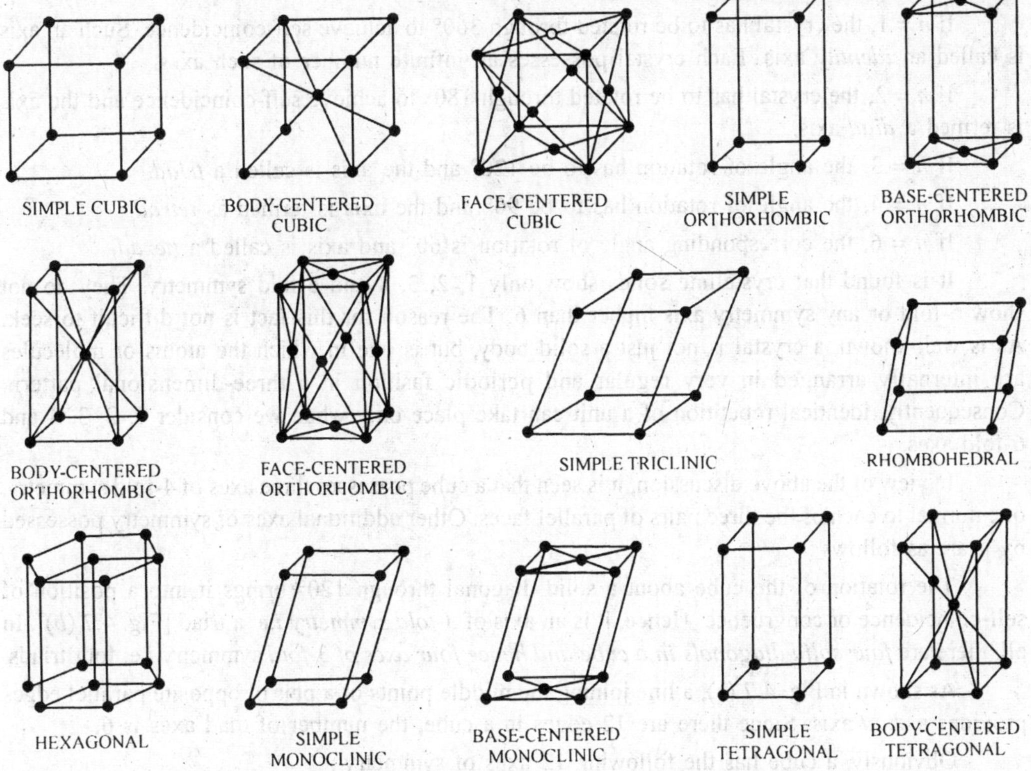

Fig. 4.5 (*b*)

As shown in Fig. 4.6, when a cube is laid flat on a table its four side faces become vertical. The intersections of these four faces are four vertical lines which are parallel to each other. It is obvious that if one of the intersection edges is made vertical, then the four faces automatically become vertical planes. This fact leads one to the conclusion that these faces are related to each other in some way. These four faces are said to constitute a *zone* and any direction parallel to the intersection edges is called the *zone axis.*

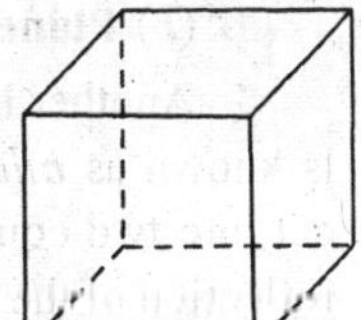

Fig. 4.6

(*a*) **Axis of Symmetry**

As shown in Fig. 4.7 (*a*), if a cube is rotated about a vertical axis (parallel to intersection edges and passing through the centre of the horizontal face), then in one complete revolution of 360°, there are found to be four positions of the cube which are coincident with its original position. In fact, when the cube is rotated through 90°, it is not possible to distinguish it from its original position. In other words, each rotation of 90° brings the cube into self-coincidence or in a congruent position.

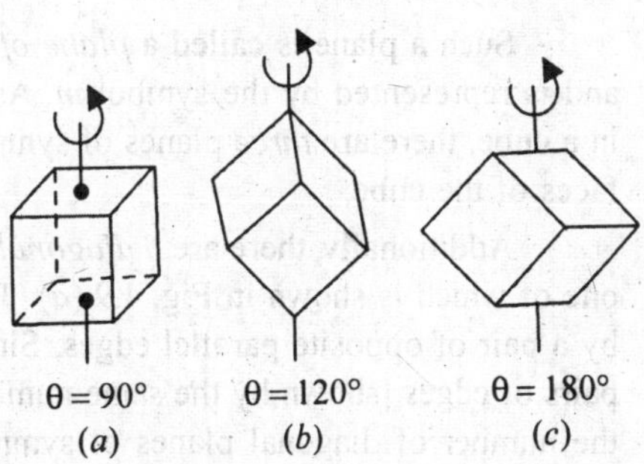

Fig. 4.7

This rotation axis is known as an *axis of symmetry* and since there are four congruent positions in one complete revolution, it is called a *four-fold* axis of symmetry. In general, if a rotation through an angle of $2\pi/n$ radians or $360/n$ degrees about an axis brings a figure into congruent position, the axis is called an n-fold axis of symmetry.

If $n = 1$, the crystal has to be rotated through 360° to achieve self-coincidence. Such an axis is called an *identity* axis. Each crystal possesses an infinite number of such axes.

If $n = 2$, the crystal has to be rotated through 180° to achieve self-coincidence and the axis is termed a *diad* axis.

If $n = 3$, the angle of rotation has to be 120° and the axis is called a *triad.*

If $n = 4$, the angle of rotation has to be 90° and the axis is termed as *tetrad.*

If $n = 6$, the corresponding angle of rotation is 60° and axis is called a *hexad.*

It is found that crystalline solids show only 1, 2, 3, 4 and 6-fold symmetry. They do not show 5-fold or any symmetry axis higher than 6. The reason for this fact is not difficult to seek. As is well-known, a crystal is not just a solid body, but is one in which the atoms or molecules are internally arranged in very regular and periodic fashion in a three-dimensional pattern. Consequently, identical repetition of a unit can take place only when we consider 1, 2, 3, 4 and 6-fold axes.

In view of the above discussion, it is seen that a cube possesses three axes of 4-fold symmetry, one normal to each of the three pairs of parallel faces. Other additional axes of symmetry possessed by it are as follows :

The rotation of the cube about a solid diagonal through 120° brings it into a position of self-coincidence or congruence. Hence, it is an axis of *3-fold symmetry i.e.* a triad [Fig. 4.7 (*b*)]. In all, there are *four solid diagonals in a cube and hence four axes of 3-fold* symmetry i.e. four triads.

As shown in Fig. 4.7 (*c*), a line joining the middle points of a pair of opposite parallel edges provides a *diad* axis. Since there are 12 edges in a cube, the number of diad axes is 6.

Obviously, a cube has the following 12 axes of symmetry :

3 tetrad axes ; 4 triad axes and 6 diad axes.

(*b*) **Planes of Symmetry**

Another kind of symmetry shown by a cube is known as *bilateral* symmetry. A crystal can be cut into two equal halves such that one half is the reflection of the other half. There is a certain plane along which if the crystal is cut and then placed on a mirror, the image reproduces the other half of the crystal.

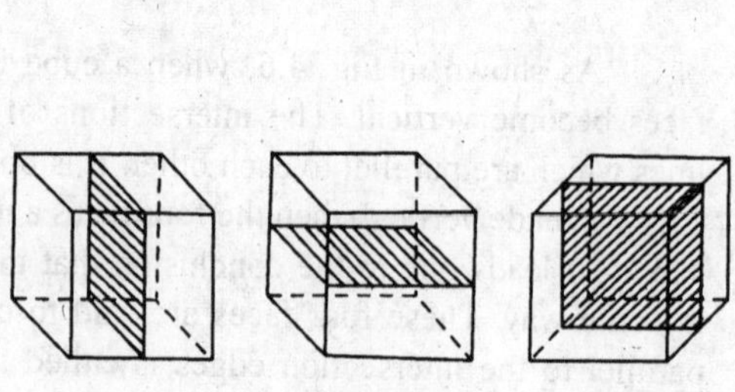

Fig. 4.8

Such a plane is called a *plane of symmetry* and is represented by the symbol m. As shown in Fig. 4.8, in a cube, there are *three* planes of symmetry parallel to the faces of the cube.

Additionally, there are *6 diagonal planes of symmetry* one of which is shown in Fig. 4.9 (*a*). This plane is formed by a pair of opposite parallel edges. Since there are 6 such pairs of edges [shown by the same numbers in Fig. 4.9 (*b*)], the number of diagonal planes of symmetry is six. Hence, total number of planes of symmetry in a cubic crystal is = (3 + 6) = 9.

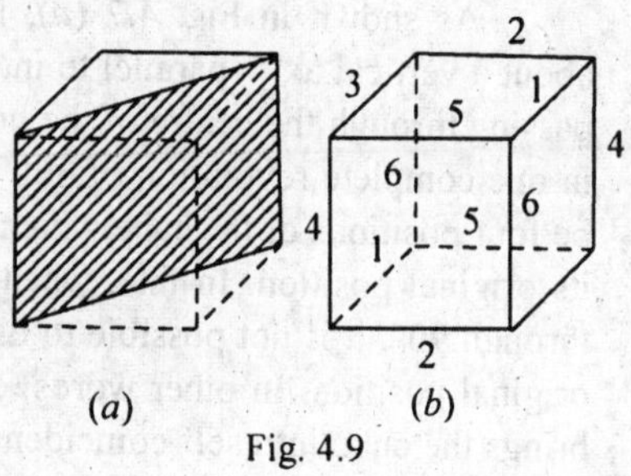

(*a*) (*b*)

Fig. 4.9

(*c*) Centre of Symmetry

It is also known as the centre of inversion (*I*). It is such a point in a crystal that if a line is drawn from any point on the crystal through this point and produced an equal distance on the other side of this centre, it meets an identical point.

It is obvious from the above definition that the centre of symmetry leads to parallel pairs of faces on opposite sides of the crystal. In fact, centre of symmetry really represents a reflection through a point instead of reflection in a plane.

In the end, it may be noted that the full crystallographic symmetry of a cubic crystal comprises of the following 23 elements of symmetry :

(*i*) 13 axes of symmetry — 3 tetrads, 4 triads and 6 diads.

(*ii*) 9 planes of symmetry — 3 planes parallel to the faces of the cube and 6 diagonal planes.

(*iii*) one centre of symmetry.

The unit cells belonging to the seven crystal systems have the following symmetry axes :

(*i*) triclinic system has no axis of symmetry.

(*ii*) the monoclinic system has one diad axis only. Moreover, there are no axes of higher degree.

(*iii*) the orthorhombic system has three diad axes only.

(*iv*) the rhombohedral system has one 4-fold axis. This unique axis is called the *c*-axis (i.e. *Z*-axis).

(*v*) the cubic system has four triad axes.

(*vi*) the tetragonal system has one triad axis. This unique axis is equally inclined to the three coordinate axes.

(*vii*) the hexagonal system has one hexad. This unique axis is at right angles to the other coordinate axes.

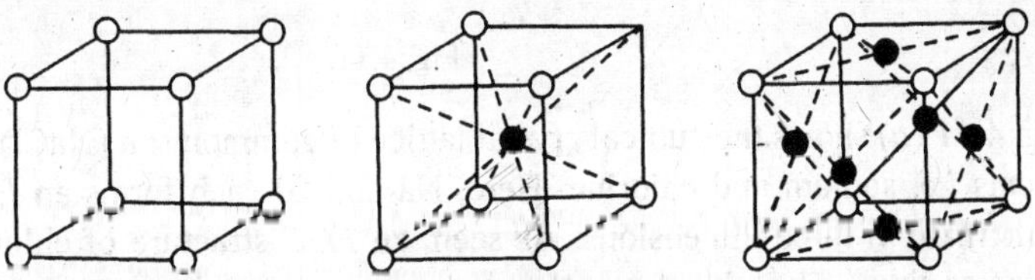

(*a*) (*b*) (*c*)

Fig. 4.10

4.8. Space Lattices of Cubic System

An mentioned earlier, cubic structure is the simplest known type of array in which the atoms take positions at the corners of the cube. However, there are three types of lattices possible in this system depending on the positions of the lattice points in the unit cell :

(*i*) simple cubic (*SC*) lattice — Fig. 4.10 (*a*)

(*ii*) body-centred cubic (*BCC*) lattice — Fig. 4.10 (*b*)

(*iii*) face-centred cubic (*FCC*) lattice — Fig. 4.10 (*c*)

Some elements as well compounds possess *FCC* lattice whereas many others crystallise into *BCC* lattice. These are listed in Table No. 4.2 which also gives the value of primitive a in angstrom units (1 Å = 10^{-10} m).

Table No. 4.2.

Elements of Cubic Crystal System

Face-centred		*Body-centred*	
element	a (in Å)	*element*	a (in Å)
Aluminium	4·049	Barium	5·025
Copper	3·615	α-iron	2·867
Gold	4·079	Rubidium	5·630
γ-iron	3·591	Sodium	4·291
Lead	4·951	Titanium	3·306
Nickel	3·524	Tungsten	3·115
Platinum	4·086	Uranium	3.474
Silver	4·086	Zirconium	3·620

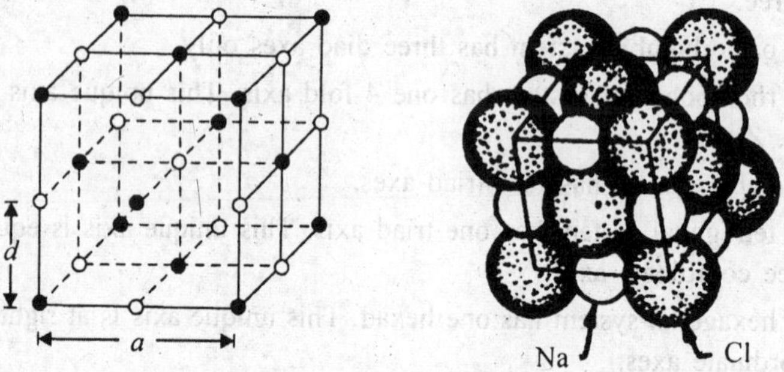

(a) (b)

Fig. 4.11

Fig. 4.11 (*a*) shows the cubical crystal lattice of common salt (NaCl) which is a compound of two elements *i.e.* sodium and chlorine. Here, Na and Cl each forms an *FCC* pattern and the two patterns alternate in three dimensions. As seen, an *FCC* structure of chlorine penetrates a similar structure of sodium. The above-mentioned three types of lattices i.e. *SC*, *BCC* and *FCC* are sometimes represented by the symbols *P*, *I* and *F* respectively. Not all such lattices are possible in all crystal systems. For example, in the cubic system, we have *P*, *I* and *F* whereas triclinic and hexagonal systems have *P* lattice only.

4.9. Calculations of Dimensions of Unit Cell

The dimensions of the unit cell or the interatomic distance in a crystal lattice can be computed from the knowledge of (*i*) molecular weight of the crystalline compound (*ii*) Avogadro's number *N* (*iii*) density of the material and (*iv*) its crystalline form. As an example, consider the following cases :

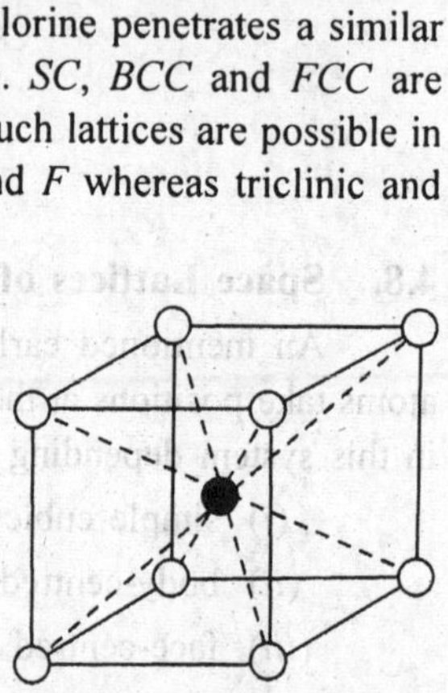

Fig. 4.12

(*i*) *Body-centred Cubic* (*BCC*) *Lattice*. The example of such lattice is afforded by α-iron crystal whose each until cell (Fig. 4.12) consists of 8 corner iron atoms and one atom at the body centre. The atom at the body centre belongs entirely to the unit cell. However, each of the other 8 corner atoms is shared by 9 adjacent unit cells so that each contributes one-eighth of an atom to the cell under consideration. Hence, total number of iron atoms associated with each unit cell = 1 + (8/8) = 2. For iron, density is 7·86 g/cm^3, atomic weight = 55·85, hence

$$a^3 \times 7{\cdot}86 = 2 \times 55{\cdot}85/6{\cdot}02 \times 10^{23}$$

$$a = 2{\cdot}87 \times 10^{-8} \text{ cm} = 2{\cdot}87 \text{ Å}$$

Alternatively, we could use the formula

$$a^3\rho = nA/N$$

where A = atomic weight.

(*ii*) *Face-centred Cubic* (*FCC*) *Lattice*. The common example is the crystal of sodium chloride (NaCl). It can be proved that each unit cell in sodium chloride has four sodium Na^+ atoms (strictly, ions) and four Cl^- atoms associated with it. As shown in Fig. 4.13, the sodium atom in the centre belongs entirely to the cell shown. Each of the other twelve sodium atoms is shared with *three* adjacent unit cells so that each contributes one-fourth of an atom to the cell under consideration. Hence, total number of Na atoms in each unit cell is

$$= 1 + \frac{1}{4}(12) = 4.$$

Now, consider chlorine atoms. There are (*i*) 8 corner atoms and (*ii*) 6 face atoms i.e. one in each face. As proved below, out of the fourteen chlorine atoms, only four are associated with the unit cell. It is easy to visualize that each one of the corner chlorine atoms is shared by 8 adjacent unit cells so that each contributes one-eighth of an atom to the cell under consideration (Art. 4.13). Hence, number of corner atoms per unit cell is = 8/8 = 1.

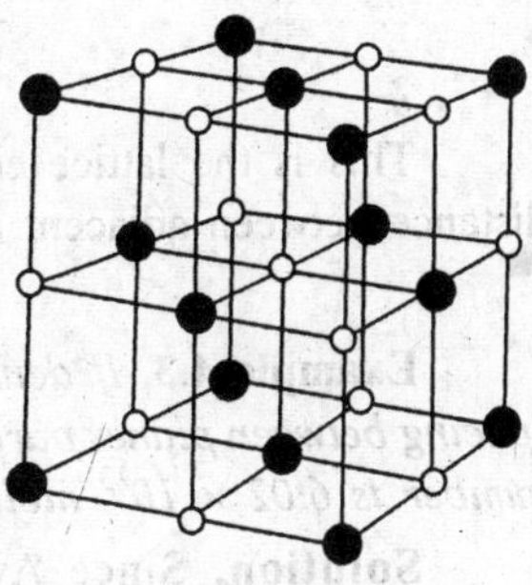

Fig. 4.13

Now, take the face-centred chlorine atoms. Each of these atoms is shared by *two* adjacent unit cells. Hence, each contributes one-half of an atom to the cell under consideration. Number of face-centred chlorine atoms associated with each unit cell = 6/2 = 3. Hence, total number of chlorine atoms belonging to each unit cell of NaCl is = 1 + 3 = 4. Taken together, it means that there are 4 *molecules* of NaCl per unit cell.

Molecular weight of NaCl = 23 + 35·5 = 58·5

Mass of each molecule = $58{\cdot}5/6{\cdot}02 \times 10^{23}$ gram

Mass of 4 molecules = $4 \times 58{\cdot}5/6{\cdot}02 \times 10^{23}$ gram

If 'a' cm is the length of the edge of the unit cell, then its volume is a^3 and since density of NaCl is 2·18 g/cm^3, mass of each unit cell = $a^3 \times 2{\cdot}18$ gram.

$$\therefore \quad a^3 \times 2{\cdot}18 = 4 \times 58{\cdot}5/6{\cdot}02 \times 10^{23}$$

$$\therefore \quad a = 5{\cdot}63 \times 10^{-8} \text{ cm} = 5{\cdot}63 \text{ Å}$$

As will be seen from Fig. 4.3 (*b*), this is the distance between two adjacent atoms of the same kind i.e. chlorine atoms in this case. Obviously, the distance between two adjacent atoms i.e. Na and Cl atoms is half of the above value *i.e.* $d = 5{\cdot}63/2 = 2{\cdot}815$ Å

Alternatively, we could use the relation

$$\alpha^3\rho = n\,M/N$$

where ρ = density of crystal; M = molecular weight of crystal material; n = No. of molecules/unit cell; N = Avogadro's number.

Example 4.1. *Classify the following unit cells into proper system :*

(*i*) $a = 10{\cdot}8$ Å, $b = 9{\cdot}47$ Å, $c = 5{\cdot}2$ Å

$\alpha = 41°$, $\beta = 83°$ and $\gamma = 93°$

(*ii*) $a = b = 10{\cdot}73$ Å, $c = 14{\cdot}3$ Å

$\alpha = \beta = 90°$ and $\gamma = 120°$

Solution. (*i*) Here, $a \neq b \neq c$ *i.e.* the three primitives are unequal. Also, no two axes are at right angles to each other. In other words, there is complete lack of symmetry which is characteristic of *triclinic* structure (Table No. 4.1). Hence, this unit cell belongs to the triclinic crystal system (Art. 4.5).

(*ii*) Here, $a = b \neq c$ and $\alpha = \beta = 90°$ and $\gamma = 120°$. As seen, from Table No. 4.1, this unit cell belongs to *hexagonal* crystal system.

Example 4.2. *NaCl crystals have f.c.c. structure. The density of sodium chloride is 2180 kg/m³. Calculate the distance between two adjacent atoms. Given that the atomic weight of sodium = 23 and that of chlorine = 35·5. Take N = 6·02 × 10²⁶ per kg-mol.*

Solution. We will use the formula : $a^3\rho = nM/N$

Now, $M = (23 + 35{\cdot}5) = 58{\cdot}5$, $n = 4$ — Art. 4–9

$\therefore$ $a^3 \times 2180 = 4 \times 58{\cdot}5/6{\cdot}02 \times 10^{26}$

$\therefore$ $a = 5{\cdot}63 \times 10^{-10}$ m $= 5{\cdot}63$ Å.

This is the lattice constant *i.e.* distance between adjacent atoms of the *same* kind. The distance between adjacent atoms *i.e.* between sodium and chlorine atoms is

$$d = a/2 = 5{\cdot}63/2 = \mathbf{2{\cdot}815\ Å}$$

Example 4.3. *If density of NaCl is 2·163 g/cm³ and its molecular weight 58·45, find the spacing between planes parallel to the cubic lattice faces of the NaCl crystal. Given that Avogadro's number is 6·02 × 10²³ molecules/g-mol.*

Solution. Since Avogadro's number represents the number of molecules in one gram-molecular-wt. of the substance, it means that there are $6{\cdot}02 \times 10^{23}$ molecules in 58·45 gram of NaCl.

No. of molecules/gram $= 6{\cdot}32 \times 10^{23}/58{\cdot}45 = 1{\cdot}03 \times 10^{22}$

No. of molecules/cm³ $= 1{\cdot}03 \times 10^{22} \times 2{\cdot}163 = 2{\cdot}23 \times 10^{22}$

Since there are two atoms in a molecule,

No. of atoms/cm³ $= 2 \times 2{\cdot}23 \times 10^{22} = 4{\cdot}46 = 10^{22}$

Since, NaCl crystal lattice consists of a multitude of cubes with atoms at the corners, there will be $(4{\cdot}46 \times 10^{22})^{1/3} = 3{\cdot}54 \times 10^7$ atoms in a row 1 cm long. The spacing between atoms and hence between rows or planes is

$$d = 1/3{\cdot}54 \times 10^7 = 2{\cdot}83 \times 10^{-8} \text{ cm} = \mathbf{2{\cdot}83\ Å}$$

4.10. Crystal Structures of Commercially Important Materials

Many materials of commercial importance have relatively simple crystal structures. Some of these are described below :

(*i*) **The hexagonal close-packed (hcp) structure.** This structure is built up by stacking close-packed planes in a simple sequence. The method of stacking can be understood with the

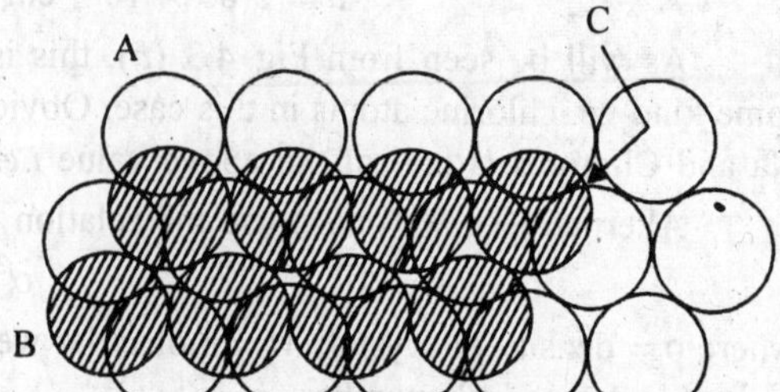

Fig. 4.14

help of Fig. 4.14 which shows two close-packed planes lying one above the other. It is seen that each atom of one plane rests *in a depression* against three atoms of the other. Also plane *B* has two possible positions : the one as indicated and the other given by position *C*. For either choice, one atom of plane *B* rests against three atoms of lower plane *A* and *vice versa*. However, the only important choice is that for the third plane which also has two choices. Let the first two planes have the sequence *AB*. Then the third plane may be added over the second in position *A*. In that case, the atoms of plane 3 would be directly above those of plane 1.

Alternatively, plane 3 may be stacked in position *C* (Fig. 4.14) so that atoms of plane 3 are directly above *interstices* of plane 1. The hcp structure has the first stacking sequence of *ABAB* ..... as shown in Fig. 4.15 (*b*). Heavy lines in Fig. 4.15 (*a*) show the unit cell of hcp crystal.

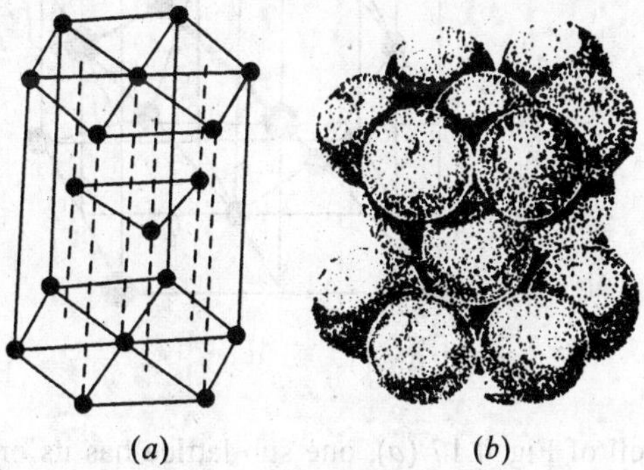

(*a*) (*b*)

Fig. 4.15

This structure is called close-packed because it possesses the maximum density of packing of hard spheres. Each atom has 12 *nearest* neighbours. In other words, coordination number is 12 (Art. 4.10).

(*ii*) **The face-centred cubic (fcc) or cubic close-packed structure.** The unit cell of this structure is shown in Fig. 4.16. It has the stacking sequence of *ABCABC* ..... . This structure is also close-packed because each atom has 12 *nearest* neighbours. As shown in Fig. 4.16 (*b*), the atoms are in contact with the diagonal atoms.

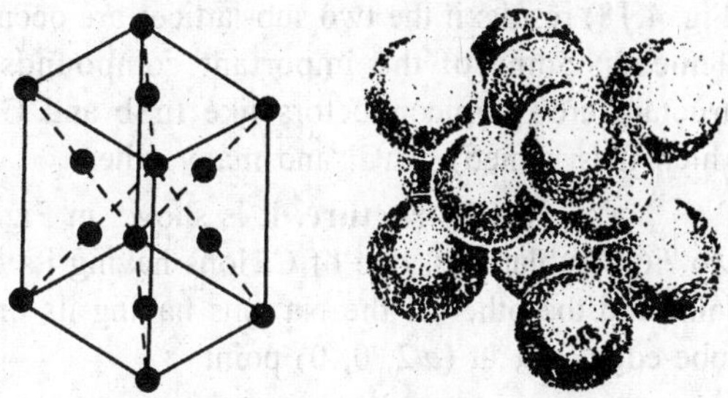

(*a*) (*b*)

Fig. 4.16

(*iii*) **The body centred cubic (bcc) structure.** The unit cell of such a structure is sketched in Fig. 4.12. It has one atom at each corner and one atom at the centre of the cube. Such a structure is not formed of close-packed planes as the above two since each atom has only eight nearest neighbours. However, the atoms are *in contact along body diagonals.*

(*iv*) **The simple cubic (sc) structure.** As shown in Fig. 4.10 (*a*), the unit cell of this structure has atoms only at the corners of the cube. Consequently, these atoms touch along cube edges. This structure is loosely-packed because each atom has only six *nearest* neighbours. Only one element *i.e.* polonium (Po) exhibits this structure (in a certain temperature region).

Many crystal structures are a variation of this simple structure. In fact, many of them consist of sets of interpenetrating simple cubic sub-lattices. For example, *BCC* lattice can be considered to consist of two identical *SC* lattices, one consisting of corner atoms and the other of the centre atoms. Similarly, an *FCC* lattice can be considered to contain four identical *SC* lattices properly disposed with respect to each other.

(*v*) **Other structures.** Although many elements solidify in one of the structures described above, several other elements and all compounds possess other structures. Some of these structures are derivates from or combination of the basic structures described above.

(*a*) **Diamond cubic structure.** Germanium (Ge), silicon (Si) and carbon (C) possess this structure which is a combination of two interpenetrating *FCC* sub-lattices. As seen from the unit

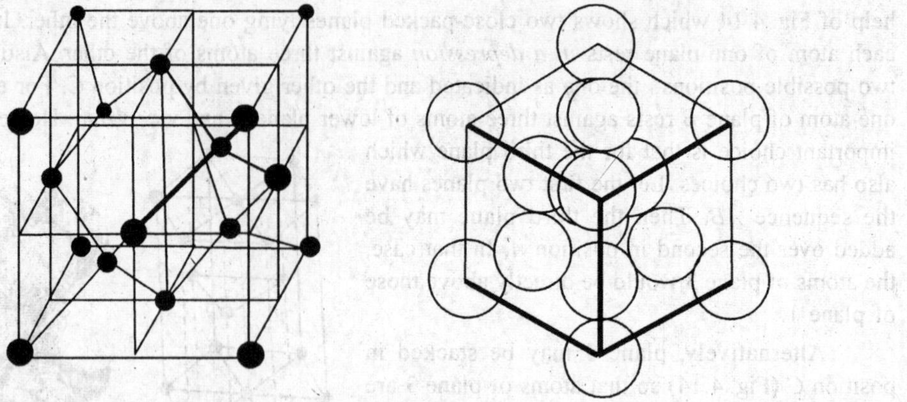

Fig. 4.17

cell of Fig. 4.17 (*a*), one sub-lattice has its origin at (0, 0, 0) and the other at a point one quarter of the way along the body diagonal *i.e.* at the (*a*/4, *a*/4, *a*/4) point. This structure is loosely-packed since each atom has only four *nearest* neighbours.

(*b*) **Zinc blende or sphalerite cubic structure.** This structure (Fig. 4.18) results if the two sub-lattices are occupied by two different elements. Some of the important compounds which possess this structure are : semiconductors like InSb and GaAs; also ZnS (after which it is named), CuCl and many others.

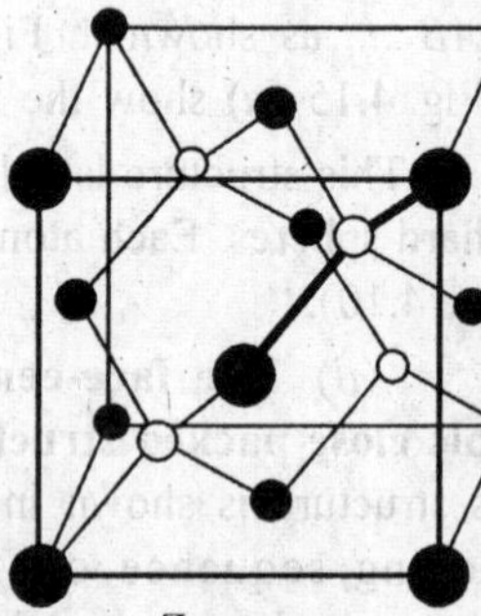

Fig. 4.18

(*c*) **NaCl structure.** It is shown in Fig. 4.11 and consists of two *FCC* sublattices, one of Cl ions having its origin at the (0, 0, 0) point and the other of the Na ions having its origin midway along a cube-edge, *say,* at (*a*/2, 0, 0) point.

4.11. Coordination Number

By coordination number, it is meant the number of *nearest* neighbours which an atom has in the unit cell of any crystal structure.

Following cases would be considered :

(*i*) **sc structure** [Fig. 4.10 (*a*)]

In this case, the coordination number is *six*. It is so because each corner atom is linked with seven other unit cells that can be imagined to be built around the unit cell containing the atom. In that case, each corner atom has four neighbours in the same plane, one vertically above and one immediately below, giving a total of six *nearest* neighbouring atoms.

(*ii*) **bcc structure** [Fig. 4.10 (*b*)]

Its coordination number is *eight*. It should be noted that in this case, the *nearest* neighbours of any corner atom are the body-centred atoms (and not the other corner atoms). Since, there are eight surrounding unit cells for any corner atom, their eight body-centred atoms form the *nearest* neighbours for any corner atom.

(*iii*) **fcc structure** [Fig. 4.10 (*c*)]

The coordination number of such a structure is *twelve*. In this case, the *nearest* neighbours of any corner atom are the face-centred atoms of the surrounding unit cells. Any corner atom has four such atoms in its own plane, four in a plane above it and four in a plane below it.

4.12. Atomic Radius

It is defined as *half the distance between nearest neighbours in a crystal of a pure element.* Usually, it is expressed in terms of the cube edge *a*. It must be remembered that *any two nearest neighbouring atoms touch each other.*

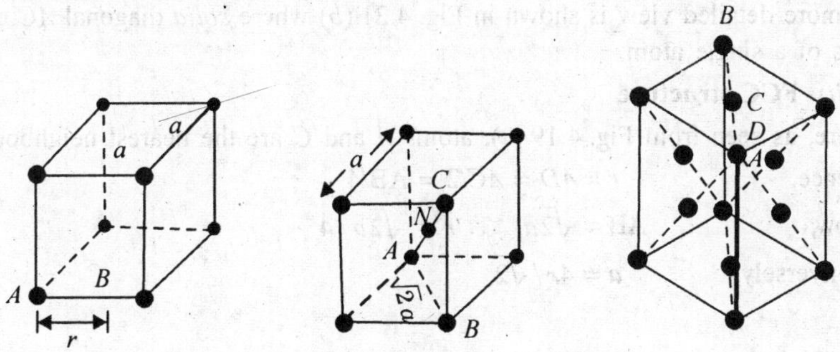

(*a*) (*b*) (*c*)

Fig. 4.19

(*i*) **SC structure.** By definition, *AB* represents the atomic radius. It is obvious from Fig. 4.19 (*a*) and 4.20 (*b*) that $r = a/2$.

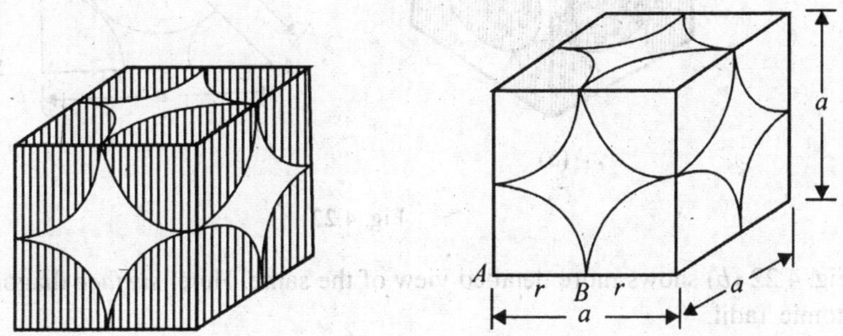

(*a*) (*b*)

Fig. 4.20

(*ii*) **BCC structure.** As seen from Fig. 4.19 (*b*), atoms *C* and *N* are the nearest neighbours. By definition $r = CN/2$.

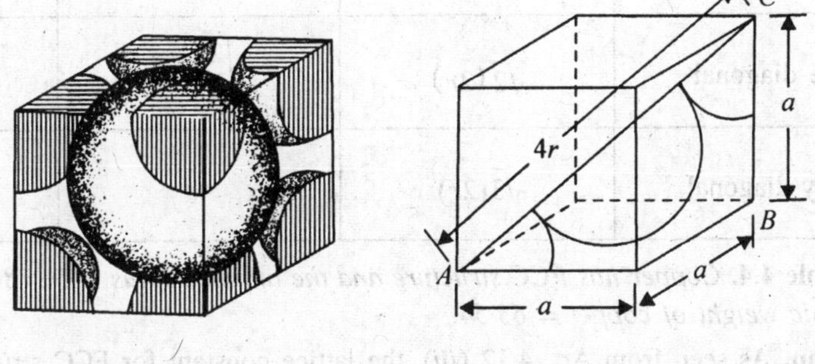

(*a*) (*b*)

Fig. 4.21

Now $AC = \sqrt{AB^2 + BC^2} = \sqrt{(\sqrt{2}a)^2 + a^2} = \sqrt{3}a$

$r = CN/2 = AC/4 = \sqrt{3}a/4$

Conversely, $a = 4r/\sqrt{3}$

A more detailed view is shown in Fig. 4.21 (*b*) where *solid* diagonal *AC* equals four times the radius of a single atom.

(*iii*) **FCC structure**

Here, as seen from Fig. 4.19 (*c*), atoms *A* and *C* are the nearest neighbours.

Hence, $r = AD = AC/2 = AB/4$

Now, $AB = \sqrt{2}a \quad \therefore r = \sqrt{2}a/4$

Conversely, $a = 4r/\sqrt{2}$

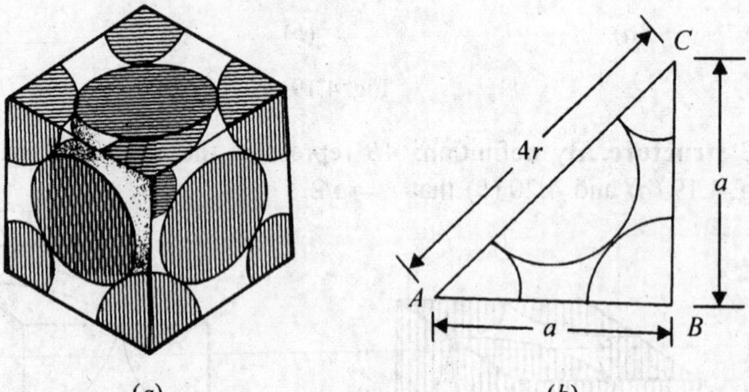

(*a*) (*b*)

Fig. 4.22

Fig. 4.22 (*b*) shows more detailed view of the same. Here, *surface* diagonal *AC* equals four times atomic radii.

The lattice constant or side of a cubic unit cell as well its face and body diagonals in terms of atomic radius are tabulated as under.

Table No. 4.3

	SC	*BCC*	*FCC*
Side	$2r$	$4r/\sqrt{3}$	$4r/\sqrt{2}$
Face diagonal	$\sqrt{2}(2r)$	$\sqrt{\frac{2}{3}}(4r)$	$4r$
Body diagonal	$\sqrt{3}(2r)$	$4r$	$\sqrt{\frac{3}{2}}(4r)$

Example 4.4. *Copper has FCC structure and the atomic radius is 1·278 A.U. Calculate its density. Atomic weight of copper = 63·54.*

Solution. As seen from Art. 4.12 (*iii*), the lattice constant for FCC structure is given by

$$a = 4r/\sqrt{2} = 4 \times 1{\cdot}278/\sqrt{2} = 3{\cdot}61 \text{ A.U.}$$

Volume of unit cell = $a^3 = (3{\cdot}61 \times 10^{-8})^3$ cm^3

$= 47 \times 10^{-24}$ cm^3

If its density is ρ, then mass of unit cell

$= \rho \times 47 \times 10^{-24}$ gram

Now, let us find the mass of the same unit cell in a different way.

Each unit cell of *FCC* structure contains 4 atoms. Obviously, their mass would represent the mass of the cell. Since 6·02 × 10 atoms have a mass of 63·54 gram, 4 atoms will have a mass

$= 4 \times 63{\cdot}54/6{\cdot}02 \times 10^{23} = 4{\cdot}22 \times 10^{-23}$ gram

Equating the two masses, we get

$$\rho \times 47 \times 10^{-24} = 4{\cdot}22 \times 10^{-23};\ \rho = \mathbf{8{\cdot}98\ g/cm^3}.$$

Alternate Solution

We could use the following formula

$$a^3\rho = nA/\text{N}$$

where A = atomic weight

$$\therefore \quad (3{\cdot}61 \times 10^{-8})^3\,\rho = 4 \times 63{\cdot}54/6{\cdot}02 \times 10^{23}$$

$$\therefore \quad \rho = \mathbf{8{\cdot}98\ g/m^3}.$$

4.13. Number of Atoms per Unit Cell

(*i*) Simple Cube

Since each corner atom is shared by *eight* surrounding cubes, share of each cube comes to one-eighth of an atom (Fig. 4.23). As there are eight corner atoms in all, each cube has 8 × 1/8 = 1 atom.

Fig. 4.23

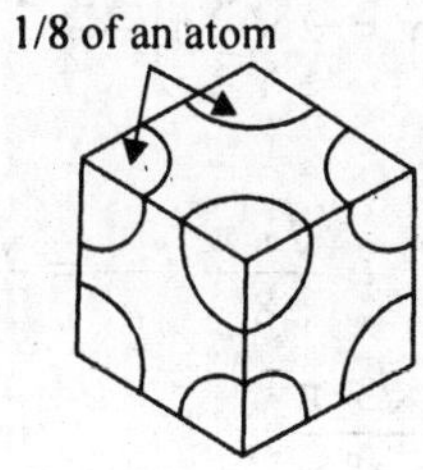

Fig. 4.24

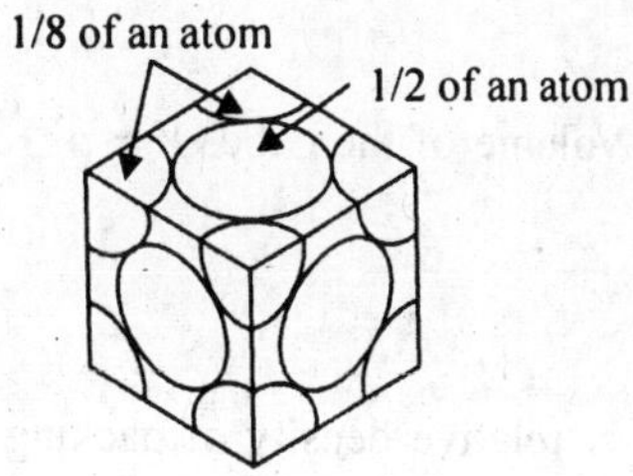

Fig. 4.25

(*ii*) Body-centred Cube

It has eight corner atoms and one centre atom (Fig. 4.24). The number of atoms which belong to this cube are

(*a*) one centre atom and

(*b*) 8 × 1/8 = 1 corner atom.

Hence, number of atoms per cube = 1 + 1 = 2.

(*iii*) Face-centred Cube (Fig. 4.25)

It has 6 face-centred and 8 corner atoms. Each of the 6 face-centred atoms is shared by the two adjoining cubes. Hence, a total of 6/2 = 3 such atoms belong to the cube. As before, one corner atom comes to the share of this cube. Hence, total number of atoms per cube is (3 + 1) = 4.

4.14. Density of Packing

The relative density of packing is *defined as the fraction of volume occupied by spherical atoms as compared to the total available volume of the structure.*

(*i*) Simple Cube

As shown earlier, this cube has one atom.

Volume of one atom $= \frac{4}{3}\pi r^3$

Now, atomic radius $r = \frac{a}{2}$

$\therefore$ Volume of the atom, $v = \frac{4}{3}\pi\left(\frac{a}{2}\right)^3 = \frac{\pi a^3}{6}$

Volume of the cube, $V = a^3$

$\therefore$ relative density of packing is $= \frac{v}{V} = \frac{\pi a^3/6}{a^3} = \frac{\pi}{6}$

(*ii*) Body-centred Cube

In this case, number of atoms per cube is 2 (Art. 4.13)

Volume occupied by *two* atoms is

$$v = 2\times\frac{4}{3}\pi r^3$$

Now $r = \sqrt{3}a/4$

$$\therefore \quad v = 2\times\frac{4}{3}\pi\left(\frac{\sqrt{3}\cdot a}{4}\right)^3 = \frac{\sqrt{3}\pi}{8}\cdot a^3$$

Volume of the cube, $V = a^3$

$$\therefore \quad \frac{v}{V} = \frac{\sqrt{3}\cdot\pi}{8}\times\frac{a^3}{a^3} = \frac{\sqrt{3}\cdot\pi}{8}$$

$\therefore$ relative density of packing $= \frac{\sqrt{3}\cdot\pi}{8}$

(*iii*) Face-centred Cube

As shown in Art. 4.13, in this case, there are 4 atoms per cube.

$\therefore$ volume occupied by *four* atom is

$$v = 4\times\frac{4}{3}\pi r^3$$

Now, $r = \sqrt{2}\cdot a/4$

$$\therefore \quad v = 4\times\frac{4}{3}\cdot\pi\left(\frac{\sqrt{2}\cdot a}{4}\right)^3 = \frac{\sqrt{2}\cdot\pi}{6}\cdot a^3$$

$$V = a^3$$

$$\therefore \quad \frac{v}{V} = \frac{\sqrt{2}\cdot\pi}{6}\cdot\frac{a^3}{a^3} = \frac{\sqrt{2}\cdot\pi}{6}$$

$\therefore$ relative density of packing $= \frac{\sqrt{2}\cdot\pi}{6}$

Some of the details of elementary cubic structure are summarised below in Table No. 4.4.

Table No. 4.4

	SC	*BCC*	*FCC*
Coordination number	6	8	12
Atomic radius (*r*)	$\frac{a}{2}$	$\frac{\sqrt{3}\cdot a}{4}$	$\frac{\sqrt{2}\cdot a}{4}$
Face diagonal	1	2	4
Body diagonal	$\frac{\pi}{6}$	$\frac{\sqrt{3}.\pi}{8}$	$\frac{\sqrt{2}.\pi}{6}$

Example 4.5. (*a*) *With reference to crystals, state what you understand by the terms* (*i*) *unit cell* (*ii*) *coordination number and* (*iii*) *factor.*

(*b*) *Show that for bcc and fcc crystal structures, the lattice constants are given by*

$$a_{bcc} = \frac{4r}{\sqrt{3}} \text{ and } a_{fcc} = \frac{4r}{\sqrt{2}}$$

where r is the atomic radius.

Solution. For solution please refer to Art. 4.3, 4.10, 4.11 and 4.12.

4.15. Important Plane Systems in a Cubic Crystal

In a cubic crystal, there are three sets of planes which are very rich in atoms. Consequently, Bragg reflections (i.e. X-ray reflections) from these planes are more intense than from others. These three sets of planes which have different spacings are shown in Fig. 4.26.

(*i*) The first set of planes consists of *surface* planes of the cube and those parallel to them such as *BFGC*, *AEHD*, *ABCD*, *EFGH* etc. [Fig. 4.26 (*a*)]. Let the distance between consecutive parallel surface planes be d_1.

(*ii*) The second set of planes consists of parallel planes like *AFGD* which are inclined at an angle of 45° to the planes mentioned above [Fig. 4.26 (*b*)]. Their spacing, as shown in Fig. 4.27, is d_2.

(*iii*) The third set of planes consists of those planes which are parallel to the plane *AFH* [Fig. 4.26 (*c*)]. Let the spacing of such planes be d_3 as shown in Fig. 4.27.

(*a*) Spacing of planes

The knowledge of spacing of these three sets of planes is essential for identifying different types of crystals. It is so because for each basic crystal structure, a definite ratio exists between the spacings of planes which are rich in atoms.

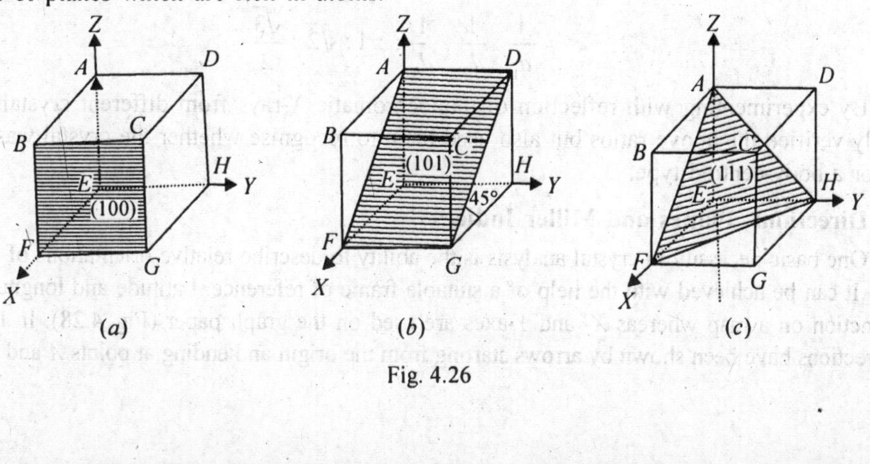

Fig. 4.26

Consider the case of a simple cubic crystal with atoms at the corners of the cube as shown in Fig. 4.26 (*a*). The spacing of face planes like *ABCD* and *EFGH* is d_1.

The spacing of planes like *AFGH* which are inclined at 45° to the first set of planes is $d_2 = d_1/\sqrt{2}$.

The triangle *AFH* represents the third set of planes whose spacing d_3 may be found by drawing the triangle *AEK* in true shape as in Fig. 4.27 (*b*).

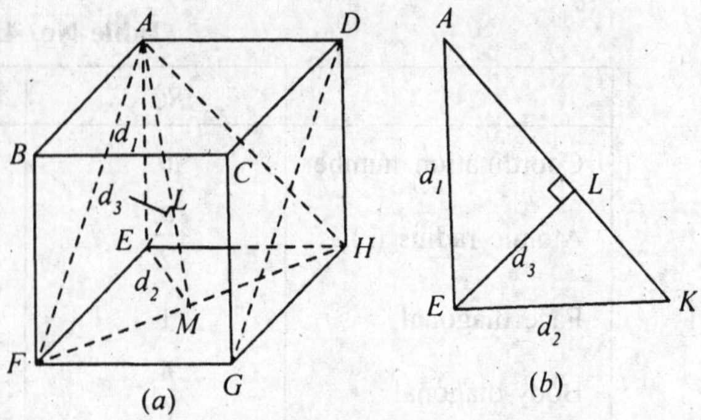

Fig. 4.27

In Fig. 4.27 (*a*), *EK* is perpendicular to *FH* and $EL = d_3$ is perpendicular to *AK*. From the two similar triangles *ELK* and *AEK* of Fig. 4.27 (*b*), we have

$$\frac{EL}{EK} = \frac{AE}{AK}$$

or $$\frac{d_3}{d_2} = \frac{d_1}{\sqrt{(d_1^2 + d_2^2)}}$$

or $$d_3 = \frac{d_1 d_2}{\sqrt{(d_1^2 + d_2^2)}}$$

Now, putting $d_2 = d_1/\sqrt{2}$, we get

$$d_3 = \frac{d_1^2}{\sqrt{2}.\sqrt{(d_1^2 + \frac{1}{2}d_1^2)}} = \frac{d_1}{\sqrt{3}}$$

$$\therefore \frac{1}{d_1} : \frac{1}{d_2} : \frac{1}{d_3} : : 1 : \sqrt{2} : \sqrt{3}$$

Similarly, it can be shown that for a body-centred lattice structure

$$\frac{1}{d_1} : \frac{1}{d_2} : \frac{1}{d_3} : : 1 : \frac{1}{\sqrt{2}} : \sqrt{3}$$

Also, for face-centred lattice structure

$$\frac{1}{d_1} : \frac{1}{d_2} : \frac{1}{d_3} : : 1 : \sqrt{2} : \frac{\sqrt{3}}{2}$$

By experimenting with reflection of monochromatic *X*-rays from different crystals, Bragg not only verified the above ratios but also used them to recognise whether the crystal was simple, cubic or a body-centred type.

4.16. Directions, Planes and Miller Indices

One basic necessity in crystal analysis is the ability to describe relative orientations of lines and planes. It can be achieved with the help of a suitable frame of reference. Latitude and longitude serve this function on a map whereas *X*- and *Y*-axes are used on the graph paper (Fig. 4.28). In Fig. 4.28, two directions have been shown by arrows starting from the origin and ending at points *A* and *B*. These directions can be satisfactorily described by giving the coordinates of the *first* whole-numbered point (x, y) through which each passes. For direction OA, it is (2, 1) and for OB , it is (1, 1).

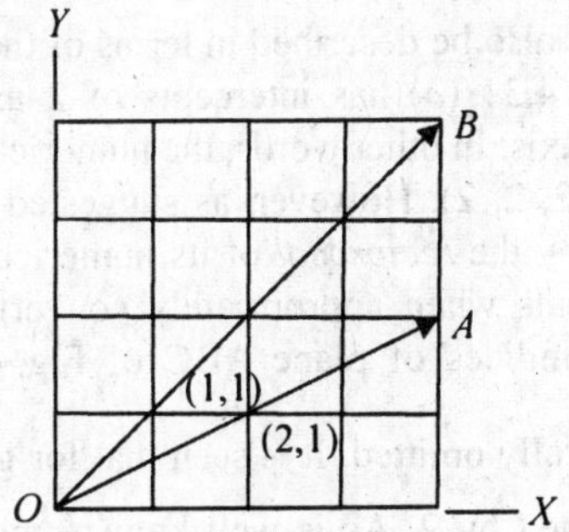

Fig. 4.28

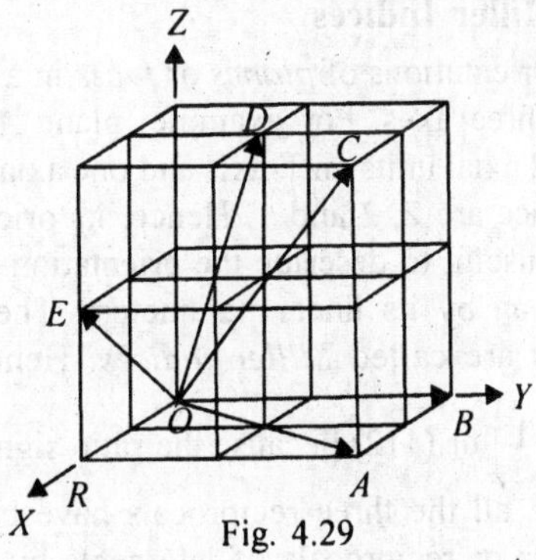

Fig. 4.29

Three-dimensional graph paper of Fig. 4.29 can be used for describing a direction in space. As before, these directions are described by giving the coordinates of the *first* whole-numbered point (x, y, z) through which they pass. The directions of various lines of Fig. 4.29 are as under :

OA [110]	*OR* [100]
OB [010]	*OD* [½, ½, 1] or [112]
OC [111]	*OE* [1, 0, ½] or [201]

The square brackets are used to indicate a *direction.* Commas are unnecessary but the digits are read as "one one zero" and not "one hundred and ten". The digits in square brackets indicate the indices of that direction. In general, the indices of a direction are written as [*uvw*]. A negative index is indicated by a bar over the digit.

For example, +*X*-axis has indices of [100] whereas –*X*-axis has [100] indices. Similarly, +*Y*-axis has indices [010] and the –*Y*-axis [010]. Positive *Z*-axis has indices [001] whereas –*Z*-axis has [001].

4.17. Unit Face and Axial Units

In Fig. 4.30, let *OX*, *OY* and *OZ* be taken as the reference axes. The faces a_1 b_1 c_1 and a_2 b_3 c_6 make certain intercepts or segments on these axes. The first face or plane intercepts segments of oa_1, ob_1 and oc_1 and the second – oa_2, ob_3 and oc_6. These segments represent the respective *linear parameters* of the two *faces.*

It is seen that the first face intercepts one spacing along each axis whereas second face intercepts two, three and six spacings along the respective axes. The first face is called *unit face* which may be defined *as the face which makes one intercept on each axis.*

Obviously, the unit face produces three values which are related to each other as 1 : 1 : 1. The second face produces three values of $2a$, $3b$ and $6c$ which are related to each other as 2 : 3 : 6. These figures of 1, 2, 3, 6 represent the *numerical parameters* of the *faces* and must always be whole numbers.

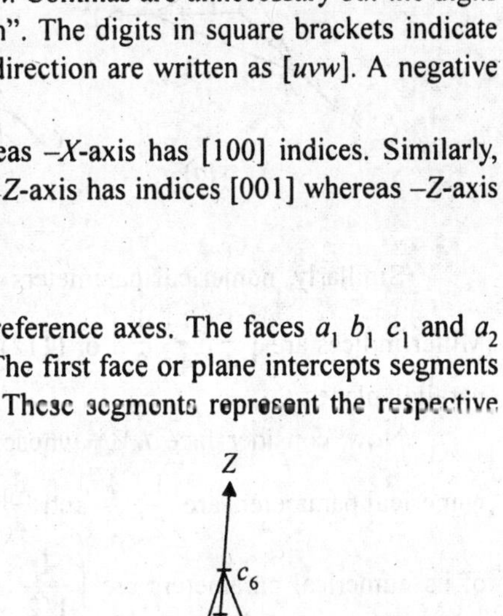

Fig. 4.30

Linear parameters intercepted by the unit *face* are called *axial units*. In the present case, these are a_1, b_1 and c_1 although they are represented generally by a, b, and c along X, Y, and Z axes respectively.

4.18. Miller Indices

Orientations of *planes* or *faces* in a crystal may also be described in terms of their intercepts on the three axes. For example, plane ABC of Fig. 4.31 (*a*) has intercepts of 2 axial units on X-axis, 2 axial units on Y-axis and one axial unit on Z-axis. In other words, the numerical parameters of the face are 2, 2 and 1. Hence, its orientation is (2, 2, 2). However, as suggested by Miller, it is more useful to describe the orientation of a plane by the *reciprocal* of its numerical parameters rather than by its linear parameters. These reciprocals when appropriately converted to whole numbers are called *Miller indices*. Hence, Miller indices of plane ABC of Fig. 4.31 (*a*) are $\left(\frac{1}{2}:\frac{1}{2}:1\right)$ or (112) because the ratio sign (:) is generally omitted. It is seen that for getting whole numbers, all the three reciprocals have been multiplied by 2. As is well known, multiplying *all* intercepts or reciprocals of intercepts by the *same* number does not change the orientation of a plane. The numbers for *planes* are written in parenthesis and not in square brackets. The general expression for indices of a plane is $(h\ k\ l)$. The symbol for a family of parallel planes is $<h\ k\ l>$.

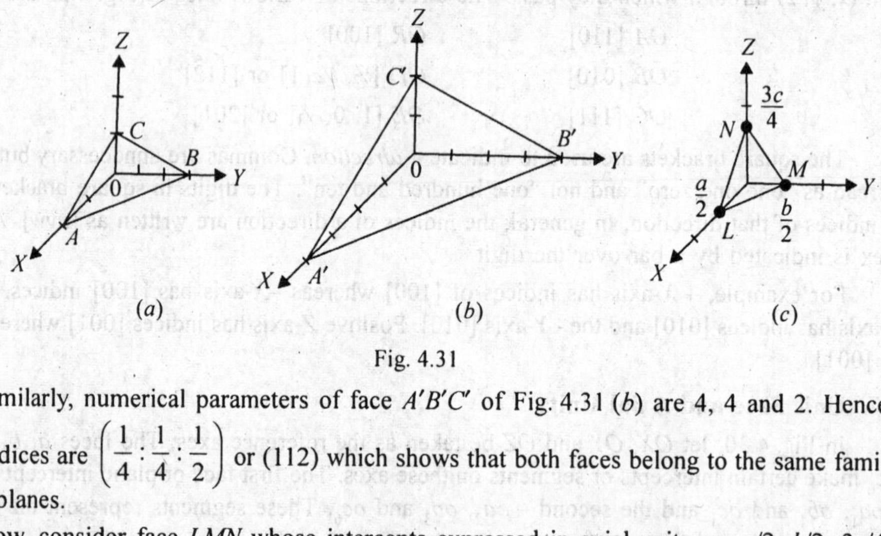

Fig. 4.31

Similarly, numerical parameters of face $A'B'C'$ of Fig. 4.31 (*b*) are 4, 4 and 2. Hence, its Miller indices are $\left(\frac{1}{4}:\frac{1}{4}:\frac{1}{2}\right)$ or (112) which shows that both faces belong to the same family of parallel planes.

Now, consider face LMN whose intercepts expressed in axial units are $a/2$, $b/2$, $3c/4$. Its numerical parameters are $\frac{1}{2}$, $\frac{1}{2}$ and $\frac{3}{4}$. Hence, its Miller indices which are nothing but reciprocal of its numerical parameters are $\left(\frac{1}{1/2}:\frac{1}{1/2}:\frac{1}{3/4}\right)$ or $\left(2\ 2\ \frac{4}{3}\right)$ or (6 6 4) or (3 3 2).

Hence, Miller index of a crystal face of plane is *given by three smallest figures which are inversely proportional to the numerical parameters of the face.*

General Case. Let the intercepts of a crystal face on the three reference axes as expressed in terms of the axial units be pa, qb and rc. Obviously, its numerical parameters are p, q and r. Then Miller indices of the face are given by three numbers h, k, l such that

$$h : k : l :: \frac{1}{p} : \frac{1}{q} : \frac{1}{r}$$

with the condition that h, k and l are the smallest possible integers satisfying the above equation.

Note. If a plane has Miller indices of (h, k, l), then its intercepts on the three axes are a/h, b/k and c/l respectively.

4.19. Miller Indices of Cubic Crystal Planes

While finding Miller indices of a plane, following points should be kept in mind :

(*i*) When a plane is parallel to one of the coordinate axes, it is said to meet that axis at infinity. Since $1/\infty = 0$, the Miller index for that axis is zero.

(*ii*) When the intercept of a plane is on the negative part of any axis, the Miller index is distinguished by a minus sign put directly over it.

As an example, consider the shaded planes of Fig. 4.26. The face plane *BFGC* of Fig. 4.26 (*a*) has Miller indices of (100). It is so because it has an intercept of unity on *X*-axis but is parallel to both *Y*- and *Z*-axes. Since the shaded plane *AFGD* of Fig. 4.26 (*b*) intercepts *X*- and *Z*-axes but not the *Y*-axis, it has indices of (101). The plane *AFH* of Fig. 4.26 (*c*) has equal intercepts on the three axes, hence it has indices of (111).

The set of parallel planes perpendicular to *Y*-axis as shown in Fig. 4.32 (*a*) has Miller indices (010). Similarly, planes normal to *X*-axis are designated (100) and are sketched in Fig. 4.32 (*b*).

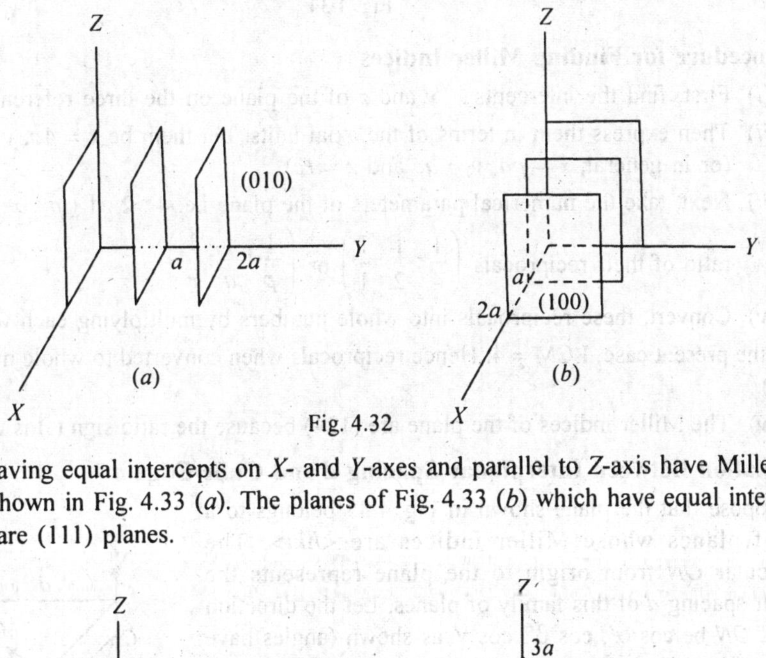

Fig. 4.32

Planes having equal intercepts on *X*- and *Y*-axes and parallel to *Z*-axis have Miller indices (110) and are shown in Fig. 4.33 (*a*). The planes of Fig. 4.33 (*b*) which have equal intercepts on the three axes are (111) planes.

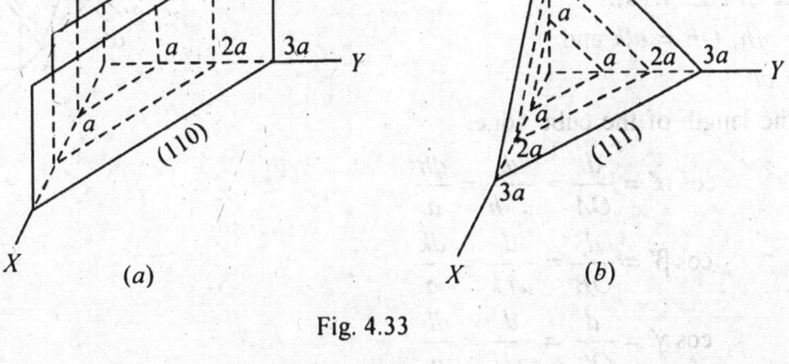

Fig. 4.33

Fig. 4.34 (*a*) shows a few planes which are designated as (112). Finally, Fig. 4.34 (*b*) shows a family of planes in which there is negative intercept on the *X*-axis. Their Miller indices are (110).

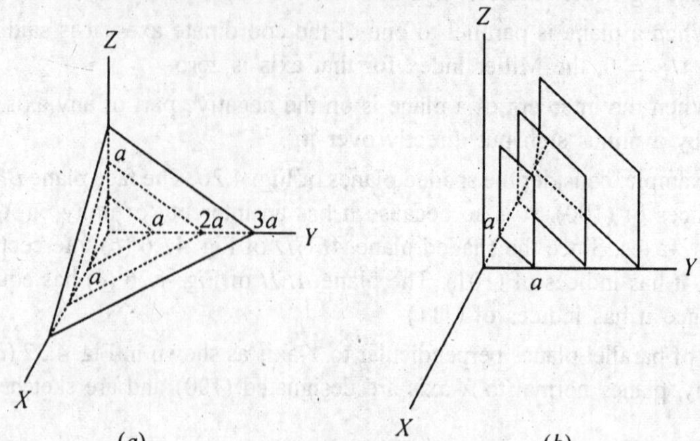

(*a*) (*b*)

Fig. 4.34

4.20. Procedure for Finding Miller Indices

(*i*) First, find the intercepts *x*, *y* and *z* of the plane on the three reference axes.

(*ii*) Then express them in terms of the axial units. Let them be $x = 4a$, $y = 2b$ and $z = c$ (or in general, $x = pa$, $y = qc$ and $z = rc$).

(*iii*) Next, take the numerical parameters of the plane i.e. 4 : 2 : 1 ($p : q : r$) and find the ratio of their reciprocals $\left(\frac{1}{4}:\frac{1}{2}:\frac{1}{1}\right)$ or $\left(\frac{1}{p}:\frac{1}{q}:\frac{1}{r}\right)$.

(*iv*) Convert, these reciprocals into whole numbers by multiplying each with their LCM.

In the present case, LCM = 4. Hence reciprocals when converted to whole numbers become (1 : 2 : 4).

(*v*) The Miller indices of the plane are (124) because the ratio sign (:) is usually omitted.

4.21. Relation Between Interplanar Spacing *d* and Cube Edge *a*

Suppose that the plane shown in Fig. 4.35 belongs to a family of planes whose Miller indices are <*hkl*>. The perpendicular *ON* from origin to the plane represents the interplanar spacing *d* of this family of planes. Let the direction cosines of *ON* be cos α′, cos β′, cos γ′ as shown (angles have been taken with a dash so as to avoid confusion with interaxial angles α, β and γ.) The intercepts of the plane on the three axes are (see Note of Art. 4.18).

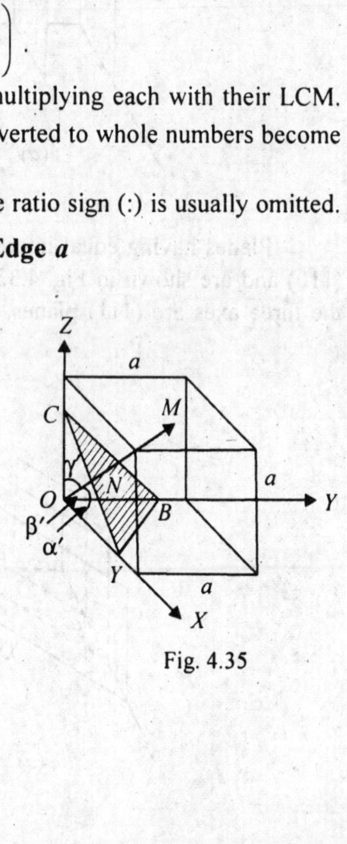

Fig. 4.35

$OA = a/h$; $OB = a/k$ and

$OC = a/l$

where *a* is the length of the cube edge.

$$\cos\alpha' = \frac{d}{OA} = \frac{d}{a/h} = \frac{dh}{a}$$

$$\cos\beta' = \frac{d}{OB} = \frac{d}{a/k} = \frac{dk}{a}$$

$$\cos\gamma' = \frac{d}{OC} = \frac{d}{a/l} = \frac{dl}{a}$$

where $d = ON$—perpendicular distance between adjacent members of the same family of planes.

Now, $\cos^2 \alpha' + \cos^2 \beta' + \cos^2 \gamma' = 1$

$$\therefore \quad \left(\frac{dh}{a}\right)^2 + \left(\frac{dk}{a}\right)^2 + \left(\frac{dl}{a}\right)^2 = 1$$

$$\therefore \quad \frac{d^2}{a^2}(h^2 + k^2 + l^2) = 1$$

$$\text{or} \quad d^2 = \frac{a^2}{h^2 + k^2 + l^2}$$

$$\text{or} \quad \boxed{d = \frac{a}{\sqrt{h^2 + k^2 + l^2}}}$$

4.22. Important Features of Miller Indices

There are many important features of the scheme of Miller indices particularly for the cubic system :

(*i*) all equally-spaced parallel planes have the same index numbers ($h\ k\ l$).

(*ii*) a plane parallel to one of the coordinate axes has an intercept of infinity and index of zero for that axis.

(*iii*) Miller indices do not define a particular plane but a set of parallel planes.

(*iv*) it is only the ratio of the indices which is of importance, *i.e.* the (422) planes are the same as (211) planes.

(*v*) a plane passing through the origin is defined in terms of a parallel plane having non-zero intercepts.

The Miller index notation is especially useful for cubic system of crystals. Some of its desirable features are :

(*a*) the angle θ between two directions (u_1, v_1 w_1) and (u_2 v_2 w_2) can be easily calculated from the expression

$$\cos\theta = \frac{u_1u_2 + v_1v_2 + w_1w_2}{\left(u_1^2 + v_1^2 + w_1^2\right)^{1/2}\left(u_2^2 + v_2^2 + w_2^2\right)^{1/2}}$$

(*b*) the normal to the plane with index number ($h\ k\ l$) is the direction [$h\ k\ l$].

This fact can be utilized to find the angle between two planes by using the above relation.

(*c*) the distance d between adjacent planes of a set of parallel planes of the indices ($h\ k\ l$) is given by

$$d = \frac{a}{(h^2 + k^2 + l^2)^{1/2}}$$

where a is the edge of the cube.

In general, it may be noted that planes with low index numbers have wide interplanar spacing compared with those having high index numbers. Moreover, low index planes have a higher density of atoms per unit area than the high index planes. In fact, it is the low index planes which play an important role in determining the physical and chemical properties of solids.

4.23. Crystal Plane and its Normal

In Fig. 4.35, let the linear parameters of the shaded plane be *pa*, *qb* and *rc* respectively. The direction cosines of the normal *ON* are cos α′, cos β′, and cos γ′. The analytical relation between the axial intercepts of a plane and direction cosines of its normal is

$$\cos\alpha' : \cos\beta' : \cos\lambda' : : \left(\frac{1}{pa}\right) : \left(\frac{1}{qb}\right) : \left(\frac{1}{rc}\right)$$

$$: : \left(\frac{h}{a}\right) : \left(\frac{k}{b}\right) : \left(\frac{l}{c}\right) \qquad \text{—Art. 4.18}$$

Now, in a simple cubic crystal, $a = b = c$

$\therefore \cos\alpha' : \cos\beta' : \cos\gamma' : : (h) : (k) : (l)$

In other words, the direction cosines of a normal to a plane are the same as the Miller indices of the plane.

Example 4.6. *Obtain the Miller indices of a plane which intercepts at a, b/2, 3c in a simple cubic unit cell.*

Solution. The linear parameters of the plane are *a*, *b*/2, 3*c*. Its *numerical* parameters (Art. 4.18) are 1, 1/2, 3. Hence, its Miller indices are

$$\left(\frac{1}{1}, \frac{1}{1/2}, \frac{1}{3}\right) \text{ or } \left(1\ 2\ \frac{1}{3}\right) \text{ or } \mathbf{(361)}$$

Example 4.7. *A certain orthorhombic crystal has axial units a : b : c : : 0·424 : 1 : 0·367. Find the Miller indices of crystal faces whose intercepts are* :

(*i*) *0·212 : 1 : 0·183*

(*ii*) *0·848 : 1 : 0·732*

(*iii*) *0·424 : ∞ : 0·123*

Solution. Here, axial units are *a* : *b* : *c* : : 0·424 : 1 : 0·367.

(*i*) Now, $pa = 0{\cdot}212$ or $p \times 0{\cdot}424 = 0{\cdot}212$

so that $p = 0{\cdot}212/0{\cdot}424 = 1/2$

Similarly, $qb = 1$ or $q \times 1 = 1$ so that $q = 1$.

Also $rc = 0{\cdot}183$ or $r \times 0{\cdot}367 = 0{\cdot}183$

so that $r = 0{\cdot}183/0{\cdot}367 = 1/2$

Hence, numerical parameters of the plane are 1/2, 1, 1/2.

$$\text{Miller indices} = \left(\frac{1}{1/2} : \frac{1}{1} : \frac{1}{1/2}\right) = \mathbf{(212)}$$

(*ii*) The numerical parameters of this plane are 2, 1, 2.

$$\text{Miller indices are } \left(\frac{1}{2}\ 1\ \frac{1}{2}\right) = \mathbf{(121)}$$

(*iii*) Here, numerical parameters are 1, ∞, 1/3

$$\therefore \text{ Miller indices} : \left(1\ \frac{1}{\infty}\ \frac{1}{1/3}\right) \text{ or } \mathbf{(103)}.$$

Example 4.8. *Sketch the (112) and (111) planes in a simple cubic cell.*

Solution. A plane whose Miller indices are (112) has the following intercepts on the three axes

$$\frac{1}{1}, \frac{1}{1}, \frac{1}{2}$$

This plane is sketched in Fig. 4.36 (*a*).

Similarly, the axial intercepts of the (111) plane are

$$\frac{1}{1}, \frac{1}{1}, \frac{1}{1}$$

This plane is shown in Fig. 4.36 (*b*).

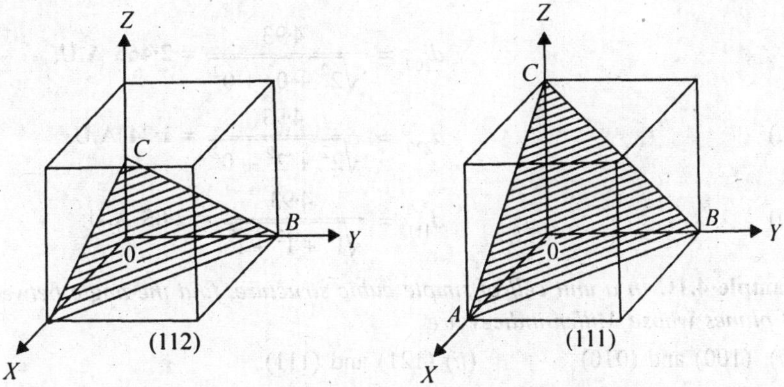

(*a*) (*b*)

Fig. 4.36

Example 4.9. *Draw the different (111) planes in the unit cell of a simple cubic structure.*

Solution. Fig. 4.37 shows the unit cell of simple cubic structure. The four (111) triangular planes are *AFH*, *CFH*, *ACH*, and *ACF*.

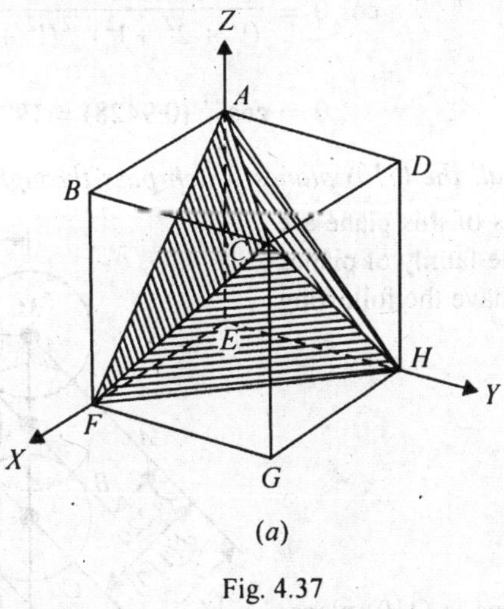

(*a*)

Fig. 4.37

Example 4.10. *Lead is face-centred cubic with an atomic radius of r = 1·746 A.U. Find the spacing of (i) (200) planes (ii) (220) planes and (iii) (111) planes.*

Solution. As stated in Art. 4.22, the interplanar spacing is given by

$$d = \frac{a}{\left(h^2 + k^2 + l^2\right)^{1/2}}$$

Also the lattice constant of the unit cell of lead (FCC) is related to its atomic radius by the relation

$$a = 4r/\sqrt{2} = 4 \times 1{\cdot}746/\sqrt{2} = 4{\cdot}93 \text{ A.U.}$$

(*i*) For (200) planes, $h = 2,\ k = 0,\ l = 0$

$$\therefore \quad d_{200} = \frac{4{\cdot}93}{\sqrt{2^2 + 0^2 + 0^2}} = \mathbf{2{\cdot}465\ A.U.}$$

(*ii*)
$$d_{220} = \frac{4{\cdot}93}{\sqrt{2^2 + 2^2 + 0^2}} = \mathbf{1{\cdot}74\ A.U.}$$

(*iii*)
$$d_{111} = \frac{4{\cdot}93}{\sqrt{1^2 + 1^2 + 1^2}} = \mathbf{2{\cdot}85\ A.U.}$$

Example 4.11. *In a unit cell of simple cubic structure, find the angle between the normal to pair of planes whose Miller indices are*

(*i*) (100) and (010) (*ii*) (121) and (111).

Solution. The directions of the two normals are [100], [010] and [121], [111] respectively. The angle between the two normals can be found by using the relation given in Art. 4.22 (*a*).

(*i*)
$$\cos\theta = \frac{1 \times 0 + 0 \times 1 + 0 \times 0}{(1^2 + 0^2 + 0^2)^{1/2}(0^2 + 1^2 + 0^2)^{1/2}} = 0$$

$$\therefore \quad \theta = \mathbf{90^\circ}$$

(*ii*)
$$\cos\theta = \frac{1 \times 1 + 2 \times 1 + 1 \times 1}{(1^2 + 2^2 + 1^2)^{1/2}(1^2 + 1^2 + 1^2)^{1/2}} = 0{\cdot}9428$$

$$\therefore \quad \theta = \cos^{-1}(0{\cdot}9428) = \mathbf{19^\circ 28'}$$

Example 4.12. *Sketch all the (110) planes which pass through an FCC unit cell.*

Solution. The intercepts of this plane are 1, 1, ∞. This plane is one of the family of planes which, as shown in Fig. 4.38, have the following intercepts on the three axes :

Plane A = 0, 0, ∞

Plane B = ½, ½, ∞

Plane C = 1, 1, ∞

Plane D = 1½, 1½, ∞

Plane E = 2, 2, ∞

Each of the above planes is (110) plane. Incidentally, it may be noted that there are four (220) interplanar spacings per face diagonal *MN* of the *FCC* cell.

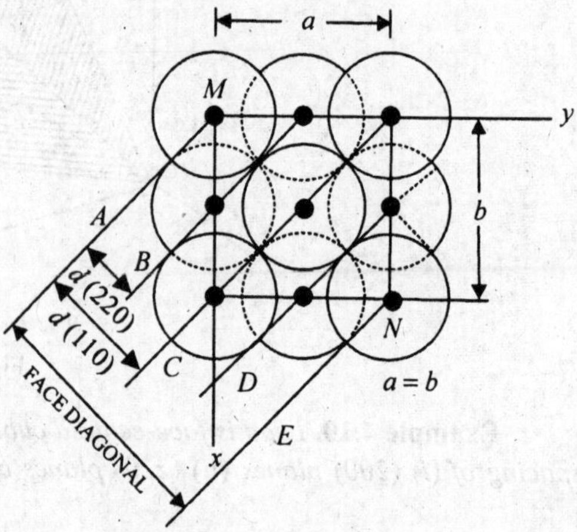

Fig. 4.38

Example 4.13. *How many atoms per mm² surface area are here in (i) (100) plane (ii) (110) plane and (iii) (111) plane for copper which has FCC structure and a lattice constant* $a = 3{\cdot}61 \times 10^{-10}$ *m.*

Solution. (*i*) Two views of the (100) plane are shown in Fig. 4.39. The number of atoms/mm² area of this plane may be calculated from the following consideration.

Surface area of the face *CDHG* is

$= a^2 = (3{\cdot}61 \times 10^{-10})^2$ m²

$= (3{\cdot}61 \times 10^{-7})^2$

$= 13 \times 10^{-14}$ mm²

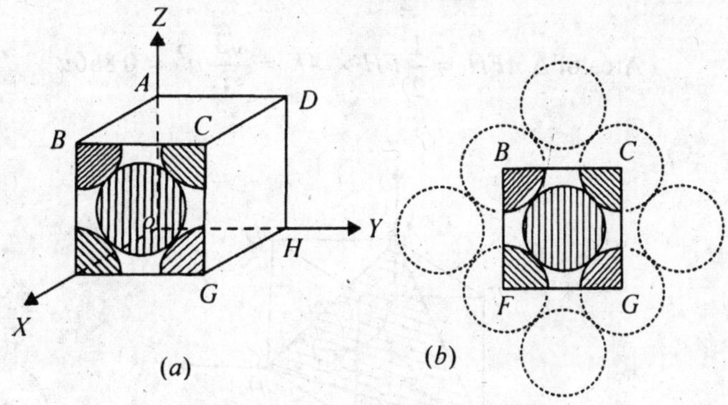

Fig. 4.39

As seen from Fig. 4.39 (*b*), the (100) plane includes one full atom and four quarter atoms. Hence, total number of atoms included = 1 + 4 (1/4) = 2. Since an area of 13×10^{-14} mm² contains 2 atoms, the number of atoms per mm² of surface area is

$$= \frac{2}{13 \times 10^{-14}} = \mathbf{1{\cdot}54 \times 10^{13}\ atoms/mm^2}$$

(*ii*) Fig. 4.40 shows the (110) plane *BFHD*. As seen, it contains four quarter atoms and two half atoms. Total number of atoms

$$= 4 \times (1/4) + 2\,(1/2) = 2$$

Area of the plane $= BF \times BD = a \times \sqrt{2}.a$

$$= \sqrt{2}a^2 = \sqrt{2}\,(3{\cdot}61 \times 10^{-7})^2 \text{ mm}^2$$

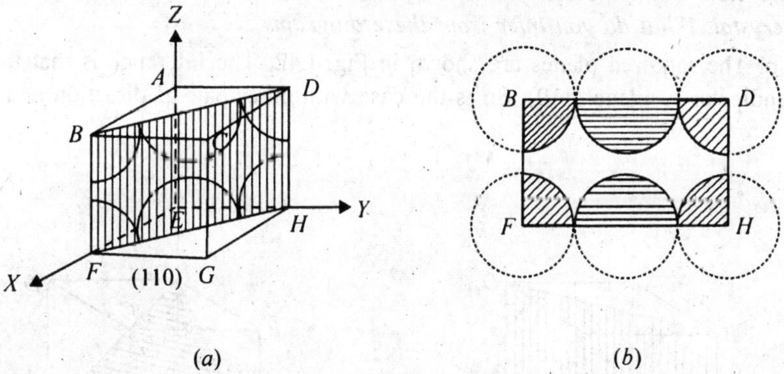

Fig. 4.40

$$\therefore \quad \text{No. of atoms/mm}^2 = \frac{2}{\sqrt{2}(3{\cdot}61 \times 10^{-7})^2}$$

$$= \mathbf{1{\cdot}08 \times 10^{13}\ atoms/mm^2.}$$

(*ii*) The two views of the 111 plane are shown in Fig. 4.41. As seen, it is an equilateral triangle having each side equal to the face diagonal of the cube.

$$AF = AH = FH = \sqrt{2}a$$

$$AK = \frac{\sqrt{3}}{2} AH = \frac{\sqrt{3}}{2} \times \sqrt{2}a$$

$$\text{Area of } \Delta\, AFH = \frac{1}{2} FH \times AK = \frac{\sqrt{3}}{3} a^2 = 0{\cdot}866a^2$$

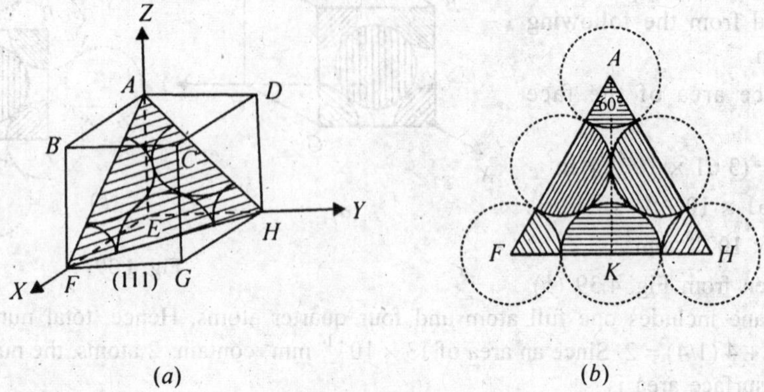

Fig. 4.41

As seen from Fig. 4.41 (*b*), it contains three one-sixths of an atom and three one-halfs. Hence, number of atoms

$$= (3 \times 1/6) + (3 \times 1/2) = 2$$

As before, the number of atoms/mm^2 surface area of (111) plane is

$$\frac{2}{0{\cdot}866 \times (3{\cdot}61 \times 10^{-7})^2} = \mathbf{1{\cdot}778\ atoms/mm^2}$$

Example 4.14. *Draw the (110) and (111) planes and the* [*110*] *and* [*111*] *directions in a simple cubic crystal. What do you infer from these diagrams ?*

Solution. The required planes are shown in Fig. 4.42. The inference is that the direction [110] is perpendicular to plane (110). So is the case with other pair of direction and plane.

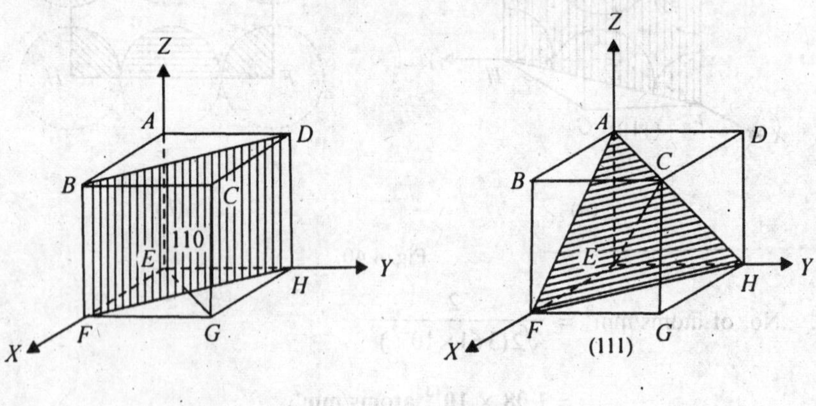

Fig. 4.42

TUTORIAL PROBLEMS

1. The atomic weight of silver is 108. It has a lattice constant of 4·077 Å. Calculate the density of silver if it has fcc lattice structure. **[10·6 g/cm^3]**
2. A KCl crystal which has fcc lattice structure has a density of $1{\cdot}98 \times 10^3$ kg/m^3. Its molecular weight is 74·55. Find the distance between adjacent atoms. **[3·15 Å]**
3. Potassium chloride (KCl) is a face-centred cubic crystal having a density of 1·98 g/cm^3. If its molecular weight is 74·6, calculate (*i*) the distance from one atom to the next atom of the same kind and (*ii*) the distance between adjacent atoms. **[(*i*) 6·3 Å (*ii*) 3·15 Å]**
4. Nickel has face-centred cubic structure and its atomic radius is $1{\cdot}243 \times 10^{-10}$ m. Calculate the interplanar spacing for (*i*) (200) planes (*ii*) (220) planes and (*iii*) (111) planes. **[(*i*) 1·76 A.U. (*ii*) 1·234 A.U. (*iii*) 2·04 A.U.]**
5. The lattice constant of a unit cell of lead is 4·93 A.U. Find the number of atoms/mm^2 of (*a*) (100) plane (*b*) (111) plane if lead has *FCC* structure. **[(*a*) $8{\cdot}2 \times 10^{12}$ (*b*) $9{\cdot}5 \times 10^{12}$]**
6. Sketch the (111) planes of Cu which has *FCC* structure and a lattice constant of 3·61 A.U. Calculate (*a*) the number of atoms/mm^2 and (*b*) interplanar spacing. **[(*a*) $1{\cdot}778 \times 10^{13}$ (*b*) 2·08 A.U.]**
7. Calculate the interplanar spacings for (111) and (321) planes in a simple cubic lattice where $a = 4{\cdot}2$ A.U. **[2·42 A.U., 1·12 A.U.]**
8. Find the Miller indices of planes shown in Fig. 4.43.

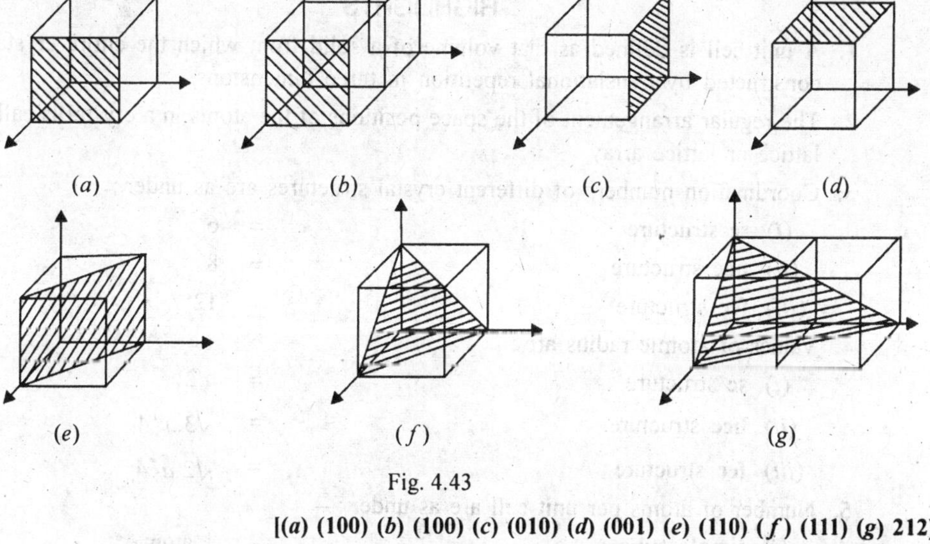

Fig. 4.43

[(*a*) (100) (*b*) (100) (*c*) (010) (*d*) (001) (*e*) (110) (*f*) (111) (*g*) 212)]

9. Compute the Miller indices of the two planes shown in Fig. 4.44. **[200, 100]**

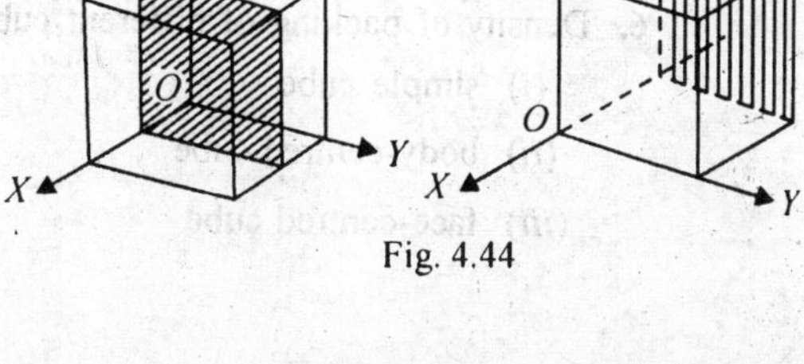

Fig. 4.44

10. Prove that the volume V of the unit cell for the seven crystal systems is given by the following formulae :

(*i*) triclinic system

$$V = abc\sqrt{1 - \cos^2\alpha - \cos^2\beta - \cos^2\gamma + 2\cos\alpha\cos\beta\cos\gamma}$$

(*ii*) monoclinic system

$$V = abc \sin\beta$$

(*iii*) orthorhombic system

$$V = abc$$

(*iv*) tetragonal system

$$V = a^2c$$

(*v*) cubic system

$$V = a^3$$

(*vi*) hexagonal system

$$V = \frac{\sqrt{3}}{2}a^2c$$

(*ii*) rhombohedral system

$$V = a^3\sqrt{1 - 3\cos^2\alpha + 2\cos^3\alpha}$$

(referred to rhombohedral coordinates).

HIGHLIGHTS

1. A unit cell is defined as that volume of a solid from which the entire crystal can be constructed by translational repetition in three dimensions.
2. The regular arrangement of the space positions of the atoms in a crystal is called space lattice or lattice array.
3. Coordination numbers of different crystal structures are as under :

   (*i*) **sc** structure = 6

   (*ii*) **bcc** structure = 8

   (*iii*) **fcc** structure = 12
4. Values of atomic radius are :

   (*i*) **sc** structure = $a/2$

   (*ii*) **bcc** structure = $\sqrt{3}.a/4$

   (*iii*) **fcc** structure = $\sqrt{2}.a/4$
5. Number of atoms per unit cell are as under :—

   (*i*) simple cube — 1 atom

   (*ii*) body-centred cube — 2 atoms

   (*iii*) face-centred cube — 4 atoms
6. Density of packing of different cubic structures are as under :

   (*i*) simple cube — $\pi/6$

   (*ii*) body-centred cube — $\sqrt{3}.\pi/8$

   (*iii*) face-centred cube — $\sqrt{2}.\pi/6$

7. Miller indices of a plane are given by the reciprocal of its three numerical parameters expressed as whole numbers.
8. The relation between the spacing of the important sets of planes in a cubic system are as under

$$\frac{1}{d_1}:\frac{1}{d_2}:\frac{1}{d_3}::1:\sqrt{2}:\sqrt{3} \qquad \text{— for } SC$$

$$::1:\frac{1}{\sqrt{2}}:\sqrt{3} \qquad \text{— for } BC$$

$$::1:\sqrt{2}:\frac{\sqrt{3}}{2} \qquad \text{— for } FCC$$

Can you Answer This ?

1. How are crystals classified ? Discuss crystal symmetry and atomic packing factor.
2. (*a*) With reference to crystals, state what you understand by the terms (*i*) crystal lattice (*ii*) unit cell (*iii*) single crystal (*iv*) coordination number (*v*) packing factor and (*vi*) atomic radius.

   (*b*) Calculate the atomic radius in case of simple cubic, b.c.c. and f.c.c. lattices.
3. Write short note on Miller indices.
4. What are Miller indices ? Draw neat diagram to indicate Miller indices of the important plane systems in a simple cube crystal. Obtain a relation between the interplanar spacings and the cube edge.
5. (*a*) What do you mean by a crystal lattice ? With the help of an example, distinguish between crystal lattice and crystal structure.

   (*b*) What do you understand by packing factor ? Show that the packing factor for simple lattice, b.c.c. and f.c.c. lattice are $\pi/6$, $\pi/8$ and $\sqrt{2}\cdot\pi/6$ respectively.
6. (*a*) Define space lattice. Name the seven types of crystal systems and give the relation of lengths of axes and the relation of angles between the axes of a unit cell in each type.

   (*b*) Establish the relation between the edge of the unit cell and the atomic radius for the b.c.c. and the f.c.c. lattices.
7. Derive the relation between the lattice constant of a cubic crystal and the density of the crystal material.
8. Obtain expressions for the values of packing density in the case of (*i*) simple cube (*ii*) body-centred cube and (*iii*) face-centred cube.
9. Establish the relation between the lattice constant of a cubic cell and the density of the crystal material.
10. (*a*) What is space lattice ? Describe briefly the seven systems of crystals.

    (*b*) Show that for b.c.c. and f.c.c. crystal structure, the lattice constants are given by

$$a_{bcc} = \frac{4r}{\sqrt{3}} \quad \text{and} \quad a_{fcc} = \frac{4r}{\sqrt{2}}$$

where r is the atomic radius.

11. Explain the term 'symmetry' in the structure of crystals. Describe the different symmetry elements.

12. Explain the following terms :

(*i*) Bravais' lattices and (*ii*) Miller indices.

13. Explain the following terms with reference to a crystal

(*i*) unit cell

(*ii*) coordination number

(*iii*) packing factor.

14. Show that in a crystal of cubic structure, the distance between the planes with Miller indices *h*, *k*, *l* is equal to

$$d = \frac{a}{\sqrt{h^2 + l^2 + k^2}}$$

where a is the lattice parameter.

15. How many atoms are there are unit cell in crystal with simple cubic, body-centred and face-centred cubic structures. Explain with suitable diagrams.

16. Explain with suitable diagram, how one can determine the orientation and symmetry of a crystal by Laue method.

17. Draw a diagram of the unit cell of the close-packed hexagonal lattice.

18. For a cubic lattice, draw the plane (6 2 3) plane showing the value of intercepts with the coordinate axes.

19. Define atomic packing fraction and coordination number and calculate these quantities for the body-centred and face-centred cubic lattices.

20. Draw a schematic diagram of the unit cell for the simple cubic, body-centred cubic and face-centred cubic lattices.

21. Explain with suitable diagrams, how in a crystal, planes are denoted by the Miller indices (*hkl*) and directions by [*hkl*].

22. (*a*) Define the terms : (*i*) single crystal (*ii*) unit cell and (*iii*) Miller indices.

(*b*) Define 'atomic packing factor' and 'ionic packing factor' and explain their importance.

23. Write short note on coordinate number.

5 QUANTUM THEORY

What is Quantum Theory — Black Body — Planck's Law of Black Body Radiations — Quantum Theory of Radiation — Properties of Photons — Collision Between Photon and Matter Particle — Photoelectric Effect — Experimental Study of Photoelectric Effect — Einstein's Photoelectric Equation — Long Wavelength Limit (λ_0) — Kinetic Energy of Photoelectrons — Laws of Photoelectric Emission — Photoelectric Work Function — Types of Photoelectric Cells — Photoemissive Cell or Phototube — Automatic Bell Alarm Circuit — Photoelectric Relay — Smoke Detector — On and Off Control Circuit — Photomultiplier Tube — Photovoltaic Cell — Photoconductive Cell — Highlights — Can You Answer This ?

5.1. What is Quantum Theory

This theory was proposed by Max Planck in 1900 when he was trying to explain the shape of the black body radiation curves obtained experimentally. According to this theory, matter is composed of a large number of oscillating particles, which vibrate with different frequencies. According to classical theory, these particles can have any value of frequency and hence any amount of vibration energy. However, according to Quantum Theory, an oscillating particle *cannot* have any energy but only those energies given by

$$E = n\,hf^{*} \qquad ...(i)$$

where f is oscillator frequency, h is a constant (now called Planck's constant) and n is a number (now called quantum number) which can take only integral values of 1, 2, 3.....etc.

The above equation shows that the energy of the oscillating particle is *quantized.*

Moreover, it was assumed by Planck that the vibrating particle does not radiate energy *continuously* but only in terms of discrete quanta or photons. These quanta of energy are emitted when an oscillating particle changes from its one quantized state of vibration to another. As seen from Eqn. (i) above, when n changes by 1 *i.e.* $\Delta n = 1$, the amount of energy radiated out is

$$\Delta E = \Delta n \cdot hf = hf$$

Hence, it is seen that even though an oscillator may possesses any amount of energy given by Eqn. (i) above, it can radiate it out only in terms of quanta of magnitude 'hf'. So long as an oscillator remains in one of its quantized states, it neither emits nor absorbs any energy.

5.2. Black Body

It is a matter of common knowledge that a black matt surface is a very good absorber of heat radiations and, also, a very good emitter of such radiations when heated. But it is still not a perfect black body which has been defined as one *that absorbs radiations of all wavelengths incident upon it.* The radiations emitted by such a body (when heated) are called black body radiations or full radiations or total radiations.

No actual body is black in the above sense, closest approach being lamp black which absorbs about 99% of incident radiations. For all theoretical considerations, ideal absorber of radiant energy

*According to later developments, $E = (n + 1/2)\,hf$.

or perfect black body is a small hole or opening in a uniform-temperature closed chamber, say, a hollow sphere as shown in Fig. 5.1 (*a*). When any radiation enters the opening, it suffers

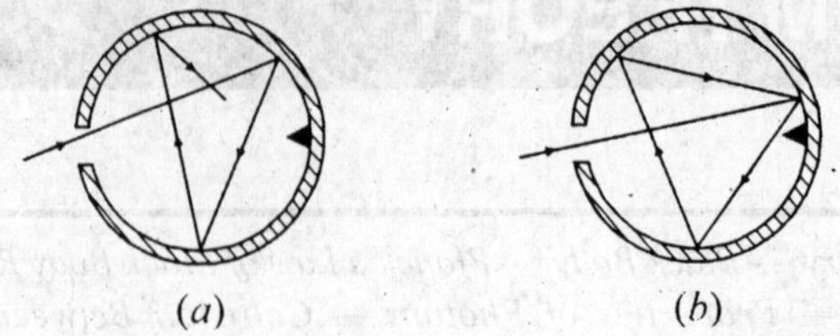

Fig. 5.1

repeated reflections from the inner walls until it is finally absorbed. Hence, the opening behaves like a perfect absorber. Conversely, the radiations emitted by the walls (when the sphere is heated) and escaping the opening will be of the same nature as emitted by a perfect black body as shown in Fig. 5.1 (*b*).

The type of black body radiator discussed above was devised by Ferry. It is a hollow thick-walled sphere painted lamp black internally and having a circular opening. Inside the sphere and directly opposite to the opening, there is a projection from the wall for preventing direct reflection of any of the radiations entering through the opening.

Conception of a black body radiator has been very useful in the theory of heat radiations. It is of importance when using radiation pyrometers. The readings of such instruments are true when they are sighted on a black body.

5.3. Planck's Law of Black Body Radiations

The quantum theory arose in the first instance out of attempts to correctly explain the energy distribution in the continuous spectrum emitted by an incandescent lamp.

When a black body is heated, it emits thermal radiations having a wide range of wavelengths. The amount of heat energy carried away by different wavelengths is different. This spectral distribution of radiant energy was first analyzed by Wien who found that energy density E_λ in wavelength range λ to $(\lambda + d\lambda)$ is given by

$$E_\lambda = \frac{c_1 \lambda^{-5}}{e^{c_2/kT}} \cdot d\lambda$$

where T is the temperature of black body radiator in °K and c_1 and c_2 are constants. The graph for Wien's law is shown in Fig. 5.2. However, it was found that the above formula fitted the experimental results fairly well in the short wavelength region but gave values which were too low elsewhere. Later on, Rayleigh and Jeans adopted a more vigorous method based on classical Physics and came up with the formula

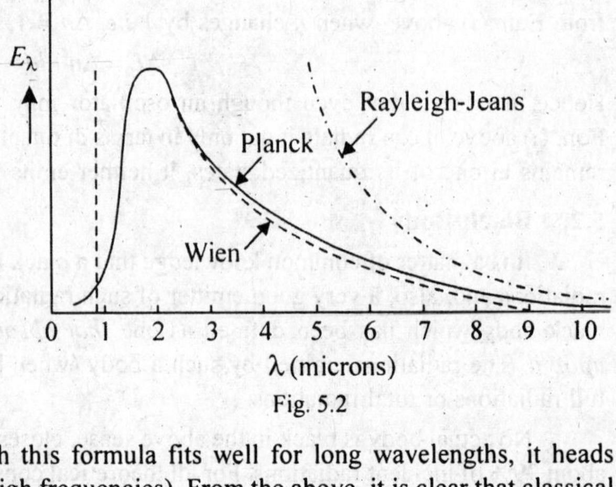

Fig. 5.2

$$E_\lambda = \frac{8\pi kT}{\lambda^4} d\lambda$$

where k is Boltzmann constant. Although this formula fits well for long wavelengths, it heads towards infinity at short wavelengths (or high frequencies). From the above, it is clear that classical Physics was unable to account for an important observation concerning the radiations emitted by a black body radiator. It was this dilemma to which Max Planck directed his attention.

In 1900, Planck made the startling non-classical assumption that radiation is not emitted continuously but in small packets of energy called *photons*. The amount of energy associated with each photon is $E = hf$ where h is Planck's constant and f is the frequency of radiations. This means that energy states of an emitter or oscillator have to be a multiple of hf *i.e.* they could be $hf, 2hf, 3hf.....nhf$, where n is an integer. On the basis of this Quantum Theory, Planck deduced the law of black body radiations as

$$E_\lambda = \frac{8\pi ch\lambda^{-5}}{(e^{ch/\lambda kT} - 1)} d\lambda$$

Alternately, using the same constants as used by Wien, the above law can be put as

$$E_\lambda = \frac{c_1\lambda^{-5}}{(e^{c_2/\lambda T} - 1)} d\lambda$$

where $c_1 = 8\pi ch$ and $c_2 = ch/k$.

This formula is found to fit the experimental curve over the whole range of wavelengths.

Following points are worth noting :

(*i*) For long wavelengths, the above equation is reduced to

$$E_\lambda = \frac{c_1\lambda T}{c_2\lambda^5} d\lambda = \frac{(c_1/c_2)T}{\lambda^4} \cdot d\lambda$$

$$= \frac{8\pi kT}{\lambda^4} d\lambda$$

which is the Rayleigh-Jeans formula.

(*ii*) For short wavelengths,

$$E_\lambda = \frac{c_1}{\lambda^5 e^{c_2/\lambda T}} d\lambda$$

which is Wien's formula.

5.4. Quantum Theory of Radiation

The Quantum Theory of radiation was first proposed by Max Planck (1858-1947) in 1901 in connection with thermal radiations emitted by a black body. It was extended by Einstein in 1905 to cover all radiations including light rays and γ-rays etc. In fact, the study of the nature of radiations was originally made in connection with the nature of light and how it travelled from one point to another. Towards the end of 17th century, Newton propounded his Corpuscular Theory according to which light consists of a stream of invisible fast-moving elastic particles of small mass called *corpuscles*. This particle view of light was superseded by Huygens' Wave Theory according to which light travels in the form of waves in a hypothetical medium called ether. This theory was followed in 1864 by Maxwell's Electromagnetic Theory according to which light consists of changes of electric and magnetic stress originating in the motion of electricity in the atoms from which light is emitted.

In 1901, Max Planck while studying black body radiations had come to the conclusion that the absorption or emission of thermal energy is *not a continuous process but takes place in discrete amounts*. In 1905, Einstein quantized the radiant energy by postulating that

(*i*) light energy is not emitted continuously but intermittently by indivisible amounts of radiant energy called *quanta*.

(*ii*) these quanta travel in space in the form of bundles or packets called *photons* and carry a definite amount of energy.

The energy carried by each photon is given by

$$E = hf$$

where h is Planck's constant and f is the frequency of radiation. Obviously, photons or quanta of high frequency such as those of γ-rays have large amounts of energy while those of low frequency such as photons of radio waves have comparatively less amount of energy. In other words, '*sizes*' of photons vary with their frequencies.

5.5. Properties of Photons

Photons are the basic constituents of any radiation and possess the following characteristic properties :

(*i*) Energy content

The energy content of a photon is always found to be in terms of the quantum represented by 'hf'. This quantum unit 'hf' is not the same for all kinds of radiations because their frequencies are different. In the original quantum theory, the energy of the photon was always considered to be an integral multiple of hf *i.e.* $E = nhf$ where $n = 1, 2, 3, 4$, etc. But according to Quantum Mechanics, $E = (n + 1/2)\, hf$ where n is an integer. This means that the limiting value of the energy of the photons is not zero but $(1/2)\, hf$.

(*ii*) Energy of the photon is independent of intensity

The energy of a photon is determined by the frequency of the radiation and not by its intensity ($\because\; E = hf$). According to Quantum Mechanics, intensity simply gives the total number of photons in a beam and has no relation whatsoever with the energy of the individual photons in the beam.

(*iii*) Mass and momentum of a photon

Since all photons travel with the velocity of light, Einstein's famous mass-energy relation $E = mc^2$ can be applied to them. Hence, the mass of a photon is given by

$$m = \frac{E}{c^2} = \frac{hf}{c^2} = \frac{h}{c\lambda}$$

The momentum of a photon is given by

$$p = mc = \frac{hf}{c^2} \times c = \frac{hf}{c} = \frac{h}{\lambda}$$

(*iv*) Non-electrical nature of photons

Photons are electrically neutral and hence are unaffected by electric or magnetic fields. They do not ionize. The ordinary methods used in the case of charged particles cannot be used to detect photons or measure their energy or intensity. In fact, all instruments used in their analysis are sensitive not to them as such but to the secondary effects produced by their interaction with matter.

5.6. Collision Between Photon and Matter Particle

According to Quantum Theory of radiation, the law of conservation of energy applies when photon collides with a matter particle. Suppose during such a collision, a radiation of frequency f has appeared or disappeared which is the same thing as saying that a photo of energy $E = hf$ has been created or destroyed (Fig. 5.3). If m is the mass of the particle and v its velocity, then kinetic energy lost or gained by it is $\frac{1}{2}mv^2$.

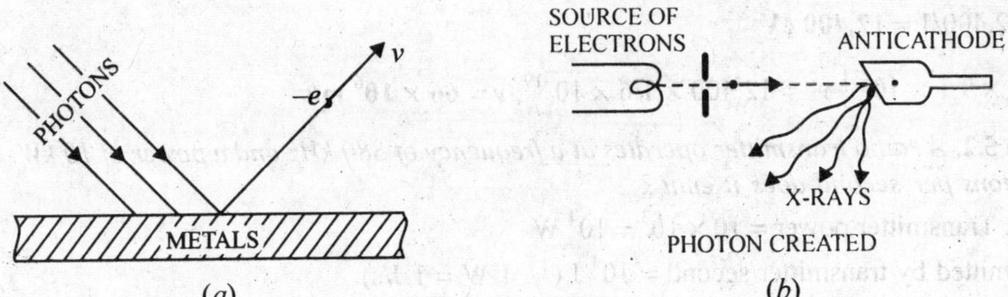

Fig. 5.3

According to the law of conservation of energy,

$$hf = \frac{1}{2}mv^2 \qquad ...(i)$$

If the material particle carries a charge of e and has been accelerated by a potential of V, then its kinetic energy is eV.

$$\therefore \qquad hf = eV \qquad ...(ii)$$

From Eqns. (*i*) and (*ii*), we get

$$\frac{1}{2}mv^2 = hf = eV \qquad ...(iii)$$

Eqn. (*iii*) has been utilized to find the following :

(*i*) the velocity of a charged material particle when accelerated through a certain potential difference. For an *electron*

$$v = \sqrt{\frac{2eV}{m}} = \sqrt{2\times1\cdot76\times10^{11}V} = 5\cdot93\times10^5\sqrt{V} \text{ m/s.}$$

It is so because for an electron

$$e/m = 1\cdot602\times10^{-19}/9\cdot1\times10^{-31} = 1\cdot76\times10^{11} \text{ C/kg}$$

(*ii*) the frequency of the radiations emitted when an accelerated charged particle hits a metal surface can be found by the relation

$$hf = eV \text{ or } \frac{hc}{\lambda} = eV$$

or
$$\lambda = hc/eV \qquad ...(iv)$$

For example, when electrons in an X-ray tube having an energy of eV are made to impinge on the anticathode (target), X-rays are emitted whose wavelength λ can be computed from the above relation.

Note. The wavelength of emitted radiations is given by

$$\lambda = \frac{hc}{eV}\frac{6\cdot62\times10^{-34}\times3\times10^8}{1\cdot602\times10^{-19}V} \text{ metre}$$

$$\lambda = \frac{12\cdot4\times10^{-7}}{V} \text{ m} = \frac{12,400}{V} \text{ Å}$$

where V is in volts and 1 Å = 10^{-10} m.

Example 5.1. *Calculate the energy (in eV) and the velocity of an electron giving X-rays of wavelength (i) 4 Å and (ii) 1 Å on impact.*

Solution. (*i*) $E = 12,400/4 =$ **3,100 eV** Art. 5.12

Also $\frac{1}{2}\times9\cdot1\times10^{-31}v^2 = 3100\times1\cdot6\times10^{-19}$, $v =$ **33 × 10⁶ m/s** (*ii*) $E = 12{,}400/1 =$ **12,400 eV**

$$\frac{1}{2} \times 9{\cdot}1 \times 10^{-31}\, v^2 = 12{,}400 \times 1{\cdot}6 \times 10^{-19},\ v = \mathbf{66 \times 10^6\ m/s}$$

Example 5.2. *A radio transmitter operates at a frequency of 880 kHz and a power of 10 kW. How many photons per second does it emit ?*

Solution. Transmitter power $= 10 \times 10^3 = 10^4$ W

Energy emitted by transmitter/second $= 10^4$ J ($\because$ 1 W = 1 J/s)

Energy carried by each photon

$$= hf = 6{\cdot}625 \times 10^{-34} \times 880 \times 10^3 = 583 \times 10^{-30}\ \text{J}$$

Hence, number of photons emitted per second is

$$= \frac{10^4}{583 \times 10^{-30}} = \mathbf{17{\cdot}15 \times 10^{30}}$$

Example 5.3. *A mercury arc is rated at 200 W. How many light quanta are emitted per second in radiation having wavelength of 6123 Å if the intensity of this line is 2% only. Assumes that 50% of power is spent for radiation.*

Solution. Radiant output of the arc is

$$= 200 \times 0{\cdot}5 = 100\ \text{W or } 100\ \text{J/s}$$

Wavelength of the radiation is $\lambda = 6123$ Å

Energy content of its quanta

$$= hf = \frac{hc}{\lambda} = \frac{6{\cdot}625 \times 10^{-34} \times 3 \times 10^8}{6123 \times 10^{-10}}$$

$$= 3{\cdot}246 \times 10^{-19}\ \text{J}$$

Since its intensity is only 2%, it means that only 2% of 100 J/s = 2 J/s are carried away by the quanta of this wavelength.

$\therefore$ No. of quanta emitted/s $= 2/3{\cdot}246 \times 10^{-19} = \mathbf{6{\cdot}16 \times 10^{18}}$

Example 5.4. *A pulse of radiation consisting of 5×10^4 photons of $\lambda = 3000$ Å falls on a photosensitive surface whose sensitivity for this wavelength region is $J = 5$ mA/W. Find the number of photoelectrons liberated by the pulse.*

Solution. Energy content of each photon

$$= hf = hc/\lambda = 6{\cdot}625 \times 10^{-34} \times 3 \times 10^8/3000 \times 10^{-10}$$

$$= 6{\cdot}625 \times 10^{-19}\ \text{J}$$

Total energy of the radiation pulse

$$= 5 \times 10^4 \times 6{\cdot}625 \times 10^{-19} = 3{\cdot}31 \times 10^{-14}\ \text{J}$$

Let the duration of the pulse be Δt second.

$\therefore$ Power incident on the photosensitive surface is

$$= 3{\cdot}31 \times 10^{-14}/\Delta t\ \text{watt}$$

Current produced is

$$I = 5 \times 3{\cdot}31 \times 10^{-14}/\Delta t\ \text{mA}$$

$$= 16{\cdot}55 \times 10^{-14} \times 10^{-3}/\Delta t\ \text{ampere}$$

$$= 16{\cdot}55 \times 10^{-17}/\Delta t\ \text{ampere}$$

Charge which flows during Δt second

$$Q = I \times \Delta t\ \text{coulomb}$$

$$= \frac{16{\cdot}55 \times 10^{-17}}{\Delta t} \times \Delta t = 16{\cdot}55 \times 10^{-17}\ \text{C}$$

Now, charge on one electron = $1 \cdot 6 \times 10^{-19}$ C

$\therefore$ No. of photoelectrons ejected by the radiation pulse

$$= 16 \cdot 55 \times 10^{-17}/1 \cdot 6 \times 10^{-19} = \mathbf{1034}$$

Example 5.5. (*a*) *How many photons of a radiation of wavelength* $\lambda = 5 \times 10^{-7}$ *m must fall per second on a blackened metallic plate in order to produce a force of* 10^{-5} *newton ?*

(*b*) *At what rate will the temperature of the plate rise if its mass is* $1 \cdot 5$ *gram and specific heat* $0 \cdot 1$ *? Take* $h = 6 \cdot 62 \times 10^{-34}$ *J-s.*

Solution. (*a*) As seen from Art. 5.5 (*iii*), momentum of a photon is

$$p = mc = \frac{hf}{c} = \frac{h}{\lambda}$$

If n is the number of photons falling per second on the plate, then total momentum per second of the incident photons is

$$p = n \times h/\lambda$$

Since plate is blackened, all these photons are absorbed by it.

According to Newton's Second Law of Motion, rate of change of momentum represents force.

$$\therefore \quad nh/\lambda = 10^{-5}$$

or $$n = 10^{-5} \times \lambda/h = 10^{-5} \times 5 \times 10^{-7}/6 \cdot 62 \times 10^{-34} = \mathbf{7 \cdot 54 \times 10^{21}}$$

(*b*) Energy of each photon = $hf = hc/\lambda$

Since n such photons fall on the plate per second, total energy absorbed by the plate in one second is

$$= n \times hc/\lambda = \left(10^{-5} \times \frac{\lambda}{h}\right)\frac{hc}{\lambda} = c \times 10^{-5} \text{ J/s}$$

$$= 3 \times 10^{8} \times 10^{-5} = 3000 \text{ J/s} = 3000/4200 \text{ kcal/s}$$

If θ is the rise in temperature in °C/s, then heat absorbed by the plate per second is

$$mS\theta = 1 \cdot 5 \times 10^{-3} \times 0.1 \times \theta$$

$$\therefore \quad 1 \cdot 5 \times 10^{-3} \times 0 \cdot 1 \times \theta = 3000/4200; \ \theta = \mathbf{4 \cdot 8 \times 10^{3}\ °C/s}$$

Example 5.6. *A blue lamp emits light of mean wavelength of 4500 Å. The lamp is rated at 150 W and 8% of the energy appears as emitted light. How many photons are emitted by the lamp per second ?*

Solution. Lamp power emitted as light

$$= 150 \times 8/100 = 12 \text{ W}$$

Light energy emitted/second = 12 J

Energy carried by one photon is

$$hf = \frac{hc}{\lambda} = \frac{6 \cdot 625 \times 10^{-34} \times 3 \times 10^{8}}{4500 \times 10^{-10}} = 44 \cdot 2 \times 10^{-20} \text{ J}$$

No. of photons emitted by the lamp per second is

$$= \frac{\mathbf{12}}{\mathbf{44 \cdot 2 \times 10^{-20}}} = \mathbf{27 \cdot 15 \times 10^{18}}$$

Example 5.7. *Frequency of an electric magnetic radiation is* 10^{12} *Hz. Calculate the number of photons that will make up 6·625 joules of energy of this radiation.*

Solution. Energy per photon

$$E = hf = 6 \cdot 625 \times 10^{-34} \times 10^{12} = 6 \cdot 625 \times 10^{-22} \text{ J}$$

∴ No. of photons required for 6·625 joules of energy

$$= 6{\cdot}625/6{\cdot}625 \times 10^{-22} = \mathbf{10^{-22}}$$

Example 5.8. *With what velocity must an electron travel so that its momentum is equal to that of a photon with a wavelength of λ = 5,200 A.*

Solution. Momentum of a photon with $\lambda = 5200$ Å $= 5200 \times 10^{-10}$ m is (Art. 5.5)

$$p = \frac{h}{\lambda} = \frac{6\cdot625\times10^{-34}}{5200\times10^{-10}} \text{ kg-m/s}$$

Momentum of the electron

$$= mv = 9{\cdot}1 \times 10^{-31}\ v$$

Equating the two, we have

$$9{\cdot}1 \times 10^{-31}\ v = 6{\cdot}625 \times 10^{-34}/5200 \times 10^{-10}$$

$$\therefore \qquad v = \mathbf{1400\ m/s}$$

5.7. Photoelectric Effect

Ejection of electrons from a metal plate when illuminated by light or any other radiation of suitable wavelength (or frequency) is called photoelectric effect.

This phenomenon was first discovered by Heinrich Hertz in 1887. One year later, Hallwachs made the important observation that when a negatively-charged zinc plate is irradiated with ultraviolet light, it loses its negative charge. Afterwards, it was discovered that alkali metals like lithium, sodium, potassium, rubidium and caesium eject electrons when visible light falls on them. Ten years later, J.J. Thomson and P. Lenard showed that the action of light was to cause the emission of free electrons from metal surface. Although these electrons are no different from all other electrons, it is customary to refer to them as *photoelectrons.*

5.8. Experimental Study of Photoelectric Effect

Photoelectric phenomenon can be studied with the help of a simple apparatus shown in Fig. 5.4. It consists of two photosensitive surfaces E and C enclosed in an evacuated quartz bulb. In the absence of light, there is no flow of current in the circuit and the ammeter A reads zero. When plate E is exposed to some source of monochromatic light, current starts flowing. However, no current is found to flow when light falls on plate C.

The explanation of the above behaviour lies in the fact that when E is irradiated with light, the incident photons eject electrons by collision with its atoms. These photoelectrons are immediately attracted by the collector plate C thereby starting current flow as indicated by the ammeter.

When C is irradiated, even then photoelectrons are produced but they are not allowed to leave plate C.

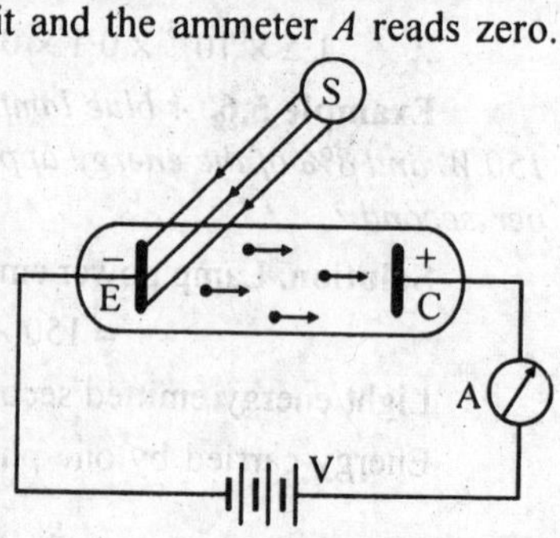

Fig. 5.4

(*i*) firstly, because of the pulling effect of positive potential of C and

(*ii*) secondly, due to repulsion from the negative plate E.

Hence, no current is found to flow in the circuit.

5.9. Einstein's Photoelectric Equation

Following Planck's idea that light consists of photons, Einstein proposed an explanation of photoelectric effect as early as 1905. According to this explanation when a single photon is incident on a metal surface, it is completely absorbed and imparts its energy hf to a single electron. The photon energy is utilized for two purposes :

(*i*) partly for getting the electron free from the atom and away from the metal surface. This energy is known as the *photoelectric work function* of the metal and is represented by W_0.

(*ii*) the balance of the photon energy is used up in giving the electron a kinetic energy of $\frac{1}{2}mv^2$.

$$\therefore \quad \boxed{hf = W_0 + \frac{1}{2}mv^2} \quad ...(i)$$

It is known as Einstein's photoelectric equation.

In case, the photon energy is just sufficient to liberate the electron only, then no energy would be available for imparting kinetic energy to the electron. Hence, the above equation would reduce to

$$hf_0 = W_0$$

where f_0 is called the *threshold* frequency. It is defined as the *minimum frequency which can cause photoelectric emission.* For frequencies lower than f_0, there would be no emission of electrons whereas for frequencies greater than f_0, electrons would be ejected with a certain definite velocity (and hence kinetic energy).

Substituting this value of W_0 in Eqn. (*i*) above, the Einstein's photoelectric equation becomes

$$hf = hf_0 + \frac{1}{2}mv^2$$

$$\text{or} \quad \boxed{hf = hf_0 + K.E.}$$

5.10. Long Wavelength Limit (λ_0)

It is the wavelength corresponding to the threshold frequency f_0. Its physical significance is that radiations having wavelength longer than λ_0 would not be able to eject electrons from a given material whereas those having $\lambda < \lambda_0$ will. In other words, it represents the upper limit of wavelength for photoelectric phenomenon. By analogy, it is also referred to as *threshold wavelength.*

$$\text{Now,} \quad c = f_0 \lambda_0 \qquad \therefore \quad \lambda_0 = \frac{c}{f_0}$$

$$\text{Also,} \quad W_0 = h f_0 \qquad \therefore \quad \frac{1}{f_0} = \frac{h}{W_0}$$

$$\therefore \quad \boxed{\lambda_0 = \frac{ch}{W_0}}$$

(*i*) When W_0 is in joules

$$\lambda_0 = \frac{3\times10^8 \times 6\cdot625\times10^{-34}}{W_0}$$

$$= \frac{19\cdot875\times10^{-26}}{W_0} \text{ metre}$$

(*ii*) When W_0 is in electron-volts (eV)

$$\lambda_0 = \frac{3\times10^8\times6\cdot625\times10^{-34}}{1\cdot602\times10^{-19}W_0} = \frac{12\cdot4\times10^{-7}}{W_0} \text{ metre}$$

$$= \frac{12{,}400}{W_0} \text{ Å}.$$

5.11. Kinetic Energy of Photoelectrons

Einstein's photoelectric equation can be used to find the velocity and hence the kinetic energy of an ejected photoelectron.

$$hf = W_0 + \frac{1}{2}mv^2$$

$$= hf_0 + K.E.$$

$\therefore$ $$\text{K.E.} = hf - hf_0 = h(f - f_0)$$

Now, $$f = c/\lambda \text{ and } f_0 = c/\lambda_0$$

$\therefore$ $$\text{K.E.} = ch\left(\frac{1}{\lambda} - \frac{1}{\lambda_0}\right)$$

$$= 3 \times 10^8 \times 6{\cdot}625 \times 10^{-34}\left(\frac{1}{\lambda} - \frac{1}{\lambda_0}\right)$$

$$= 19{\cdot}875 \times 10^{-26}\left(\frac{1}{\lambda} - \frac{1}{\lambda_0}\right) \text{ joules} \qquad \text{—}\lambda \text{ and } \lambda_0 \text{ in metres}$$

$$= 19{\cdot}875 \times 10^{-16}\left(\frac{1}{\lambda} - \frac{1}{\lambda_0}\right) \text{ joules} \qquad \text{—}\lambda \text{ and } \lambda_0 \text{ in Å}$$

$$= \frac{19{\cdot}875 \times 10^{-16}}{1{\cdot}602 \times 10^{-19}}\left(\frac{1}{\lambda} - \frac{1}{\lambda_0}\right) \text{ eV} \qquad \text{—}\lambda \text{ and } \lambda_0 \text{ in Å}$$

$\therefore$ $$\boxed{\text{K.E.} = 12{,}400\left(\frac{1}{\lambda} - \frac{1}{\lambda_0}\right) \text{ eV}} \qquad \text{—}\lambda \text{ and } \lambda_0 \text{ in Å}$$

Incidentally, it may be noted that this also represents the maximum value of the kinetic energy a photoelectron can have.

$\therefore$ $$E_{max} = h(f - f_0) = h\,\Delta f \text{ joules}$$

$$= 12{,}400\left(\frac{1}{\lambda} - \frac{1}{\lambda_0}\right) \text{ eV} \qquad \text{—}\lambda \text{ and } \lambda_0 \text{ in Å}$$

5.12. Photoelectric Work Function

As explained in Art. 5.9, it is defined as *the energy which is just sufficient to liberate electrons from a body with zero velocity.* Its value is given by

$$W_0 = hf_0 = \frac{ch}{\lambda_0} = \frac{3 \times 10^8 \times 6{\cdot}625 \times 10^{-34}}{\lambda_0}$$

$$= \frac{19{\cdot}875 \times 10^{-26}}{\lambda_0} \text{ joules} \qquad \text{—}\lambda_0 \text{ in metres}$$

$$= \frac{19{\cdot}875 \times 10^{-16}}{\lambda_0} \text{ joules} \qquad \text{—}\lambda_0 \text{ in Å}$$

$$= \frac{19{\cdot}875 \times 10^{-16}}{1{\cdot}602 \times 10^{-19}\lambda_0} \text{ eV} \qquad \text{—}\lambda_0 \text{ in Å}$$

$\therefore$ $$\boxed{W_0 = \frac{12{,}400}{\lambda_0} \text{ eV}} \qquad \text{—}\lambda_0 \text{ in Å}$$

5.13. Laws of Photoelectric Emission

The apparatus shown in Fig. 5.5 was used by Millikan to study the photoelectric effect and the various laws governing it. *S* is a source of radiations of variable but known frequency *f* and

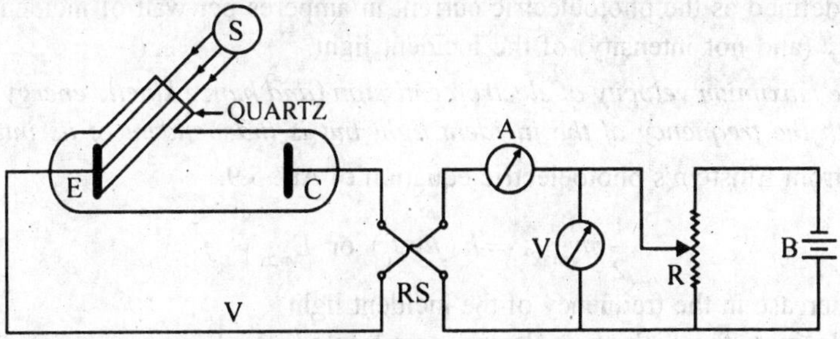

Fig. 5.5

intensity *I*. *E* is the emitting electrode made of the photosensitive material and *C* is the collecting electrode. Both these electrodes are enclosed in an evacuated glass envelope with a quartz window that permits the passage of ultraviolet and visible light. As shown, any potential difference can be established between the two electrodes. A reversing switch *RS* enables the polarities of the two electrodes to be reversed. If the tube is in the dark, then no photoelectrons are emitted and the microammeter *A* reads zero. However, if ultraviolet or visible light is allowed to fall on the emitting electrode, electrons are liberated and circuit current is set up.

From the experimental data collected by Richardson and Compton in 1912, photoelectric emission was found to be governed by the following laws :

(*i*) *Photoelectric current* (*i.e. number of electrons emitted per second*) *is directly proportional to the intensity of the incident light* (*or radiation*).

This can be verified by increasing the intensity of light and measuring the corresponding photoelectric current while holding the *frequency* of the incident light and the accelerating potential *V* of the collecting electrode *C* constant. The graph is similar to one shown in Fig. 5.6. Increase in intensity means more photons and hence ejection of electrons.

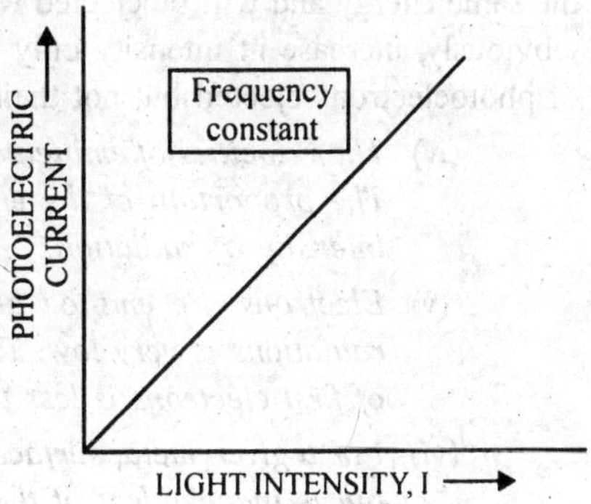

Fig. 5.6

(*ii*) *For each photosensitive surface, there is a minimum frequency of radiation* (*called threshold frequency*) *at which emission begins.*

Light of lower frequency (or longer wavelength), however strong, will not be able to produce any electron emission. This fact can be verified by keeping the light intensity constant while varying the frequency. The graph so obtained is shown in Fig. 5.7. The current is found to increase (though non-linearly) with the frequency of the incident light.

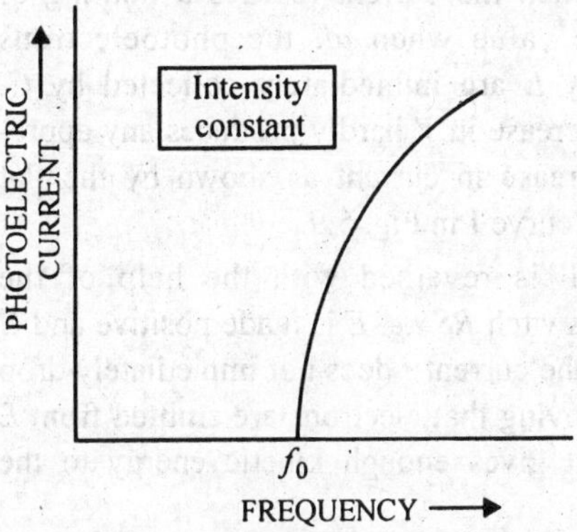

Fig. 5.7

Moreover, it is seen that there is a limiting or critical frequency below which no photoelectrons are emitted (Art. 5.9). It is called threshold frequency and its value depends on the nature of the material irradiated because for each material there is a certain minimum energy necessary to liberate an electron. This energy is known as *photoelectric work function* or threshold energy W_0. As seen from Art. 5.9, $f_0 = W_0/h$.

The wavelength corresponding to the threshold frequency f_0 is called threshold wavelength or long wavelength limit. No photoelectrons are emitted for wavelength greater than λ_0, no matter how long the light falls on the surface or how greater is its intensity. The photoelectric or quantum yield (which is defined as the photoelectric current in amperes per watt of incident light) depends on the frequency (and not intensity) of the incident light.

(*iii*) *The maximum velocity of electron emission (and hence kinetic energy) varies linearly with the frequency of the incident light but is independent of its intensity.*

As seen from Einstein's photoelectric equation of Art. 5.9,

$$\frac{1}{2}mv_{max}^2 = h(f - f_0) \text{ or } E_{max} \propto f$$

Hence, increase in the frequency of the incident light increases the velocity with which photoelectrons are ejected. The same fact is illustrated by Fig. 5.8. Incidentally, it may be noted that the slope of the curve gives the value of Planck's constant h.

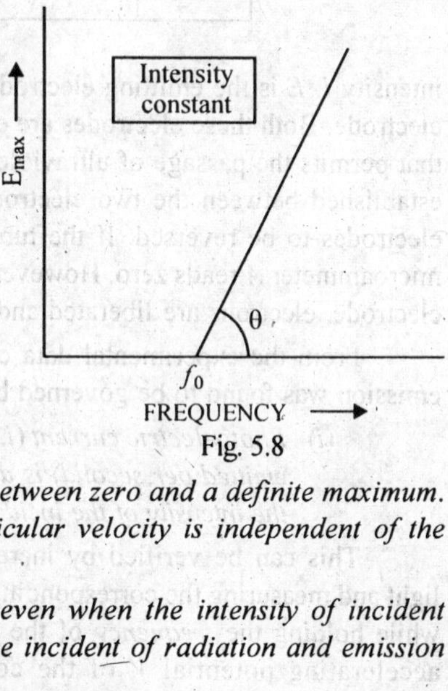

Fig. 5.8

If the *intensity* of the incident light is increased (keeping *frequency* constant), more photons will be incident on the metal surface, each photon having the same energy. Hence, more photoelectrons will be ejected. Since an electron can absorb only one photon, each photoelectron will have the same energy and will be ejected with the same velocity. Obviously, increase in intensity only increases the number of photoelectrons ejected but not their kinetic energy.

(*iv*) *The velocities of emitted electrons have values between zero and a definite maximum. The proportion of the electrons having a particular velocity is independent of the intensity of radiation.*

(*v*) *Electrons are emitted almost instantaneously even when the intensity of incident radiations is very low. The time lag between the incident of radiation and emission of first electrons is less than 10^{-8} second.*

(*vi*) *For a given metal surface, stopping potential V_0 is directly proportional to frequency but is independent of the intensity of the incident light.*

Suppose in Fig. 5.9, the frequency and intensity of incident light are held constant but the potential difference V between the two electrodes E and C is increased. Up to some stage as this p.d. is increased, photoelectric current is also increased. However, soon some value of V is reached when the current reaches a limiting or saturation value when *all* the photoelectrons emitted by E are immediately collected by C. Further increase in V hardly produces any appreciable increase in current as shown by the flat portion of curve I in Fig. 5.9.

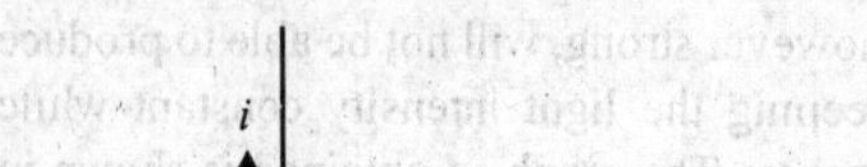

Fig. 5.9

If V is reversed with the help of the reversing switch RS *i.e.* E is made positive and C negative, the current i does not immediately drop to zero proving that electrons are emitted from E with some definite velocity. This velocity is such that it gives enough kinetic energy to the electrons so as to surmount the retarding electric field between the two electrodes. Hence, some electrons do succeed in reaching C despite the fact that the electric field opposes their motion.

When reversed V is made large enough, a value V_0 (called stopping or inhibiting potential) is reached when current is reduced to zero.

Stopping potential is *that value of the retarding potential difference between the two electrodes which is just sufficient to halt the most energetic photoelectrons emitted.*

As seen from curve II of Fig. 5.9, doubling the intensity of the incident light merely doubles the current but does not affect the value of V_0.

Now, if v_{max} is the maximum velocity of emission of a photoelectron and V_0 the stopping potential, then

$$\frac{1}{2}mv_{\max}^2 = eV_0 \text{ or } v_{max} = \sqrt{\frac{2 \cdot eV_0}{m}}$$

$$v_{max} = \sqrt{2 \times 1 \cdot 76 \times 10^{11} \times V_0}$$

$$= 5 \cdot 93 \times 10^5 \sqrt{V_0} \text{ m/s}$$

Obviously, $\quad E_{max} = eV_0$ joules

$$= V_0 \text{ electron-volt}$$

If, however, the experiment is repeated by varying the frequency of the light, it is found that the stopping potential varies linearly with frequency as shown in Fig. 5.10. Below threshold frequency, no electrons are emitted, hence stopping potential is zero for that reason. But as frequency is increased above f_0, the stopping potential varies linearly with the frequency of incident light.

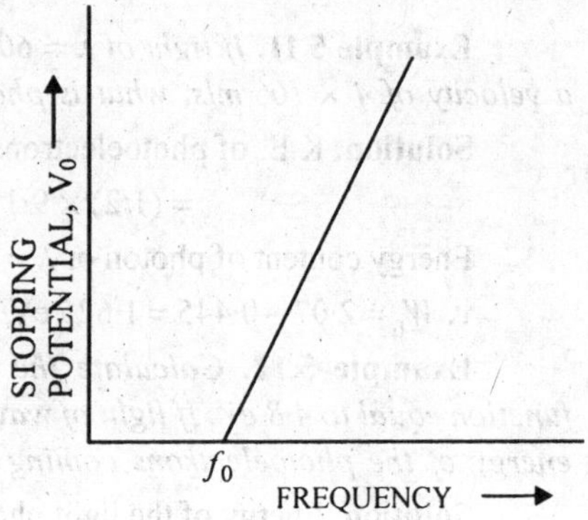

Fig. 5.10

Einstein's photoelectric equation may be expressed in terms of stopping potential as given below :

$$hf = W_0 + \frac{1}{2}mv_{\max}^2$$

Now, $\quad W_0 = hf_0$ and $\frac{1}{2}mv_{\max}^2 = eV_0$

$\therefore \quad hf = hf_0 + eV_0$ or $V_0 = \dfrac{h(f - f_0)}{e}$

Now, $\quad f = \dfrac{c}{\lambda}$ and $f_0 = \dfrac{c}{\lambda_0}$

$$V_0 = \frac{ch}{e}\left(\frac{1}{\lambda} - \frac{1}{\lambda_0}\right)$$

$$= 12 \cdot 4 \times 10^{-7} \left(\frac{1}{\lambda} - \frac{1}{\lambda_0}\right) \text{ volt} \qquad \text{—}\lambda \text{ and } \lambda_0 \text{ in metres}$$

$$\therefore \quad V_0 = 12{,}400 \left(\frac{1}{\lambda} - \frac{1}{\lambda_0}\right) \text{ volt} \qquad \text{—}\lambda \text{ and } \lambda_0 \text{ in A}$$

Example 5.9. *Photoelectrons with a maximum speed of 7×10^5 m/sec are emitted from a metal surface when light of frequency 8×10^{14} Hz falls on it. Calculate the threshold frequency of the surface.*

Solution. As seen from Art. 5.11

$$E_{max} = h(f - f_0)$$

$$\therefore \quad \frac{1}{2} m v_{max}^2 = h(f - f_0)$$

$$\text{or} \quad \frac{1}{2} \times 9{\cdot}1 \times 10^{-31} \times (7 \times 10^5)^2 = 6{\cdot}625 \times 10^{-34} (8 \times 10^{-14} - f_0)$$

$$\therefore \quad f_0 = \mathbf{4{\cdot}635 \times 10^{14}\ Hz.}$$

Example 5.10. *A tungsten cathode whose threshold wavelength is 2300 Å is irradiated by ultraviolet light of wavelength 1800 Å. Calculate (i) the maximum energy of the photoelectrons emitted and (ii) the work function for tungsten, both in electron-volts.*

Solution. As seen from Art. 5.10 and 5.11,

(*i*) $W_0 = \frac{12,400}{\lambda_0} = \frac{12,400}{2300} = \mathbf{5{\cdot}4\ eV}$

(*ii*) As seen from Art. 5.11

$$E_{max} = 12,400 \left(\frac{1}{\lambda} - \frac{1}{\lambda_0} \right) \text{eV}$$

$$= 12,400 \left(\frac{1}{1800} - \frac{1}{2300} \right) = \mathbf{1{\cdot}5\ eV}$$

Example 5.11. *If light of λ = 6000 Å falls on a metal surface and emits photoelectrons with a velocity of 4×10^5 m/s, what is photoelectric threshold wavelength ?*

Solution. K.E. of photoelectrons

$$= (1/2) \times 9{\cdot}1 \times 10^{-31} \times (4 \times 10^5)^2 / 1{\cdot}6 \times 10^{-19} = \mathbf{0{\cdot}445\ eV}$$

Energy content of photon or λ = 6000 Å = 12,400/6,000 = 2.07 eV

$\therefore W_0 = 2{\cdot}07 - 0{\cdot}445 = 1{\cdot}625$ eV, $\lambda_0 = 12,400/1{\cdot}625 =$ **7,631 Å**

Example 5.12. *Calculate the threshold frequency for gold having photoelectric work function equal to 4·8 eV. If light of wavelength 2220 Å falls on gold, what will be maximum kinetic energy of the photoelectrons coming out ?*

Solution. Energy of the light photon = 12,400/2220 = 5·58 eV.

Out of this, 4·8 eV would be used for dislodging the electron and the balance would represent its kinetic energy.

$$E_{max} = 5{\cdot}58 - 4{\cdot}8 = \mathbf{0{\cdot}78\ eV}$$

Alternatively, λ_0 = 12,400/4·8 = 2583 Å. Hence, we may use

$$\therefore \quad E_{max} = 12,400 \left(\frac{1}{\lambda} - \frac{1}{\lambda_0} \right) \text{eV}$$

Example 5.13. *When violet light of λ = 4000 AU strikes the cathode of a photocell, a retarding potential of 0·4 V is required to stop emission of electrons. Calculate (i) light frequency (ii) photon energy (iii) work function (iv) threshold frequency and (v) net energy after the electron leaves the surface.*

Solution. (*i*) $f = c/\lambda = 3 \times 10^8 / 4000 \times 10^{-10}$

$$= \mathbf{7{\cdot}5 \times 10^{14}\ Hz}$$

(*ii*) $E = hf = 6{\cdot}625 \times 10^{-34} \times 7{\cdot}5 \times 10^{14}$

$$= 4{\cdot}95 \times 10^{-19}\ \text{J} = \mathbf{3{\cdot}1\ eV}$$

(*iii*) $W_0 = hf - \text{K.E.} = hf - V_0 = 3{\cdot}1 - 0{\cdot}4 = \mathbf{2{\cdot}7\ eV}$

(*iv*) $f_0 = W_0/h = 2{\cdot}7 \times 1{\cdot}6 \times 10^{-19}/6{\cdot}625 \times 10^{-34}$

$= \mathbf{6{\cdot}5 \times 10^{14}\ Hz}$

(*v*) Net energy $= hf - W_0 = 3{\cdot}1 - 2{\cdot}7 = 0{\cdot}4\ \text{eV} = \mathbf{6{\cdot}4 \times 10^{-20}\ J}$

Example 5.14. *A photon of wavelength 3310 Å falling on a photo cathode ejects an electron of energy 3 × 10^{-19} J and one of wavelength 5000 Å ejects an electron of energy 0·972 × 10^{-19} J. Calculate the value of Planck's constant and the threshold wavelength for the photo cathode.*

Solution. As seen from Art 5.9

$$hf = W_0 + \text{K.E.} \quad \text{or} \quad \frac{hc}{\lambda} = W_0 + \text{K.E.}$$

In the first case, $$\frac{h \times 3 \times 10^8}{3310 \times 10^{-10}} = W_0 + 3 \times 10^{-19}$$

In the second case, $$\frac{h \times 3 \times 10^8}{5000 \times 10^{-10}} = W_0 + 0{\cdot}972 \times 10^{-19}$$

Subtracting one from the other,

$$h = \mathbf{6{\cdot}62 \times 10^{-31}\ J\text{-}s.}$$

Subtracting this value of h, $W_0 = 3 \times 10^{-19}$ J.

$\therefore \quad \lambda_0 = ch/W_0 = 3 \times 10^8 \times 6{\cdot}62 \times 10^{-34}/3 \times 10^{-19}$

$= \mathbf{6620 \times 10^{-10}\ m}$

$v_{max} = \mathbf{7{\cdot}12 \times 10^5\ m/s}$

Example 5.15. *A certain metal has a threshold wavelength of 6525 Å. Find the stopping potential when the metal is irradiated with*

(*a*) *monochromatic light having a wavelength of 4000 Å*

(*b*) *light having twice the frequency and three times the intensity of that in* (*a*) *above.*

(*c*) *If a material having double the work function were used, what would be the answers to* (*a*) *and* (*b*) *above ?*

Solution. (*a*) $$V_0 = 12{,}400 \left(\frac{1}{4000} - \frac{1}{6525}\right) = \mathbf{1{\cdot}2\ volt}$$

(*b*) As explained in Art. 5.12, stopping potential is independent of the intensity of the incident light but varies directly as frequency f provided it is more than f_0.

Since frequency is twice, the wavelength of the light is half *i.e.* 2000 Å.

$$V_0 = 12{,}400 \left(\frac{1}{2000} - \frac{1}{6525}\right) = \mathbf{4.3\ V}$$

(*c*) If work function is double, then λ_0 is reduced to half *i.e.* $\lambda_0 = 6525/2 = 3262{\cdot}5$ Å. Since the incident light has $\lambda = 4000$ Å, it would not be able to produce photoemission.

In the second case, $\lambda = 2000$ Å

$$V_0 = 12{,}400 \left(\frac{1}{2000} - \frac{1}{3262.5}\right) = \mathbf{2.4\ volt}$$

Example 5.16. *A certain metallic surface is illuminated by monochromatic light of variable wavelength. No photoelectrons are emitted above a wavelength of 5000 Å. With an unknown wavelength, a stopping potential of 3·1 V is necessary to stop photoelectric current. Find the unknown wavelength.*

Solution. Here, $\lambda_0 = 5000$ Å, $V_0 = 3{\cdot}1$ V, $\lambda = ?$

Now, $V_0 = 12{,}400\left(\frac{1}{\lambda} - \frac{1}{\lambda_0}\right)$

$\therefore$ $3{\cdot}1 = 12{,}400\left(\frac{1}{\lambda} - \frac{1}{5000}\right)$

$\therefore$ $\lambda = \mathbf{2{,}222}$ **Å**

Example 5.17. *Light of wavelength 2000 Å falls on an aluminium surface. In aluminium, 4·2 eV are required to remove an electron. Determine (i) K.E. of the fastest emitted photoelectron (ii) K.E. of the slowest emitted photoelectron (iii) Stopping potential and (iv) cutoff wavelength for aluminium.*

Solution. Photon energy of the incident light is

$$= 12{,}400/2000 = 6{\cdot}2 \text{ eV}$$

(*i*) $E_{max} = (6{\cdot}2 - 4{\cdot}2)$ eV

$= 2 \text{ eV} = 2 \times 1{\cdot}6 \times 10^{-19} = \mathbf{3{\cdot}2 \times 10^{-19}}$ **J**

(*ii*) $E_{min} = 0$

(*iii*) $V_0 = E_{max} = \mathbf{2}$ **V**

(*iv*) $\lambda_0 = 12{,}400/W_0 = 12{,}400/4.2 = \mathbf{2952{\cdot}4}$ **Å**

Example 5.18. *The stopping potential is 4·6 V for light of frequency 2 × 10*^{15} *Hz. When light of frequency 4 × 10*^{15} *Hz is used, the stopping potential is 12.9 V. Calculate the value of Planck's constant.*

Solution. As seen from Art. 5.12 (*vi*) if V_0 is the stopping potential, then

$$eV_0 = h(f - f_0)$$

Substituting the two given values, we get

$$4{\cdot}6\, e = h(2 \times 10^{15} - f_0)$$

$$12{\cdot}9\, e = h(4 \times 10^{15} - f_0)$$

Subtracting one from the other, we have

$$8{\cdot}3\, e = 2h \times 10^{15}$$

$$8{\cdot}3 \times 1{\cdot}6 \times 10^{-19} = 2h \times 10^{15}$$

$$h = \frac{8{\cdot}3 \times 1{\cdot}6 \times 10^{-19}}{2 \times 10^{15}} = \mathbf{6{\cdot}44 \times 10^{-34}\ \text{J-s.}}$$

5.14. Types of Photoelectric Cells

Following three main types are worth considering :

(*i*) Photoemissive cell — it depends on the emission of electrons from a *metal* cathode when it is exposed to light or other radiation.

(*ii*) Photovoltaic cell — here sensitive element is a semiconductor (not metal) which generates voltage in proportion to the light or any radiant energy incident on it.

(*iii*) Photoconductive cell — it uses a semiconductor material whose resistance changes in accordance with the radiant energy received.

Further details of all these cells are given below :

5.15. Photoemissive Cell or Phototube

(*a*) **Construction.** It consists of two metallic electrodes *i.e.* a cathode and an anode supported in an evacuated glass bulb fitted with a base like a thermionic valve as shown in Fig. 5.11. The cathode is either V-shaped or semi-cylindrical and is made of a metal coated with an emissive material.

The anode is in the form of a thin wire facing the cathode or a small button so placed as not to obscure the light source from the cathode.

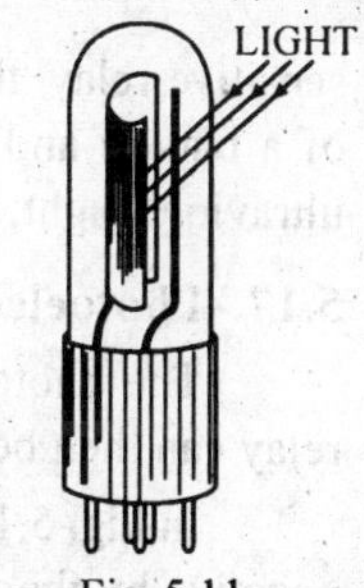

Fig. 5.11

Various types of photocathodes in use are as under :

(*i*) Pure metals like Ni, Zn, Al etc. for ultraviolet radiations.

(*ii*) Alkali metals for visible light.

(*iii*) Composite (atomic film) cathodes for infra-red radiations. Example of such a cathode being caesium-on-oxygen-on-silver.

(*iv*) Alloy cathodes made up of thin films of antimony-caesium.

(*b*) **Working.** When cathode is illuminated, electrons are emitted by it which are attracted by the positive anode. Hence, a current is produced whose magnitude, for a given cathode, depends on

(*i*) intensity of incident radiation and

(*ii*) anode-to-cathode voltage.

The variations are shown in Fig. 5.12. As seen, the current produced by such a cell is of the order of a few μA which has to be further amplified by an amplifier before being out to use.

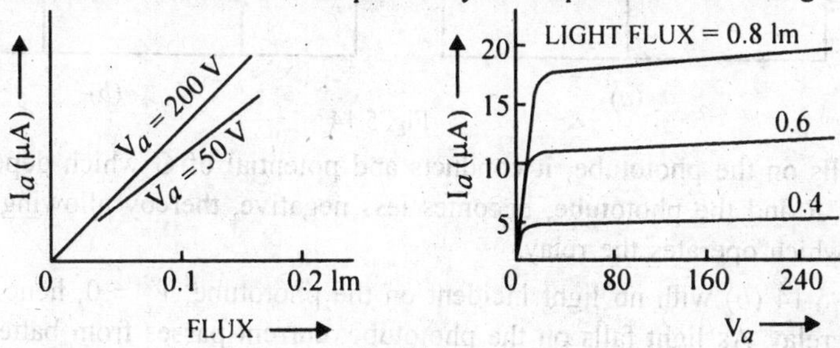

Fig. 5.12

(*c*) **Uses.** It finds wide use in the field of photometry and calorimetry, for sound reproduction from a motion-picture film, for 'on and off' circuits and in other circuits concerning the counting or sorting of objects on a conveyer belt, the automatic opening of a door as it is approached and fire-alarm etc.

5.16. Automatic Bell Alarm Circuit

In Fig. 5.13 is shown the circuit in which a bell rings up whenever light beam between the lamp and phototube is interrupted. Light falling on the phototube causes a small current to flow from battery *B* through resistance *R*. As a result, some voltage drop = *iR* is developed across *R* with point *G* being negative with respect to cathode. This negative bias is applied to the grid of a

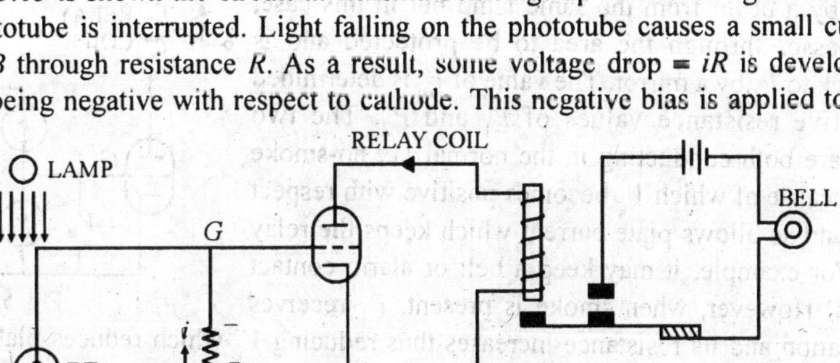

Fig. 5.13

triode valve, hence very little or no anode current flows. Should the light weaken or be cut off, the grid bias reduces and allows more current to flow. The anode current flows through the coil of a sensitive relay, the armature of which is attracted thus closing the contacts of a circuit consisting of a battery and an electric bell or buzzer. If photocathode is made sensitive to the infra-red or ultraviolet light, then visible light need not be used.

5.17. Photoelectric Relay

The photoelectric cell can be made to control a relay with the help of a triode valve. The relay can then be used to switch on an alarm or operate a counter or start or stop an electric motor.

In Fig. 5.14 (*a*) with *no light incident* on the phototube, valve grid *G* is maintained sufficiently negative by the battery *B* to cut off plate current so that the relay remains inoperative. However,

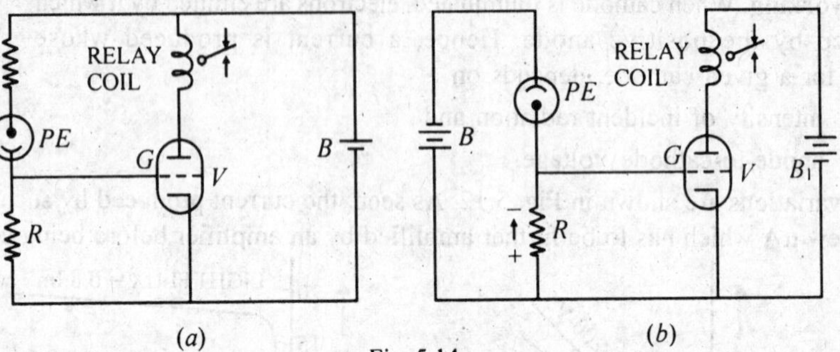

Fig. 5.14

when light falls on the phototube, it conducts and potential of *G* which depends on the relative resistances of *R* and the phototube, becomes less negative, thereby allowing the triode valve to pass current which operates the relay.

In Fig. 5.14 (*b*) with no light incident on the phototube, $V_c = 0$, hence current flows thus operating the relay. As light falls on the phototube, current passes from battery *B* through *R* and the phototube, thereby making grid *G* sufficiently negative with respect to cathode, so as to cut off the plate current. It is obvious that so long as the phototube is illuminated, the relay remains inoperative but it starts operating immediately on interruption of the light.

5.18. Smoke Detector

A simple form of a smoke detector employing two similar phototubes P_1 and P_2 is shown in Fig. 5.15. P_2 is illuminated directly by a lamp. P_1 is also illuminated by a beam from the same lamp but in this case, the beam passes through the area to be protected and is reflected back to P_1 by a mirror. The value of V_g is determined by the relative resistance values of P_1 and P_2. The two phototubes are both conducting in the normal *i.e.* no-smoke condition because of which V_g becomes positive with respect to cathode and so allows plate current which keeps the relay operative. For example, it may keep a bell or alarm contact switch open. However, when smoke is present, P_1 receives less illumination and its resistance increases thus reducing V_g which reduces plate current so that the relay releases the switch. Hence, a bell starts ringing. Smoke detectors are generally used as part of fire prevention scheme.

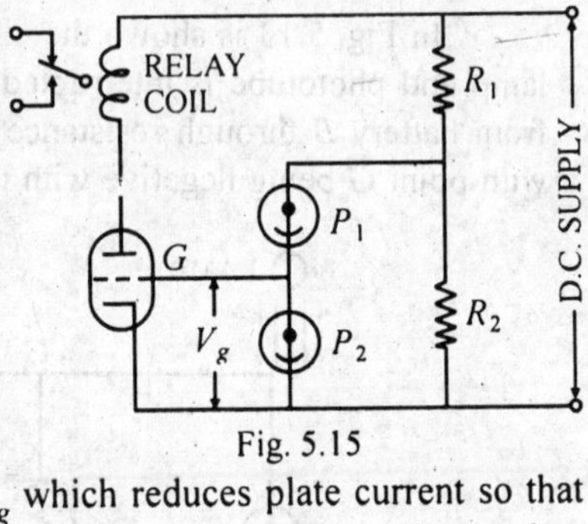

Fig. 5.15

5.19. On and Off Control Circuit

Suppose it is required to light an electric lamp by means of a match-stick and to extinguish it by waving the hand. The circuit for the purpose is shown in Fig. 5.16 which employs a phototube in conjunction with a thyratron. The rating of thyratron must be high enough to supply the lamp current. The resistor *R* acts as load resistor for phototube but grid resistor for thyratron. The plate circuit of thyratron is connected to an a.c. supply. As long as phototube remains unilluminated, no current flows through R and hence no voltage develops across R and bias voltage B_2 is large enough to prevent firing of thyratron. When a match-stick is lit in front of the phototube, it is illuminated by the light so produced and hence becomes conducting so that B_1 passes current through R across which a certain voltage is dropped. The net cathode-grid voltage becomes greater than the critical grid voltage of thyratron for some point on the alternating plate voltage wave. Hence, lamp lights up and if its light falls on the phototube, the lamp remains on. If hand is interposed between the phototube and the lamp so that phototube receives no light from the lamp for a short while, the lamp will remain extinguished.

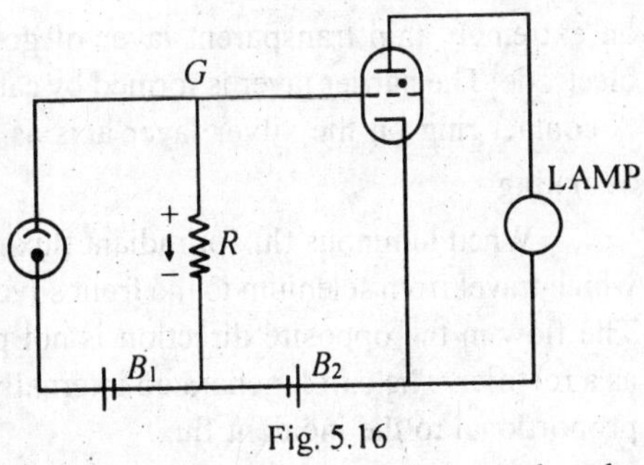

Fig. 5.16

5.20. Photomultiplier Tube

It is a phototube in which current is amplified by electron multiplication through secondary emission. Cross-section of one form of such a tube is shown in Fig. 5.17. It consists of a 9 electrodes (called dynodes) which are maintained at increasing potentials in sequence from photocathode O to the anode marked 10 in the diagram. When light falls on the cathode, photoelectrons are emitted which are accelerated towards dynode No. 1 since it is at a higher potential. On colliding with this dynode, they liberate secondary electrons which are attracted by dynode No. 2 since it is at a relatively still higher potential. These electrons release further secondary electrons from it which are attracted by dynode No. 3. This process of secondary emission from different dynodes continues till the electrons are finally collected by the anode. It is obvious that at each dynode, the number of secondary electrons keeps on multiplying.

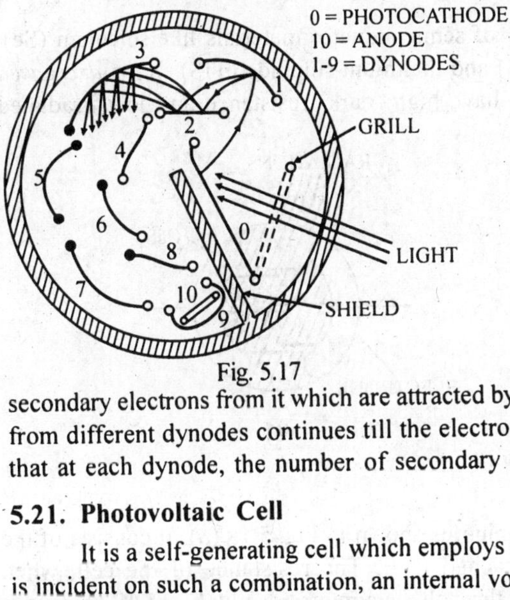

Fig. 5.17

5.21. Photovoltaic Cell

It is a self-generating cell which employs semiconductor contacts against metals. When light is incident on such a combination, an internal voltage is generated which causes the current to flow through internal circuit even though no external battery is applied. The e.m.f. generated is in proportion to the radiant energy received as in solar batteries.

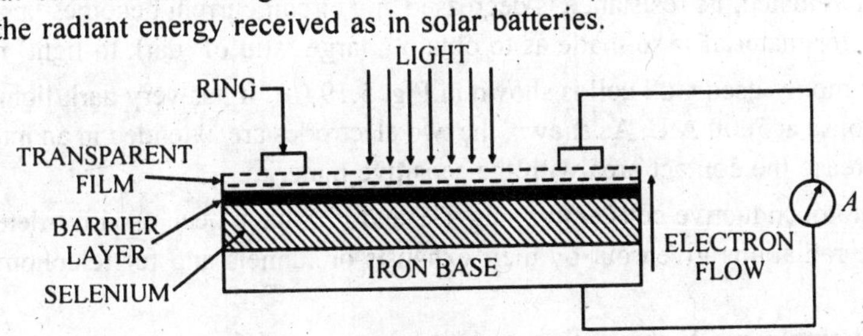

Fig. 5.18

Construction

The most commonly used photovoltaic cells are of the barrier layer type like iron-selenium cells or Cu-CuO_2 cells. In the iron-selenium cell, selenium layer is placed on an iron disc and then an extremely thin transparent layer of gold or silver is formed on the selenium to act as a front electrode. The barrier layer is formed by cathode-sputtering the semi-transparent film on the selenium. A contact ring on the silver layer acts as one electrode and the iron base as the other.

Working

When luminous flux or radiant flux falls on the semiconductor *i.e.* selenium, it ejects electrons which travel from selenium to the front silver electrode through the barrier layer as shown in Fig. 5.18. The flow in the opposite direction is not permitted by the (boundary) barrier layer because it acts as a rectifier. The e.m.f. generated internally between silver electrode and selenium is almost directly proportional to the incident flux.

Uses. The main advantage of a photovoltaic cell is that it *requires no external battery for its own operation i.e.* it is self-generating. Moreover, the internal e.m.f. and hence current generated by it are *large* enough to be measured on a pointer galvanometer. Hence, such cells are used in devices like portable exposure meters, direct-reading illumination metres and with low-resistance relays for 'on/off' operations and other monitoring operations in industry.

5.22. Photoconductive Cell

It is based on the principle that resistivity of semiconductor materials like selenium (Se), cadmium sulphide (CdS) and lead sulphide (PbS) and thalmium-sulphide (TIS) etc. is *decreased* when irradiated. In other words, such materials have high 'dark' resistance and low irradiated resistance.

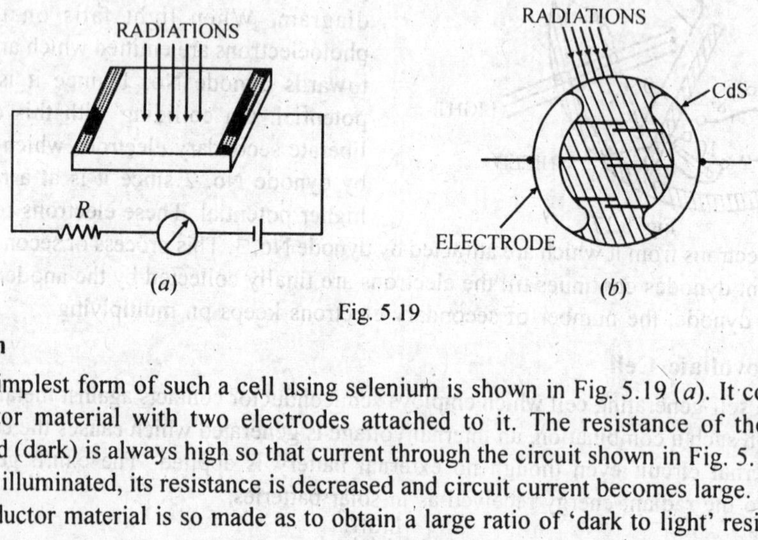

Fig. 5.19

Construction

The simplest form of such a cell using selenium is shown in Fig. 5.19 (*a*). It consists of the semiconductor material with two electrodes attached to it. The resistance of the cell, when unilluminated (dark) is always high so that current through the circuit shown in Fig. 5.19 (*a*) is low. When cell is illuminated, its resistance is decreased and circuit current becomes large. The shape of the semiconductor material is so made as to obtain a large ratio of 'dark to light' resistance.

The commonly-used CdS cell is shown in Fig. 5.19 (*b*). It has very dark/light ratio and gives maximum response at 5000 A.U. As shown, the two electrodes are extended in an interdigital pattern in order to increase the contact area with the sensitive material.

Uses. Photoconductive cells employing TIS and PbS have been used for detecting ships and aircrafts by the radiations given out by their exhausts or funnels and for telephony by modulated infrared light.

A circuit employing such a cell and which may be used for low-speed counting or as a fire or burglar alarm is shown in Fig. 5.20 (*a*). Fig. 5.20 (*b*) shows a similar circuit using a thyratron and an a.c. supply. In each case when sufficient light falls on the cell, the value conducts and energies the relay. Resistor *R* can be changed to adjust the light level for relay operation.

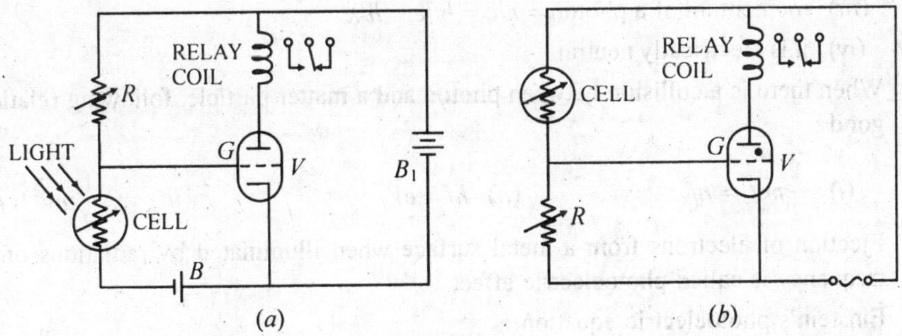

Fig. 5.20

TUTORIAL PROBLEMS

1. Find the energy content of ultraviolet light photons of wavelength 3000 Å in (*i*) joules and (*ii*) electron-volt. **[(*i*) 6·62 × 10^{-19} J (*ii*) 4·13 eV]**
2. What is the energy of photons of a spectral emission line whose wavelength is 21 cm? **[5·9 × 10^{-6} eV]**
3. Earth receives solar radiations at the rate of 19·86 kcal/m²-min. If the average wavelength of the solar radiation is assumed as 5500 Å, convert the above into photons/m²-min. Take J = 4200 joules/kcal. **[23·1 × 10^{22}]**
4. Calculate the frequency and amount of energy (in eV) associated with a quantum of electromagnetic radiation at each of the following wavelengths (*i*) 1 m (*ii*) 1 mm (*iii*) 5000 Å (*ic*) 1 Å. **[(*i*) 3 × 10^8 Hz; 1·25 × 10^{-6} eV (*ii*) 3 × 10^{11} Hz; 1·25 × 10^{-3} eV (*iii*) 6 × 10^{14} Hz; 1·25 eV (*iv*) 3 × 10^{18} Hz; 1·25 × 10^4 eV]**
5. Silver has a photoelectric function of 4·5 eV. Find (*i*) its threshold wavelength (*ii*) maximum velocity of ejected electrons with ultraviolet light of wavelength 2000 Å (*iii*) stopping potential for the most energetic photoelectrons. **[(*i*) 2760 Å (*ii*) 7·71 × 10^5 m/s (*iii*) 1·7 V]**
6. The threshold wavelength for tungsten is 2300 Å. Find the energy of the emitted electrons from the surface by ultraviolet light of wavelength 1800 Å. **[2·4 × 10^{-19} J]**
7. With what velocity must an electron travel so that its kinetic energy is equal of a photon with a λ = 5,200 Å. **[9·2 × 10^5 m/s]**
8. The long wavelength limit for photoelectric effect for a certain metal is 2750 Å. Find (*i*) work function W_0 (*ii*) the maximum velocity of electrons ejected from the metal by radiation of λ = 1800 Å (*iii*) maximum K.E. of these electrons. **[(*i*) 4·5 eV (*ii*) 9·1 × 10^5 m/s (*iii*) 3·8 × 10^{-19} J]**
9. Calculate the frequency of the incident radiations which eject from a metal surface electrons fully retarded by a reverse potential of 3 V. The threshold frequency for this metal is 6 × 10^{14} Hz. Also find the work function of this metal. **[13·2 × 10^{14} Hz; 2·48 eV]**
10. A ray of ultraviolet light of wavelength 3000 Å falling on the surface of a material whose work function is 2·82 eV ejects an electron. What will be the velocity of the emitted electrons? **[8·1 × 10^5 J]**
11. If a photon of energy 4 eV falls on a surface with work function of 1·9 eV, what will be the maximum kinetic energy in joule of the electron emitted? **[3·26 × 10^{-19} J]**

HIGHLIGHTS

1. According to Quantum Theory :
   (*i*) radiant energy is not emitted continuously but intermittently in the form of photons.
   (*ii*) these quanta or photons travel in space with the velocity of light.
2. A photon has the following properties :
   (*i*) it has energy = hf
   (*ii*) mass of photon = $E/c^2 = hf/c^2 = h/c\lambda$

(*iii*) momentum of a photon = $mc = hf/c = h/\lambda$

(*iv*) it is electrically neutral.

3. When there is a collision between photon and a matter particle, following relations hold good :

(*i*) $\frac{1}{2}mv^2 = hf$ (*ii*) $hf = eV$ (*iii*) $\therefore\ \frac{1}{2}mv^2 = hf = eV.$

4. Ejection of electrons from a metal surface when illuminated by radiations of suitable frequency is called photoelectric effect.

5. Einstein's photoelectric equation is

$$hf = W_0 + \frac{1}{2}mv^2$$

6. Photoelectric work function W_0 is defined as the energy which is just sufficient to liberate electrons from a metal surface with zero velocity.

$$W_0 = hf_0 = \frac{ch}{\lambda} = \frac{12,400}{\lambda_0}\ \text{eV} \qquad \text{—}\lambda_0 \text{ in Å}$$

where f_0 is threshold frequency and λ_0 is threshold wavelength.

7. Threshold frequency f_0 is defined as the minimum frequency of the incident radiation which can cause photoelectric emission.

8. Laws of photoelectric emission are :

(*i*) Number of electrons emitted per second (*i.e.* current) is directly proportional to the intensity of the incident radiation.

(*ii*) For each photosensitive surface, there is minimum frequency of incident radiation at which electron emission begins.

(*iii*) The maximum velocity of electron emission varies linearly with the frequency of the incident radiation but is independently of its intensity.

(*iv*) The velocities of emitted electrons have value between zero and a definite maximum.

(*v*) Electrons are emitted almost instantaneously.

(*vi*) For a given metal surface, stopping potential is directly proportional to frequency but is independent of the intensity of the incident radiation.

9. Photoelectric cell working depends on the fact that electrons are emitted when a metal cathode is irradiated.

10. Photovoltaic cell is based on the principle that a sensitive semiconductor material generates voltage in proportion to the radiant energy incident on it.

11. Photoconductive cell utilizes that fact that resistance of certain semiconductor materials decreases when irradiated.

Can You Answer This ?

1. What is photoelectric effect ? Explain how has this been explained by Einstein. Discuss practical applications of this effect.

2. Give in detail with neat diagrams the experimental procedure for the verification of Einstein's photoelectric equation.

3. In an experiment on photoelectric effect for measuring the energy and number of photoelectrons, what happens if :

(*i*) the target material is changed, frequency and intensity of light being kept constant.

(*ii*) the frequency of light is changed, target material and intensity of light being kept constant.

(*iii*) the intensity of light is changed, frequency of light and target material being kept constant ?

4. Write short note on photoconduction cell.
5. Give laws of photoelectric emission of electrons and Einstein's explanation for the photoelectric effect.
6. (*a*) What is photoelectric effect ? Define photoelectric work function and threshold frequency.

   (*b*) Derive in a simple way Einstein's photoelectric equation.
7. Write short note on photoelectric cells.
8. What is photoelectric effect ?

   State the characteristics of photoelectrons. How are they studied experimentally ?
9. What are the fundamental laws of photoelectric emission ? State Einstein's photoelectric equation and show how it explains the laws.
10. (*a*) State and explain the laws of photoelectric effect.

    (*b*) Describe Millikan's method of verifying Einstein's photoelectric equation.

    (*c*) Calculate the velocity of the particle at which the kinetic energy of particle is equal to its rest energy.
11. State Einstein's law of photoelectric effect. Describe an experiment designed to verify this law.
12. Briefly outline two applications of photoelectric effect in modern life.
13. Describe in details the experimental arrangement and the procedure to study the photoelectric effect. State the conclusions.

BONDS IN SOLIDS 6

Solid State of Matter — Ionic Bonds — Covalent Bonds — Metallic Bonds — Molecular Bonds — Characteristics of Ionic Solids — Characteristics of Covalent Solids — Metallic Compounds — Molecular Solids — Highlights — Can You Answer This ?

6·1. Solid State of Matter

As compared to gaseous and liquid states, the atoms and molecules in a solid state are more closely-packed and are held together by strong mutual forces of attraction. These interatomic forces are basically electrostatic in nature and are a direct consequence of the electronic structure of matter. When atoms come close and finally unite to form molecules, their electrons *rearrange themselves in such a way as to achieve a stable configuration.* This rearrangement of electrons gives rise to different types of bonds which hold the atoms together in a solid state. The principal types of bonds in solids are :

(*i*) ionic or electrovalent bonds

(*ii*) covalent or homopolar bonds

(*iii*) metallic bonds and

(*iv*) molecular bonds (Van der Waals' forces)

As said above, these bonds are electrostatic in origin and, in many cases, determine the characteristic physical properties of a solid. Depending on the type of bonding force, the different solids may be categorized as follows :

(*a*) ionic or electrovalent solids : NaCl, MgO, CaF_2

(*b*) Covalent solids : C, Si, Ge

(*c*) Metallic solids : Na, Al, Cu, Ag

(*d*) Molecular solids : solid A, solid Kr and solid Xe.

6·2. Ionic Bonds

Such bonds are mainly formed in *inorganic compounds* like NaCl and KOH etc. and never in pure elements. These are *very strong* bonds and develop between those two different atoms one of which has low ionization energy and the other has high electron affinity. As explained below, these bonds are formed due to the electrostatic attraction between two stable ions which are themselves produced due to the *complete* transfer of electrons from one atom to another.

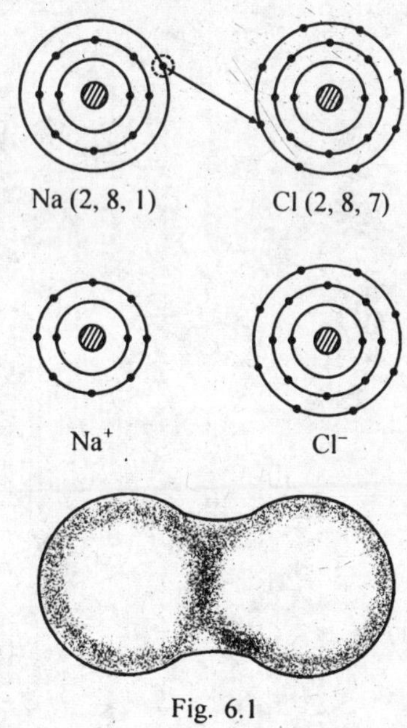

Fig. 6.1

Consider the formation of ionic bonds in NaCl. As shown in Fig.6.1, Na has one 3*s* valence electron in its outermost sub-orbit which it can lose easily *because Na atom has low ionization energy*. Chlorine, on the other hand, has seven valence electrons and is ever ready to accept one electron to complete its 3*p* sub-orbit. When Na atom loses one electron, it becomes a positive ion (*i.e.* anion) and as Cl gains one electron, it becomes a negative ion (*i.e.* cation). The two ions attract each other with coulombic forces which lead to the formation of an electrovalent or ionic bond between two neighbouring Na and Cl atoms. At the same time, the two ions achieve *noble gas structure i.e.* they have their shells filled. For example, Na^+ achieves neon structure (2, 8) and Cl^- achieves argon structure (2, 8, 8).

Since the electrostatic field of an ion extends in all direction, the above process of Na^+ ion attracting a Cl^- ion does *not stop at two ions only*. As shown in Fig. 6.2 (*a*), a positive Na ion will

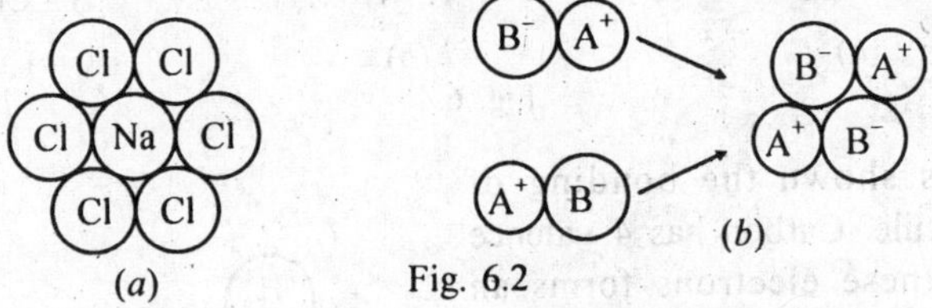

Fig. 6.2

be surrounded by negative Cl ions. In the same way, the negative Cl ion will be surrounded by positive Na ions. This leads to the crystalline structure of NaCl belonging to the cubic system having *FCC* lattice.

In general, in the formation of an ionic compound, a positive ion A^+ combines with a negative ion B^- to give a molecule AB as shown in Fig. 6.2 (*b*). When two such molecules or ion pairs come closer, force of attraction between the ions of one pair and the oppositely-charged ions of the *other* pair is set up. However, at the same time, force of repulsion between A^+—A^+ and B^-—B^- is also brought into play. Since this repulsive force is negligible, the process of clustering of ions continues unchecked. Hence, crystal of an ionic compound may be looked upon as a cluster of ions in which a positive ion is surrounded spatially by a certain number of negative ions while a negative ion is similarly surrounded by a number of positive ions.

6·3. Covalent Bonds

A covalent bond is formed when two similar or dissimilar atoms achieve stability by *sharing* valence electrons between themselves. This bonding force arises due to the presence of *shared* electrons. It should be clearly understood that in a covalent bond, electrons are *shared* by both atoms; the electrons do not become the *exclusive* property of any one atom as is the case of an ionic bond. By sharing the electrons, the two atoms, in many cases, acquire a noble gas configuration. The compound so formed is known as *covalent* or *non-polar* compound.

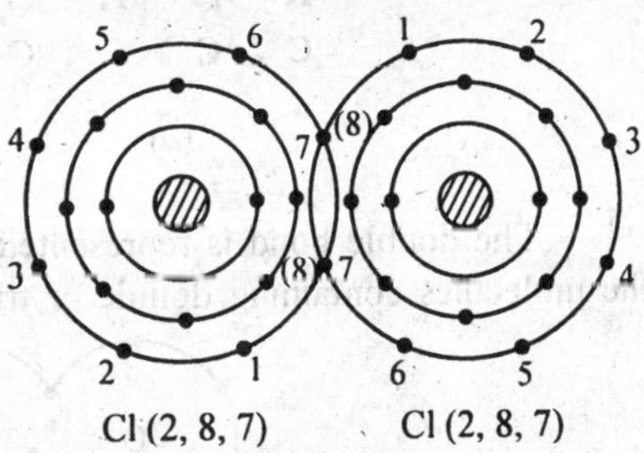

Fig. 6.3

Consider the case of Cl atoms. As shown in Fig. 6.3, each Cl atom has 7 valence electrons. Covalent bond is established between the two atoms because each contributes one electron and the electron *pair* is shared by the two atoms so that each has 8 electrons in its valence shell thereby achieving stability. As seen, each chlorine atom achieves argon structure. For clarity sake, the two electrons have been shown by dots and crosses in Fig. 6.4 (*a*). Similarly, covalent bond between hydrogen atoms is created due to their sharing their 1*s* electrons [Fig. 6.4 (*b*)]. The two hydrogen atoms acquire He structure. Fig. 6.4 (*c*) shows covalent bond between hydrogen and chlorine atoms. In this case, hydrogen achieves He structure whereas Cl achieves the argon structure.

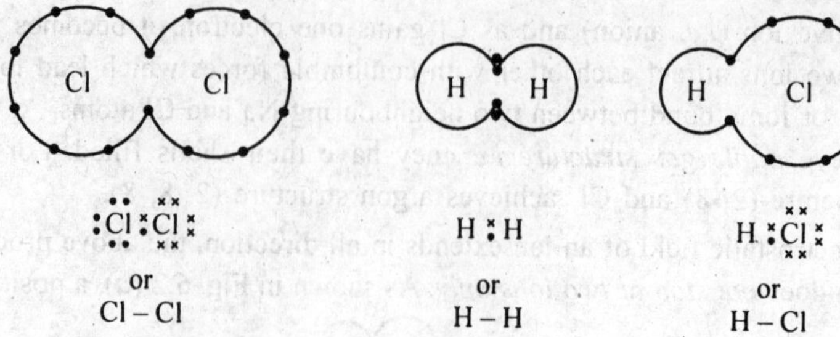

(a) (b) (c)

Fig. 6.4

In Fig. 6.5 is shown the bonding of methane (CH_4) molecule. Carbon has 4 valence electrons. Each of these electrons forms an electron pair with the 1*s* electrons of the four surrounding hydrogen atoms.

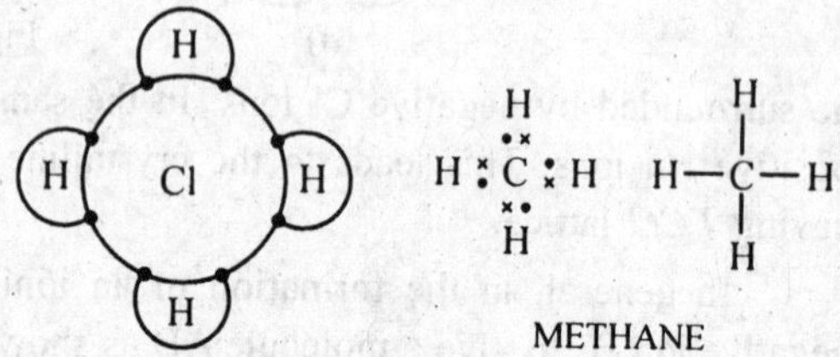

Fig. 6.5

Diamond is another example of crystal structure in which carbon atoms are linked by covalent bonds. As seen from Fig. 6.6, each carbon atom shares its four valence electrons with those of neighbouring atoms thus forming single covalent bonds with each of them. When more than one pair of electrons is shared between two atoms, a *double* or *triple* covalent bond is said to be formed.

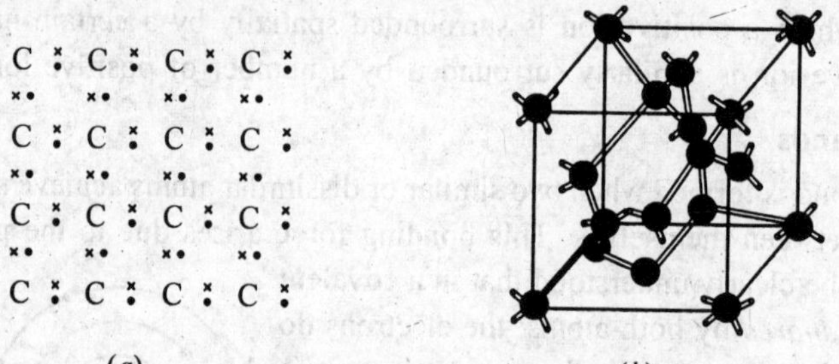
(a) (b)

Fig. 6.6

The double bond is represented by double lines and a triple bond by three lines. Some of the molecules containing double or triple bonds are shown in Fig. 6.7.

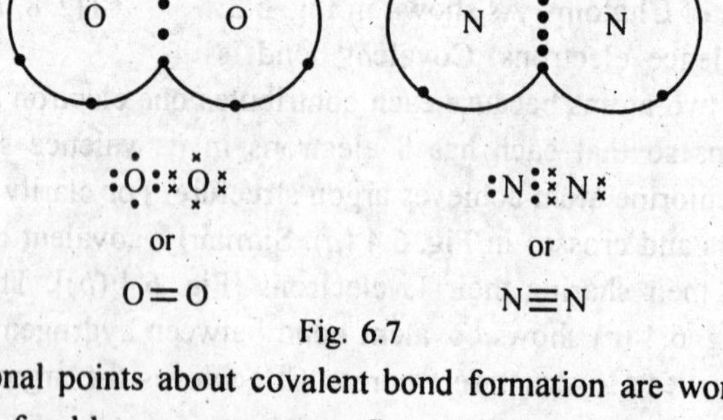

Fig. 6.7

Following additional points about covalent bond formation are worth noting:

(*i*) attainment of noble gas or octet configuration is not essential in such a bond.

(*ii*) the essential feature of such a bond is the fact that it involves the *pairing of two electrons with opposite spin.*

(*iii*) a covalent bond may be either polar or non-polar depending on the fact whether the electron pair is shared unequally or equally between the two atoms.

6·4. Metallic Bonds

Such a bond results from the sharing of *variable* number of electrons by a *variable* number of atoms. Such bonds are responsible for holding the atoms together in metals and their alloys. It has been found that in metals, each atom loses all its valence electrons and becomes a positively-charged ion. These free and mobile electrons form a kind of electron *cloud* or *gas* which permeates all atoms. In fact, one may look upon a metal as consisting of an array of closely-packed ions immersed in a sea of free electrons. The valence electrons are not bonded directly to an individual atom but they move freely in the sphere of influence of other atoms and are bound *to different atoms at different times* and that too for part of the time.

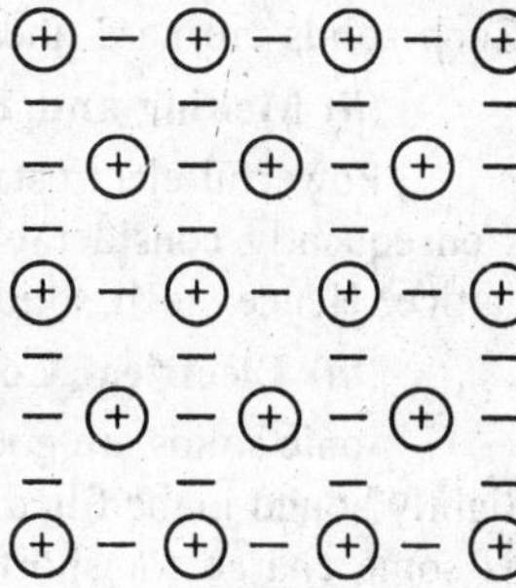

Fig. 6.8

A metallic bond may be looked upon as an *unsaturated* covalent bond which is due to the electrostatic attraction between the negative electron cloud and positive ion 'cores'. Fig. 6.8 shows the ion 'cores' of a metal (consisting of the nucleus and the non-valence electrons) surrounded by mobile electrons. The cohesion of a metallic crystal is due to the attraction of the positive nuclei and the valence electrons *passing between them.*

The metallic bond is comparatively weaker than the ionic and covalent bonds which are called *saturated bonds.*

6·5. Molecular Bonds

This bonding occurs for those elements or compounds whose electron configuration is such that little electron transfer takes place between their atoms. Consequently, in their case, the three primary bonds considered above cannot be formed. Molecular bonds are formed by the weak Van der Waals' forces of attraction which exist between various atoms as explained below.

In a noble gas like helium, the outer shell with its two 2*s* electrons is complete. Similarly, other noble gases like neon and argon have full compliment of 8 electrons in their outermost shells. In such stable situations, the primary bonds (discussed earlier) cannot be formed because they all require adjustment of valence electrons which is not possible in the case of noble gases. Hence, atoms of all noble gases have very little attraction for each other and consequently they remain monoatomic under ordinary temperature and pressure conditions. These gases condense only at extremely low temperatures when the thermal agitation of their atoms is considerably reduced. Obviously, the condensation could not have taken place if there were no interatomic forces, however, weak. These interatomic forces of attraction are called Van der Waals' forces. Their importance lies in the fact that in cases like the one considered above, they happen to be the only forces which operate in the process of interatomic bonding.

Van der Walls' forces are due to the electrostatic attraction between the nucleus of one atom and the electrons of the other. This is largely but not completely neutralized by the electrostatic repulsion of electrons of one atom by the electrons of the other and repulsion of the nucleus of one atom by the nucleus of the other. The resultant weak attraction between the two atoms is called Van der Waals' force.

6·6. Characteristics of Ionic Shells

Such solids are found to possess the following characteristics :

(*i*) Crystal Structure

As revealed by X-ray diffraction patterns, such solids usually have crystalline structure. Moreover, their compounds are *ions* not molecules. For example, in a NaCl crystal, each Na^+ ion is surrounded by 6 evenly-spaced negative Cl^- ions and every Cl^- ion is surrounded by $6Na^+$ ions. Such solids are hard and brittle.

(*ii*) Melting and Boiling Points

Powerful electrostatic forces exist between the ions in the crystal lattice of an ionic solid. Consequently, considerable energy is required to overcome these forces and break down the crystal lattice. Hence, such solids have high melting and boiling points.

(*iii*) Electrical Conductivity

Ionic solids are good insulators. There is no electronic conduction because the electrons are tightly bound in the filled shells of the two types of ions involved. However, in ionic crystals, there is some charge transport by the *charged ions themselves*. In other words, ions are the charge carriers (not electrons as is the case with metals). Since ionic mobility increases with temperature, the electrical conductivity of ionic solids also increases with temperature (unlike metals whose electrical conductivity decreases with rise in temperature).

(*iv*) Solubility

Ionic solids are readily soluble in solvents like water (H_2O) and liquid ammonia (NH_3) because their molecules interact strongly with the crystal ions. Dissolution is also favoured by the decrease of the electrostatic force between crystal ions because of the high permittivity of the solvents.

(*v*) Ionic Reactions

Ionic reactions are practically *instantaneous* (*i.e.* precipitation). For example, Cl^- ions present in NaCl and $BaCl_2$ give a white precipitate of AgCl as soon as Ag^+ ions in the form of $AgNO_3$ solution are mixed.

6·7. Characteristics of Covalent Solids

Some of the characteristics of such solids are as under :

(*i*) Physical State

Covalent compounds may be solids, liquids or gases. At normal temperature and pressure, they exist as gases or solids. For example, chlorine (molecular weight = 71) is gas, bromine (molecular weight = 160) is liquid whereas iodine (molecular weight = 254) is a solid.

(*ii*) Crystal Structure

The crystals of covalent compounds are of three types :

(*a*) those in which molecules are small and held together by weak forces. Such crystals are soft and easily fusible. Examples are : sulphur and iodine.

(*b*) those in which each atom is united with the other by covalent links resulting in the formation of *giant* molecules. Examples are : diamond and silicon carbide. In diamond, each atom is united by covalent bonds with four neighbouring carbon atoms held at the corners of a regular tetrahedron.

(*c*) those which consist of separate *layers* such as graphite. Here, carbon atoms are arranged in regular hexagons in flat parallel layers, such that each atom is linked by the neighbouring atoms. However, there is *no strong bonding between different layers*, which are, therefore, easily separable from each other. This is the cause of softness and lubricating action of graphite.

(iii) Melting and Boiling Points

Since covalent bonds are not as powerful as ionic bonds, the covalent compounds have comparatively low melting and boiling points.

(iv) Electrical Conductivity

All covalent crystals are basically insulators because ordinarily no electrons are available for conduction. However, some covalent crystals like those of diamond are excellent insulators whereas others like germanium (Ge) are semiconductors. The conductivities of such conductors can be increased considerably by suitable doping (Art. 7.17).

(v) Solubility

Covalent compounds are soluble in non-polar solvents such as benzene and carbon tetrachloride.

6·8. Metallic Compounds

(*i*) They have a crystalline structure.

(*ii*) Their electrical conductivity is extremely high due to the availability of unlimited number of free electrons.

(*iii*) Since metallic bonds are not very strong, such solids have usually moderate to high melting temperatures.

(*iv*) They have high thermal conductivity which is again due to the presence of free electrons which serve as carriers of heat.

(*v*) Since free electrons in a metal absorb light energy, all metals are opaque to light.

6·9. Molecular Solids

(*i*) Molecular structures can be both crystalline and non-crystalline. Their structural units are molecules. Moreover, as these crystals are formed of elements with low atomic numbers, they have low densities.

(*ii*) Because of weak molecular bonds arising out of Van der Waals' forces, such solids have low melting points.

(*iii*) Because in such solids no valence electrons are available, they are good insulators.

(*iv*) They are usually transparent to light.

(*v*) They are soluble in both polar and non-polar liquids.

HIGHLIGHTS

1. Ionic bonds are formed due to electrostatic attraction between two stable atoms which become positive and negative ions because of complete transfer of electrons between them. Such bonds are sufficiently strong.
2. Covalent bonds are formed between two atoms when they share one or more pair of electrons between them.
3. Metallic bonds result from the sharing of variable number of electrons by a variable number of atoms.
4. Molecular bonds are formed by weak Van der Waals' forces which exist between two atoms.
5. Ionic solids are crystalline, have high melting points, are insulators and easily soluble in water and liquid ammonia.
6. Covalent solids may exist in a gaseous, liquid or solid state, have comparatively low melting points, are semiconductors and are soluble in non-polar solvents.

7. Metallic compounds have crystalline structure, are very good conductors of electricity, have moderate melting points, have high thermal conductivity and are opaque to light.
8. Molecular solids may be crystalline or non-crystalline, have low melting points, are insulators, are transparent to light and are easily soluble.

Can You Answer This ?

1. What are bonds in a solid ?
2. How are metallic bonds explained ? In what way do metallic compounds differ from ionic compounds ?
3. (*a*) What are primary and secondary bonds in a solid ? Explain each in detail.

   (*b*) How are ionic compounds compared with covalent compounds ?
4. Write note on bonds in solids.
5. Discuss the various types of bondings in crystals giving examples.
6. (*a*) Explain the following terms with examples :

   (*i*) covalent bond and (*ii*) ionic bond.

   (*b*) State four important distinctions between ionic and covalent compounds.
7. (*a*) Give a qualitative explanation for the formation of bonds in solids.

   (*b*) How can one explain the electrical properties of a metal, semiconductor and insulator in terms of their bond picture.
8. Explain : (*i*) ionic and (*ii*) covalent bonding in solids.
9. (*a*) Explain the atomic bonds, their nature and classification.

   (*b*) Deduce the expression for bond energy in general.
10. (*a*) Explain the nature of ionic bond.

    (*b*) Enumerate the typical properties of ionic crystals.
11. Write a short note on ionic and covalent bonding.
12. Explain the various types of bondings in crystals. Illustrate with examples.
13. Explain the type of bonding in a pure germanium crystal. What physical properties are associated with it ?
14. Distinguish between covalent and ionic bonds in solids. Illustrate with examples.

CLASSIFICATION OF SOLIDS 7

Energy Bands in Solids — Spacing Between Energy Levels of a Band — Energy Bands in Lithium — Valence and Conduction Bands — Conductors, Semiconductors and Insulators — Model Energy Diagrams for Typical Metals — Energy Diagram of Group IV Elements — Electron Conductivity of a Metal — Atomic Bonding in Semiconductors — Intrinsic Semiconductors — Hole Formation in Semiconductors — Fermi Level in an Intrinsic Semiconductor — Extrinsic Semiconductors — Insulators and Their Uses — Electron Conductivity of a Metal — Current Carriers in Semiconductors — Conductivity When Intrinsic Charge Carrier Densities are Neglected — Conductivity of Pure and P-type Germanium — P-N Junction Diode — Depletion Layer — Junction Voltage — Junction or Diffusion Capacitance — Preparation of Semiconductor Materials — Production of P- and N-type Crystals — Junction Triode Transistors — Working of P-N-P Transistor in Common-base Configuration — Working of a N-P-N Transistor in Common-base Configuration — Highlights — Can You Answer This ?

7.1. Energy Bands in Solids

In the case of an isolated single atom, there are single energy levels as shown for a hydrogen atom in Fig. 3.12. But there are significant changes in the energy levels when atoms are brought close together as in solids.

It is found that each of the energy levels of an atom splits into N levels of energy where N is the number of atoms in the crystal. Each original energy level becomes a *band* of very closely-spaced levels of slightly different energy. The individual energies within the band are so close together that, for many purposes, the energy band may be considered to be continuous.

Fig. 7.1 (*a*) shows the splitting of K, L and M levels as the distance between different atoms is reduced. At first, only valence level or M-level is affected as shown by dotted vertical line marked A, then as separation is reduced, inner shells also become affected as indicated by dotted vertical line B.

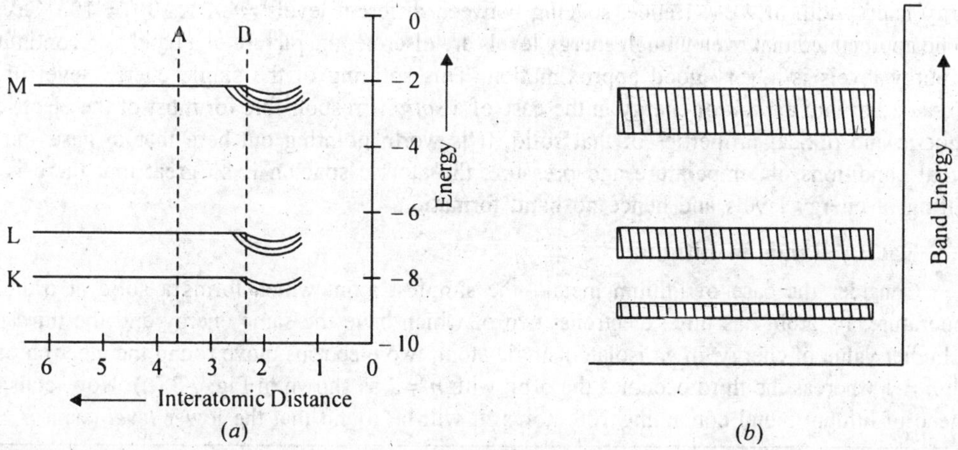

Fig. 7.1

Consider the case of Na crystal which consists of an ordered array of many closely-packed sodium atoms usually referred to as crystal lattice. Each Na atom has 11 electrons arranged in different shells and sub-shells as shown in Fig. 7.2 (*a*). As seen, the 1*s*, 2*s* and 2*p* sub-shells are filled but 3*s* sub-shell is incomplete and could hold one more electron. The electrons in each sub-shell occupy specific energy levels as shown in Fig. 7.2 (*b*)*.

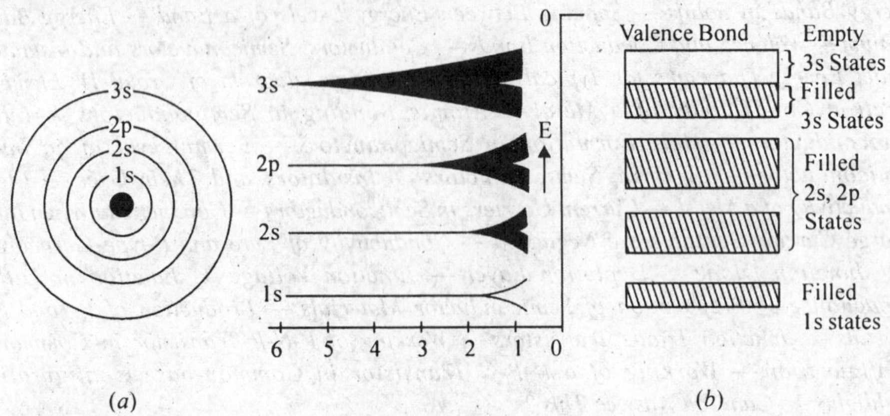

Fig. 7.2

For a small sodium crystal containing 10^{20} atoms, the band formed by splitting of *s*-subshell will have 2×10^{20} levels of slightly different energy because the *s*-subshell has 2 electronic levels (one with spin up and the other with spin down). Similarly, in a *p*-band there will be 6×10^{20} closely-packed levels because there are 6 electrons in a filled *p*-subshell.

In general, in an assembly of *N* atoms, the number of possible energy states is *N*. Since only two electrons of opposite spin can occupy the same state (as per Pauli's Exclusion Principle discussed earlier), the maximum number of electrons which these *N* states can occupy is 2*N*.

7.2. Spacing Between Energy Levels of a Band

It will be quite interesting as well as instructive to calculate the spacing between different energy levels in an energy band. A crystal weighing one milligram contains about 10^{19} atoms. If we assume the valency band to be a *s*-band, it will contain 2×10^{19} levels. Suppose the width of the energy band is 2 eV. Then, it is obvious that 2×10^{19} levels per milligram spread over an energy band width of 2 eV. Hence, spacing between different levels $= 2/2 \times 10^{19} = 10^{-19}$ eV. It will be appreciated that even though energy levels are discrete, the picture of a band as a continuum of energy levels is a very good approximation. This splitting of the single energy level of an *isolated* atom into a *band* of energy in the case of a *solid* is responsible for most of the electrical, magnetic and optical properties of that solid. It is worth pointing out here that in gases under normal conditions of temperature and pressure, the atomic spacing is so great that there is no splitting of energy levels and hence no band formation.

7.3. Energy Bands in Lithium

Consider the case of lithium metal—the simplest atom which forms a solid at ordinary temperature. Its atom has three electrons, two of which have the same energy and the third one has higher value of energy. In an isolated single atom, two electrons move round the electron orbit with $n = 1$ whereas the third occupies the orbit with $n = 2$ as shown in Fig. 7.3 (*a*). Now, consider a piece of lithium metal containing 100 atoms. It will be found that the lower *level* (with $n = 1$)

*The stability of crystal requires that the energies of these levels pass through a minimum and then rise again when the interatomic distance is further reduced as shown in Fig. 7.1 (*a*) but here they are shown simply branching off in a brush-like fashion.

forms a *band* of 200 electrons occupying 100 different energy states. The higher level (with $n = 2$) forms a wider band also of 100 energy states which could, as before, accommodate 200 electrons. But as there are only 100 electrons available (one from each atom), this energy band remains half-filled [Fig. 7.3 (*b*)].

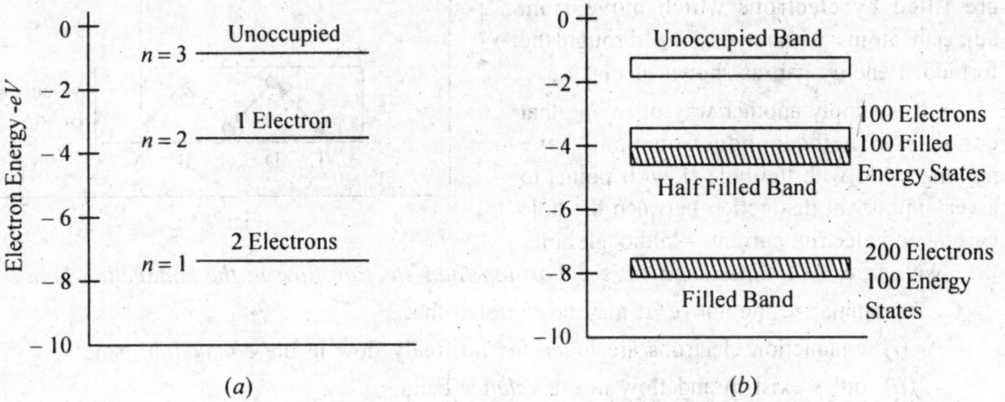

Fig. 7.3

7.4. Valence and Conduction Bands

The outermost electrons of an atom *i.e.* those in the shell furthermost from the nucleus are called *valence* electrons and have the *highest energy**. It is these electrons which are most affected when a number of atoms are brought very close together as during the formation of a solid. The states of lower-energy electrons orbiting in shells nearer to the nucleus are little, if at all, affected by this atomic proximity.

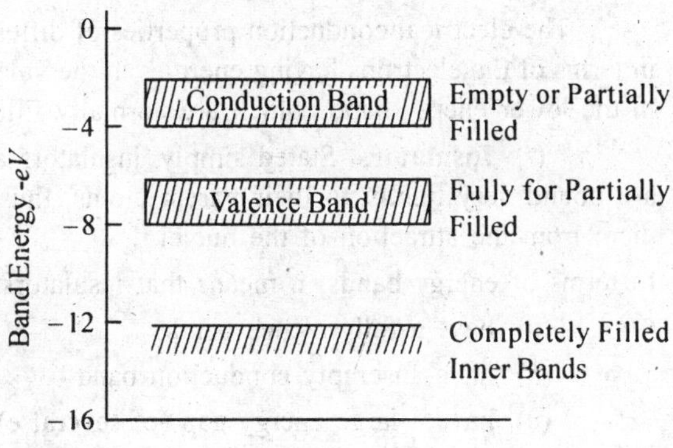

Fig. 7.4

The band of energy occupied by the valence electrons is called the *valence band* and is, obviously, the *highest occupied band.* It may be completely filled or partially filled with electrons.

The next higher permitted energy band is called the *conduction* band and may either *be empty or partially filled* with electrons. In fact, it may be defined *as the lowest unfilled energy band.*

In conduction band, electrons can move freely and hence are known as *conduction* electrons. The gap between these two bands is known as the forbidden energy gap.

It may be noted that the covalent forces of the crystal lattice have their source in the valence band. If a valence electron happens to absorb enough energy, it jumps across the forbidden energy gap and enters the conduction band. An electron in the conduction band can jump to an adjacent conduction band more readily than it can jump back to the valence band from where it had come earlier. However, if a conduction electron happens to radiate too much energy, it will suddenly reappear in the valence band once again.

When an electron is ejected from the valence band, a covalent bond is broken and positively-charged hole is left behind. This hole can travel to an adjacent atom by acquiring an

*Algebraically — not numerically.

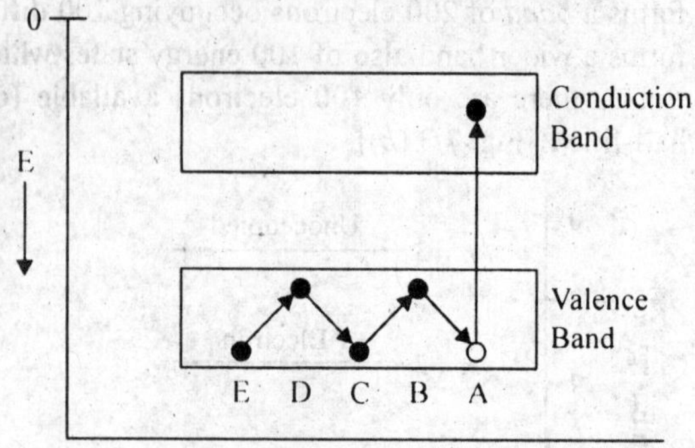

Fig. 7.5

electron from that atom which involves breaking an existing covalent band and then re-establishing a covalent bond by filling up the hole. It is to be noted carefully that holes are filled by electrons which move from adjacent atoms without passing through the forbidden energy gap as shown in Fig. 7.5.

It is simply another way of saying that conditions in the conduction band have nothing to do with the hole flow. It points to a very important distinction between the hole current and electron current — although holes flow with case, they *experience more opposition than electron flow in the conduction band.*

To summarize the above, it may be repeated that :

(*i*) conduction electrons are found in and freely flow in the *conduction* band

(*ii*) holes exist in and flow in the *valence* band

(*iii*) conduction electrons move almost twice as fast as the holes.

7.5. Conductors, Semiconductors and Insulators

The electrical conduction properties of different elements and compounds can be explained in terms of the electrons having energies in the valence and conduction bands. The electrons lying in the lower energy band, which are normally filled, play no part in the conduction process.

(*i*) **Insulators.** Stated simply, insulators are those materials in which valence electrons are bound very tightly to their parent atoms, thus requiring very large electric field to remove them from the attraction of the nuclei.

In terms of energy bands, it means that insulators

(*a*) have a full valence band

(*b*) have an empty conduction band

(*c*) have a large energy gap (of several eV) between them and

(*d*) at all ordinary temperatures, the probability of electrons from full valence band gaining sufficient energy so as the surmount energy gap and becoming available for conduction in the conduction band is light.

This is shown in Fig. 7.6 (*a*). For conduction to take place, electrons must be given sufficient energy to jump from the valence band to the conduction band. Increase in temperature enables some electrons to go to the conduction band which fact accounts for the negative resistance-temperature coefficient of insulators.

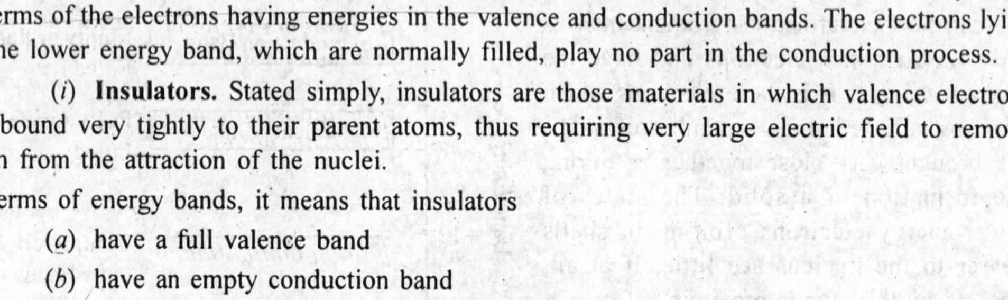

Fig. 7.6

(*ii*) **Conductors.** Put in a simple way, conducting materials are those in which plenty of free electrons are available for electric conduction.

In terms of energy bands, it means that electrical conductors are those which have overlapping valence and conduction bands as shown in Fig. 7.6 (*b*).

In fact, there is no physical distinction *between the two bands*. Hence, the availability of a large number of conduction electrons.

Another point worth nothing is that in the absence of forbidden energy gap in good conductors, there is no structure to establish holes. *The total current in such conductors is simply a flow of electrons*. It is exactly for this reason that the existence of holes was not discovered until semiconductors were studied thoroughly.

(*iii*) **Semiconductors.** A semiconductor material is one whose electrical properties lie in between those of insulators and good conductors. Examples are : germanium and silicon.

In terms of energy bands, semiconductors can be defined as those materials which have almost an empty conduction band and almost filled valence band with a very narrow energy gap (of the order of 1 eV) separating the two.

At 0°K, there are no electrons in the conduction band and the valence band is completely filled. However, with increase in temperature, width of the forbidden energy band is decreased so that some of the electrons are liberated into the conduction band. In other words, conductivity of semiconductors increases with temperature. Moreover, such departing electrons leave behind positive holes in the valence band (Fig. 7.5). Hence, semiconductor current is the sum of electron and hole currents flowing in opposite directions (Art. 7.15).

Semiconductors are of two types :

(*a*) pure or intrinsic semiconductors and

(*b*) impure or extrinsic semiconductors.

7.6. Model Energy Diagrams for Typical Metals

As said earlier, metals are those solids which have vacant electron energy states *immediately* above the highest-filled level of the valence band. Now, this can happen in the following two ways :

(*i*) In the first case, the valence band is only partially-filled as shown in Fig. 7.7 (*a*). When an electric field is applied, electrons can acquire sufficient energy and hence velocity so as to jump to higher levels. Monovalent metals such as alkali metals (Li, Na and K) have one electron per atom in their outermost sub-shell. Consequently, their outermost bands are half-filled. So is the case with trivalent elements like Al etc.

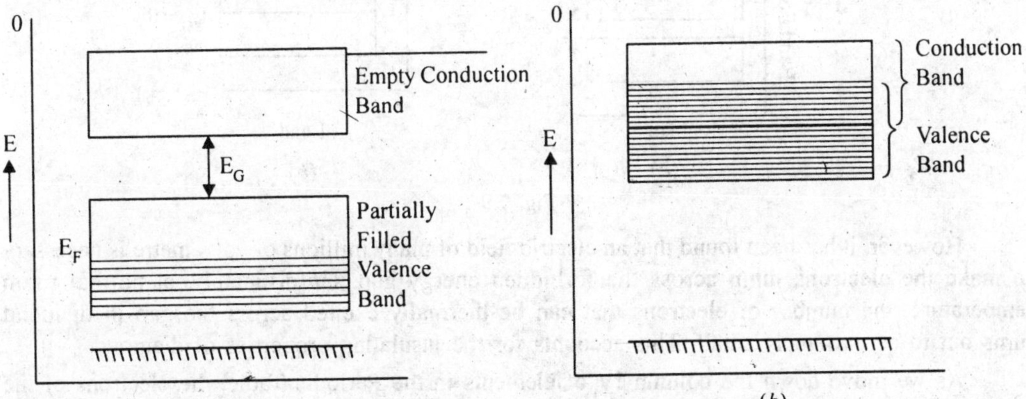

Fig. 7.7

(*ii*) In the second case, a full valence band overlaps conduction band as shown in Fig. 7.7 (*b*). Obviously, here the forbidden energy gap is zero. Examples of such elements are Mg and Be which conduct *even if their valence bands are full.*

7.7. Energy Diagrams of Group IV Elements

Some of the elements which belong to this group are carbon, silicon, germanium, tin and lead. The electrons in their outermost orbits are *even* in number as in divalent metals mentioned above. But usually there is no overlap between their valence and conduction bands. In the case of diamond (a form of carbon), the energy gap between the two bands is very large (nearly 6 eV) as shown in Fig. 7.8 (*a*).

The only way electrons can conduct electric current is by transition across the forbidden gap into the conduction band. This can be brought about in the following two ways :

(*i*) by applying a very strong electric field or

(*ii*) by thermal excitation.

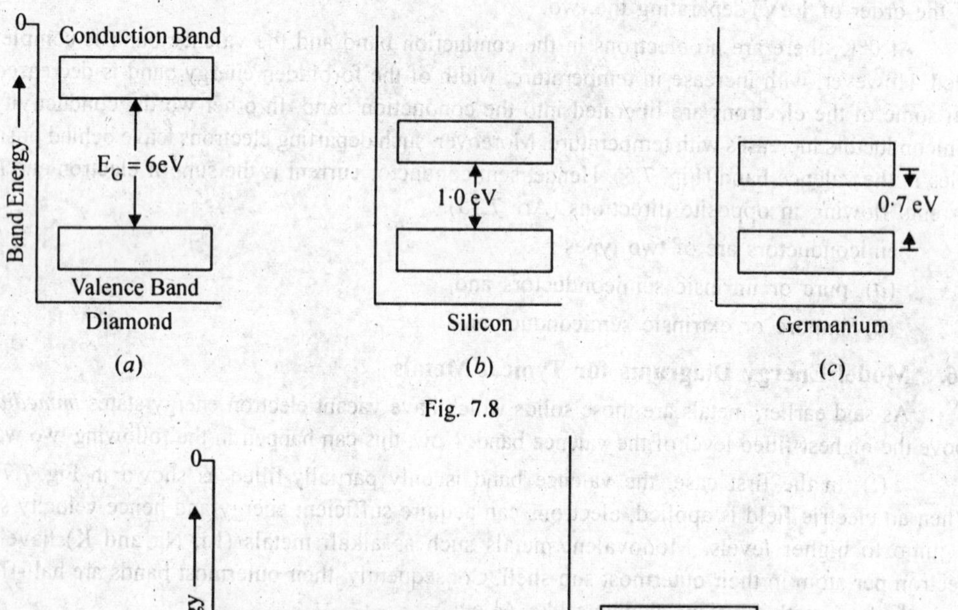

Fig. 7.8

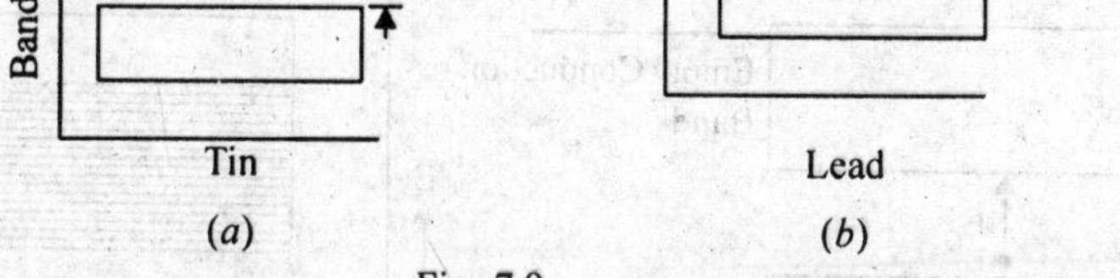

Fig. 7.9

However, it has been found that an electric field of many millions of volts/metre is necessary to make the electrons jump across the forbidden energy gap. Unfortunately, at normal room temperature, the number of electrons that can be thermally excited across the gap in diamond turns out to be extremely small. This accounts for the insulating properties of diamond.

As we move down the column IV of elements in the Periodic Table, the electrons of the outermost shells go farther away from the nucleus. This reduces the force of attraction between these electrons and the nucleus. This weakening effect predominates over the effect of increasing charge on the nucleus with the result that the electrons become less tightly-bound to their nuclei in the case of those elements. It means that energy gap between the valence and conduction bands is reduced. As shown in Fig. 7.8 and Fig. 7.9, it is reduced from 6 eV for diamond to 1.0 eV in silicon, 0.7 eV in germanium and a mere 0.08 eV in grey tin. In lead, the forbidden energy gap becomes zero [Fig. 7.9 (*b*)].

7.8. Atomic Binding in Semiconductors

Semiconductors like germanium and silicon, have crystalline structure. Their atoms are arranged in an ordered array known as crystal lattice. Both these materials are tetravalent *i.e.* each has four valence electrons in its outermost shell.* The neighbouring atoms form *covalent* bonds by sharing four electrons with each other so as to achieve inert gas structure (*i.e.* 8 electrons in the outermost orbit). A two-dimensional view of the germanium crystal lattice is shown in Fig. 7.10 (*b*) in which circles represent atom cores consisting of the nuclei and inner 28 electrons. Each pair of lines represents a covalent band. The dots represent the valence electrons. It is seen that each atom has 8 electrons under its influence.

A 3-dimensional view of germanium crystal lattice is shown in Fig. 7.10 (*a*) where each atom is surrounded symmetrically by four other atoms forming a tetrahedral crystal. Each atom shares a valence electron with each of its four neighbours, thereby forming a stable structure.

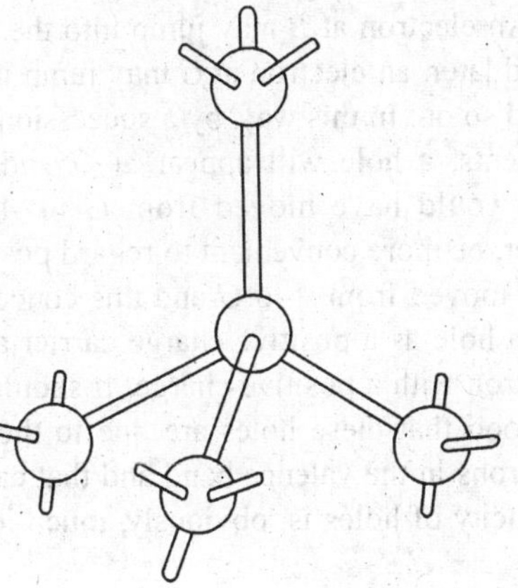

(*a*)

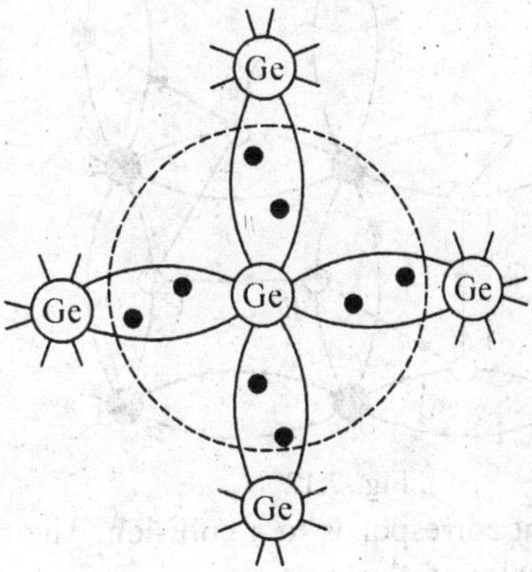

(*b*)

Fig. 7.10

In the case of pure (intrinsic) germanium, the covalent bonds have to be broken to provide electrons for conduction. There are many ways of rupturing the covalent bond and hence setting the electron free. One way is to increase the crystal temperature above 0°K.

It may be noted that covalent crystals are characterised by their hardness and brittleness. Their brittleness is due to the fact that in such crystals, adjacent atoms must remain in accurate alignment since the bond is strongly directional and formed along a lines joining the atoms. The hardness is due to the great strength of the covalent bond itself.

7.9. Intrinsic Semiconductors

An intrinsic semiconductor is *one which is made of the semiconductor material in its extremely pure form.*

---

*A single germanium atom has 32 electrons out of which only four electrons take part in electrical properties of germanium, the remaining 28 electrons being tightly bound to the nucleus. The four electrons revolve in the outermost shell and are called valence electrons.

Common examples of such semiconductors are : pure germanium and silicon which have forbidden energy gaps of 0.72 eV and 1.1 eV respectively. The energy gap is so small that even at ordinary room temperature, there are many electrons which possess sufficient energy to jump across the small energy gap from the valence to the conduction band. However, it is worth noting that for each electron liberated into the conduction band, a positively-charged hole is created in the valence band. When an electric field is applied to an intrinsic semiconductor at a temperature greater than 0°K, conduction electrons move to the anode and the holes in the valence band move to the cathode. Hence semiconductor current consists of movement of electrons and holes in opposite directions.

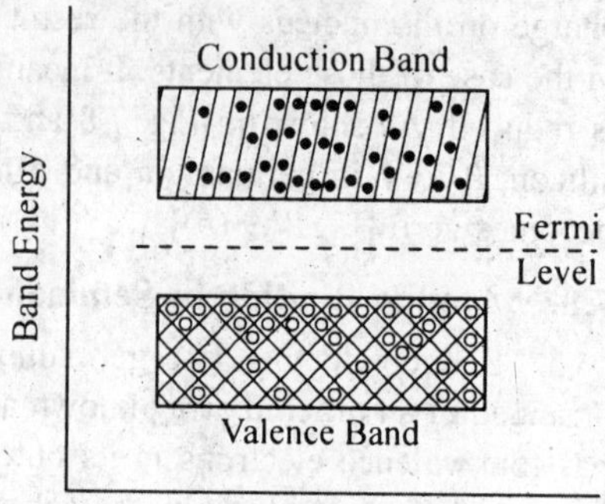

Fig. 7.11

7.10. Hole Formation in Semiconductors

The formation of '*hole*' which is a positive charge carrier is explained below :

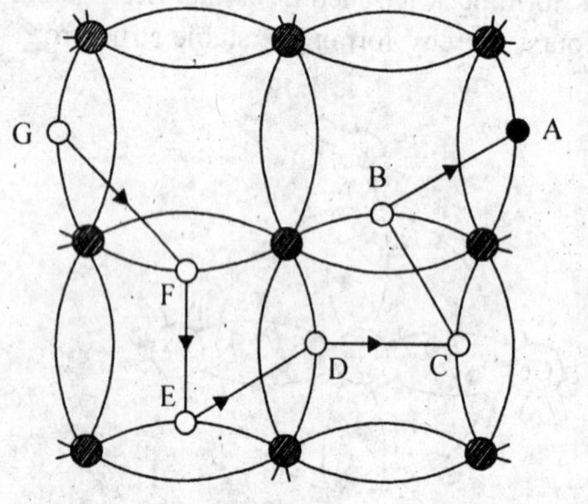

Fig. 7.12

As shown in Fig. 7.12 suppose the covalent bond is broken at *A* and the electron has moved through the crystal lattice leaving behind a hole in the covalent bond. An electron at *B* may jump into the vacant hole at *A* and later, an electron at *C* may jump into the hole at *B* and so on. In this way, by a succession of electron movements, a hole will appear at *G* and a negative charge would have moved from *G* to *A*. It would, however, be more convenient to regard positive charge to have moved from *A* to *G* and this conception gives rise to a hole as a positive charge carrier as if it were an electron with a positive charge. It should be clearly understood that these holes are due to the movement of electrons in the valence band and that each electron movement corresponds to a collision. The drift velocity of holes is, obviously, much less than the drift velocity of electrons.

Alternatively, an intrinsic semiconductor may be defined as one *in which the number of conduction electrons is equal to the number of holes.*

Schematic energy band diagram of an intrinsic semiconductor at room temperature is shown in Fig. 7.11. Only two bands *i.e.* valence and conduction bands have been shown since lower filled bands are not of any consequence. Here, Fermi level* lies exactly in the middle of the forbidden energy gap.

7.11. Fermi Level in an Intrinsic Semiconductor

It can be proved that in an intrinsic semiconductor, Fermi energy level E_F lies in the middle of the energy gap *i.e.* midway between the conduction and valence bands.

Let at any temperature T°K

n_c = No. of electrons in the conduction band

n_v = No. of electrons in the valence band

$N = n_c + n_v$

= No. of electrons in both bands

*For the present discussion, Fermi level may be defined as the energy which corresponds to the centre of gravity of conduction electrons and holes weighted according to their energies.

Further, let us make the following simplifying assumptions :

(*i*) Widths of energy bands are small as compared to forbidden energy gap between them

(*ii*) Since band widths are small, all levels in a band have the same energy.

(*iii*) Energy of all levels in valence band is zero as shown in Fig. 7.13.

(*iv*) Energy of all levels in conduction band is = E_g.

In Fig. 7.13, the zero-energy reference level has been arbitrarily taken at the top of valence band.

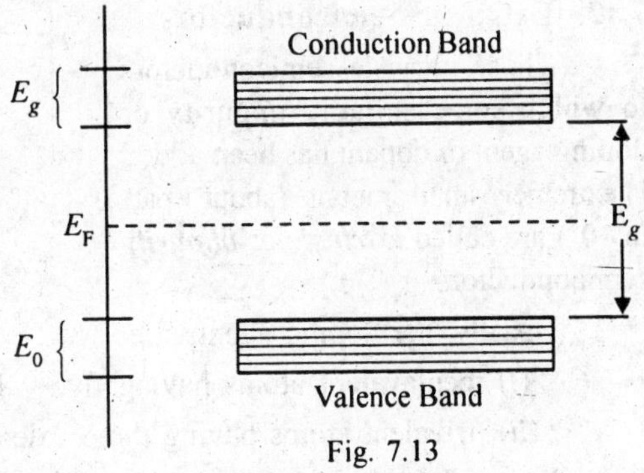

Fig. 7.13

Now, number of electrons in the conduction band is

$$n_c = N \cdot P(E_g)$$

where $P(E_g)$ is the probability of an electron having energy E_g. Its value can be found from Fermi-Dirac probability distribution function given by

$$P(E) = \frac{1}{1 + e^{(E - E_F)/kT}}$$

where $P(E)$ is the probability of finding an electron having any particular value of energy E and E_F is Fermi level.

$$P(E_g) = \frac{1}{1 + e^{(E_g - E_F)/kT}}$$

$$\therefore \quad n_c = \frac{N}{1 + e^{(E_g - E_F)/kT}}$$

Now, number of electrons in valence band is

$$n_v = N \cdot P(0)$$

The probability of an electron being found in the valence band with zero energy can again be calculated by putting $E = 0$ in the Fermi-Dirac probability distribution function.

$$\therefore \quad P(0) = \frac{1}{1 + e^{(0 - E_F)/kT}} = \frac{1}{1 - e^{-E_F/kT}}$$

$$\therefore \quad n_v = \frac{N}{1 + e^{E_F/kT}}$$

Now, $N = n_c + n_v$

or

$$N = \frac{N}{1 + e^{(E_g - E_F)/kT}} + \frac{N}{1 + e^{-E_F/kT}}$$

$$\therefore \quad 1 - \frac{1}{1 + e^{-E_F/kT}} = \frac{1}{1 + e^{(E_g - E_F)/kT}}$$

which on simplification gives

$$\boxed{E_F = \frac{E_g}{2}}$$

Hence, it shows that in an intrinsic semiconductor, the Fermi level lies midway between the conduction and valence bands. That this conclusion is physically plausible can be best seen with the help of Fig. 7.14.

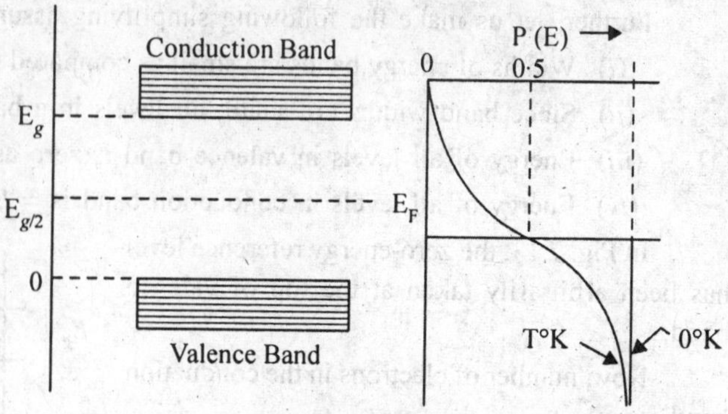

Fig. 7.14

7.12. Extrinsic Semiconductors

Those intrinsic semiconductors to which some suitable impurity or doping agent or dopant has been added in extremely small amounts (about 1 part in 10^8) are called *extrinsic* or *impurity* semiconductors.

Usually, the doping agents are

(*i*) pentavalent atoms having five valence electrons (arsenic, antimony, phosphorus) or

(*ii*) trivalent atoms having three valence electrons (gallium, indium, aluminium, boron).

Pentavalent doping atom is known as *donor* atom because it donates or contributes one electron to the conduction band of pure germanium. The trivalent atom, on the other hand, is called *acceptor* atom because it accepts one electron from the germanium atom.

Depending on the type of doping material used, extrinsic semiconductors can be divided into two classes :

(*i*) *N*-type semiconductors and (*ii*) *P*-type semiconductors.

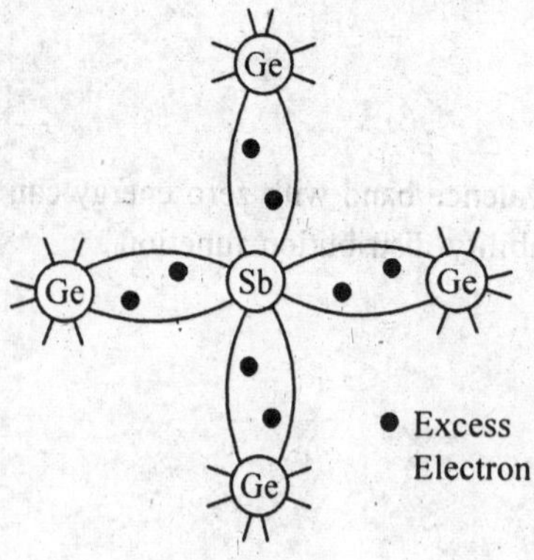

(*a*)

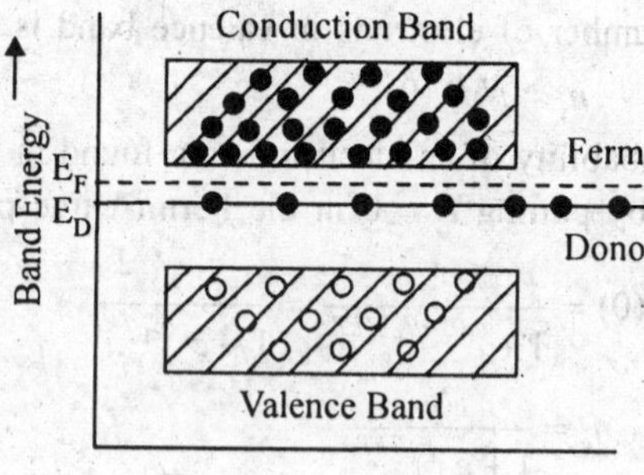

(*b*)

Fig. 7.15

(*a*) ***N*-type Extrinsic Semiconductor.** This type of semiconductor is obtained when a pentavalent material like antimony (Sb) is added to pure germanium crystal. As shown in Fig. 7.15 (*a*), each antimony atom forms covalent bonds with the surrounding four germanium atoms with the help of four of its five electrons. The fifth electron is superfluous and is loosely bound to the antimony atom. Hence, it can be easily excited from the valence band to the conduction band by the application of electric field or increase in thermal energy. Thus, practically every antimony atom introduced into the germanium lattice contributes one conduction electron *without creating a positive hole*. Antimony is called donor impurity and makes the pure germanium *N*-type (*N* for negative) extrinsic semiconductor.

It may be noted that by giving away its one valence electron, the donor atom becomes a positively-charged ion. But it cannot take part in conduction because it is firmly fixed or tied into the crystal lattice. As seen, in addition to the electrons and holes *intrinsically available in germanium*, the addition of antimony greatly increases the number of conduction electrons. Hence, concentration of electrons in the conduction band is increased and exceeds the concentration of holes in the valence band. Because of this, Fermi level shifts upwards towards the bottom of the conduction band as shown in Fig. 7.15 (*b*).*

In terms of energy levels, the fifth antimony electron has an energy level (called donor level) just below the conduction band. Usually, the donor level is 0·01 eV below conduction band for germanium and 0.054 eV for silicon.

It is seen from the above description that in an *N*-type semiconductor, electrons *are the majority carriers* while holes constitute the minority carriers. Hence, *N*-type semiconductor conducts principally by electrons in the nearly empty conduction band and the process is called '*excess*' conduction.

Another point worth noting is that even though *N*-type semiconductor has excess of electrons, still it is electrically neutral. It is so because by addition of donor impurity, number of electrons available for conduction purposes becomes more than the number of holes available intrinsically. But the total charge of the semiconductor does not change because the donor impurity brings in as much negative charge (by way of electrons) as positive charge (by way of protons).

(*ii*) ***P*-type Extrinsic Semiconductor.** This type of semiconductor is obtained when traces of a trivalent impurity like boron (B) are added to a pure germanium crystal.

In this case, the three valence electrons of boron atom form covalent bonds with four surroundings germanium atoms but one bond is left incomplete and gives rise to a hole as shown in Fig. 7.16 (*a*).

Thus, boron which is called an acceptor impurity causes as many positive holes in a germanium crystal as there are boron atoms thereby producing a *P*-type (*P* for positive) extrinsic semiconductor.

In this type of semiconductor, conduction is by means of holes in the valence band. Accordingly, holes form the majority carriers *whereas electrons constitute minority carriers*. The process of conduction is called 'defect' or 'deficit' conduction.

Since concentration of holes in the valence band is more than the concentration of electrons in the conduction band, Fermi level shifts nearer to the valence band [Fig. 7.16 (*b*)] The acceptor level lies immediately above the Fermi level. Conduction is by means of hole movement of the top of valence band, the acceptor level readily accepting electrons from the valence band.

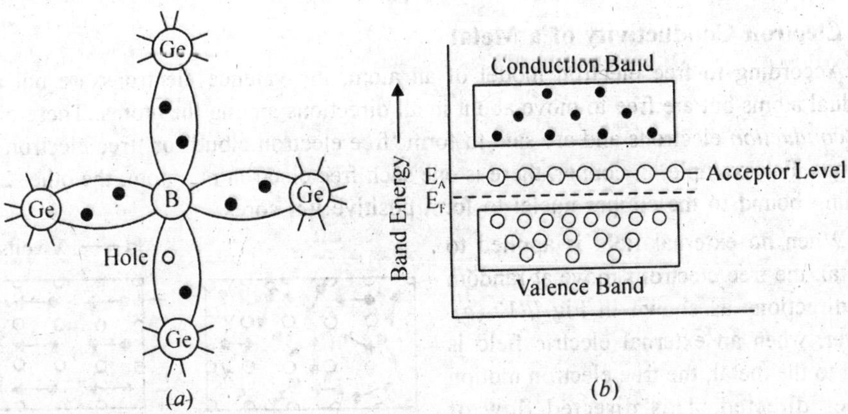

Fig. 7.16

*Since the number of electrons as compared to the number of holes increases with temperature, the position of Fermi-level also changes considerably with temperature.

Again it may be noted that even though *P*-type semiconductor has excess of holes for conduction purposes, as a whole it is electrically neutral for the same reasons as given above.

7.13. Insulators — Their Properties and Uses

Insulators are those materials which, on application of a potential difference pass such extremely small current that it is generally considered negligible. More comprehensive definition has already been given in Art. 7.5.

The main function of insulators is to confine the path of electric current along conductors. It is achieved either by covering these conductors or else by supporting them by an insulating material.

The different properties desired in a good insulation material are as under:

(*a*) permanence — it is the most important quality but one least easily attained.

(*b*) high dielectric strength *i.e.* ability to resist electrical breakdown under considerable electrical stress.

(*c*) mechanical strength and

(*d*) fairly high insulation resistance.

Some of the commonly-used insulators along with their special fields of application are given below:

(*i*) Asbestos — for covering wires in high-power machines.

(*ii*) Bitumen (vulcanised) — for low-voltage mining cables and also as a cable-box filling compound.

(*iii*) Cotton — for covering electric wires.

(*iv*) Ebonite — for making covers for resistance boxes.

(*v*) Empire cloth — for wrapping armature coils.

(*vi*) Gutta-percha — for covering submarine cables.

(*vii*) Micanite — for insulating commutator segments and as slot lining for high-voltage machines.

(*viii*) Paper and paraffin wax — for cable insulation when oil impregnated and for covering transformer conductors.

(*ix*) Porcelain — for insulators and use in overhead transmission lines.

(*x*) Shellac — in the form of insulating varnish.

7.14. Electron Conductivity of a Metal

According to free electron model of an atom, the valence electrons are not attached to individual atoms but are free to move about in all directions among the atoms. These electrons are called *conduction* electrons and are said to form 'free electron cloud' or 'free electron gas' or the Fermi gas. For example, in copper, there is one such free electron per atom, the other 28 electrons remaining bound to the copper nuclei to form positive ion cores.

When no external field is applied to the metal, the free electrons move at random in all directions as shown in Fig. 7.17 (*a*). However, when an external electric field is applied to the metal, the free electron motion becomes directed. This directed flow of electrons results in a net charge displacement in a definite direction. This type of motion is known as *drift* and the phenomenon is

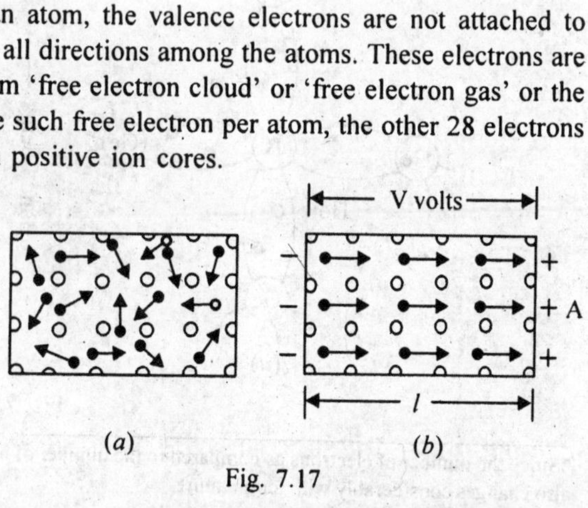

Fig. 7.17

referred to as process of conduction by drift charge. The drift velocity (v) of the electrons is dependent upon the electron mobility (μ_e) and the applied electric field E. The actual relation is $v = \mu_e E$.

Let, e = electron charge

v = electron drift velocity

A = conductor cross-section

n = number of free electrons per unit volume of the conductor *i.e.* electron density

l = length of the conductor

V = voltage applied across the two ends of the conductor

E = applied electric field

$= V/l$

Now, electric current flowing in any conductor is given by the amount of charge which flows in one second across any plane of the conductor (Fig. 7.18). The *total* number of electrons which cross the plane P of cross-section A in one second

$= n \times (v \times A)$

Charge carried by them is

$= env\ A$

$\therefore \quad I = env\ A$

Substituting the value of v, we get

$I = ne\ A\mu_e\ E = nAe\mu_e \cdot V/l$

$$\therefore \quad R = \frac{V}{I} = \frac{l}{A}\left(\frac{1}{ne\mu_e}\right) = \rho\frac{l}{A}$$

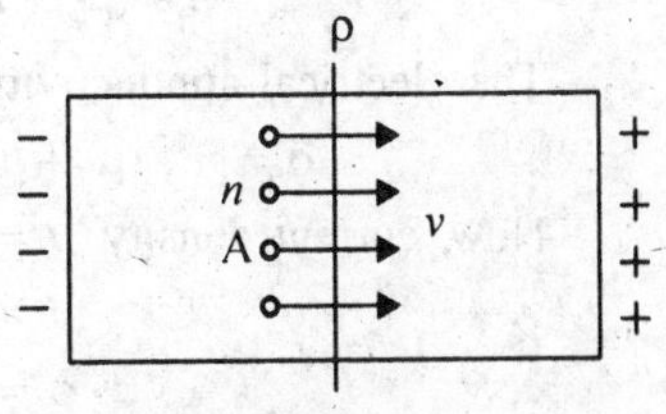

Fig. 7.18

$\therefore$ resistivity $\boxed{\rho = 1/ne\mu_e}$ ohm-m

conductivity $\boxed{\sigma = ne\mu_e}$ mho/m

Incidentally, it may be noted that conductivity of a semiconductor differs from that of a metal in one important respect *i.e.* in a semiconductor, charge carriers are both holes as well as electrons whereas in metals electrons are the only charge carriers (Art. 7.15).

7.15. Current Carriers in Semiconductors

We will consider the following cases separately :

(*a*) **Intrinsic Semiconductors.** In their case, current flow is due to the movement of electrons and holes in opposite directions. Even though the number of electrons equals the number of holes, hole mobility μ_h is practically half of electron mobility μ_e.

As shown in Fig. 7.19, the total current flow which is due to the sum of electron flow and hole flow is given by

$$I = I_e + I_h = n_1 e v_e A + p_1 e\ v_h\ A$$

where v_e = drift velocity of electrons

v_h = drift velocity of holes

n_1 = density of free electrons in an intrinsic semiconductor

p_1 = density of holes in an intrinsic semiconductor

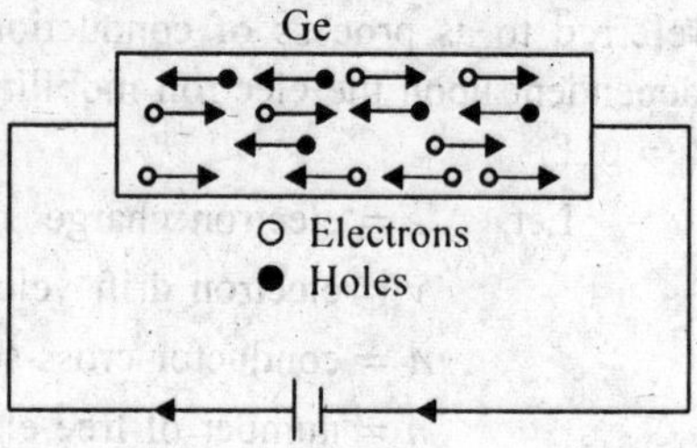

Fig. 7.19

e = electron charge

A = cross-section of the semiconductor

Since in an intrinsic semiconductor $n_1 = p_1$

$$I = n_1 e\,(v_e + v_h)\,A$$
$$= n_1 e\,(\mu_h + \mu_e) = EA$$

where μ_e = electron mobility = v_e/E

μ_h = hole mobility = v_h/E

E = electric field

Since $E = V/l$ where l is the length of the intrinsic semiconductor,

$$\therefore \quad I = n_1 e\,(\mu_h + \mu_e)\,AV/l$$

$$\text{Now,} \quad \frac{V}{I} = \frac{l}{A} \cdot \frac{1}{n_1 e\,(\mu_e + \mu_h)} = \rho\,\frac{l}{A}$$

where ρ is the resistivity of the semiconductor. It is given by

$$\rho = \frac{1}{n_1 e\,(\mu_e + u_h)}$$

The electrical conductivity σ which is the reciprocal of resistivity is given by

$$\sigma = n_1 e\,(\mu_e + \mu_h)$$

Now, current density $J = I/A$

$$\therefore \quad J = n_1 e\,(\mu_e + \mu_h)\,E = \sigma E$$

$$\therefore \quad \sigma = J/E$$

Obviously, conductivity of semiconductors depends on two factors (*i*) number of current carriers present per unit volume and (*ii*) the mobility of the current carriers.

(*b*) **Extrinsic Semiconductors.** The general expression for current density in the case of an extrinsic semiconductor when an electric field is employed is

$$J = (ne\mu_e + pe\mu_h)\,E \qquad \text{...}(i)$$

1. If it is *N*-type semiconductor, then the above expression becomes

$$J = e\,(n_n\mu_e + p_n\mu_h)\,E$$

where n_n and p_n represent the electron and hole densities in the *N*-type semiconductor after doping.

2. If it is a *P*-type semiconductor, then

$$J = e\,(n_p\mu_e + p_p\mu_h)\,E$$

where n_p and p_p represent similar quantities in a *P*-type semiconductor after doing.

The conductivity is given by $\sigma = J/E$

$$\therefore \quad \sigma = ne\mu_e + peu_h$$

or $\quad \sigma_n = e\,(n_n\mu_e + p_n\mu_h)$ — for *N*-type

and $\quad s_p = e\,(n_p m_e + p_p m_h)$ — for *P*-type

(*i*) In *N*-type semiconductors, electrons form the majority carriers although holes are also available as minority carriers.

The current density in such a conductor is given by Eq. (*i*) above. However, since electron density in such extrinsic semiconductors is much more than hole density *i.e.* $n_n \gg p_n$, the above expressions are simplified to

$$\boxed{J = n_n e\mu_e E \text{ and } \sigma_n = n_n e\mu_e} \qquad \text{...}(ii)$$

(*ii*) In *P*-type semiconductors, conduction is by means of holes in the valence band which form majority carriers in this case although electrons are available as minority carriers.

Since in such extrinsic semiconductors, $n_p \ll p_p$, the above expressions become

$$\boxed{J = p_v e \mu_h E \text{ and } \sigma_p = p_p e \mu_h} \quad ...(iii)$$

7.16. Conductivity When Intrinsic Charge Carrier Densities are Neglected

In case, density of charge carriers available intrinsically is negligible as compared to the added impurity atoms (whether of donor or acceptor type), then the formulae for conductivity in Eq. (*ii*) and (*iii*) above will be changed as follows :

(*a*) For *N*-type semiconductor

As seen from Eq. (*ii*) above, the conductivity is given by

$\sigma_n = n_n e \mu_e$

where n_n = electron density after doping.

In this relation, intrinsic hole density has already been neglected. This electron density is made up of two components :

(*i*) intrinsic electron density due to electrons available in *pure* semiconductor

(*ii*) electron density N_d contributed by the added donor impurity.

However, if we further neglect the intrinsic electron density, then

$$\boxed{\sigma_n = N_d e \mu_e}$$

Obviously, in this approximate formula, the conductivity solely depends on the density of electrons supplied by the donor impurity.

(*b*) For *P*-type semiconductor

As seen from Eq. (*iii*) above, the conductivity is given by

$$\sigma_p = p_p e \mu_h$$

where p_p = hole density after doping.

Again, in this relation, intrinsic electron density has been neglected.

This hole density again consists of two components :

(*i*) intrinsic hole density due to holes available intrinsically in pure semiconductor

(*ii*) hole density (N_a) contributed by the added acceptor impurity.

However, if we further neglect the intrinsic hole density, then

$$\boxed{\sigma_p = N_a e \mu_h}$$

7.17. Conductivity of Pure and *P*-type Germanium

As shown in Art. 7.15, the conductivity of pure germanium is given by

$$\sigma = n_1 e (\mu_e + \mu_h) = p_1 e (\mu_e + \mu_h)$$

when germanium is doped by a trivalent impurity like indium, it becomes a *P*-type semiconductor. After doping, its conductivity depends on the number of charge carriers available in it. The law of Mass Action can be used for finding this number. For acceptor impurity, the law may be stated as follows :

$$n_p p_p = n_1 p_1 = n_1^2 = p_1^2$$

where n_p and p_p represent the 'free' electron and hole densities respectively in the semiconductor *after* doping and n_1 and p_1 the electron and hole densities *before* doping *i.e.* in an intrinsic semiconductor. Put in another way, it simply means that, at constant temperature, the product of the number of electron carriers and the number of hole carriers is independent of the density of acceptor atoms. In physical terms, it means that the introduction of *P*-type impurity fills some of the electron levels produced by thermal action.

By calculating n_p and p_p from above and knowing μ_e and μ_h, conductivity after doping can be found out as illustrated by the following examples.

(*a*) **Pure Germanium**

$$\sigma = n_1 e\,(\mu_e + \mu_h)$$
$$= p_1 e\,(\mu_e + \mu_h)$$

Now,
$$n_1 = p_1 = 2 \times 10^{19} \text{ per m}^3$$
$$\mu_e = 0.36 \text{ m}^2/\text{V-s}$$
$$\mu_h = 0.17 \text{ m}^2/\text{V-s};\ e = 1.6 \times 10^{-19} \text{ C}$$
$$\sigma = 2 \times 10^{19} \times 1.6 \times 10^{-19}\,(0.36 + 0.17)$$
$$= 1.69 \text{ mho/m}$$

(*b*) ***P*-type Germanium**

$$\sigma_p = e\,(n_p\mu_e + p_p\mu_h)$$

Suppose, we add 10^{22} atoms/m^3 of indium and that $n_1 = 2 \times 10^{19}$ charge carriers (either electrons or holes) per m^3. Then

$$n_p p_p = n_1^2 = (2 \times 10^{19})^2 = 4 \times 10^{38} \text{ and } p_p - n_p = 10^{20}$$

$$\therefore \quad p_p - 4 \times 10^{38}/p_p = 10^{20}$$

$$p_p^2 - 10^{20}\, p_p - 4 \times 10^{38} = 0$$

Solving the above quadratic equation and taking positive value only,

$$p_p = 1.04 \times 10^{20} \text{ and } n_p = 0.04 \times 10^{20}$$

$$\therefore \quad \sigma_p = 1.6 \times 10^{-19}\,(0.04 \times 10^{20} \times 0.36 + 1.04 \times 10^{20} \times 0.17) = 3.1 \text{ mho/m}$$

It is seen that conductivity is almost doubled.

Example 7.1. *Calculate the energy gap in silicon if it is given that it is transparent to radiation of wavelength greater than 11,000 Å but it absorbs radiation of shorter wavelengths.*

Solution. Energy gap in a semiconductor represents the energy required to shift the conduction electron from top of valence band to bottom of conduction band. It represents minimum energy needed for electron shift. Obviously, photons of radiations with $\lambda > 11{,}000$ Å do not have this minimum energy and hence are not absorbed by the semiconductor for shifting electrons.

$$\therefore \quad E_g = 11{,}000/12{,}400 = \mathbf{0{\cdot}887\ eV}$$

Example 7.2. *Find the mobility of electrons in copper assuming that each atom contributes one free electron for conduction. For Cu, resistivity = 1.7×10^{-6} ohm-cm, density = 8·96 g/cc and atomic wt. = 63.5. Avogadro's number = 6.02×10^{23}/g-mol.*

Solution. $\rho = 1/\eta e \mu_e$ ...Art. 7.14

Since 63.5 gram contain 6.02×10^{23} atoms, 8.96 g contain $8.96 \times 6.02 \times 10^{23}/63.5 = 8.49 \times 10^{22}$ atoms. It represents the number of Cu atoms as well as electrons per cc *i.e.* $n = 8.49 \times 10^{22}$ per c.c.

$$\therefore \quad \mu_e = 1/1.7 \times 10^{-6} \times 1.6 \times 10^{-19} \times 8.49 \times 10^{22} = \mathbf{43.3\ cm^2/V\text{-}s.}$$

Example 7.3. *A specimen of pure germanium at 300°K has a density of charge carriers of $2.5 \times 10^{19}/m^3$. It is doped with donor impurity atoms at the rate of one impurity atom for every 10^6 atoms of germanium. All impurity atoms may be supposed to be ionized. The density of germanium atoms is 4.2×10^{28} atoms/m^3. Find the resistivity of the doped germanium if electron mobility is 0.36 m^2/V-s.*

Solution. Density of added impurity atoms

$$N_d = 4.2 \times 10^{28}/10^6 = 4.2 \times 10^{22} \text{ atoms/m}^3$$

As seen, it is extremely large as compared to the intrinsic charge carrier density of $2.5 \times 10^{19}/\text{m}^3$ which will, therefore, be neglected. Now, as seen from Art. 7.16.

$$\sigma_n = N_d e \mu_e$$
$$= 4.2 \times 10^{22} \times 1.6 \times 10^{-19} \times 0.36 = 2.42 \times 10^3 \text{ mho/m}$$
$$\therefore \quad \rho_n = 1/\sigma_n = 1/2.42 \times 10^3 = 0.413 \times 10^{-3} \ \Omega\text{-m}$$

Example 7.4. *In germanium the energy gap is about 0.75 eV. What is the wavelength at which germanium starts to absorb light?*

Solution. $\lambda = 12{,}400/0.75 =$ **16,533 Å**

7.18. *P-N* Junction Diode

(*i*) **Construction.** It consists of a piece of germanium, one-half of which is doped with *P*-type material and the other half with *N*-type material as shown in Fig. 7.20. The plane dividing the two zones is known as *junction.*

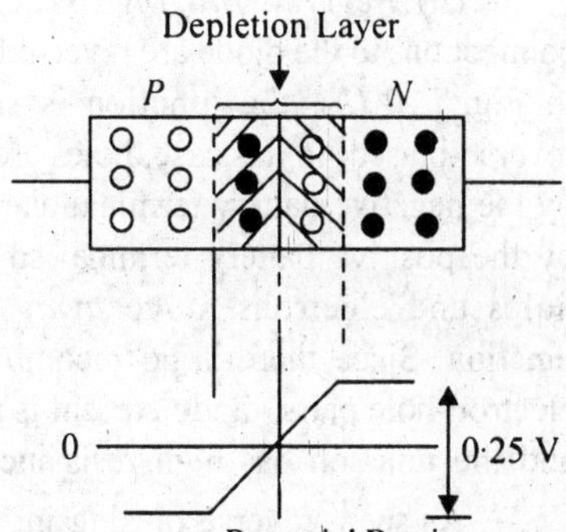

Fig. 7.20

Diffusion causes some electrons and holes to migrate across the junction thereby forming a depletion layer on either side of the junction by neutralization of holes in the *P*-region and of free electrons in the *N*-region. This diffusion of holes and electrons across the junction continues till a potential barrier is developed in the depletion layer which then prevents further diffusion.

This potential barrier or hill is either decreased or increased by applying an external voltage.

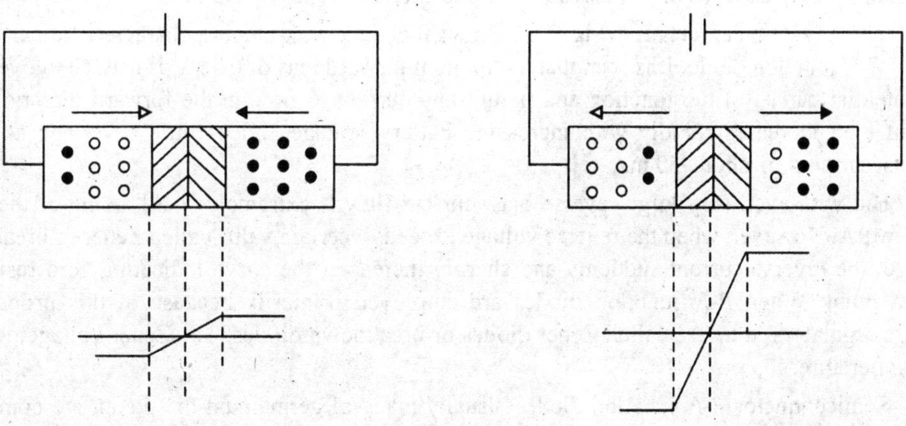

Fig. 7.21

(*ii*) **Rectifying Action.** Let us see what happens when an external voltage is applied across the diode terminals.

(*a*) *Forward Biasing*. Suppose the positive battery terminal is connected to the *P*-type region and the negative battery terminal to the *N*-type region as shown in Fig. 7.21 (*a*). When battery is so connected, the junction is said to be biased in the *forward* direction because it permits flow of current across the *P-N* junction.

It will be realized that holes will be repelled by the positive battery terminal and the electrons by the negative battery terminal with the result that both holes and electrons will be driven *towards* the junction where they will recombine. Hence, large current flows so long as the battery voltage is applied. Another way of looking at this process is that they forward bias lowers the potential barrier across the depletion layer thereby allowing more current to flow across the junction.

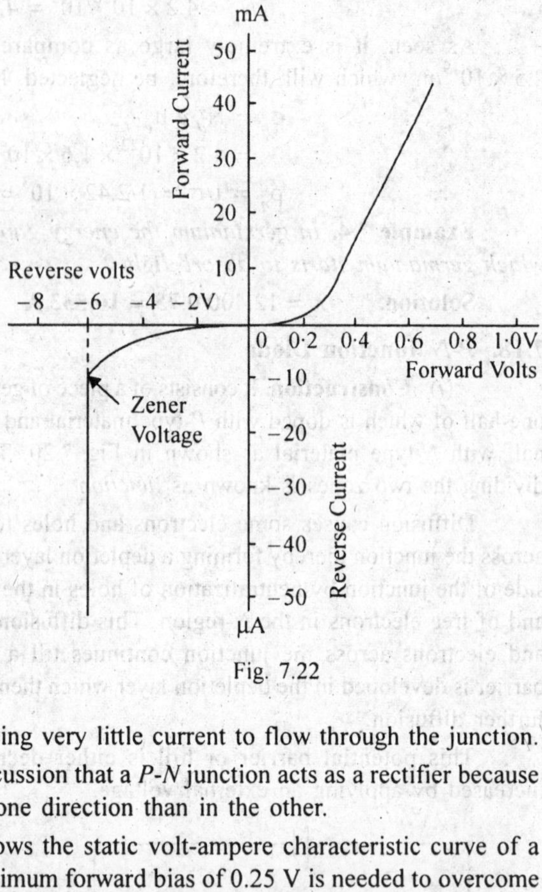

Fig. 7.22

(*b*) *Reverse Biasing*. When battery connections to the diode are reversed as shown in Fig. 7.21 (*b*), the junction is said to be reverse-biased. In this case, holes are attracted by the negative battery terminal and electrons by the positive battery terminal so that both holes and electrons move *away* from the junction. Since there is no recombination of electron-hole pairs, diode current is negligible and the junction has high resistance.

As seen, reverse biasing increases the potential barrier at the junction, thereby allowing very little current to flow through the junction.

It should be obvious from the above discussion that a *P-N* junction acts as a rectifier because of its ability to pass current more readily in one direction than in the other.

(*iii*) ***V/I* characteristics.** Fig. 7.22 shows the static volt-ampere characteristic curve of a typical *P-N* junction diode. It is seen that a minimum forward bias of 0.25 V is needed to overcome the potential barrier at the junction and permit any current to flow in the forward direction. The current then increases rapidly with increasing battery voltage and as little a voltage as 1 volt permits currents of about 50 mA.

But with even very large reverse bias, current flow is extremely small, being of the order of a few μA. However, when the reverse voltage exceeds a certain value called Zener or breakdown voltage, the reverse current suddenly and sharply increases, the curve indicating zero resistance at that point. When *P-N* junction diodes are employed primarily because of this property as voltage regulators, they are called Zener diodes or breakdown diodes. The Zener voltage depends on temperature.

Semiconductor *P-N* junction diodes, usually made of germanium or silicon, are commonly used as power rectifiers.

7.19. Depletion Layer

As Shown in Fig. 7.20, depletion layer is the narrow region centred around the junction plane of a *P-N* diode. A simple way to account for the formation of this layer or charge-free region is to postulate that at the middle junction plane, equal number of oppositely-charged impurity atoms exist which cancel each other out.

However, a more accurate picture involves *dynamic* rather than a static concept of charge behaviour near the *P-N* junction. Suppose that a *P-N* junction has just been formed. There is a natural tendency for the electrons and holes (*provided by the impurity atoms*) to diffuse across the junction and terminate their existence by recombination.

However, this process is accompanied by a secondary effect *i.e.* by the formation of ions. This happens because the impurity atoms which provide these migratory electrons and holes are left behind in an ionized state bearing a charge which is opposite to that of the departed carrier. Also these impurity ions, like germanium atoms, are fixed in their positions in the crystal lattice in the *P*- and *N*-regions of the diode. Hence, as shown in Fig. 7.23 (*a*) they form parallel rows or 'plates' of opposite charges facing each other across the depletion layer. Obviously, the row of fixed *positive* ions in the *N*-region is produced by the migration of electrons from the *N*- to *P*-region. Similarly, the row of fixed *negative* ions in the *P*-region is produced by the migration of holes from the *P*- to *N*-region.

If a *majority carrier* (either an electron or a hole) tries to cross into the depletion layer, it can suffer the following two fates :

(*i*) either it can be trapped or captured by the row of fixed impurity ions of opposite sign which guard its own region. For example, a hole trying to approach the depletion layer may be neutralized by the row of fixed negative ions situated in the *P* region itself at the edge of the depletion layer. So will be the case with the electrons trying to approach the depletion layer from *N*-region.

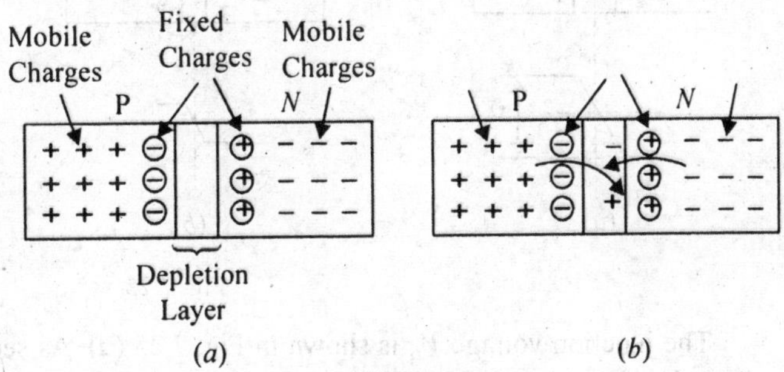

Fig. 7.23

(*ii*) or it may succeed in entering the depletion layer where it will be repelled by the row of similarly charged impurity ions guarding the *other* region. But its life will be cut short by recombination with a majority carrier of opposite sign which has similarly entered the depletion layer from the other half of the diode.

At the same time, the rows of oppositely-charged ions on the two sides of the depletion layer are themselves recombining and reforming. The whole process is one of confusion and chaos so far as individual carriers are concerned, but considering the whole process in its entirety there is produced some net effect. The overall effect is the *establishment and maintenance* of depletion layer and prevention of mass annihilation of the mobile charges. Instead, there prevails a limited rate of majority-carrier recombination in the depletion layer or to put it more correctly,

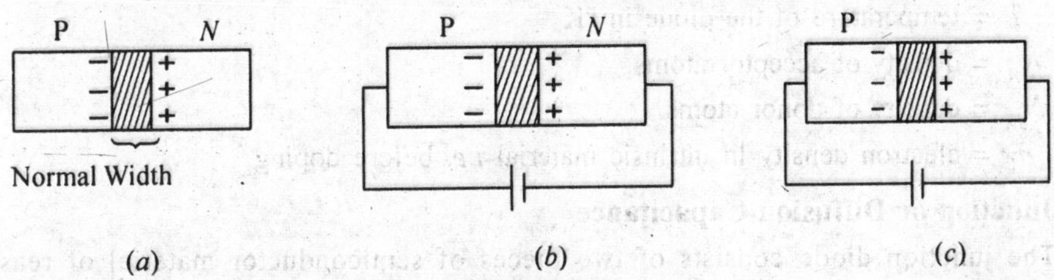

Fig. 7.24

the depletion layer exists by virtue of this limited recombination. Meanwhile, in the body of both regions of the *P-N* diode, charge replenishment of ionized impurity atoms is constantly provided by thermally-generated minority carriers.

Fig. 7.24 (*a*) shows the normal width of the depletion layer. With forward bias, width is decreased [Fig. 7.24 (*b*)] whereas with reverse bias, it is increased as shown in Fig. 7.24 (*c*). As discussed below, this variation in the width of the depletion layer affects the junction capacitance.

7.20. Junction Voltage

As seen from above, although depletion layer is cleared of majority free charges, it has fixed atomic ions on its both sides. Because of this charge separation, an electric potential V_B develops across the junction under equilibrium conditions *i.e.* with the junction externally isolated. It gives rise to a strong electric field in the narrow depletion layer whose thickness varies from 10^{-6} to 10^{-8} m. For example, with $V_B = 0{\cdot}1$ V and depletion layer thickness of 10^{-6} m, $E = 0.1/10^{-6} = 100{,}000$ V/m.

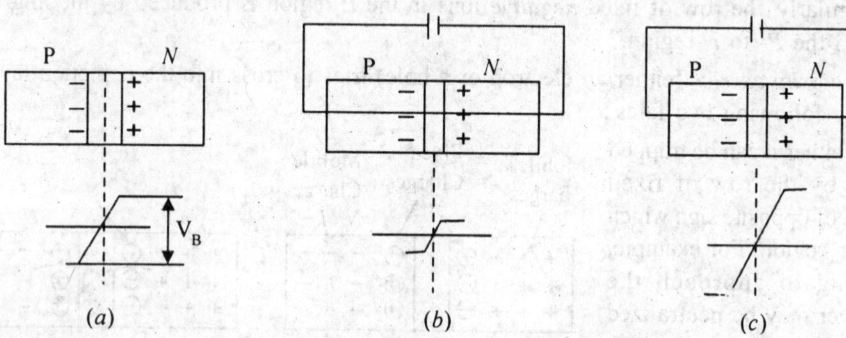

Fig. 7.25

The junction voltage V_B is shown in Fig. 7.25 (*a*). As seen from the same figure, V_B (which is also called barrier voltage) is decreased with forward bias and increased with reverse bias of the *P-N* junction.

Creation of this junction voltage (or internal voltage) can be explained in terms of energy band diagram. Fig. 7.26 (*a*) shows the condition before the junction is formed. The doping impurities are so sparsely distributed even in heavily-doped material that impurity atoms do not alter the basic levels.

Thus the energies at the valence and conduction band edges, *i.e.* E_e and E_v are the same in both *P* and *N* regions of the semiconductor. After junction formation, the Fermi-level is constant throughout the system in equilibrium as shown in Fig. 7.26 (*b*).

It can be shown that

$$V_B = kT \log N_A N_D / n_1^2$$

where k = Boltzmann constant

T = temperature of the diode in °K

N_A = density of acceptor atoms

N_D = density of donor atoms

n_1 = electron density in intrinsic material *i.e.* before doping.

7.21. Junction or Diffusion Capacitance

The junction diode consists of two pieces of semiconductor material of reasonable conductivity separated by an insulating (depletion) layer. It resembles a parallel plate capacitor which consists of two conducting plates with a dielectric in between. In the case of a *P-N* junction, the two parallel rows of charged impurity ions act as charged plates of a capacitor whose capacitance is inversely proportional to their separation *i.e.* thickness of the depletion layer. The value of this capacitance is of the order of a few picofarads and is known as *junction or diffusion capacitance*. Its value can be increased by applying a forward bias to the diode because in that case, thickness of the depletion layer is decreased and, consequently, capacitance is increased. Since reverse bias increases the thickness of depletion layer, junction capacitance is reduced. Obviously, by varying the applied voltage and hence the thickness of the depletion layer, diode capacitance can be varied. Hence, the *P-N* diode can be used as a variable capacitor whose capacitance can be controlled by applied voltage.

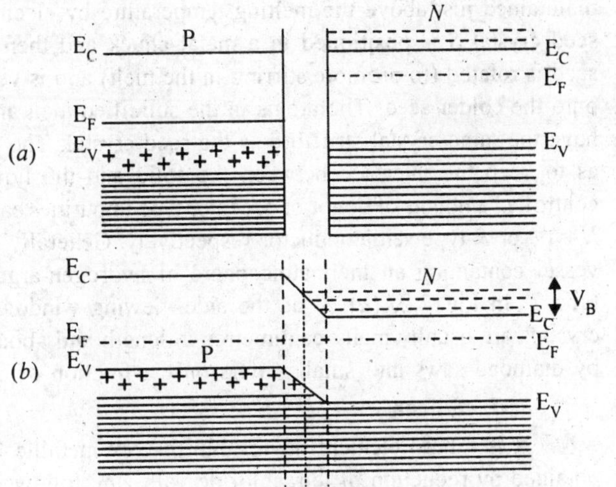

Fig. 7.26

However, a reverse-biased *P-N* diode works better as a capacitor because its capacitance varies with the applied voltage according to the formula $1/V_r^{1/3}$ to $1/V_r^{1/2}$ where V_r is the reverse voltage. Capacitive diodes called varactors are especially designed to have the maximum capacitance range. Varactors are used in switching, limiting, harmonic generation and parametric amplification applications.

7.22. Preparation of Semiconductor Materials

The two most commonly-used semiconductor materials are germanium and silicon whose details of preparation are given below.

(*i*) Germanium

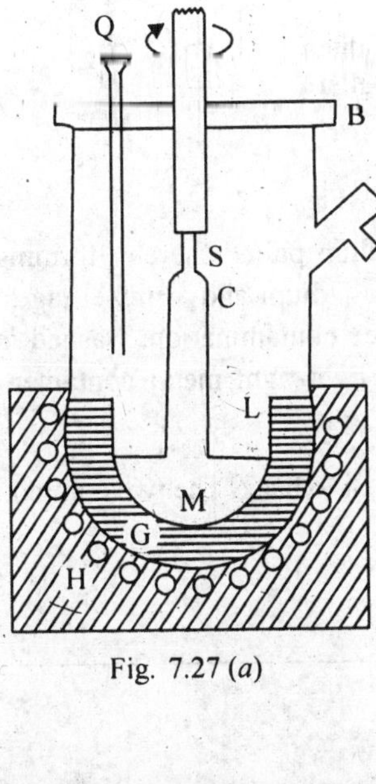

Fig. 7.27 (*a*)

It was discovered in 1886 and is a hard dense element obtained as a by-product of zinc refining or from the flue dust of certain coals. It has grey metallic lustre. Commercially, it is available in the oxide form (GeO_2) which becomes reduced germanium powder when heated at about 650°C in an atmosphere of hydrogen. When it is further heated above the melting point of 937°C, germanium is obtained in the form of bars of low impurity. Further purification is achieved by the process of zone-refining in inert gas atmosphere.

Seed crystals are obtained by taking a portion of an ignot of germanium obtained from zone-refining process. The desired single crystals are grown by dipping a small seed crystal into a bath of molten germanium and withdrawing it uniformly as it grows. The bath is maintained at an accurately-controlled temperature under an inert gas atmosphere. Fig. 7.27 (*a*) shows a cross-sectional view of a typical piece of equipment used for the purpose of crystal pulling. Here, the germanium melt *M* is kept in a graphite crucible *G* lined by quartz liner *L* and is maintained just above the melting temperature by electrical heater *H*. To start the growth, a tiny seed crystal *S* is positioned in a metal chuck and then immersed in the liquid germanium. The seed is rotated (to promote stirring in the melt) and is very slowly withdrawn as the liquid freezes onto the colder seed. The atoms in the solidified melt arrange themselves in such a way that they have the same crystal structure as the seed crystal. The growing crystal *C* is raised at such a rate as to keep the interface between the solid and the liquid at the surface of the melt. Carefully controlled amounts of donor or acceptor type impurities can be added via quartz tube *Q* for producing *N*-type or *P*-type semiconductor respectively. Generally, the entire apparatus is placed in a sealed vessel containing an inert atmosphere of hydrogen-argon which can be introduced via the side tube *T*. In Fig. 7.27 (*a*), *W* is the side viewing window and *B* is the brass support plate. Final crystals are usually many centimetres in length and about 2 to 3 cm in diameter. These are sliced by diamond saws into small wafers only a fraction of a mm in thickness.

(*ii*) Silicon

It is a hard element having bluish-grey metallic lustre and melting point of 1420°C. It is obtained by reduction of tetrachloride with zinc followed by use of hydrogen and other reducing agents. Required purification is obtained by zone-refining process. Single crystals are grown by a method similar to that used for germanium.

7.23. Producing of *P*- and *N*-type Crystals

A junction made by simply placing *P*- and *N*-type germanium (or silicon) in contact has no useful electrical properties. The junction has to be grown in a single crystal. One of the following two methods is usually employed :

(*a*) A pure germanium crystal is pulled up very slowly from an *N*-type melt. Then a *P*-type impurity is added to the melt in such quantity as is sufficient to make holes the majority carriers and the crystal growing is continued. In this way, a junction between *N* and *P* is formed which can be located by etching the crystal in hydrofluoric acid. Diamond cutters are used to slice the crystal so produced into large number of pieces (Art. 7.22)

(*b*) In this method, an *N*-type germanium crystal is sliced into thin wafers and an indium (a trivalent acceptor material) is placed on each slice as shown in Fig. 7.27 (*b*). Many such wafers are heated at the same time in a hydrogen atmosphere. The indium melts (M.P. = 156°C) at a lower temperature than germanium (M.P. = 1420°C) and thus dissolves some germanium from the slice to form *P*-type material. A strict temperature control is essential during the process. On cooling, the contaminated germanium solidifies first and then the rest of the indium pallet, thereby forming the *P-N* junction. As shown in Fig. 7.27 (*c*) a wire is soldered to the indium and a nickel tag to the germanium. The *P-N* diode so formed is etched to remove surface contamination, washed in deionized water, dried and mounted in water-repellent grease in a light-tight metal container.

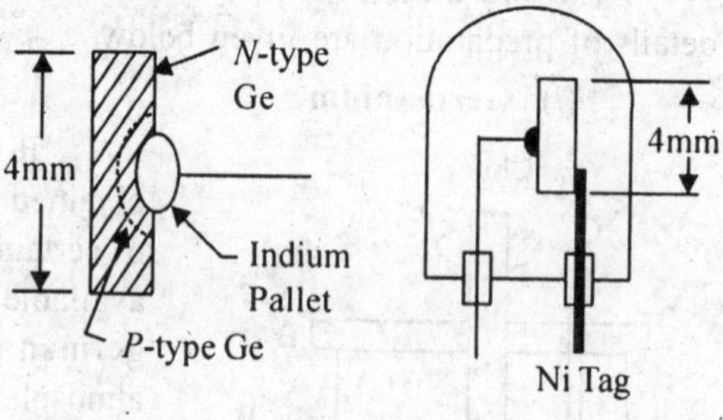

Fig. 7.27

7.24. Junction Triode Transistors*

A triode transistor consists of two *P-N* junction diodes placed back to back. The two common types of junction transistors are

(*i*) grown junction type and

(*ii*) alloy-junction type.

*The word transistor is derived from two words : *transfer* and *resistor*.

Fig. 7.28 (*a*) shows a grown *P-N-P* junction triode transistor whereas Fig. 7.28 (*b*) shows the form of *N-P-N* junction transistor. The alloy-junction transistor is shown in Fig. 7.28 (*c*). In the manufacture of grown junction transistors shown in Fig. 7.28 (*a*) and (*b*), the single crystal-growing process is employed. As the crystal is grown, the molten portion is changed so as to form *P*-type, *N*-type and *P*-type regions in succession thereby producing *P-N-P* transistor. Likewise, changes in the molten material which form *N*-type, *P*-type and *N*-type portions in succession lead to a *N-P-N* transistor. The left-hand section or region is called the *emitter* whereas the right-hand section is known as *collector*. The middle section called base region or *base* is extremely thin as compared to either the emitter or collector and is lightly doped.

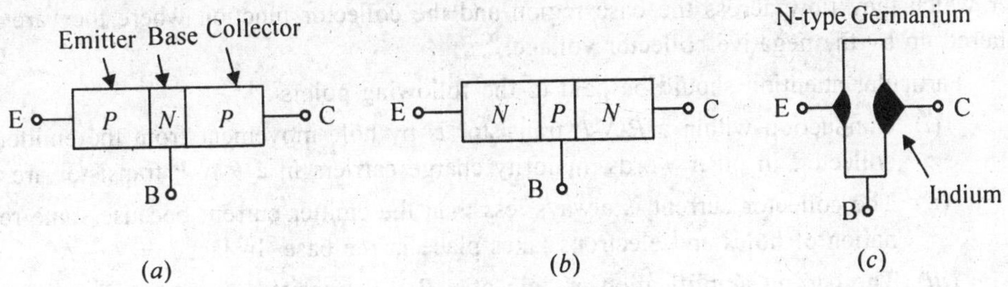

Fig. 7.28

Function of the emitter is to inject majority charge carriers into the base and that of the collector is to collect or attract these carriers through the base.

Fig. 7.28 (*c*) shows an alloy-junction transistor which consists of two beads of indium metal alloyed on the opposite sides of a thin slice of an *N*-type germanium. Collector is larger in size than the emitter. These transistors may be also of the *P-N-P* type or *N-P-N* type.

7.25. Working of *P-N-P* Transistor in Common-base Configuration

The basic mechanism of transistor operation can be easily understood if the following facts are kept in mind.

Since emitter is to provide charge carriers, it is *always forward-biased.* First letter of transistor type indicates the polarity of the emitter voltage with respect to base. Similarly, collector's job is to collect or attract those carriers through the base, hence it is *always reverse-biased.* The second letter of transistor type indicates the polarity of collector voltage with respect to the base.

The above statement applies both to *P-N-P* and *N-P-N* transistors.

Fig. 7.29 shows a *P-N-P* transistor connected in the common-base (or grounded-base) configuration. It is so called because both the emitter and collector are returned to the base terminals.

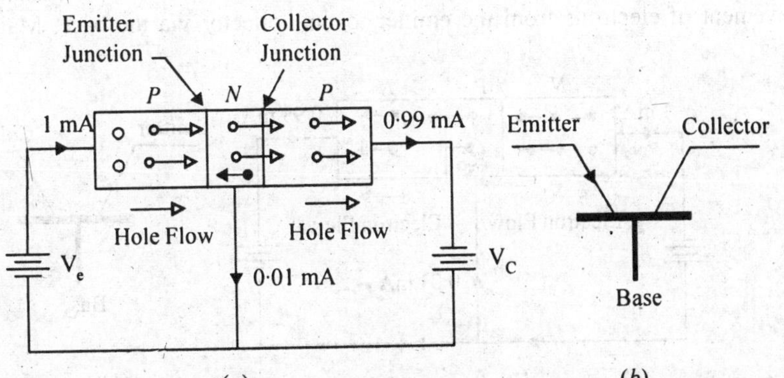

Fig. 7.29

As seen, the emitter junction is forward-biased whereas the collector junction is reverse-biased. The holes in the emitter are repelled by the positive battery terminal towards the *P-N* or emitter junction. The potential barrier at the junction is reduced (Art. 7.18) due to the forward-bias, hence holes cross the junction and enter the *N*-type base.

Since the base is thin and lightly-doped, majority of the holes (about 95%) are able to drift across the base without meeting electrons to combine with. The balance of 5% of holes are lost in the base region due to recombination with electrons. The holes which after crossing the *N-P* or collector junction enter the collector region are swept up by the negative collector voltage V_C. It is seen from the above discussion that holes are continuously injected into the emitter junction after which they flow across the base region and the collector junction where they are quickly gathered up by the negative collector voltage.

Particular attention should be paid to the following points :

(*i*) Conduction within a *P-N-P* transistor is by hole movement from the emitter to the collector. In other words, majority charge carriers in a *P-N-P* transistor are holes.

(*ii*) The collector current is always less than the emitter current because some recombination of holes and electrons takes place in the base. In fact $I_c = I_e - I_b$.

(*iii*) The current amplification or gain of a *P-N-P* transistor for steady conditions when connected in common base configuration is given by

$$\alpha = \frac{\text{collector current}}{\text{emitter current}} = \frac{I_c}{I_e}$$

Obviously, $\alpha < 1$.

(*iv*) Emitter arrow shows the direction of flow of *conventional* current. Obviously, electron flow will be in the opposite direction.

7.26. Working of a *N-P-N* Transistor in Common-base Configuration

Fig. 7.30 (*a*) shows battery connections for *N-P-N* junction transistor, Here, again the emitter is forward-biased and the collector reverse-biased.*

The electrons in the emitter region are repelled by the negative battery terminal towards the emitter or *N-P* junction. Since potential barrier at the junction is reduced due to forward bias, electrons cross over into the *P*-type base region. Since the base is thin and lightly-doped, most of the electron (about 95%) cross over the collector junction and enter the collector region where they are readily swept up by the positive collector region where they are readily swept up by the positive collector voltage V_C. Only about 5% of the emitter electrons combine with the holes in the base and are lost as charge carriers. Obviously, conduction through a *N-P-N* transistor takes place by the movement of electrons from the emitter to the collector via the base. Main points worth noting are :

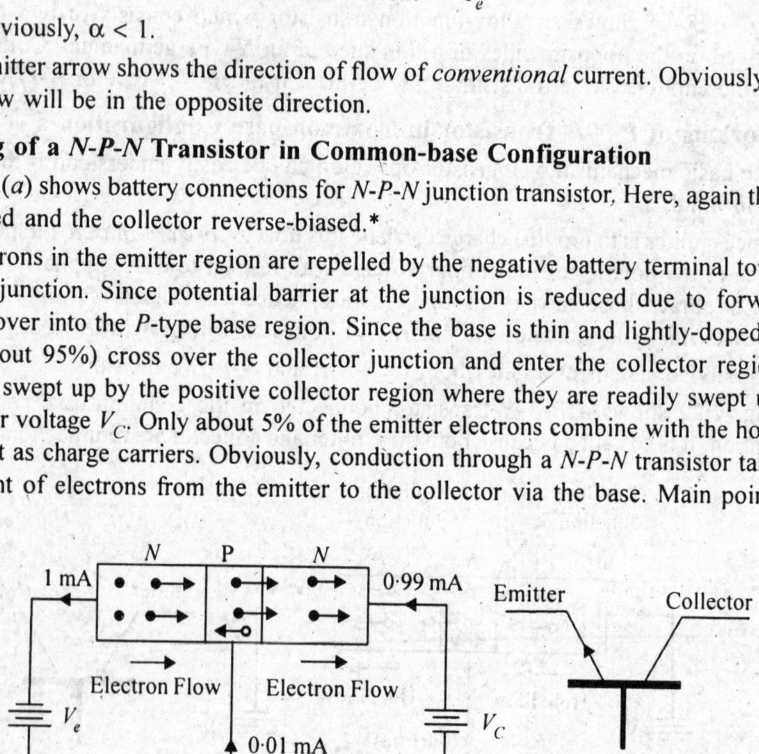

(*a*) (*b*)

Fig. 7.30

(*i*) in an *N-P-N* transistor, electrons constitute the majority charge carriers.

(*ii*) collector current is less than the emitter current so that $\alpha < 1$.

It may be noted again that emitter arrow in Fig. 7.30 (*b*) shows the direction of flow of conventional current (and not electronic current).

HIGHLIGHTS

1. Valence electrons are those
   (*i*) which occupy the outermost shell of an atom
   (*ii*) which have the highest energy among the electrons
   (*iii*) which are most affected when a number of atoms are brought together as in the formation of a solid.
2. Conduction electrons are those valence electrons which have gained enough energy to take part in conduction of electricity through a solid.
3. Valence band is the band of energy occupied by valence electrons
   (*i*) it is the highest occupied band
   (*ii*) it may be completely or partially filled with electrons.
4. Conduction band is the next higher energy band to the valence band
   (*i*) it is occupied by conduction electrons
   (*ii*) it may be empty or partially filled
   (*iii*) it is the lowest unfilled or unoccupied energy band.
5. Insulators are those materials which
   (*i*) have full valence band
   (*ii*) have an empty conduction band and
   (*iii*) have a large energy gap between the valence and conduction bands.
6. Conductors are those materials which have overlapping valence and conduction bands. Conduction takes place with the help of conduction electrons.
7. Semiconductors are those materials which
   (*i*) have almost empty conduction band
   (*ii*) have almost filled valence band
   (*iii*) have narrow energy gap between the two.
8. Intrinsic semiconductors are those which are made of the semiconductor material in its extremely pure form. Their current is due to the movement of electrons and holes whose number is equal.

   Current $I = n_1 e\,(\mu_e + \mu_h)\,EA$

   $= n_1 e\,(\mu_e + \mu_h)\,AV/l$

   Conductivity, $\sigma = n_1 e\,(\mu_e + \mu_h)$

   Current density $J = n_1 e\,(\mu_e + \mu_h)\,E = \sigma E$
9. Extrinsic or impurity semiconductors are those intrinsic semiconductors to which some suitable impurity or doping agent has been added in extremely small amount (about one part in 100 million)

   Extrinsic semiconductors can be of two types depending on the doping agent.

   (*i*) *N*-type extrinsic semiconductors are those intrinsic or pure crystals which have been doped by a pentavalent element like antimony etc. In such semiconductors, conduction is mostly by electrons.

$\sigma_n = e\,(n_n\mu_e + p_n\mu_h)$ ¾ $n_n e\mu_h$ – neglecting hole density

$J = n_n e\mu_h E.$

(*ii*) *P*-type extrinsic semiconductors are those intrinsic semiconductors which have been doped by a trivalent element like boron. In such semiconductors, conduction is mostly by means of hole movement :

$\sigma_p = e\,(n_p\mu_e + p_p\mu_h)$ ¾ $p_p e\mu_h$ – neglecting electron density

$J = p_p e\mu_h E.$

10. *P-N* junction diode consists of a piece of germanium, one half of which is doped with *P*-type material and the other half with *N*-type material.

11. A transistor consists of two *P-N* junction diodes placed back-to-back. It may be of the *P-N-P* type or *N-P-N* type.

Can You Answer This ?

1. How have the solids been classified as metals, semiconductors and insulators ?
2. How are valence, conduction and forbidden bands explained ?
3. Write note an semiconductors.
4. (*a*) What is the difference between *N* and *P* types of semiconductors ? Draw diagram to explain them.
   (*b*) Describe with suitable diagram, the construction and action of a *P-N* junction diode. Discuss the physical processes that take place in and near the depletion layer.
5. Write notes on conductors and semiconductors.
6. (*a*) Define and explain the functions of (*i*) valence and (*ii*) conduction band and (*iii*) forbidden gap.
   (*b*) What is the difference between *N*- and *P*-types of semiconductors ? Draw diagram to explain them.
7. (*a*) Explain the formation of different energy bands in a solid and hence distinguish between an insulator, a semiconductor and a conductor.
   (*b*) With the help of energy band diagram, explain the difference between an intrinsic and an extrinsic semiconductor giving examples.
8. (*a*) What is the difference between *N*- and *P*-types of semiconductors ? Explain with the help of diagrams.
   (*b*) Explain the operation of *PNP* and *NPN* transistors.
9. (*a*) Explain, in brief, the different energy bands in a solid.
   (*b*) Obtain expressions for the (*i*) total conductivity of a semiconductor and (*ii*) conductivity of *N*- and *P*-type semiconductors.
10. (*a*) Explain the difference between the band structure of a semiconductor and that of a metal. Why does a semiconductor act as an insulator at 0°K and why does its conductivity increase with increasing temperature ?
    (*b*) With the help of energy band diagrams, explain with examples the distinction between an intrinsic and an extrinsic semiconductor.
11. Explain why a contact difference of potential develops across an open-circuited *P-N* junction. How can you explain the presence of this potential ?
12. Write short note on band theory of solids.

13. (*a*) Explain how a *P-N* junction can act as a semiconductor rectifier.

    (*b*) Explain what intrinsic and extrinsic (impurity) semiconductors are.

14. (*a*) Draw schematically the energy bands of sodium, diamond and silicon. Compare their electrical properties in terms of the band picture.

    (*b*) Draw band diagram of silicon when the following impurities are added to it (*i*) arsenic and (*ii*) gallium

15. Write short note on semiconductor.

16. (*a*) Define the terms : Valence band, conduction band and forbidden gap.

    (*b*) Describe in short the formation of energy bands in solids and hence explain how it helps to classify the materials into conductors and insulators.

17. (*a*) Describe with suitable diagrams the construction and action of *P-N* junction diode.

    (*b*) Discuss the physical processes that take place in and near the depletion layer.

    (*c*) How can the *P-N* diode be used as a variable capacitor ?

18. Write a note on intrinsic and extrinsic semiconductors.

19. What are valence and conduction bands ? Distinguish between insulators, conductors and semiconductors in terms of their energy bands.

X-RAYS 8

Production of X-rays — Origin of X-rays — X-ray Spectrum — Moseley's Law — Importance of Moseley's Law — Formulae for K- and L-series of X-ray Spectrum — Intensity Measurement of X-rays — Absorption of X-rays — Definition of Absorption Coefficient (μ) — Wave Nature of X-rays — Explanation of Laue's Spots — Uses of Laue's Spots — Bragg's X-ray Spectrometer — Theory of Bragg's Diffraction — Bragg's Law — Bragg's Law and Crystal Structure — Laue Spot Method for Study of Crystal Structure — Powder Crystal Method — Experimental Arrangement — Compton Scattering of X-rays — Properties of X-rays — Practical Applications of X-rays — Highlights — Can You Answer This ?

8.1. Production of X-rays

X-rays were accidentally discovered by Wilhelm Röntgen in 1985 during the course of some experiments with a discharge tube. At present, it is well known that these rays are produced whenever fast-moving electrons strike a high atomic weight solid like tungsten kept in vacuum.

(*a*) X-ray Tube

The essential elements of a modern Coolidge X-ray vacuum tube which is widely used for commercial and medical purposes are shown in Fig. 8.1. Electrons are produced thermionically from a tungsten filamentary cathode F which is heated to incandescence either by a storage battery (Fig. 8.1) or by a low-voltage alternating current from a step-down transformer T_2 (Fig. 8.2). These electrons are focussed on the target T with the help of a cylindrical shield S which surrounds F and is maintained at a negative potential. The electrons are accelerated to very high speeds (up to 10% velocity of light) by the d.c. potential difference (of about 50 kV-100 kV) applied between F and the anode (also called anticathode). This high d.c. potential is obtained from a step-up transformer T_1 whose output is converted into direct current by full-wave rectifier and a suitable filter.

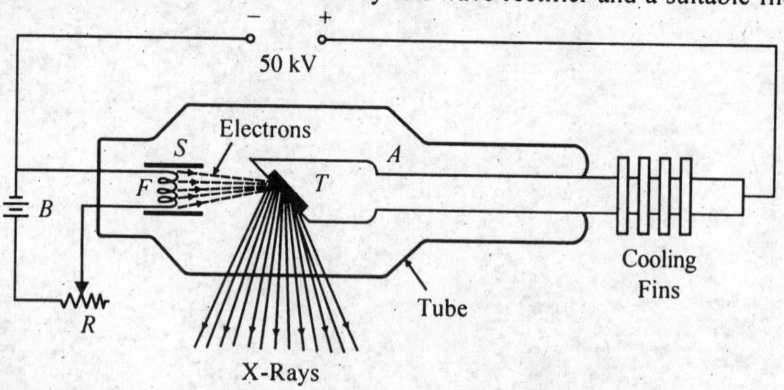

Fig. 8.1

The target T usually employed in X-ray tubes is a massive block of tungsten or in many cases, a molybdenum plug embedded in the face of a solid copper anode. The face of the copper anode is sloped at about 45° to the electron beam. Being very good conductor of heat, copper helps to conduct heat efficiently to the external cooling fins or the water-cooling system. Under the territic bombardment of the target by so many electrons, most metals will melt*. That is why metals

*About 99.8 percent of the energy of the incident beam of electrons is converted into heat and only 0.2 percent into energy of X-rays.

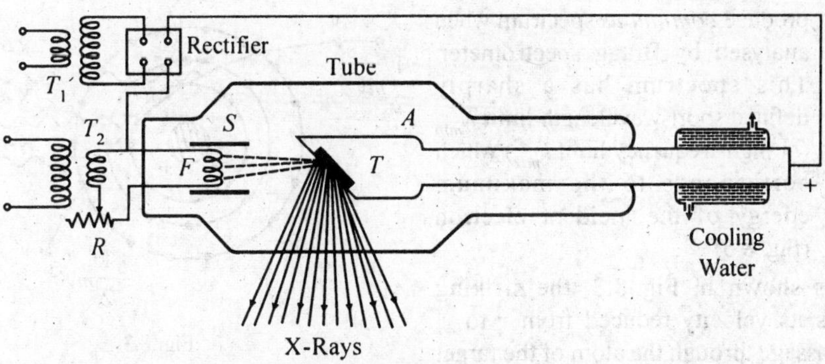

Fig. 8.2

like tungsten, platinum and molybdenum etc. are used which have high melting points and who have a high atomic weight (which is essential for abundant production of X-rays).

When the electrons strike the tungsten target, they give up their kinetic energy and thereby produce X-rays.

(*b*) Control of Intensity and Quality

The intensity of X-rays depends on the *number* of electrons striking the target. This number is determined by the temperature of the electron-emitting filament which itself is proportional to the heater current. Hence, by controlling the filament current with the help of a rheostat *R*, thermionic emission and hence intensity of X-rays can be controlled.

The *quality* of X-rays is measured in terms of their penetrating power which is dependent on the potential difference between filamentary cathode and the anode. Greater this accelerating voltage, height the speed of the striking electrons and, consequently, more penetrating the X-rays produced. It is customary to refer to highly penetrating X-rays (*i.e.*, those possessing high frequency) as *hard* X-rays and to those less penetrating (*i.e.*, of low frequency) as *soft* X-rays. Obviously, the quality of penetrating power of X-rays can be controlled by varying the potential difference between the cathode and anode.

It will be noticed from above explanation that in Coolidge X-ray tube, it is possible to achieves separate control of the intensity and quality of X-rays independent of each other.

It has been found that apart from the intensity and quality of X-rays, their abundance depends on the atomic weight of the target material. Target materials of higher atomic weights yield a greater abundance of X-rays than those of lower atomic weights.

8.2. Origin of X-rays

X-rays are produced when high-speed electrons strike some material object. However, majority of the electrons that strike a solid target, do nothing spectacular at all ! Most of them undergo glancing collisions with the matter particles, lose their energy a little bit at a time and thus merely increase the average kinetic energy of the particles of the target material. It is found that nearly 99.8 percent of the energy of the electron beam goes into heating the target.

But a small number of the bombarding electrons produce X-rays by losing their kinetic energy in the following *two ways* :

(*i*) Some of the high-velocity electrons penetrate the interior of the atoms of the target material and are attracted by the positive charge of their nuclei. As an electron passes close to the positive nucleus, it is deflected from its path as shown in Fig. 8.3. The electron experiences deacceleration during its deflection in the strong field of the nucleus. The energy lost during this deacceleration is given off in the form of X-rays of continuously varying wavelength (and hence frequency). These X-rays produce *continuous* spectrum when analysed by Bragg spectrometer. This spectrum has a sharply defined short-wavelength limit λ_{min} (or high-frequency limit f_{max}) which corresponds to the maximum energy of the incident electron (Fig. 8.6).

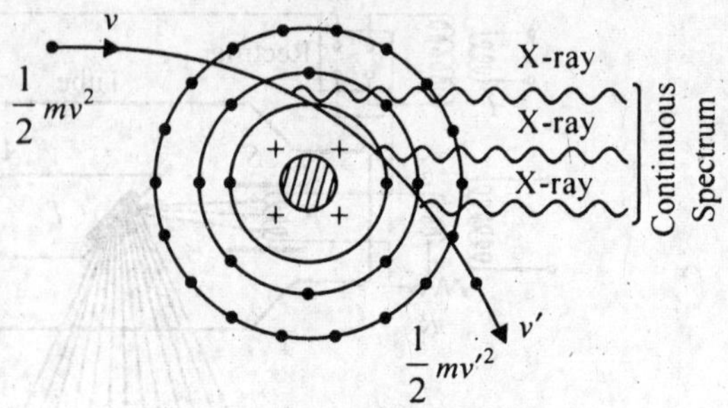

Fig. 8.3

If, as shown in Fig. 8.3, the striking electron has its velocity reduced from v to v' during its passage through the atom of the target material, then its loss of energy is $= \left(\frac{1}{2}mv^2 - \frac{1}{2}mv'^2\right)$. This must equal the energy of the X-ray photons emitted.

$$\therefore \qquad \frac{1}{2}m(v^2 - v'^2) = hf$$

The highest or maximum frequency of the emitted X-rays corresponds to the case when the electron is *completely stopped, i.e.*, when $v' = 0$. In that case

$$\frac{1}{2}mv^2 = hf_{max} \qquad ...(i)$$

If the electron is accelerated through a potential of V volts, then

$$\frac{1}{2}mv^2 = eV \qquad ...(ii)$$

From (*i*) and (*ii*), we get

$$hf_{max} = eV; \qquad f_{max} = eV/h$$

Now, $$hf_{max} = hc/\lambda_{min} \qquad (\because c = f \cdot \lambda)$$

$$\therefore \qquad hc/\lambda_{min} = eV \quad \text{or} \quad \lambda_{min} = hc/eV$$

Substituting the values of $e = 1{\cdot}602 \times 10^{-19}$ C, $h = 6{\cdot}62 \times 10^{-34}$ J-s and $c = 3 \times 10^8$ m/s, we get

$$\lambda_{min} = \frac{6{\cdot}62 \times 10^{-34} \times 3 \times 10^8}{1{\cdot}602 \times 10^{-19}\, V} = \frac{1{\cdot}24 \times 10^{-6}}{V} \text{ m}$$

or $$\boxed{\lambda_{min} = \frac{12{,}400}{V} \text{ Å}} \qquad (\because 1 \text{ Å} = 10^{-10} \text{ m})$$

Such X-rays are very aptly called '*braking*' radiations because they are due to *braking* or *slowing down* of high velocity electrons in the positive field of a nucleus. These radiations constitute, as said earlier, the *continuous* spectrum of the X-rays because they consists of a series of uninterrupted wavelengths having a sharply-defined short-wavelength limit λ_{min}. These X-rays are independent of the nature of the target material but are determined *by the potential difference between the cathode and anode of the X-ray tube.*

(*ii*) Some of the high-velocity electrons while penetrating the interior of the atoms of the target material, knock off the *tightly*-bound electrons in the innermost shells (like *K*, *L*-shells etc.) of the atoms*. When electrons from outer orbits jump to fill up the vacancy so produced, the energy difference is given out in the form of X-rays of *definite* wavelength (and frequency). These wavelengths constitute the *line* spectrum which is *characteristic of the material of the target.*

*Not the valance electrons in the outermost shells when jumps produce less energetic visible radiations.

Fig. 8.4 (*a*) shows the case when the high-velocity incident electron knocks off the one electron from the *K*-shell. As shown in Fig. 8.4 (*b*), this vacancy in *K*-shell is filled up by a nearby electron in the *L*-shell. During this jump an *X*-ray radiation is emitted whose frequency is given by

$$E_k - E_l = hf$$

where E_k is the energy required to dislodge an electron from the *K*-shell and E_l is that required for *L*-shell. Since this energy difference is comparatively very large, the X-rays emitted have very large energy content and hence are highly penetrating.

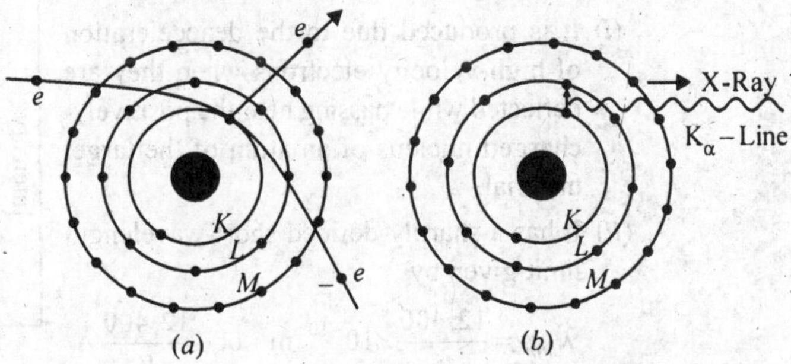

Fig. 8.4

If, however, this vacancy in *K*-shell is filled up by an electron jumping from *M*-shell, the X-rays emitted would be still more energetic and would consequently posses still higher frequency because $\Delta E = (E_k - E_m)$ is more than $\Delta E = (E_k - E_l)$. Such X-rays arising from millions of atoms produce the *K*-lines as shown in Fig. 8.5. Usually, two lines K_α and K_β of this series are detected although there are many more.

Similarly, when the incident electron carries somewhat lesser amount of energy, it dislodges an electron from the *L*-orbit and an electron either from *M*-orbit or other outer orbits takes its lace so that X-rays of frequency lower than that of the *K*-series are produced. This gives the *L*-series of the X-ray spectrum as shown K_α, L_β and L_γ lines in Fig. 8.5 (*a*). Spectral lines of *M*-series are produced in a similar way as shown in the energy-level diagram of Fig. 8.5 (*b*).

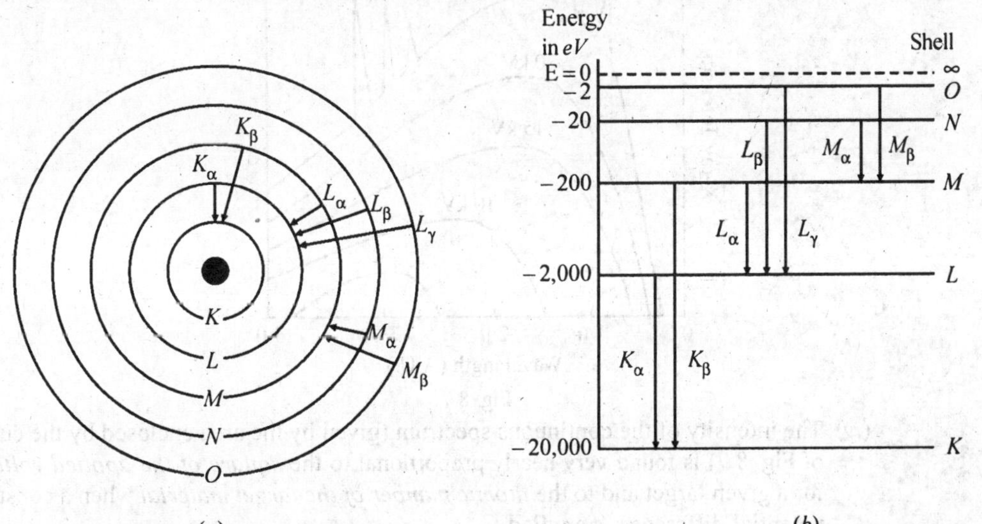

Fig. 8.5

As stated earlier, these *K*-, *L*- and *M*- series constitute the *line* spectra of the X-rays which *are characteristic of the material* used as target in the X-ray tube.

Hence, the X-rays produced by an X-ray tube consist of two parts :

(*i*) one part consists of a series of uninterrupted wavelengths having a short cut-off wavelength λ_{min}. This constitutes the *continuous spectrum* and

(*ii*) the other part consists of a number of distinct and discrete wavelengths which constitute the line or discontinuous spectrum of the X-rays.

8.3. X-ray Spectrum

As explained in Art. 8.2, X-ray spectrum consists of

(*i*) continuous spectrum and (*ii*) line spectrum. These two are shown in Fig. 8.6.

(*a*) Continuous Spectrum

(*i*) It is produced due to the deacceleration of high-velocity electrons when they are deflected while passing near the positively-charged nucleus of an atom of the target material.

(*ii*) It has a sharply-defined short-wavelength limit given by

$$\lambda_{min} = \frac{12,400}{V} \times 10^{-10} \text{ m} \quad \text{or} \quad \frac{12,400}{V} \text{ Å}$$

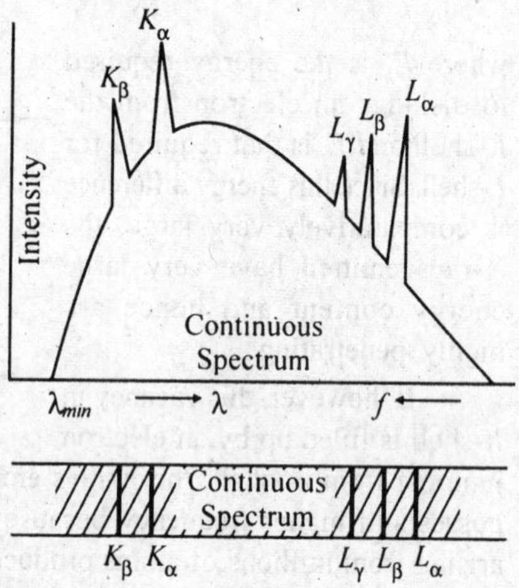

Fig. 8.6

(*iii*) The cut-off wavelength λ_{min} is independent of the nature of the target material but is inversely proportional to the potential difference between the cathode and anode of an X-ray tube. The value of λ_{min} decreases as this potential difference is increased (Fig. 8.7).

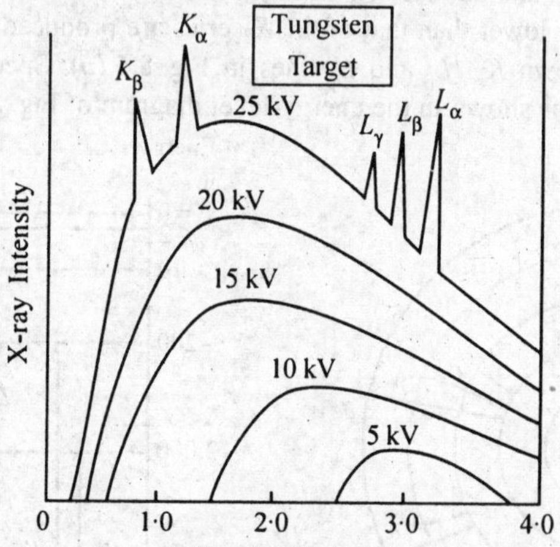

Fig. 8.7

(*iv*) The intensity of the continuous spectrum (given by the area enclosed by the curve of Fig. 8.7) is found very nearly proportional to the *square of the applied voltage* for a given target and to the *atomic number of the target material* when a constant potential difference is applied.

(*v*) There is a shift of the maximum intensity position towards the short wavelength side as voltage is increased.

(*b*) Line Spectrum

(*i*) It is produced when electrons are dislodged from the innermost orbits of the atoms of the target material followed by electron jumps from outer orbits.

(*ii*) It consists of discrete spectral lines which constitute *K*-series, *L*-series and *M*-series etc. *K*-series consists of those lines for which electron jumps end at *K*-level.

Similarly, *L*-series and *M*-series are produced when electron jumps end at *L*- and *M*-levels (Fig. 8.5).

(*iii*) *K*-series being most energetic constitute the hard X-rays whereas *L*- and *M*- series form the 'soft' X-rays.

(*iv*) Line spectrum is *characteristic of the target material used.* In fact, X-rays constituting the line spectrum are known as *characteristic X-rays*. The *number* of lines present in the spectrum depends both *on the nature of target material and the excitation voltage.*

(*v*) There is a regular shift towards shorter wavelength in the *K-spectrum* as the atomic number of the target is increased (Fig. 8.8). The exact relationship, as found by Moseley, is

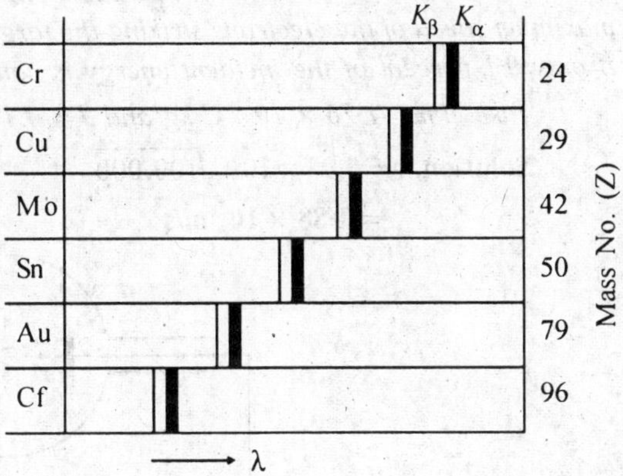

Fig. 8.8

$$\frac{\lambda_2}{\lambda_1} = \frac{f_1}{f_2} = \frac{(Z_1 - 1)^2}{(Z_2 - 1)^2}$$

where f_1 is the frequency of the K_α line for a target material having an atomic number of Z_1 and f_2 and Z_2 are similar quantities for some different target material.

Example 8.1. *An X-ray tube works on 60,000 V. What will be the wavelength of X-ray emitted in it.*

Solution. As seen from Art. 8.3

$$\lambda_{min} = \frac{12,400}{V} \text{ Å}$$

Here, $V = 60,000$ V

$$\therefore \quad \lambda_{min} = \frac{12,400}{60,000} = \mathbf{0{\cdot}2\ Å.}$$

Example 8.2. *If the potential difference applied across an X-ray tube is 12·4 kV and the current through it is 2 mA, calculate*

(*i*) *the number of electrons striking the target per second*

(*ii*) *the speed with which they strike it*

(*iii*) *the shortest wavelength emitted*

Take $e = 1{\cdot}6 \times 10^{-19}$ *C and* $m = 9{\cdot}1 \times 10^{-31}$ *kg.*

Solution. (*i*) If *n* is the number of electrons striking the anode per second, then

$$I = ne$$

$$\therefore \quad n = I/e = 2 \times 10^{-3}/1{\cdot}6 \times 10^{-19}$$

$$= \mathbf{1{\cdot}25 \times 10^{16}\ s^{-1}}$$

(*ii*) As seen from Art. 5.6

$$v = 5{\cdot}93 \times 10^5 \sqrt{V} = 5{\cdot}93 \times 10^5 \sqrt{12,400}$$

$$= \mathbf{6{\cdot}6 \times 10^7\ m/s}$$

(*iii*) $\lambda_{min} = 12,400/V = 12,400/12,400 = \mathbf{1\ Å.}$

Example 8.3. *Calculate the minimum applied potential required to produce X-rays of 1 Å wavelength.*

Solution. $\lambda_{min} = 12{,}400/V$ Å

$\therefore \quad V = 12{,}400/\lambda_{min} = 12{,}400/1 = \mathbf{12 \cdot 4\ kV.}$

Example 8.4. *An X-ray tube passes 5 mA at a potential difference of 100 kV. Calculate the maximum speed of the electrons striking the target and the rate of production of heat at the target if only 0·1 percent of the incident energy is converted into X-radiations.*

Take e/m = 1·76 × 10¹¹ C/kg and J = 4·18 joules/cal.

Solution. $v = 5 \cdot 93 \times 10^5 \sqrt{100{,}000}$

$= \mathbf{1 \cdot 88 \times 10^5\ m/s}$

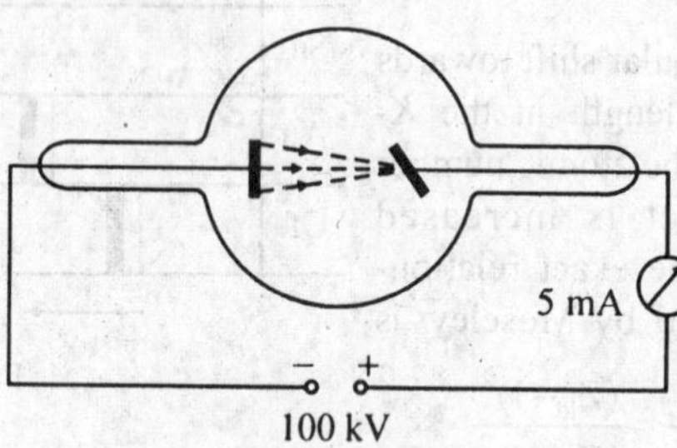

Fig. 8.9

Incident power = $100{,}000 \times 5 \times 10^{-2} = 500$ W

Power converted into heat = 99·9% of 500 = 499·5 W

Heat produced/second = 499·5/4·18 = **119 cal/s.**

Example 8.5. *An X-ray tube operated at 30 kV emits a continuous X-ray spectrum with a short-wavelength limit λ_{min} = 0·414 Å. Calculate Planck's constant h, if e = $1 \cdot 602 \times 10^{-19}$ C and c = 3×10^8 m/s.*

Solution. As shown in Art. 8.2,

$$\lambda_{min} = \frac{ch}{eV}$$

$$\therefore \quad h = \frac{eV\lambda_{min}}{c}$$

$$= \frac{1 \cdot 602 \times 10^{-19} \times 30 \times 10^3 \times 0 \cdot 414 \times 10^{-19}}{3 \times 10^8} = \mathbf{6 \cdot 63 \times 10^{-34}\ J\text{-}s.}$$

8.4. Moseley's Law

In 1913-14, Moseley carried out a systematic study of the characteristic X-ray spectra of various elements used as targets in an X-ray tube. By using Bragg's spectrometer for the purpose, he found that the characteristic X-ray spectra of different elements are remarkably similar to each other in the sense that each consists of *K*-, *L*- and *M*-series. However, there is one very important difference. The frequency of lines (in every series) produced from an element of higher atomic number is greater than that produced by an element of lower atomic number. It is due to the fact that binding energy of electrons increases as we go from one element to another of higher atomic number. Because there is greater positive charge on the nucleus of an element of higher atomic number, larger amount of energy is required to liberate an electron from the *K*, *L* and *M* shells of that element.

Consider the K_α line of the characteristic X-ray spectrum of any element. It is found that higher the atomic number of the target material, higher is the frequency of the K_α line produced by it. The exact mathematical relationship between frequency and atomic number is given by

$$f \propto (Z - b)^2$$

or

$$\sqrt{f} \propto (Z - b) \quad \text{or} \quad \sqrt{f} = a\,(Z - b)$$

where Z is the atomic number of the element and a and b are constants for a particular series but vary from one series to another, *i.e.*, their values for K-series are different from those for L-series etc. The constant b is known as *nuclear screening constant.* For lines of K-series, $b = 1$*. Its value for lines of L-series is more. The above relation is known as Moseley's Law for the characteristic or line X-ray spectrum. It may be stated as follows :

The frequency of a spectral line in the characteristic X-ray spectrum varies directly as the square of the atomic number of the element emitting it.

Fig. 8.10 shows Moseley's diagram for K_α and K_β lines which is obtained by plotting $\sqrt{f}$ versus atomic number of different elements of the periodic table. As expected, the graph is linear.

An exact form of Moseley's Law is

$$\frac{1}{\lambda} = R\,(Z - \sigma)^2 \left(\frac{1}{n_1^{\,2}} - \frac{1}{n_2^{\,2}}\right)$$

where R is Rydberg's constant, Z is the atomic number, σ a correction factor and n_1 and n_2 the principal quantum numbers of the energy levels between which the transition occurs.

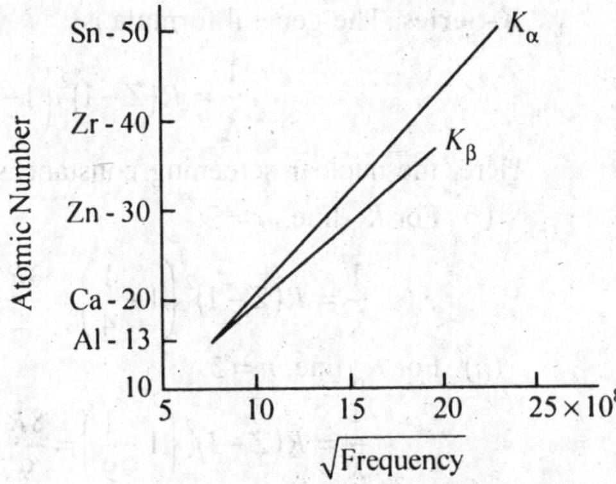

Fig. 8.10

8.5. Importance of Moseley's Law

The great significance of Moseley's Law lies in the fact that it proves for the first time that it is the atomic *number* and not the atomic *weight* of an element which determines its characteristic properties (both physical and chemical). It provides the proper guideline that elements must be arranged in the periodic table according to their atomic *numbers* and not their atomic *weights*.

Accordingly, Moseley's Law has been used to place elements in their proper sequence in the periodic table in certain questionable cases. For example, if we go by the atomic weight, potassium ($_{19}K^{39}$) should come before argon ($_{18}A^{40}$) and similarly, nickel ($_{28}Ni^{58.7}$) should precede cobalt ($_{27}Co^{58.9}$). But Moseley's Law dictates that as per their atomic numbers, their order should be just opposite of the above. This fact is further supported by the chemical properties of these elements.

Moseley's Law has led to the discovery of new elements like hafnium (72), promethium (61), technetium (43) and rhenium (75) etc. by the indication of gaps in Moseley's diagram. This law has been also helpful in determining the atomic number of rare earths thereby fixing their position in the periodic table.

It can be shown that Moseley's Law is in accordance with Bohr's theory of spectral emission from atoms. As shown in Art. 3.13, when an electron jumps from an orbit n_2 to the orbit n_1, the frequency of the radiation given out is,

$$f = \frac{me^4}{8\varepsilon_0^{\,2} h^3} \cdot Z^2 \left(\frac{1}{n_1^{\,2}} - \frac{1}{n_2^{\,2}}\right)$$

---

*Hence, for K_α and K_β lines the relation becomes that the frequency is proportional to the square of the atomic number of the element reduced by one

$$\therefore \quad \sqrt{f} \propto (Z - 1)^2 \quad \text{or} \quad \sqrt{f} = (Z - 1)$$

This may be put as

$$f = \left[\left(\frac{me^4}{8\varepsilon_0{}^2 h^3}\right)\left(\frac{1}{n_1{}^2} - \frac{1}{n_2{}^2}\right)\right] Z^2$$

or $$f \propto Z^2 \text{ or } \sqrt{f} \propto Z$$

Bohr did not take into account the screening effect of electrons whereas Moseley did. That is why the expression becomes

$$\sqrt{f} \propto (Z - b)$$

8.6. Formulae for K- and L-series of X-ray Spectrum

The frequencies of the various lines in the *K*- and *L*-series of the X-ray line spectrum are given by the following empirical formulae.

K-Series. The general formula is

$$\frac{1}{\lambda} = R(Z-1)^2\left(1 - \frac{1}{n^2}\right) \quad \text{where } n = 2, 3 \text{ etc.}$$

Here, the nuclear screening constant is unity.

(*i*) For K_α line, $n = 2$.

$$\therefore \quad \frac{1}{\lambda} = R(Z-1)^2\left(1 - \frac{1}{4}\right) = \frac{3R}{4}(Z-1)^2$$

(*ii*) For K_β line, $n = 3$.

$$\frac{1}{\lambda} = R(Z-1)^2\left(1 - \frac{1}{9}\right) = \frac{8R}{9}(Z-1)^2$$

L-Series. The general formula is

$$\frac{1}{\lambda} = R(Z - 7{\cdot}4)^2\left(\frac{1}{4} - \frac{1}{n^2}\right) \quad \text{where } n = 3, 4 \text{ etc.}$$

Here, screening constant is 7·4.

(*i*) For H_α line, $n = 3$.

$$\therefore \quad \frac{1}{\lambda} = R(Z - 7{\cdot}4)^2\left(\frac{1}{4} - \frac{1}{9}\right) = \frac{5R}{36}(Z - 7{\cdot}4)^2$$

(*ii*) For H_β line, $n = 4$.

$$\frac{1}{\lambda} = R(Z - 7{\cdot}4)^2\left(\frac{1}{4} - \frac{1}{16}\right) = \frac{3R}{16}(Z - 7{\cdot}4)^2$$

Example 8.6. *Find the nuclear screening constant for the L-series of X-rays if it is known that X-rays with a wavelength of* $\lambda = 1{\cdot}43$ Å *are emitted when an electron in a tungsten atom* ($Z = 74$) *is transferred from the M-level to L-level. Take Rydberg constant* $= 10{\cdot}97 \times 10^6\ m^{-1}$.

Solution. When electron jumps from *M*- to *L*-level, the first member of the *L*-series, *i.e.*, L_α line is given out. Its wavelength as given by Moseley's Law is

$$\frac{1}{\lambda} = R(Z-b)^2\left(\frac{1}{2^2} - \frac{1}{3^2}\right) = \frac{5R}{36}(Z-b)^2$$

Substituting the given values, we have

$$\frac{10^{10}}{1{\cdot}43} = 10{\cdot}97 \times 10^6 \times \frac{5}{36}(74-b)^2$$

$$\therefore \quad (74 - b)^2 = 4589{\cdot}8; \ (74 - b) = 67{\cdot}75; \ b = \mathbf{6{\cdot}25}.$$

Example 8.7. *The Kα and Lα absorption edges of copper occur at wavelengths 1·380 Å and 11·288 Å respectively. Calculate the atomic number of copper.*

Solution. It should be remembered that absorption edges are found in the absorption spectrum of X-rays. The absorption edge of each series (of line spectrum) represents the limit of that series. In other words, the short-wavelength limit of each series is called its absorption edge and is written as λ_∞. It may be obtained by putting $n = \infty$ in the formulae of Art. 8.6. Moreover, for *K*-series, the value of screening constant for K_∞ member is 3·3 (instead of 1 for other members). Corresponding value for *L*-series is 11 (instead of 7.4)

K-Series : $$\frac{1}{\lambda_\infty} = R(Z-3{\cdot}3)^2\left(1-\frac{1}{\infty^2}\right) = R(Z-3{\cdot}3)^2$$

$$\therefore \quad \frac{10^{10}}{1{\cdot}38} = (Z-3{\cdot}3)^2$$

L-Series : $$\frac{1}{\lambda_\infty} = R(Z-11)^2\left(\frac{1}{4}-\frac{1}{\infty^2}\right) = \frac{R}{4}(Z-11)^2$$

$$\therefore \quad \frac{10^{10}}{11{\cdot}288} = \frac{R}{4}(Z-11)^2$$

Dividing one by the other, we get

$$\frac{(Z-3{\cdot}3)^2}{(Z-11)^2} = \frac{11{\cdot}238}{4\times 1{\cdot}38} = 2{\cdot}045$$

or $$(Z-3{\cdot}3)/(Z-11) = 1{\cdot}43 \quad \text{or} \quad \mathbf{Z = 29.}$$

Example 8.8. *An impure tungsten target emits a strong Kα line of λ = 0·21 Å and a weak Kα line of λ = 1·537 Å. Can you identify the impurity taking the nuclear screening constant as unity. Given for tungsten, Z = 74, for Ni = 28.*

Solution. The wavelength for K_α line is given by the relation

$$\frac{1}{\lambda} = R(Z-1)^2\left(1-\frac{1}{4}\right) = \frac{3}{4}R(Z-1)^2$$

For tungsten :

$$\frac{1}{0{\cdot}21\times 10^{-10}} = \frac{3}{4}R(74-1)^2$$

$$= \frac{15{,}987\,R}{4}$$

For impurity :

$$\frac{1}{1{\cdot}537\times 10^{-10}} = \frac{3}{4}R\,(Z-1)^2$$

Dividing one by the other, we get

$$\frac{1{\cdot}537}{0{\cdot}21} = \frac{5329}{(Z-1)^2}; \quad \mathbf{Z = 28}$$

Obviously, impurity is **nickel** whose atomic number is 28.

8.7. Intensity Measurement of X-rays

The following three techniques are available for measuring the intensity of X-rays :

(*i*) *By measuring fluorescent light*. X-rays are found to produce fluorescence when incident on certain materials. This fact has been used for quantitative measurement of X-ray intensity by measuring the fluorescent light produced.

(*ii*) *By using a photographic plate*. X-rays are found to blacken a photographic plate when incident on it. The density of the black image produced due to X-ray exposure as measured by a densitometer can be related to the incident X-rays.

(*iii*) *By measuring ionization current*. X-rays ionize a gas while passing through it. These ions can be collected and measure quantitatively with the help of an ionization chamber. Stronger the ionization current, greater the intensity of the X-rays. One form of an ionization chamber is shown schematically in Fig. 8.11. The chamber itself is made of a metallic box into which an electron AB is inserted through an insulating plug. The electrode AB is maintained at a suitable positive potential so that it may attract all the free electrons available in the chamber. These electrons are produced when X-rays are allowed to enter the chamber through the window W. When collected by electrode AB, these electrons constitute a small electric current which is proportional to the intensity of the incident X-rays. This current after amplification is measured by an electrometer which is very sensitive and responds quickly to any changes in the X-ray intensity.

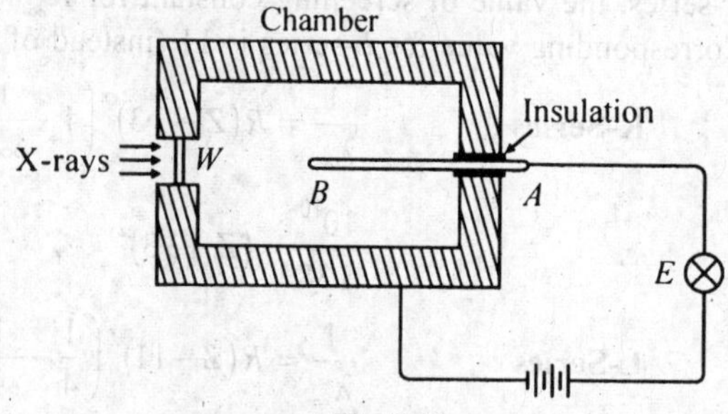

Fig. 8.11

8.8. Absorption of X-rays

When a narrow and monochromatic beam of X-rays passes through matter, part of it is absorbed and the remaining part is transmitted. Absorption of X-rays can be studied with the help of the apparatus shown in Fig. 8.12. The X-rays produced by an X-ray tube are first made into a well-defined narrow beam by passing them through two fine slits S_1 and S_2 in the two lead plates. The beam is then monochromatised by Bragg reflection from a chamber which measures the ionization current. The strength of the ionization current is a measure of the intensity of the X-ray.

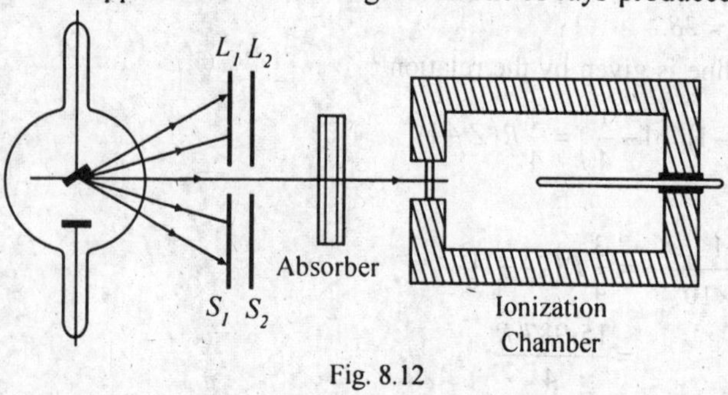

Fig. 8.12

Next, a sheet of the absorbing material is interposed in the path of the X-ray beam before it enters the ionization chamber. It is found that the ionization current and hence the intensity of X-rays is reduced by their passage through the absorber sheet. The ratio I/I_0 of the two ionization currents can be used to measure the absorption coefficient of the material.

Let I_0 be the initial intensity of a homogeneous X-ray beam incident normally on an absorber sheet and I the intensity after the beam has travelled a thickness x of the absorber.

If dI is the further decrease in intensity over a thickness dx of the absorber (Fig. 8.13), then $-dI/dx$ gives the rate of decrease of intensity with thickness. Assuring that this rate is proportional to the intensity I, we have

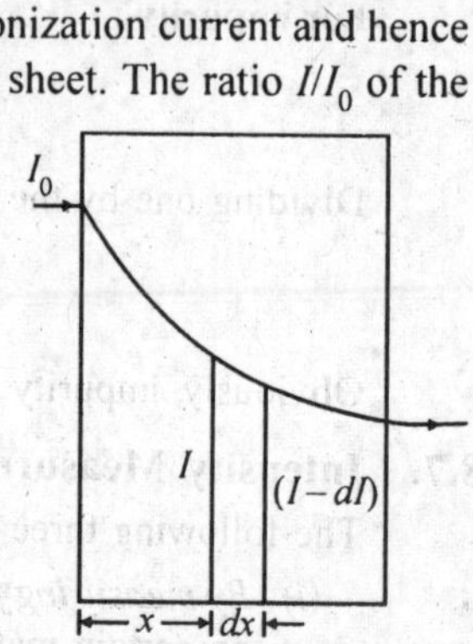

Fig. 8.13

$$-dI/dx \propto I \quad \text{or} \quad dI/dx = -\mu I$$

where μ is a constant of proportionality and is called the *linear absorption coefficient* of the absorber (it is also known as macroscopic absorption coefficient or linear attenuation coefficient).

Now, $$dI/I = -\mu \cdot dx \quad ...(i)$$

Integrating both sides of the above equation, we get

$$\int dI/I = -\mu \int dx$$

∴ $$\log_e I = -\mu x + K \quad ...(ii)$$

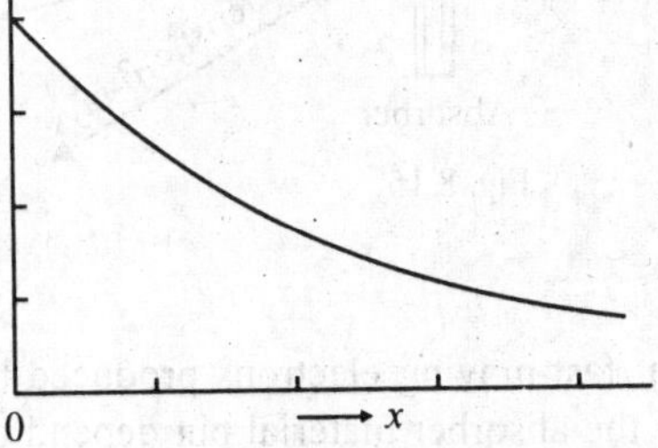

Fig. 8.14

The value of the integration constant *K* can be found from the known initial conditions which are that when $x = 0$, $I = I_0$. Substituting these values in Eq. (*ii*) above, we have

$$\log_e I_0 = K$$

Hence, Eq. (*ii*) becomes

$$\log_e I = -\mu x + \log_e I_0$$

or $$\log_e I/I_0 = -\mu x$$

or $$I/I_0 = e^{-\mu x}$$

or $$I = I_0 e^{-\mu x} \quad ...(iii)$$

It is seen that intensity of the X-ray beam decreases exponentially with the thickness of the absorbing material as shown in Fig. 8.14.

8.9. Definition of Absorption Coefficient (μ)

By rearranging Eq. (*i*) above, we get

$$\mu = \frac{-dI/I}{dx}$$

Hence, μ may be defined as *equal to the fractional decrease in the intensity of the X-rays per unit thickness of the absorber.*

Value of μ depends on the wavelength of the X-rays and on the absorbing material. Very nearly, $\mu \propto \lambda^3$. The value of μ can be determined by finding the thickness of the absorber sheet which reduces the intensity of the X-ray beam to half its initial value. Such a thickness is known as *half-value-lyer T* or *hvl.*

Putting $I/I_0 = \frac{1}{2}$ and $x = T$ in Eq. (*iii*), we get

$$I/I_0 - e^{-\mu T} = \frac{1}{2} \quad \text{or} \quad e^{\mu T} = 2$$

∴ $$\mu T = \log_e 2 = 2{\cdot}3 \log_{10} 2 = 0{\cdot}693$$

∴ $$\boxed{\mu = \frac{0{\cdot}693}{T}}$$

In practice, it is found that a quantity more important than μ is μ/ρ which is called the *mass absorption coefficient* μ_m. It may be defined as *equal to the fractional decrease in intensity of a beam of unit cross-section in travelling unit mass of the absorbing material.*

It is found that μ_m also depends on the wavelength of the X-rays and the atomic number of the absorber material, the exact relation being $\mu_m = k\lambda^3 Z^3$.

In the above analysis of absorption of X-rays, it has not been considered as to what becomes of the portion of the X-rays which is absorbed. It is found that the absorbed portion gives rise to a number of rays collectively known as *secondary* radiations. These secondary radiation are found to consist of the following four distinct and important types of rays as shown in Fig. 8.15.

(i) Scattered X-rays

These are just the same as primary or incident X-rays which have merely changed their direction by passing through the absorbing material. These rays are independent of the nature of the absorbing material but like the primary rays they are also heterogeneous.

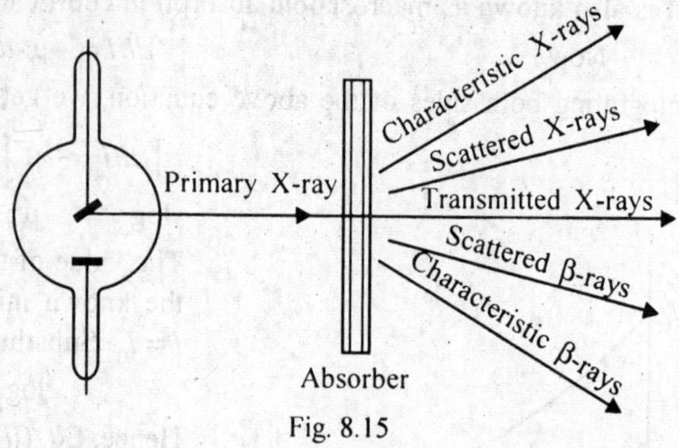

Fig. 8.15

(ii) Characteristic X-rays

As their name implies, they are typical of the nature of the absorbing material. They are homogeneous and do not depend on the *quality* of the primary rays provided they posses certain minimum hardness.

(iii) Scattered β-rays

These consist of β-particles which are nothing else but fast-moving electrons produced by photoelectric process. These are independent of the nature of the absorber material but depend on the *quality* of the primary rays. Their intensity is governed by the relation

$$I_\theta = I_{\pi/2}\,(1 + \cos^2\theta)$$

where I_θ is the intensity of the scattered rays in a direction making an angle of θ with the direction of the primary rays and $I_{\pi/2}$ is the intensity at right angles to the direction of the primary rays.

(iv) Characteristic β-rays

These rays are similar to (*iii*) above except that they are characteristic of the absorber material.

Example 8.9. *The mass absorption coefficient for aluminium for X-rays having λ = 0·32 A is 0·6 cm²/g. If density of aluminium is 2·7 g/cm³, find (i) the linear absorption coefficient of aluminium (ii) the hvl and (iii) the thickness of the absorber needed to cut down the intensity of the beam to 1/20 of the initial value.*

Solution. (*i*) $\mu_m = \dfrac{\mu}{\rho}$ or $\mu = \rho\mu_m = 2{\cdot}7 \times 0{\cdot}6 = \mathbf{1{\cdot}62\ cm^{-1}}$

(*ii*) $T = \dfrac{0{\cdot}693}{\mu} = \dfrac{0{\cdot}693}{1{\cdot}62} = \mathbf{0{\cdot}428\ cm.}$

(*iii*) $\dfrac{I}{I_0} = e^{-\mu x}$

Here, $\dfrac{I}{I_0} = \dfrac{1}{20}$ $\qquad \mu = 1{\cdot}62$

$\therefore \quad \dfrac{1}{20} = e^{-1{\cdot}62x}$ $\quad$ or $\quad 20 = e^{1{\cdot}62x}$

$\therefore \quad \log_e 20 = 1{\cdot}62x$ $\quad$ or $\quad 1{\cdot}62x = 2{\cdot}3\log_{10} 20$

or $\quad 1{\cdot}62x = 2{\cdot}3 \times 1{\cdot}301;$ $\qquad \mathbf{x = 1{\cdot}85\ cm.}$

8.10. Wave Nature of X-rays

The physical nature of X-rays remained in doubt for many years after their discovery by Röntgen in 1895. Some scientists thought them to be high speed particles like cathode rays but more penetrating whereas others regarded them as electromagnetic waves of extremely high frequency. Many experiments were performed to test these two hypotheses but without success. X-rays showed no deflection in an electric or magnetic field neither in earlier experiments, they could be reflected or refracted. To test their wave nature, attempts were made to produce interference as well as diffraction effects similar to those observed with light waves. The wavelength of X-rays is, however, considerably shorter than that of light waves. Hence, it was soon realized that to produce the same order of diffraction with a transmission grating, the X-ray grating would require 40 million rulings per cm ! Clearly, such a figure is outside the bounds of practical possibility.

In 1912, German theoretical physicist Max Von Laue suggested the use of a natural crystal as a closely-spaced three-dimensional grating. Since the structure of a crystal consists of a 3-dimensional array of regularly-spaced atoms, Laue reasoned that the equally-spaced layers of atoms would act like a diffraction grating.*

At Laue's suggestion, the crucial experiment of using crystals as diffraction gratings was performed by W. Friedrich and P. Knipping with the help of the apparatus shown in Fig. 8.16. The X-ray beam from a Coolidge tube is collimated into a fine pencil of rays by a pinhole in each of the two lead screens L_1 and L_2. Next, this beam is incident on a thin crystal of zinc sulphide (ZnS) along one of its axes of symmetry. Finally, the X-ray beam falls on a photographic film placed beyond the crystals.

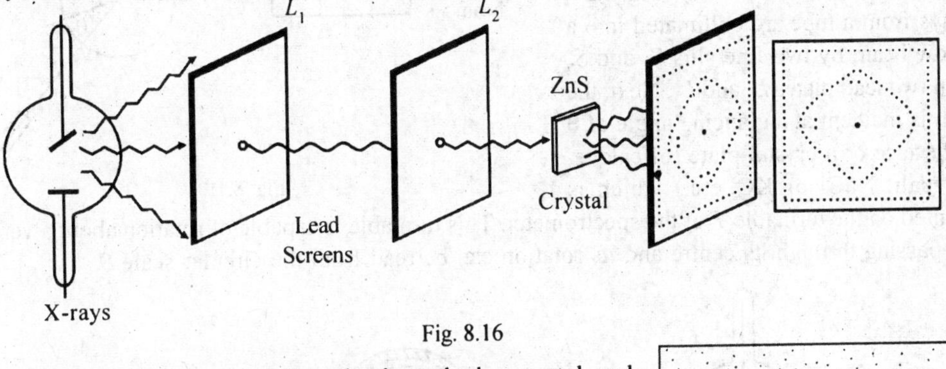

Fig. 8.16

Most of the X-rays go directly through the crystal and produce a black spot at the centre of the film. But there are many other weak diffracted beams which emerge in different directions and produce a series of dark spots on the same film. Hence, the diffraction pattern consists of a central dark spot surrounded by a series of smaller spots arranged in symmetrical order. *This symmetrical pattern of spots is known as Laue pattern* and proves that X-rays are waves.

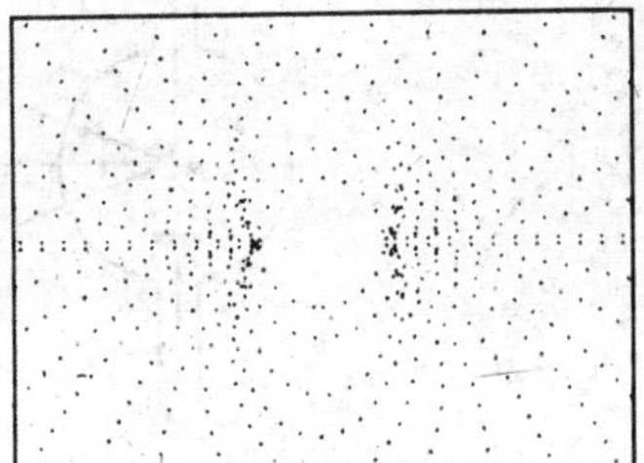

Sugar ($C_{12}H_{22}O_{11}$)

Fig. 8.17

The small number of Laue spots in the case of ZnS crystal (Fig. 8.16) is an indication of its simple crystal structure whereas very large number of spots as for sugar crystals (Fig. 8.17) signifies a relatively complex crystal structure.

8.11. Explanation of Laue's Spots

Origin of Laue's pattern can be explained with the help of the 2-dimensional view of the 3-dimensional array of sodium chloride crystal atoms (Fig. 8.18). It is seen that each Laue spot arises from the reflection of some of the incident X-rays from one of the various sets of parallel crystal planes which are particularly rich in atoms. Two such planes are shown in the figure by double lines. The X-rays obey the law of reflection in as much as the angle of reflection is equal to the angle of incidence.

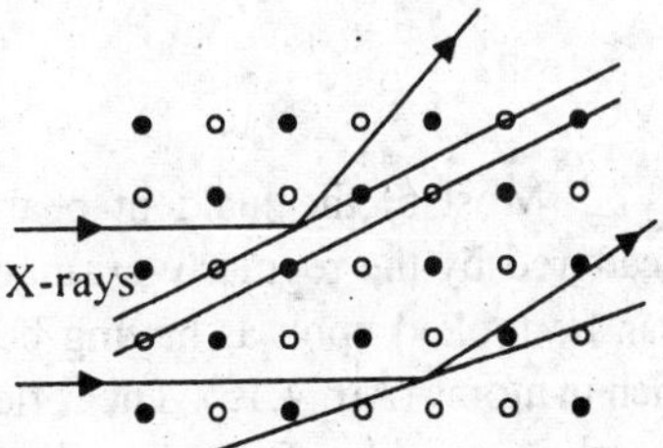

Fig. 8.18

* For details of diffraction grating and its working, please refer to Appendix A towards the end of your book.

8.12. Uses of Laue's Patterns

X-ray diffraction patterns are widely used for the following purposes :

(*i*) in determining the microscope structure of matter in solid state, as for example, in metallurgy for the study of mechanical processes of rolling, hardening and annealing.

(*ii*) in locating as well as determining angular relationship between different planes in a crystal lattice which are rich in atoms.

8.13. Bragg's X-ray Spectrometer

Soon after Laue's experiments, further study of X-ray diffraction by crystals was taken up by British father-and-son team of Sir W.H. Bragg and Sir Lawrence Bragg. For this purpose, they devised an X-ray spectrometer in which a crystal is used not as *transmission grating* (as in Laue's experiment) but as a *reflection grating*.

A schematic diagram of Bragg's X-ray spectrometer is shown in Fig. 8.19 whereas Fig. 8.20 gives a more detailed view. In Fig. 8.19, *M* represents collimator and *IC* stands for ionization chamber. As shown in Fig. 8.20, X-rays from a tube are collimated into a narrow beam by two fine slits S_1 and S_2 cut in two lead plates L_1 and L_2. Next, the beam is incident at a glancing angle of θ on the face of a crystal plate (of calcite, rock salt, NaCl or KCl etc.) which is mounted on the turntable *T* of the spectrometer. This turntable is capable of rotation about a vertical axis passing through its centre and its rotation can be read from the circular scale *S*.

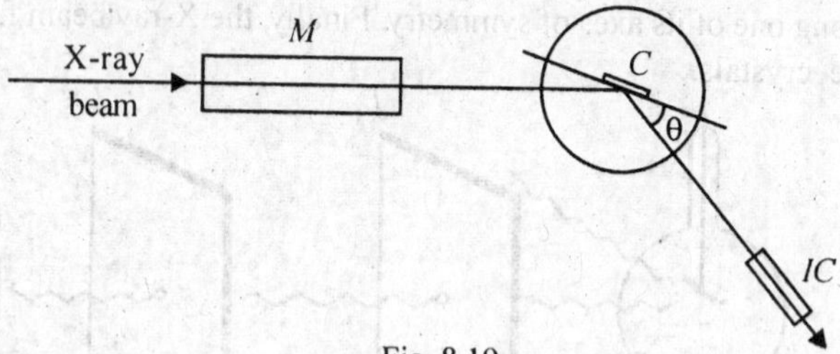

Fig. 8.19

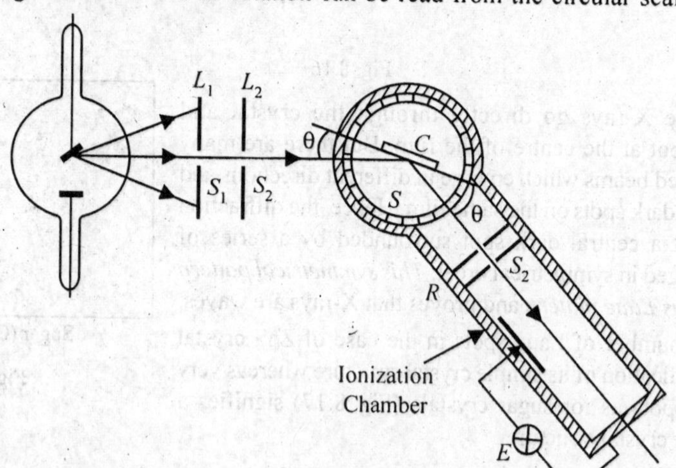

Fig. 8.20

Most of the incident beam passes straight through the crystal plate but some X-rays are scattered by the regularly-arranged atoms lying in different crystal planes. These scattered X-rays can be looked upon as having been reflected from the crystal planes particularly those which are rich in atoms (Art. 4.15). The reflected X-ray beam enters an ionization chamber carried by an arm *R* which is capable of rotation about the same axis as the turntable.

The turntable and arm *R* are so linked together that when the turntable (and hence the crystal) rotates through an angle θ, the arm *R* (and hence the ionization chamber) turns through double the angle, *i.e.*, 2θ. In this, way the beam is always reflected into the ionization chamber whatever its incident or glancing angle at the crystal surface. The ionization current produced by the reflected X-ray beam can be measured by a sensitive electrometer E (or the reflected beam can be recorded on a photographic film in which case the spectrometer is known as spectrograph).

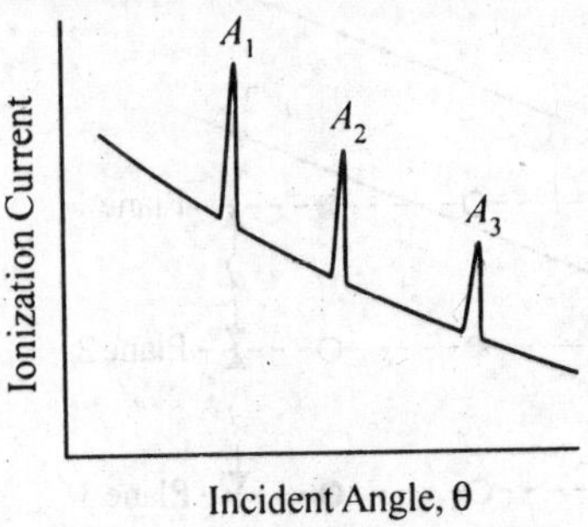

Fig. 8.21

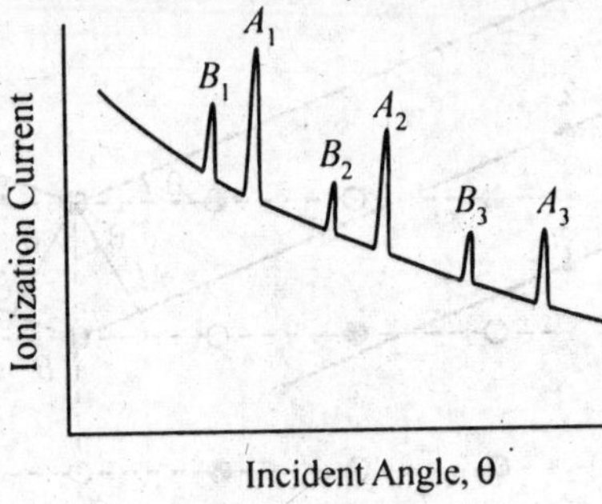

Fig. 8.22

By plotting the electrometer readings corresponding to different glancing angles, it is possible to delineate the X-ray diffraction spectrum. Fig. 8.21 shows such a spectrum when X-ray beam is monochromatic whereas Fig. 8.22 represents the case when the X-ray beam consists of two wavelengths λ_1 and λ_2.

8.14. Theory of Bragg's Diffraction

When a beam of monochromatic X-rays falls on a crystal, it is scattered by the individual atoms which are arranged in sets of parallel layers. Each atom becomes a source of scattered radiations. As already discussed in Art. 4.15, in every crystal, there are certain sets of planes which are particularly rich in atoms. The combined scattering of X-rays from these planes can be looked upon as *reflections* from these *planes*. Because, of this, Bragg *scattering* is usually referred to as Bragg *reflection* and these planes are known as *Bragg planes*. It is dues to the presence of such sets of parallel planes that a crystal acts as reflection grating.

For certain glancing angles, reflections from these sets of parallel planes are in phase with each other. Hence, different reflected X-rays reinforce each other so that the resulting reflection is exceptionally strong. Consequently, ionization current produced is also large. For some other values of θ, the X-ray reflection from different planes are antiphase with each other so that the resulting reflection is either zero or extremely feeble.

Obviously, as the angle of incidence or glancing angle is changed by rotating the turntable T of the spectrometer, a series of alternate maxima and minima of intensity are obtained.

8.15. Bragg's Law

Fig. 8.23 gives a 3-dimensional view of how a beam of monochromatic X-rays undergoes Bragg's reflection from different planes in a NaCl crystal. Fig. 8.24 gives a 2-dimensional view of the same diagram. It shows a beam of monochromatic X-rays incident at a glancing angle θ* on a set of parallel planes of NaCl crystal. The beam is successive layers rich in atoms. Ray No. 1 is *reflected* from atom A in plane 1 whereas Ray No. 2 is *reflected* from atom B lying in plane 2 immediately below atom A. Whether two reflected rays will be in phase or antiphase with each other will depend on their path difference. This path difference can

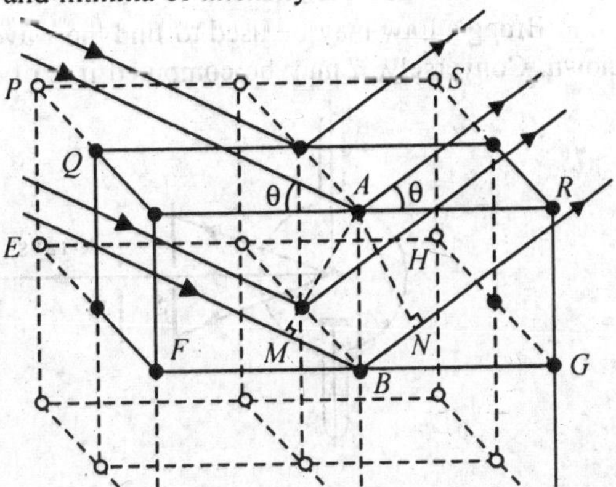

Fig. 8.23

* It may be noted that in Optics, it is usual to measure the angles of incidence and reflection between the rays and the normal to the surface. However, the Bragg angle is measured between the ray and crystal plane.

be found by drawing perpendiculars *AM* and *AN* on ray No. 2. Since the two rays travel the same distance from points *A* and *N* onwards, it is obvious that ray No. 2 travels an extra distance

$$= MB + BN$$

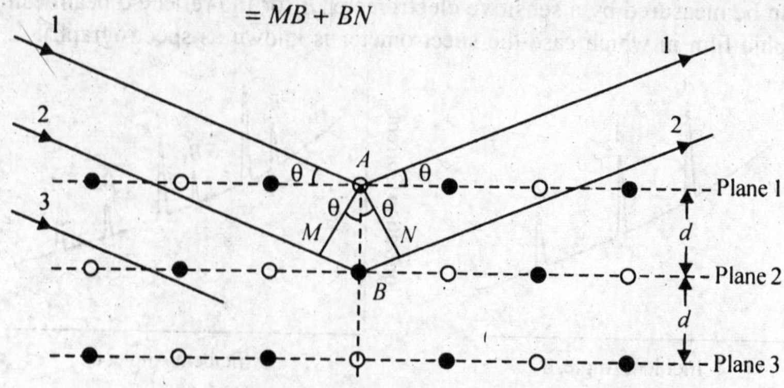

Fig. 8.24

Hence, the path difference between the two reflected beams is

$$= MB + BN = d \sin \theta + d \sin \theta = 2d \sin \theta$$

where *d* is the interplanar spacing, *i.e.*, vertical distance between two adjacent planes belonging to the same set.

The two reflected beams will be in phase with each other* if this path difference equals an integral multiple of λ and will be antiphase if it equals an odd multiple of λ/2.

Hence, the condition for producing maxima becomes

$$\boxed{2d \sin \theta = n\lambda}$$

where $n = 1, 2, 3$, etc. for the first-order, second-order and third-order maxima respectively. This relation is known as Bragg's Law.

Different directions in which intense reflections will be produced can be found by giving different values of θ in the above equation.

For the first maxima, $\sin \theta_1 = \lambda/2d$

For the second maxima, $\sin \theta_2 = 2\lambda/2d$

For the third maxima, $\sin \theta_3 = 3\lambda/2d$ etc.

As seen from Fig. 8.22, the intensities of maxima go on decreasing as the order of the spectrum increases. Moreover, it will be seen from Fig. 8.22 that the separation between maxima for λ_1 and λ_2 goes on increasing as higher orders are considered.

Bragg's Law may be used to find the wavelength λ of the X-rays if interplanar spacing *d* is known. Conversely, *d* may be computed if λ of X-rays is known from some other experiment. In

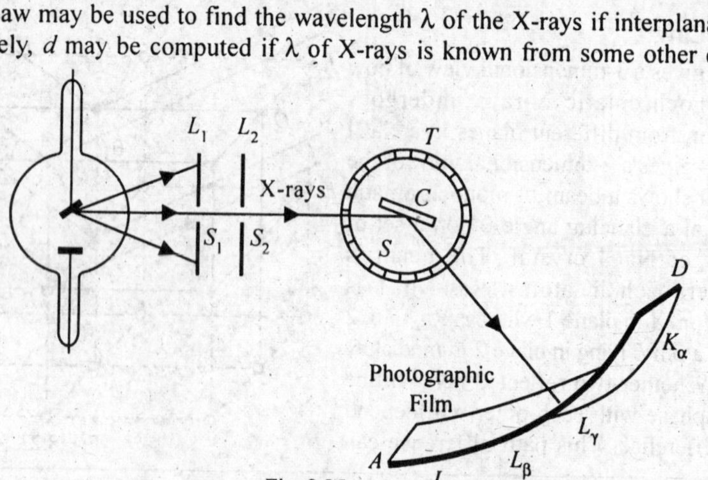

Fig. 8.25

*Same condition will apply to other parallel rays like ray No. 3 shown in the figure.

modern X-ray spectrometers, the ionization chamber is replaced by a photographic plate set at right angles to the reflected beam as shown in Fig. 8.25. Such an instrument is known as X-ray spectrograph and takes much less time for obtaining the complete spectrum.

If we substitute the value of *d* in terms of the Miller indices of the planes for a cubic system, we get

$$2\frac{a}{\sqrt{h^2+k^2+l^2}}\sin\theta = n\lambda$$

or

$$\boxed{\sin^2\theta = \frac{\lambda^2}{4a^2}(h^2+k^2+l^2)} \qquad -n=1$$

8.16. Bragg's Law and Crystal Structure

Bragg used a KCl crystal (which is cubic) and found first maxima of reflected X-rays to occur at values of θ equal to 5·22°, 7·30° and 9·05° respectively when the three planes *BFGC*, *AFGD* and *AFH* (Fig. 4.26) are used in turn as reflecting planes.

Now, for first-order spectrum,

$$2d\sin\theta = \lambda \qquad (\because n=1)$$

$$\frac{1}{d} = \frac{2\sin\theta}{\lambda}$$

$$\therefore \quad \frac{1}{d_1}:\frac{1}{d_2}:\frac{1}{d_3} :: \sin 5{\cdot}22° : \sin 7{\cdot}30° : \sin 9{\cdot}05°$$

$$:: 0{\cdot}091 : 0{\cdot}127 : 0{\cdot}157$$

$$:: 1 : 1{\cdot}4 : 1{\cdot}73$$

These figures agree remarkably well with those derived geometrically in Art. 4.15 which confirms that KCl crystal has a simple cubic crystal structure.

Suppose in the case of some crystal, we get

$$\frac{1}{d_1}:\frac{1}{d_2}:\frac{1}{d_3} :: 1:\frac{1}{\sqrt{2}}:\sqrt{3}$$

then we are sure that is has a body-centred type structure (Art. 4.15). In this way, Bragg's Law can be utilized to analyse different types of crystal structures.

Example 8.10. *X rays from a tube undergo first-order reflection at a glancing angle of 12° from the face of a calcite crystal. The grating space of the calcite is 3·04 × 10*^{8} *cm. Calculate the wavelength of the X-rays. At what angle will be third-order reflection take place from the crystal ?*

Solution. Here, $\theta = 12°, n = 1, d = 3{\cdot}04 \times 10^{-10}$ m

$$\therefore \quad 2\times 3{\cdot}04\times 10^{-10}\sin 12° = 1\times\lambda$$

$$\therefore \quad \lambda = 0{\cdot}79\times 10^{-10}\text{ m} = \mathbf{0{\cdot}79\ Å}$$

For third-order reflection,

$$2\times 3{\cdot}04\times 10^{-10}\sin\theta_3 = 3\times 0{\cdot}79\times 10^{-10}$$

$$\therefore \quad \sin\theta_3 = 0{\cdot}3898;\ \theta_3 = \mathbf{22{\cdot}9°}.$$

Example 8.11. *For a certain b.c.c. crystal, the 110 planes have separation of 1·181 Å. These 10 planes are irradiated with X-rays of wavelength 1·540 Å. How many orders of Bragg reflections can be observed in this case ?*

Solution. According to Bragg's Law; $2\,d\sin\theta = n\lambda$

Now, maximum possible value of $\sin\theta = 1$

$\therefore\ 2d = n\lambda$ or $n = 2d/\lambda$ or $n = 2 \times 1{\cdot}181/1{\cdot}54 = 1{\cdot}53$.

Since fraction is meaningless, only first-order reflection is possible.

Example 8.12. *An X-ray analysis of a crystal is made with monochromatic X-rays of wavelength 0·58 Å. Bragg's reflections are obtained at angles of (a) 6·45° (b) 9·15° and (c) 13°. Calculate the interplanar spacing of the crystal.*

Solution. Bragg's Law of reflection is $2d \sin\theta = n\lambda$

$$\therefore \quad \frac{d}{n} = \frac{\lambda}{2\sin\theta}$$

$$(a) \quad \frac{d}{n} = \frac{\lambda}{2\sin\theta} = \frac{0{\cdot}6\times10^{-10}}{2\times\sin 6{\cdot}45^\circ}$$

$$= \frac{0{\cdot}6\times10^{-10}}{2\times0{\cdot}1132} = 2{\cdot}65\times10^{-10}\text{ m}$$

$$(b) \quad \frac{d}{n} = \frac{0{\cdot}6\times10^{-10}}{2\times\sin 9{\cdot}15^\circ} = \frac{0{\cdot}6\times10^{-10}}{2\times0{\cdot}16} = 1{\cdot}875\times10^{-10}\text{ m}$$

$$(c) \quad \frac{d}{n} = \frac{0{\cdot}6\times10^{-10}}{2\times\sin 13^\circ} = \frac{0{\cdot}6\times10^{-10}}{2\times0{\cdot}2264} = 1{\cdot}32\times10^{-10}\text{ m}$$

Since value of d/n in (*a*) is twice that in (*c*), it shows that angles 6·45° and 13° represent the first and second-order reflection maxima from one set of parallel planes with a spacing which may be found from (*a*) by putting $n = 1$ or from (*c*) by putting $n = 2$.

Taking case (*a*), we have

$$\frac{d}{1} = 2{\cdot}65\times10^{-10}\text{ m} \quad \therefore \quad d = 2{\cdot}65\times10^{-10}\text{ m} = \mathbf{2{\cdot}65\ Å}.$$

Case (*b*) has a different value of d/n and must be the first-order reflection from some other set of parallel planes in the crystal. Their interplanar spacing is

$$\frac{d}{1} = 1{\cdot}875\times10^{-10}\text{ m} \quad \therefore \quad d = 1{\cdot}875\times10^{-10}\text{ m} = \mathbf{1{\cdot}875\ Å}.$$

Example 8.13. *An X-ray beam of wavelength 3 AU is diffracted from (100) planes of a cubic crystal. The first-order diffraction is obtained at an angle of 40°. Determine the spacing between the (100) planes and the volume of the unit cell.*

Solution. According to Bragg's Law, $2d\sin\theta = n\lambda$

Here, $n = 1, \theta = 40^\circ, \sin\theta = 0{\cdot}643$

$\lambda = 3\times10^{-10}$ m, $d = ?$

$$\therefore \quad 2d\times0{\cdot}643 = 1\times3\times10^{-10}$$

$$\therefore \quad d = 2{\cdot}33\times10^{-10}\text{ m} = \mathbf{2{\cdot}33\ AU}.$$

As seen from Fig. 8.24, it represents plane separation. However, lattice constant, $a = 2d = 2\times2{\cdot}33\times10^{-10} = 4{\cdot}66\times10^{-10}$ m.

$\therefore$ Volume of unit cell

$$= a^3 = (4{\cdot}66\times10^{-10})^3 = \mathbf{1{\cdot}01\times10^{-28}\ m^3}.$$

Example 8.14. *The K_α X-rays from palladium are reflected in the first-order from a certain crystal at the following glancing angles : 5·4° from the (100) planes; 7·6° from (110) planes and 9·4° from (111) planes. Show how could this data be used to determine the type of cubic lattice structure possessed by the crystal.*

Solution. From Bragg's formula, we have

$$\frac{1}{d} = \frac{2\sin\theta}{\lambda} \qquad (\because n = 1)$$

$$\frac{1}{d_1} = \frac{2\sin 5{\cdot}4^\circ}{\lambda} = \frac{2\times 0{\cdot}0941}{\lambda} \quad \text{— for (100) plane}$$

$$\therefore \quad \frac{1}{d_2} = \frac{2\sin 7{\cdot}6^\circ}{\lambda} = \frac{2\times 0{\cdot}1323}{\lambda} \quad \text{— for (110) plane}$$

$$\frac{1}{d_3} = \frac{2\sin 9{\cdot}4^\circ}{\lambda} = \frac{2\times 0{\cdot}1633}{\lambda} \quad \text{— for (111) plane}$$

$$\therefore \quad \frac{1}{d_1}:\frac{1}{d_2}:\frac{1}{d_3} :: 0{\cdot}0941 : 0{\cdot}1323 : 0{\cdot}1633$$

$$:: 1 : \sqrt{2} : \sqrt{3}$$

It shows that the given crystal has a simple cubic lattice structure (Art. 4.15).

Example 8.15. *The radiations from an X-ray tube operated at 50 kV are diffracted by a cubic KCl crystal of molecular mass 74·6 and density $1{\cdot}99 \times 10^3$ kg/m^3. Calculate (i) the short wavelength limit of the spectrum from the tube and (ii) glancing angle for first-order reflection from the principal planes of the crystal for that wavelength.*

Solution. (*i*) $$\lambda_{min} = \frac{12{,}400}{50{,}000} = \mathbf{0{\cdot}248\ \text{Å}}.$$

(*ii*) Let us first find the interplanar spacing d before using Bragg's formula $2d \sin\theta = n\lambda$.

Now, $a^3\rho = n\,M/N$ Art. 4.9

Here, $\rho = 1{\cdot}99 \times 10^3$ kg/m^3, $M = 74{\cdot}6$

$n = 4$; $N = 6{\cdot}02 \times 10^{26}$/kg-mole

$\therefore \quad a^3 \times 1{\cdot}99 \times 10^3 = 4 \times 74{\cdot}6/6{\cdot}02 \times 10^{26}$

$\therefore \quad a = 6{\cdot}28 \times 10^{-10}$; $d = a/2 = 3{\cdot}14 \times 10^{-10}$ m

Now, $2d \sin\theta = n\lambda$

$n = 1$, $\lambda = 0{\cdot}248 \times 10^{-10}$ m

$d = 3{\cdot}14 \times 10^{-10}$ m

$\therefore \quad 2 \times 3{\cdot}14 \times 10^{-10} \sin\theta = 0{\cdot}248 \times 10^{-10}$

$\sin\theta = 0{\cdot}0395$; $\theta = \mathbf{2°12'}$.

Example 8.16. *The K_α X-ray line for copper has $\lambda = 1{\cdot}54$ Å and gives first-order Bragg reflection from a cleavage face of NaCl at an angle of 15·9°. Find (i) lattice spacing of NaCl (ii) Avogadro's number if density of NaCl is 2·164 g.cm^3 and molecular weight 58·45.*

Solution. (*i*) For first-order reflection, $2d \sin\theta = \lambda$

$d = \lambda/2 \sin\theta = 1{\cdot}54/2 \times \sin 15{\cdot}9°$

$= 1{\cdot}54/2 \times 0{\cdot}274 = \mathbf{2{\cdot}81\ \text{Å}}$

(*ii*) Let N be Avogadro's number. Then

No. of molecules/gram $= N/58{\cdot}45$

$$\text{No. of molecules/cm}^3 = \frac{N}{58{\cdot}45} \times 2{\cdot}164$$

Since each NaCl molecule contains two atoms (one of sodium and other of chlorine),

$$\text{No. of atoms/cm}^3 = \frac{2N \times 2{\cdot}164}{58{\cdot}45}$$

Since NaCl crystal lattice is a simple cube with atoms at the corners,

No. of atoms in a row one cm long is

$$= \left(\frac{2N \times 2{\cdot}164}{58{\cdot}45}\right)^{1/3}$$

Spacing between atoms and hence rows is

$$d = \frac{1}{(2N \times 2{\cdot}164/58{\cdot}45)^{1/3}}$$

$$\therefore \quad \left(\frac{2N \times 2{\cdot}164}{58{\cdot}45}\right)^{1/3} = \frac{1}{2{\cdot}81 \times 10^{-8}}$$

$$\therefore \quad N = \frac{58{\cdot}45}{2 \times 2{\cdot}164 \times 2{\cdot}81^3 \times 10^{-24}} = \mathbf{6{\cdot}08 \times 10^{23}}$$

8.17. Laue Method for Study of Crystal Structure

It is one of the principal methods of crystal structure analysis. In this method, a bundle of *polychromatic* X-rays impinge at 90° on the plane of a crystal under study. On their way through the crystal, the rays encounter Bragg's planes with different spacing d. These different families of planes make different angles θ with the direction of the X-rays. Some combinations of d, θ and λ satisfy the conditions of Bragg's formula with the resulting increase in intensity of diffracted X-rays.

As seen from Fig. 8.26, primary rays pass through the crystal and produce black spot at point C on the photographic plate PP. Different wavelengths included in the primary incident X-ray will produced spots though somewhat less pronounced around this central spot as explained in Art. 8.11.

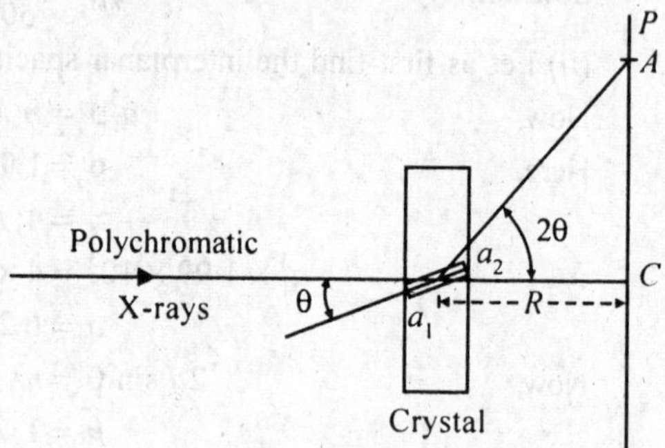

Fig. 8.26

In Fig. 8.26, supposed a_1, a_2 represents the position of one of the possible Bragg's plane, θ the corresponding angle and A the spot on the photographic plate. Then as seen

$$AC = R \tan 2\theta$$

If AC and R are measured, the θ can be found for the corresponding plane.

8.18. Powder Crystal Method

This crystallographic technique is used when *single crystals of reasonably large size are not available**. Hence, it is a standard technique when the crystalline substance is available either in the form of very small crystals or in powder form. One form of this technique known as Debye-Sherrer Method (after the name of its inventors) requires that the crystalline substance be finely ground and powdered so that its tiny crystals (*i.e.*, crystallites) assume random orientations. A small specimen of the crystalline powder is taken either in a small capillary tube of non-diffracting material or is just stuck on a hair by means of gum and interposed in the path of a narrow monochromatic beam of X-rays as shown in Fig. 8.27.

*If they are available, then Laue or Bragg's method can be used.

The basic principle underlying this powder technique is that since millions of tiny crystals in the powder have completely random orientations, all possible diffraction planes will be available for Bragg reflection to take place. Such reflections will take place from many sets of parallel planes lying at different angles to the incident X-ray beam. Moreover, each set will give not only first-order reflections but those of higher orders as well. Since all orientations are equally likely, the reflected rays will form a cone whose axis lies along the direction of the incident beam and whose semi-vertical angle is twice the glancing angle for that particular set of planes (Fig. 8.26). A more detailed view of one such cone is shown in Fig. 8.28 where C stands for the crystal. For each set of planes and for each order, there will be such a cone of reflected X-rays. Their intersections with a photographic film (set with its plane normal to the incident beam form) a series of concentric circular rings. Radii of these rings are recorded and the film can be used to find the glancing angle and hence the interplanar spacing of the crystalline substance.

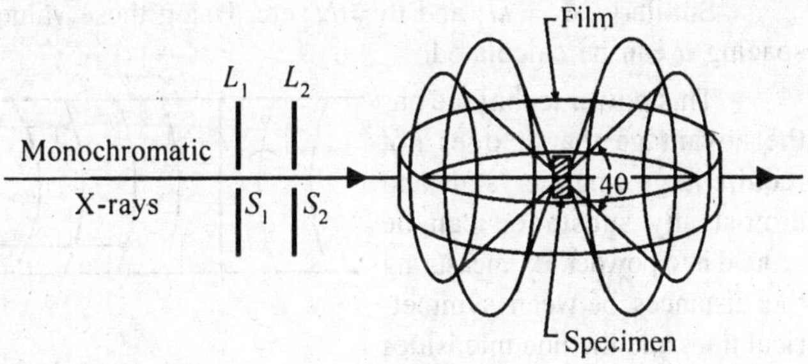

Fig. 8.27

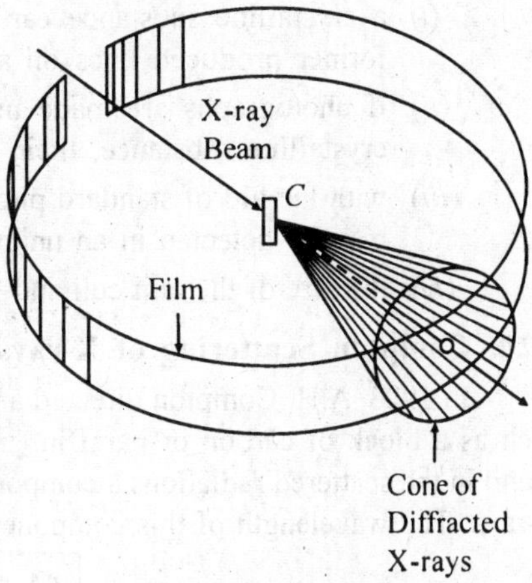

Fig. 8.28

8.19. Experimental Arrangement

As shown in Fig. 8.29, X-rays from a tube are monochromatised by passing them through a filter F which absorbs all but one wavelength of the incident X-rays. After being collimated by two fine slits S_1 and S_2 cut in two lead plates L_1 and L_2, the beam falls on the powder specimen C stuck on hair by means of gum and suspended along the axis of the cylindrical camera called Debye-Sherrer camera. The photo film fits round the inner surface of the camera and almost completely surrounds C in order to receive rays diffracted up to 180°. Since the film is of narrow width, only parts of the circular rings register on the film. Their appearance is somewhat like that shown in Fig. 8.30 when the film is laid flat. The curvature of these area reverses after the angle of diffraction exceeds 90°.

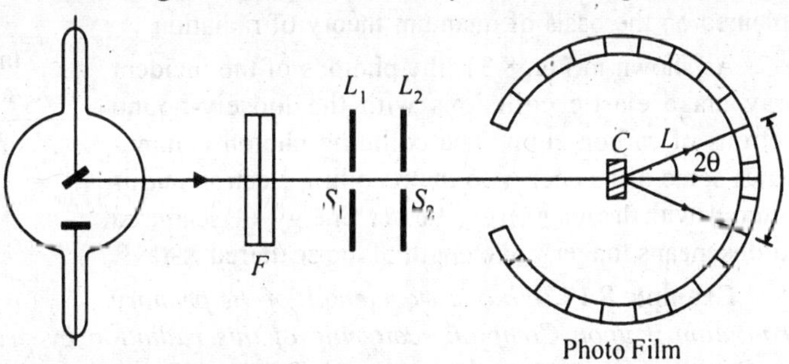

Fig. 8.29

If distances between symmetrical lines on the stretched photograph are l_1, l_2, l_3 etc. and diameter of the cylindrical film is D, then

$$\frac{l_1}{\pi D} = \frac{4\theta_1}{360°}$$

$$\theta_1 = \frac{90°}{\pi D} \cdot l_1 = k l_1 \quad \text{where } k = \frac{90°}{\pi D}$$

Similarly, $\theta_2 = kl_2$ and $\theta_3 = kl_3$ etc. Using these values of θ in Bragg's formula, interplanar spacing d can be calculated.

This power technique has the advantage that it does not require large single crystals and almost any substance can be ground into powder. By measuring the distances between symmetrical lines and the line intensities with the help of a diffractionmeter, following information can be obtained :

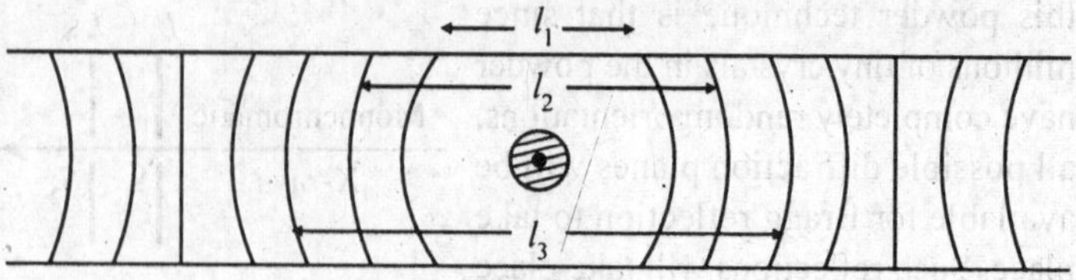

Fig. 8.30

(*i*) a crystalline substance can be distinguished from an amorphous one because the former produced lines on a photograph, the latter does not.

(*ii*) if photographs are made before and after treatment (thermal or mechanical) of a crystalline substance, their comparison can reveal any alteration in the lattice.

(*iii*) with the aid of standard photographs of known chemical compounds, their presence may be detected in an unknown mixture.

(*iv*) the size of the unit cell and the type of the lattice can be determined in simple cases.

8.20. Compton Scattering of X-rays

In 1923, A.H. Compton directed a beam of monochromatic X-rays on a scattering substance such as a block of carbon or paraffin (in which electrons are loosely bound to their atoms) and found in the scattered radiations a component whose wavelength is slightly longer than the incident X-rays. The wavelength of this component is found to be longer by an amount of $\Delta\lambda$ given by

$$\boxed{\Delta\lambda = 0{\cdot}024\,(1 - \cos\theta)\ \text{Å}}$$

where θ is the angle between the direct and scattered beams and wavelength is measured in angstrom units ($1\ \text{Å} = 10^{-10}$ m).

This is known as Compton scattering and has been explained on the basis of quantum theory of radiation.

As shown in Fig. 8.31, the photons of the incident X-ray make elastic collisions with the loosely-bound electrons of carbon atom. The colliding photon communicates some of its energy to the recoiling electron and is scattered with lesser energy. Lesser energy of scattered photons means longer wavelength of the scattered X-rays.

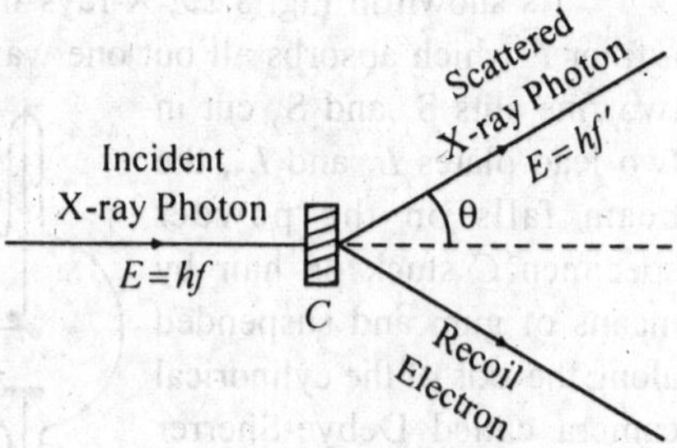

Fig. 8.31

Example 8.17. *What is wavelength of the primary X-radiation if upon Compton scattering of this radiation by graphite at an angle of 60°, the wavelength of the scattered radiation is $2{\cdot}54 \times 10^{-11}$ m ?*

Solution. As seen from Art. 8.20, the amount by which the wavelength of the scattered X-ray is increased is

$$\Delta\lambda = 0{\cdot}024\,(1 - \cos\theta)\ \text{Å}$$

$$= 0{\cdot}024\,(1 - \cos 60°) = 0{\cdot}012\ \text{Å}$$

Now, $$\lambda + d\lambda = 0{\cdot}254 \times 10^{-10}\ \text{m} = 0{\cdot}254\ \text{Å}$$

$$\lambda = 0{\cdot}254 - 0{\cdot}012 = \mathbf{0{\cdot}242\ Å.}$$

Example 8.18. *A powder pattern is obtained for lead with radiations of $\lambda = 1{\cdot}54$ Å. The (220) reflection is observed at Bragg angle $\theta = 32°$. What is the lattice parameter of lead and the radius of the atom ?*

Solution. It will be assumed that the given reflection is the first-order reflection.

Now, $$2d \sin\theta = n\lambda$$

$\therefore$ $$2d \sin 32° = 1 \times 1{\cdot}54 \times 10^{-10}\ \text{m}$$

or $$2d \times 0{\cdot}53 = 1{\cdot}54 \times 10^{-10};\ d = 1{\cdot}45 \times 10^{-10}\ \text{m}$$

This interplanar spacing and the lattice constant of lead are connected by the relation

$$d = \frac{a}{(h^2 + k^2 + l^2)^{1/2}}$$

$\therefore$ $$a = 1{\cdot}45 \times 10^{-10}\,(2^2 + 2^2 + 0^2)^{1/2} = \mathbf{4{\cdot}1 \times 10^{-10}\ m.}$$

Now, lead has face-centred cubic (*FCC*) lattice and its lattice constant is related to its atomic radius by the relation

$$a = 4r/\sqrt{2} \quad \text{or} \quad r = \sqrt{2}\cdot a/4$$

$\therefore$ $$r = \sqrt{2} \times 4{\cdot}1 \times 10^{-10}/4 = \mathbf{1{\cdot}45 \times 10^{-10}\ m.}$$

8.21. Properties of X-rays

Main properties of X-rays may be summarised as under :

(*i*) Like visible light, X-rays consist of electromagnetic waves of very short wavelength (or of very high frequency) and show reflection, refraction, interference, diffraction and polarization etc.

(*ii*) They are not deflected by electric and magnetic fields.

(*iii*) They possess high penetrating power and can pass through many solids which are opaque to visible light. The transparency depends on the density of the material. Higher the density of the substance, the less transparent it is to the X-rays. For example, a sheet of lead 1 cm thick can absorb X-rays whereas aluminium sheet of same thickness cannot.

The penetrating power of X-rays depends upon (*a*) the voltage applied across the cathode and anode of the X-ray tube and (*b*) the atomic number of the material of the cathode. Greater the accelerating potential of the X-ray tube and higher the atomic number of its target material, the more penetrating the X-rays produced.

(*iv*) They ionize a gas and also eject electrons from metals on which they fall.

(*v*) They cause fluorescence in many substances like barium, cadmium, tungstate and zinc sulphide etc.

(*vi*) They suffer Compton scattering.

(*vii*) They have a destructive effect on living tissue. Exposure of human body to X-rays causes the reddening of skin and surface sores.

8.22. Practical Applications of X-rays

On account of their diverse and distinctive properties, X-rays have been put to many uses in different fields of our daily life, *i.e.*, in industry, medicine and research.

(*a*) Industrial applications. Some of these applications are as under :

(*i*) to detect and photograph defects within a body, *i.e.*, in its internal structure such as metals, machine parts intended for withstanding high pressures, cracks in wood, porcelain and other insulators, defects in diamonds and other precious stones, in moulds, forgings and castings etc.

(*ii*) to analyse the structure of alloys and other composite bodies by determining the crystal form in an ingot with the help of diffraction of X-rays. In this way, alloys like cobalt-nickel, steel, bronze, duraluminium, artificial pearls and old paintings have been analysed.

(*iii*) to study the structure of rubber, cellulose and plastics. The diffraction of X-rays by these substances leads to valuable information about their molecular grouping.

(*b*) Applications for pure scientific research

(*i*) for investigating the structure of the atom.

(*ii*) for studying the structure of the crystalline solids and alloys (X-ray crystallography).

(*iii*) for identification of chemical elements including determination of their atomic numbers.

(*iv*) for analysing the structure of complex organic molecules by examining their X-ray diffraction patterns.

(*c*) Medical applications. These can be broadly divided into two classes; one for diagrams (radiography) and the other for curative purposes (X-ray therapy).

(*i*) *Radiography*

X-rays are being widely used for detecting fractures, tumours, the presence of foreign matter like bullets etc. in the human body as well as diseased organs of the body. It is due to differential absorption of X-rays between bones, tissues and metals. Radiographs or X-ray photos are used for this purpose. Since bones are more dense and hence more opaque to X-rays than flesh, a contrasting radiograph of the human body can be obtained for leisurely study by interposing it between the X-ray tube and photo film. Where the organs do not provide contrast as, for example, the intestines or other fleshy parts of the human body, artificial means are adopted to create sufficient contrast between them. In such cases, before taking radiograph, barium or bismuth meal is given to the patient. This meal consists of milk to which some amount of barium sulphate or bismuth carbonate has been added. A few hours after the meal has been taken, these powders settle in the gastrointestinal tract and if a radiograph is taken at that time, the intestines stand out in sharp contrast to the surrounding tissues due to the fact that absorption coefficient of barium is greater on account of its high atomic number. In this way, peptic ulcers and ruptures etc. in the internal organs of the human body can be accurately located. Similarly, radiographs are routinely used for the diagnosis of tuberculosis, stones in kidneys and gall-bladders etc.

(*ii*) *X-ray or Röntgen Therapy*

Many types of skin diseases, malignant sores, cancers and tumours have been cured by controlled exposure to X-rays of suitable quality. This curative power of X-rays is due to the fortunate fact that diseased tissue is more susceptible to destruction than the surrounding healthy tissue. Unnecessary long exposure of human body to X-rays produces many injurious effects including the loss of white cells in the blood, sterility and harmful genetic changes.

X-rays have been used for the identification of different types of cells and tissues and for bringing about genetic mutations.

TUTORIAL PROBLEMS

1. Find the shortest wavelength λ_{min} that can be emitted by the sudden stoppage of a fast-moving electron when it strikes the screen of a TV tube operating at 10 kV. What is the spectral region in which this wavelength lies ? **[1·24 Å; X-ray]**

2. What is the minimum wavelength and the corresponding maximum frequency of the X-rays that can be excited when 12·4 kV is applied to an X-ray tube ? **[1·0 Å; 3×10^{18} Hz]**

3. An X-ray tube passes 10 mA current at a potential difference of 150 kV. Calculate (*i*) the rate at which the target is being heated if only 1% of the incident energy is converted into X-rays (*ii*) the average rise in temperature of the target if it weighs 300 g and has a specific heat of 0·035. **[(*i*) 335 cal/s (*ii*) 33·8 °C/s]**

4. The wavelength of K_α X-ray line for tungsten is 0·21 mμ. What would be its value of copper ? Atomic number of tungsten is 74 and that of Cu is 29. Hint : $\sqrt{f} \propto (Z-1)$ **[0·143 mμ]**
5. X-ray spectrum of a cobalt target ($Z = 27$) contains strong K_α line of wavelength 1·785 Å and weak K_α lines having wavelength of 2·285 Å and 1·537 Å due to impurities. Using Moseley's Law, calculate the atomic numbers of the two impurities and identify their nature. The screening constant b for K series may be taken as unity. **[24 and 29; Cr and Cu]**
6. Prove that tenth-value layer of an absorber equals 2·3/μ where μ is the linear absorption coefficient of the absorber material.
7. The value of μ for aluminium is 13·9 cm^{-1} for K_α line emitted by a tungsten target in an X-ray tube. Calculate what percentage of the intensity of this line passes through a 5 mm thick plate of aluminium ? **[70·7 percent]**
8. X-rays of wavelength 1·6 Å are diffracted by a Bragg crystal spectrometer at an angle of 14·2° in the first order. What is the spacing of atomic layers in the crystal ? **[2·15 Å]**
9. The K_α line of molybdenum ($\lambda = 0{\cdot}712$ Å) undergoes Compton scattering in a carbon block. If the scattering takes place from electrons, calculate (*i*) the Compton change in wavelength at an angle of 90° and (*ii*) the scattered wavelength. **[(*i*) 0·024 Å (*ii*) 0·736 Å]**
10. A γ-ray of $\lambda = 1{\cdot}06 \times 10^{-2}$ Å from cobalt undergoes Compton scattering from a block of graphite. Find the fractional changes ($\Delta\lambda/\lambda$) in wavelength for the scattered radiation on the assumption that ejected electron has no binding energy in carbon. **[2·28]**
11. The glancing angle for the first-order spectrum is found to be 5°30′ in Bragg's X-ray spectrometer. If the crystal spacing $d = 2{\cdot}65 \times 10^{-10}$ m, calculate the wavelength of X-ray. **[0·5 AU]**

HIGHLIGHTS

1. X-rays are produced when high-speed electrons strike a metal of high atomic number.
2. X-ray spectrum consists of two spectra :

   (*i*) continuous spectrum. It is due to the continuous deacceleration of the electron as it passes through the field of the nucleus.

   Its short-wavelength limit is given by

   $$\lambda_{min} = 12{,}400/V \text{ Å}$$

   (*ii*) lines or characteristics spectrum. It is produced when electrons of the target material are dislodged from their orbits by the striking electrons. The spectrum consists of K-, L- and M-series etc. depending on where electron jump ends.
3. Moseley's Law states that the frequency of a spectral line in the characteristic X-ray spectrum varies directly as the square of the atomic number of the element emitting it.
4. Absorption of X-rays follows the exponential law

   $$I = I_0\, e^{-\mu x}$$

   where μ is the absorption coefficient.
5. The linear absorption coefficient μ may be defined as the fractional decrease in the intensity of the X-rays per unit thickness of the absorber material.
6. According to Bragg's Law, the diffraction maxima are given by the equation

   $$2d \sin\theta = n\lambda$$
7. Due to Compton scattering, change in wavelength of the scattered X-rays is given by

   $$\Delta\lambda = 0{\cdot}024\,(1 - \cos\theta) \text{ Å}$$
8. Practical applications of X-rays include (*i*) industrial applications (*ii*) scientific research and (*iii*) medical applications including radiography and X-ray therapy.

Can You Answer This ?

1. Write short note on industrial applications of X-rays.
2. (*a*) Derive an expression for Bragg's Law.
   (*b*) Explain Moseley's Law.
3. (*a*) Deduce Bragg's Law for the diffraction of X-rays in crystals.
   (*b*) Explain how Bragg's Method is used to determine the crystal structure of KCl and NaCl crystals.
   (*c*) Write short note on Moseley's Law.
4. (*a*) Describe with the help of a diagram the essential parts of an X-ray equipment. Discuss the factors that determine (*i*) the intensity and (*ii*) the quality of X-rays in an X-ray tube.
   (*b*) Describe Moseley's work stressing its importance in interpreting the structure of the atom.
5. What are continuous and characteristic of X-ray spectra ? Explain their origin.
6. (*a*) Describe and explain how wavelength of X-rays can be determined with the help of a Bragg's spectrometer.
   (*b*) Show that the minimum wavelength λ_{min} of rays produced in an X-ray tube operating at V kilovolt is given by $12{\cdot}415/V$ angstrom units.
7. Explain Bragg's Law. How can with the help of such a law crystal structure be determined ?
8. Write short note on Coolidge tube for production of X-rays.
9. (*a*) Describe the method of producing X-rays with a Coolidge tube.
   (*b*) Discuss the origin of X-rays.
10. Explain Bragg's Law in X-ray diffraction. Describe Bragg's X-ray spectrometer and explain its use in determining the wavelength of X-rays.
11. (*a*) Describe the method of producing X-rays with a Coolidge tube.
    (*b*) Describe Bragg's X-ray diffraction method of determining the crystal structure.
12. Write note on X-ray spectrometer.
13. Explain how an X-ray spectrometer can be used to study the structure of crystals.
14. (*a*) Give theory of production of X-rays and describe one form of an X-ray tube. What are characteristic X-rays ?
    (*b*) Name the rays which are more penetrating than X-rays.
15. State Bragg's Law for X-ray reflection from crystals.
16. Write short note on Coolidge X-ray tube.
17. (*a*) Explain the principle and working of an X-ray spectrometer.
    (*b*) State Moseley's Law and explain its importance.
18. Describe the X-ray powder diffraction camera and explain how it is used to determine crystal structure.
19. Write short note on applications of X-rays in industry.
20. Describe in details the construction and working of Coolidge X-ray tube.
21. Write short note on Moseley's Law.

WAVES AND PARTICLES 9

Wave and Particle Duality of Matter — De Broglie's Waves — De Broglie's Electron Waves — Properties of Matter Waves — Particle Nature of X-rays — Experimental Study of Matter Waves — The Davisson-Germer Experiment — Electron Diffraction Patterns — Bohr's Principle of Complementarity — Electron Microscope — Electron Waves Within the Atom — The New Atomic Picture — Heisenberg's Principle of Indeterminacy or Uncertainty — Free Electron Theory of Metals — Quantization of Free Electron Energy — Energy Distribution Among Free Electrons — Conduction Process in a Metal Under Electric Field — Schrödinger Equation — Applications of Schrödinger Equation — Highlights — Can You Answer This ?

9.1. Waves and Particle Duality of Matter

The concept of wave nature of matter arose from the dual character of radiation which sometimes behaves as a wave and at other times as a particle.

(*i*) Radiations including visible light, infrared, ultraviolet and X-rays etc. behave as waves in *propagation* experiments based on interference and diffraction. These experiments conclusively prove the wave nature of these radiations because they require the presence of two waves at the *same position at the same time.* Obviously, it is impossible for the two particles to occupy the same position at the same time.

(*ii*) However, radiation behaves as a particle in *interaction* experiments which include black-body radiations, photoelectric effect and Compton effect. Here, radiation interacts with matter in the form of photons or quanta.

Of course, radiation cannot exhibit its particle and wave properties *simultaneously*. Still, this dual nature of radiation was not easily accepted because of the apparently contradictory aspects of the two natures.

(*a*) A wave is specified by its (*i*) frequency (*ii*) wavelength λ (*iii*) phase or wave velocity u (*iv*) amplitude and (*v*) intensity. Moreover, a wave spreads out and occupies a relatively large region of space.

(*b*) A particle is specified by (*i*) mass m (*ii*) velocity v (*iii*) momentum p and (*iv*) energy E. Moreover, a particle occupies a definite position in space and hence is very small

In view of the above, it is rather difficult to accept the conflicting ideas that radiation is a *wave* which is spread out over space and also a *particle* which is localized at a point in space. However, this acceptance is essential if one has to satisfactorily explain the result of experiments which can be performed with radiation.

9.2. De Broglie's Matter Waves

In 1924, the French theoretical physicist de Broglie made a very bold and novel suggestion that like radiation, matter also has dual (*i.e.*, particle-like and wave-like) characteristic. His suggestion was partly based on his intuitive feeling nature is symmetrical and if radiation like light can act like wave some time and like a particle at other times, then things like electrons which are considered particles should also act as waves at times. According to his hypothesis, all matter particles like electrons, protons, neutrons, atoms or molecules have an associated wave with them which is called *matter* wave or *pilot* wave or de Broglie's wave.

The wavelength of the matter wave is given by

$$\boxed{\lambda = \frac{h}{p} = \frac{h}{mv}}$$

where m is the mass of the material particle, v its velocity and p its momentum.*

The above relation is known as de Broglie's wave equation.

9.3. De Broglie's Electron Waves

Consider the case of an electron which has been accelerated through a p.d. of V volts. Its velocity can be found from the relation

$$\frac{1}{2}mv^2 = eV \quad \text{or} \quad mv = \sqrt{2emV}$$

Hence, wavelength of electron wave is

$$\lambda = \frac{h}{mv} = \frac{h}{\sqrt{2emV}} \text{ metre}$$

Substituting $h = 6{\cdot}62 \times 10^{-34}$ J-s; $e = 1{\cdot}602 \times 10^{-19}$ C and $m = 9{\cdot}1 \times 10^{-31}$ kg, we get

$$\lambda = \frac{6{\cdot}62\times10^{-34}}{\sqrt{2V\times1{\cdot}602\times10^{-19}\times9{\cdot}1\times10^{-31}}} = \frac{12{\cdot}25\times10^{-10}}{\sqrt{V}} \text{ metre}$$

$$\therefore \quad \lambda = \frac{12{\cdot}25}{\sqrt{V}} \text{ Å}$$

(*i*) *Wave number*. The wave number (ν) of a wave is given by the reciprocal of its wavelength.

$$\therefore \quad \nu = \frac{1}{\lambda} = \frac{\sqrt{2emV}}{h} = \sqrt{\frac{2emV}{h^2}} \text{ m}^{-1}$$

(*ii*) *Wave velocity*. The velocity u of the matter wave can be found from the photon analogy.

$$E = hf$$

$$\therefore \quad f = \frac{E}{h} = \frac{\frac{1}{2}\cdot mv^2}{h} = \frac{m^2v^2}{2mh}$$

$$= \frac{1}{2m}\cdot h\cdot\frac{m^2v^2}{h^2}$$

Now, $\quad \frac{1}{2}\cdot mv^2 = eV \quad \therefore \quad m^2v^2 = 2emV$

$$\therefore \quad f = \frac{h}{2m}\cdot\frac{2emV}{h^2} = \frac{h}{2m}\cdot\frac{1}{\lambda^2}$$

$$\therefore \quad u = f\lambda = \frac{h}{2m\lambda^2}\times\lambda = \frac{h}{2m\lambda} \text{ m/s}$$

---

*In case, particle velocity is comparable with the velocity of light, then relativistic mass must be used in the above equation. Since, relativistic mass is

$$m = \frac{m_0}{\sqrt{1-(v/c)^2}} = \gamma m_0$$

$$\lambda = \frac{h}{\gamma m_0 v} \quad \text{— where } m_0 \text{ is the rest mass:}$$

$$\boxed{u = \frac{h}{2m\lambda}}$$

If $V = 100$ volts, then $\lambda = 12{\cdot}25/\sqrt{100} = 1{\cdot}225$ Å which is of the order of the interplanar distance in crystals. This fact suggests that the existence of electron wave might be shown by using crystals as diffraction gratings as is done for X-rays. Experiments based on these ideas were performed in 1927 by Davisson and Germer in USA and by G.P. Thomson in UK.

Note. If V exceeds 20,000 volts, then relativistic mass of electron must be used in the above relations.

Example 9.1. *What is the de Broglie's wavelength of an electron accelerated through 20,000 V?*

Solution. As seen from Art. 9.3

$$\lambda = \frac{12{\cdot}25}{\sqrt{V}}$$

$$= \frac{12{\cdot}25}{\sqrt{20{,}000}} = \mathbf{0{\cdot}087\ Å.}$$

Example 9.2. *An electron initially at rest is accelerated through a p.d. of 5000 V. Compute (i) the momentum (ii) the de Broglie's wavelength and (iii) the wave number of the electron wave.*

Also calculate the Bragg angle for its first order reflection from the (111) planes of nickel which are 2·04 Å apart.

Solution. (*i*) Momentum of the electron is

$$p = mv = \sqrt{2emV}$$

$$= (2 \times 1{\cdot}602 \times 10^{-19} \times 9{\cdot}1 \times 10^{-31} \times 5000)^{1/2}$$

$$= \mathbf{3{\cdot}82 \times 10^{-23}\ kg\text{-}m/s}$$

(*ii*) $\lambda = \dfrac{12{\cdot}25}{\sqrt{5000}} = \mathbf{0{\cdot}175\ Å}$

(*iii*) Wave number, $\nu = \dfrac{1}{\lambda} = \dfrac{1}{0{\cdot}175 \times 10^{-10}} = \mathbf{5{\cdot}71 \times 10^{10}}$.

Applying Bragg's formula (Art. 8.15) for reflections at the atomic planes, we have

$$2d \sin\theta = n\lambda$$

$$\therefore \quad 2 \times 2{\cdot}04 \times 10^{-10} \sin\theta = 1 \times 0{\cdot}175 \times 10^{-10}$$

$$\sin 0 = 0{\cdot}043;\ \theta = \mathbf{2{\cdot}5°}.$$

Example 8.3. *An electron initially at rest is accelerated through a p.d. of 54 volts. Compute (i) the velocity of the electron (ii) de Broglie's wavelength and (iii) phase velocity of the electron wave.*

Solution. (*i*) $\dfrac{1}{2}mv^2 = eV$

$$v = \sqrt{2eV/m} = 5{\cdot}93 \times 10^5 \sqrt{V}$$

$$= 5{\cdot}93 \times 10^5 \times \sqrt{54} = \mathbf{4{\cdot}36 \times 10^6\ m/s}$$

(*ii*) $\lambda = \dfrac{12{\cdot}25}{\sqrt{V}}\ Å = \dfrac{12{\cdot}25}{\sqrt{54}} = \mathbf{1{\cdot}67\ Å}$

(*iii*) The wave or phase velocity of the matter wave associated with the electron is

$$u = \frac{h}{2m\lambda} = \frac{6{\cdot}62 \times 10^{-34}}{2 \times 9{\cdot}1 \times 10^{-31} \times 1{\cdot}67 \times 10^{-10}}$$

$$= \mathbf{2{\cdot}18 \times 10^6\ m/s.}$$

Note. It is seen that wave velocity u is exactly half of the particle velocity v which is true of all electrons not acted on by external forces.

Example 9.4. *Compute the de Broglie's wavelength of a photon whose kinetic energy is equal to the rest energy of an electron. Mass of a photon is 1836 times that of the electron.*

Solution. According to Einstein's mass energy equation, the rest energy of an electron is

$$m_0c^2 = 9{\cdot}1 \times 10^{-31} \times (3 \times 10^8)^2 = 81{\cdot}9 \times 10^{-15}\ \text{J}$$

$$\text{Mass of proton} = 1836 \times 9{\cdot}1 \times 10^{-31} = 16{\cdot}71 \times 10^{-28}\ \text{kg}$$

$$\therefore \quad \frac{1}{2}mv^2 = 81{\cdot}9 \times 10^{-15}$$

$$\therefore \quad (mv)^2 = 2 \times 81{\cdot}9 \times 10^{-15} \times 16{\cdot}71 \times 10^{-28}$$

$$\therefore \quad mv = 16{\cdot}54 \times 10^{-21}\ \text{kg-m/s}$$

$$\lambda = \frac{h}{mv} = \frac{6{\cdot}62\times10^{-34}}{16{\cdot}54\times10^{-21}} = 4 \times 10^{-14}\ \text{m}$$

$$= 4 \times 10^{-4}\ \text{Å} = \mathbf{0{\cdot}0004\ Å.}$$

Example 9.5. *What is the de Broglie's wavelength of neutron whose energy is 1 eV? Given* e= *1·6 × 10*$^{-19}$ *C, h = 6·622 × 10*$^{-34}$ *J-s and mass of neutron = 1·676 × 10*$^{-27}$ *kg.*

Solution. Kinetic energy of the neutron is

$$= 1\ \text{eV} = 1 \times 1{\cdot}6 \times 10^{-19}\ \text{J}$$

$$= 1{\cdot}6 \times 10^{-19}\ \text{J.}$$

$$\therefore \quad \frac{1}{2}mv^2 = 1{\cdot}6 \times 10^{-19}$$

or

$$v = (2 \times 1{\cdot}6 \times 10^{-19}/1{\cdot}676 \times 10^{-27})^{1/2}$$

$$= 1{\cdot}38 \times 10^4\ \text{m/s}$$

$$\lambda = \frac{h}{mv} = \frac{6{\cdot}622\times10^{-34}}{1{\cdot}676\times10^{-27}\times1{\cdot}38\times10^4}$$

$$= 2{\cdot}86 \times 10^{-11}\ \text{m} = \mathbf{0{\cdot}286\ Å.}$$

9.4. Properties of Matter Waves

(*i*) Since, $\lambda = h/\sqrt{2emV}$, it is seen that larger the mass of the matter particle, shorter its associated wavelength.

(*ii*) These waves are not electromagnetic waves but are a *new kind* of wave.

(*iii*) They are *pilot* waves in the sense that their only function is to pilot or guide the matter particle as shown in Fig. 9.1.

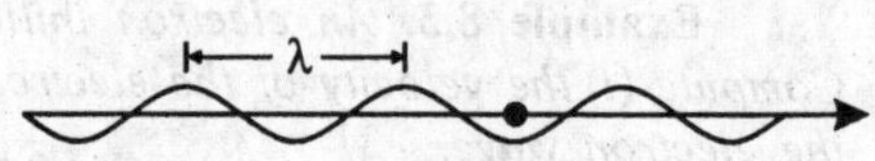

Fig. 9.1

(*iv*) These waves can *travel faster than light* as shown below :

The phase of wave velocity of the matter wave is

$$u = f\lambda, \text{ where } f \text{ is the frequency.}$$

Now, by photon analogy,

$$E = hf \quad \text{or} \quad f = E/h$$

Also,

$$E = mc^2 \quad \therefore \quad f = \frac{mc^2}{h}$$

$$\text{Now,} \qquad \lambda = \frac{h}{mv} \quad \therefore \quad u = f\lambda = \frac{mc^2}{h} \times \frac{h}{mv} = \frac{c^2}{v}$$

Since the speed v of the material particle cannot exceed the velocity of light, it is obvious that the phase velocity of the matter wave associated with it is more than the velocity of light,

(*v*) Whereas the velocity of electromagnetic radiation is constant, that of the matter wave is not since it depends on the velocity of the material particle.

(*vi*) The phase velocity u of the matter wave is also given by $u = h/2m\lambda$.
It shows that phase velocity *depends inversely on wavelength* λ *even if the material particle is moving in vacuum*. This is quite different behaviour from that of light, which moves with the same velocity c regardless of the wavelength, for all waves of the electromagnetic spectrum. This dependence of wave velocity u on wavelength in vacuum is one of the basic differences between electron waves and light waves.

(*vii*) The matter wave is not a physical phenomenon. It is rather a symbolic representation of what we know about the particle. It is a wave of probability.

9.5. Particle Nature of X-rays

Following two experiments which belong to the class known as *interaction* experiments prove the particle-like nature of X-rays.

(*a*) Photoelectric Effect of X-rays

When a beam of X-rays is incident on the surface of a thin gold sheet, many different phenomena may take place. The X-rays may act like waves and produced diffraction pattern (Fig. 8.16). Alternatively, they may act like particles and so collide with gold atoms and produce photoelectrons as in the photoelectric effect (Art. 5.7). Even if the X-ray photons may be all of the same energy, not all the ejected photoelectrons have the same velocity. Instead, they are divided into several well-defined groups as shown schematically in Fig. 9.2 (*a*) by the lengths of the arrows.

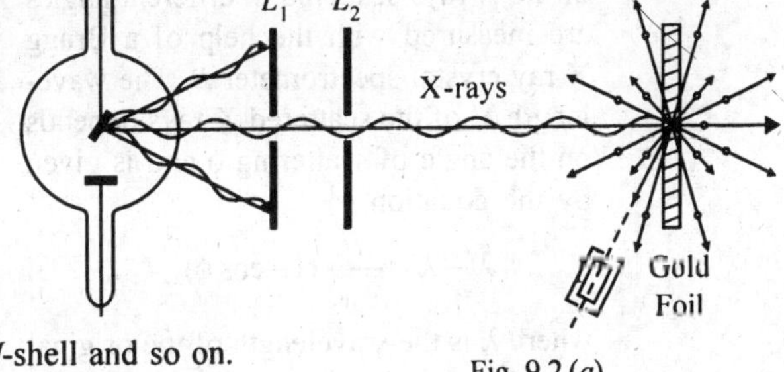

Fig. 9.2 (*a*)

The different velocity groups are associated with various electronic orbits (or shells) within the gold atoms. The slowest electrons having a velocity of v_K are ejected from K-shell, the next faster group with a velocity v_L from the L-shell, the next group with a velocity v_M from M-shell and so on.

The velocity of photoelectrons in each group is given by Einstein's photoelectric equation (Art. 5.9)

$$hf = W_0 + \frac{1}{2}mv^2$$

where W_0 is the work function and represents the energy required to free an electron from any one of the different electron shells.

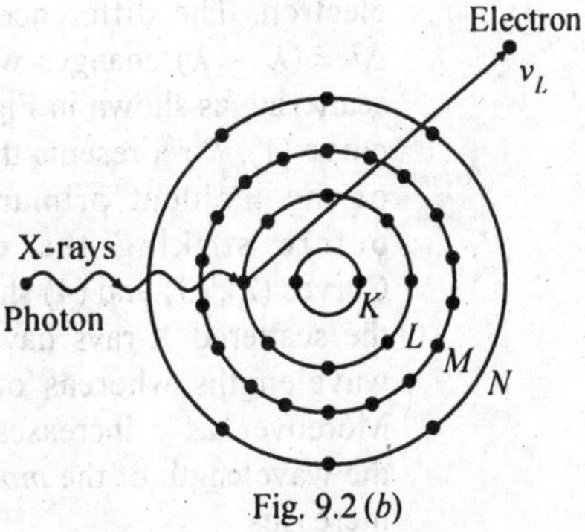

Fig. 9.2 (*b*)

Since value of W_0 for K-shell is more, v_K is less. The electrons liberated from L-shell have relatively higher velocity v_L and hence greater kinetic energy, because value of W_0 for L-shell is less and so on. It is worth noting that energy W_0 used up in ejecting a photoelectron is not lost by the atom. Instead, it is again given out in the form of X-rays of various frequencies. For example, when *K*-electron is ejected, its place may be occupied by an *L*-electron which emits a *K*-series X-rays during its jump from *L*-shell to *K*-shell. This may be followed by an *M*-electron jumping into the vacated *L*-shell, thereby emitting *L*-series X-rays and so on.

(*b*) The Compton Effect

X-rays are scattered by matter in two different ways :

(*i*) *Thomson scattering or coherent scattering* — in which the X-rays are scattered by electrons (of the atoms) without any change in their wavelength. Its effect is to change the direction of the incident X-rays.

(*ii*) *Compton scattering* — in which the scattered X-rays consist of two components : one component has the same wavelength λ as the original incident X-ray and the other component has a slightly longer wavelength λ′.

This phenomenon in which ther is change in wavelength of the scattered X-rays is called Compton Effect. It was discovered by A.H. Compton in 1923 during the course of his studies of the scattering of X-rays by matter. Compton scattering is observed when X-rays of high energy fall on solid matter.

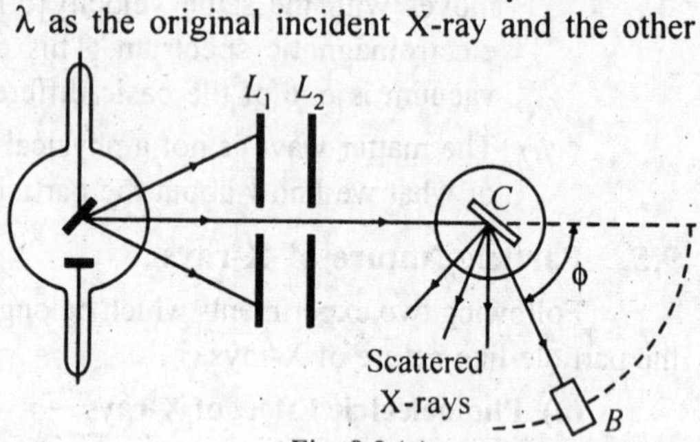

Fig. 9.3 (*a*)

As shown in Fig. 9.3 (*a*), a monochromatic beam of high energy X-rays is made to fall on one face of a small carbon block *C* which scatters out in various directions. Because of its loosely-bound electrons, carbon acts as a good scatterer. The wavelengths of the X-rays scattered at different angles are measured with the help of a Bragg X-ray crystal spectrometer *B*. The wavelength λ′ of the scattered X-rays depends on the angle of scattering φ and is given by the equation

$$\lambda' - \lambda = \frac{h}{m_0 c}(1 - \cos\phi)$$

where λ is the wavelength of the original X-rays and m_0 the rest mass of the electron. The difference in wavelength $\Delta\lambda = (\lambda' - \lambda)$ changes with the angle of scattering as shown in Fig. 9.3 (*b*). In top curve (1), *P* represents the wavelength λ of the incident primary X-ray beam before striking the carbon block. Curves (2), (3) and (4) show that some of the scattered X-rays have changed their wavelengths whereas others have not. Moreover, as φ increases, the change in the wavelength of the *modified* X-rays *M* increases.

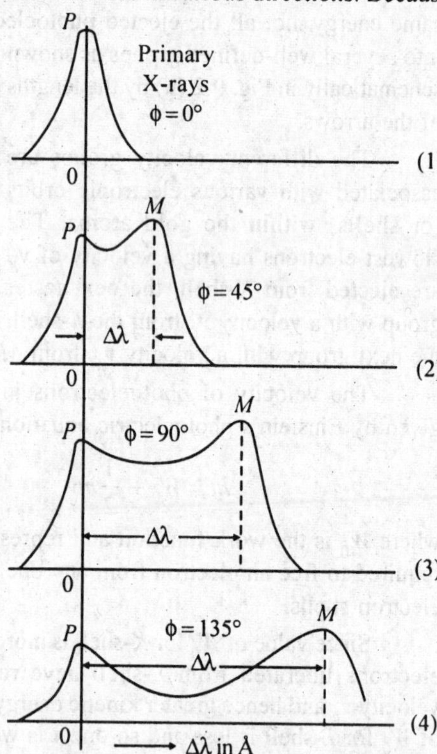

Fig. 9.3 (*b*)

The quantity h/m_0c is called *Compton wavelength* and equals $2{\cdot}43 \times 10^{-12}$ m.

$$\Delta\lambda = 2{\cdot}43 \times 10^{-12} (1 - \cos\phi) \text{ metre}$$
$$= 0{\cdot}0243 (1 - \cos\phi) \text{ A}$$
$$= 0{\cdot}0486 \sin^2 \phi/2 \text{ Å}.$$

Example 9.6. *A beam of monochromatic X-rays with $\lambda = 0{\cdot}09$ Å is incident on a carbon scatterer. The scattered rays are observed at an angle of 54° with the direction of the incident beam. Find (a) the wavelength of the scattered X-rays and (b) the energies of the incident and scattered photons.*

Solution. (*a*)
$$\Delta\lambda = 0{\cdot}0243 (1 - \cos 54°) = 0{\cdot}01 \text{ Å}$$
$$\lambda' = \lambda + \Delta\lambda = 0{\cdot}09 + 0{\cdot}01 = \mathbf{0{\cdot}1 \text{ Å}}$$

(*b*) Energy of incident photon is
$$= hf = hc/\lambda = 6{\cdot}624 \times 10^{-34} \times 3 \times 10^8/0{\cdot}09 \times 10^{-10}$$
$$= 2{\cdot}208 \times 10^{-14} \text{ J} = \mathbf{0{\cdot}138 \text{ MeV}}$$

Energy of the scattered photon is
$$= hf' = hc/\lambda' = 6{\cdot}624 \times 10^{-34} \times 3 \times 10^8/0{\cdot}1 \times 10^{-10}$$
$$= \mathbf{0{\cdot}124 \text{ MeV}}.$$

9.6. Experimental Study of Matter Waves

According to de Broglie's hypothesis, a beam of material particles must possess wave-like characteristics and should undergo reflection, refraction, interference, diffraction and polarization etc. as the more familiar light waves do. This conclusion was put to an experimental test and found to be true. The following two experiments conclusively prove the wave-like character of electrons. Selection of electrons for this purpose was based on two considerations (*i*) they can be easily produced in fairly intense beams with a definite velocity and (*ii*) on account of their small mass, the electron waves have wavelengths of the same order as X-rays and hence can be measured by methods already in use for X-rays.

9.7. The Davisson-Germer Experiment

The first experimental proof of wave nature of atomic particles was provided in 1927 by two American physicists C.L. Davisson and L.H. Germer. Electrons from a hot filament *F* (Fig. 9.4) were accelerated by a small voltage and allowed to strike a single crystal of nickel. These electrons, acting like waves, were diffracted off in certain preferred directions as given by Bragg's diffraction formula for X-rays, *i.e.*, $2d \sin\theta = n\lambda$. These directions were located with the help of an electron detector which could be rotated around the nickel crystal through a large angle. For a given accelerating voltage *V*, the scattering curve showed a peak or maximum in a particular direction. As shown in Fig. 9.5 (*a*), for 54 V electron, most pronounced reflection occurred at an angle of 50°. As shown in Fig. 9.5 (*b*), the reflecting planes in the crystal are inclined at 25° to the top surface of the crystal. Since interatomic distance for a nickel crystal is known to be $2{\cdot}15 \times 10^{-10}$ m, the interplanar distance $d = 2{\cdot}15 \times 10^{-10} \sin 25° = 0{\cdot}909 \times 10^{-10}$ m. Using Bragg formula*, we have

$$2 \times 0{\cdot}909 \times 10^{-10} \sin (90° - 25°) = 1 \times \lambda$$
$$\therefore \quad \lambda = 1{\cdot}65 \times 10^{-10} \text{ m} = 1{\cdot}65 \text{ Å}$$

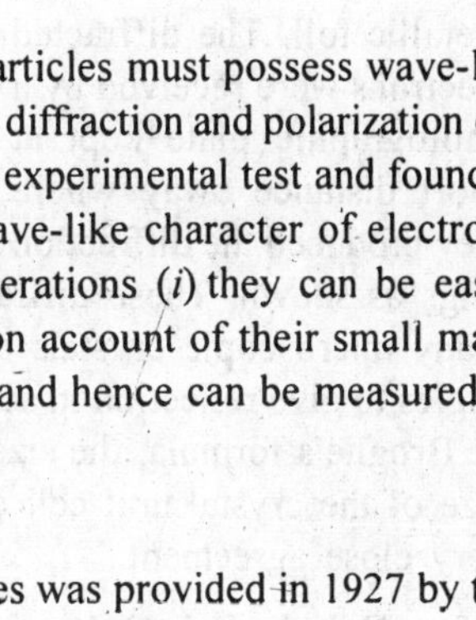

Fig. 9.4

*In this formula, the angle θ is the angle between the incident beam and the metal surface.

The theoretical value of de Broglie's electron wave is given by

$$\lambda = \frac{12{\cdot}25}{\sqrt{54}} = 1{\cdot}67\ \text{Å}$$

The close agreement between the two values confirms de Broglie's hypothesis of matter waves.

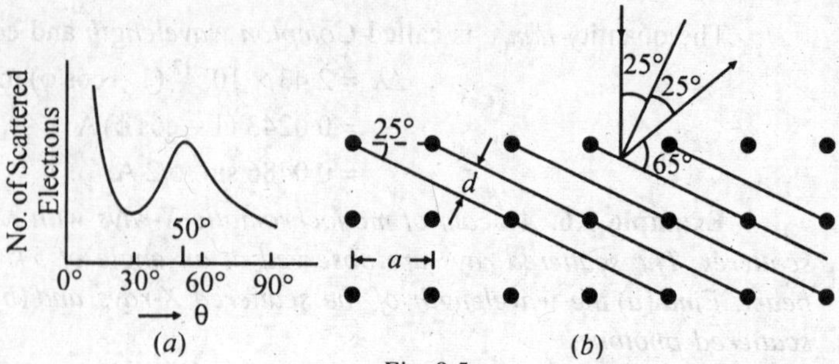

Fig. 9.5

9.8. Electron Diffraction Patterns

Experiments similar to Von Laue's X-ray diffraction experiments were performed by the English physicist G.P. Thomson and A. Reid in 1928 and independently by a Japanese physicist Kikuchi. A schematic diagram of the apparatus used by them is shown in Fig. 9.6.

A narrow beam of high-velocity electrons from an electron gun was projected on the surface of a thin metallic foil. The diffracted electrons were received by a photographic plate kept at short distance away where they produced the diffraction rings as shown. These diffraction patterns were obtained due to the fact that metals consist of many microscopic crystals arranged in random fashion so that some are always at the proper angles to give reflection in accordance with Bragg formula. Knowing the electron wavelength from de Broglie's formula, the size of the crystal unit cell was calculated from the ring diameters. The size of the crystal unit cell obtained by X-rays and by electron diffraction were found to be in very close agreement.

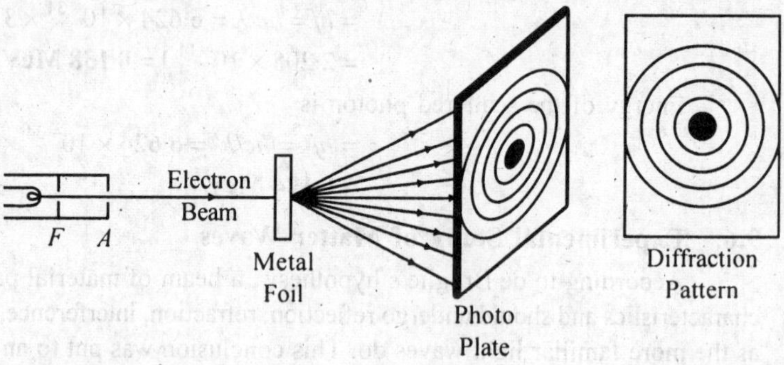

Fig. 9.6

9.9. Bohr's Principle of Complementarity

Compton scattering (Art. 9.5) has shown that waves have particle aspect whereas de Broglie's hypothesis of matter waves has shown that particles have wave characteristics. However, it has been found that both these aspects have never been shown in a single experiment. The above fact has been given the name of Bohr's Principle of Complementarity according to which the wave and particle aspects are complementary.

In simple words, it means that in no single experiment does a photon show both wave and particle properties nor does a particle *simultaneously* show both particle and wave properties. In view of the above duality, it was humorously suggested by Eddington that primordial entity seems to be in reality a 'wavicle'.

9.10. Electron Microscope

Like an optical microscope, its purpose is to magnify extremely minute objects. The resolving power of an instrument is inversely proportional to the wavelength of the radiation used for illuminating the object under study. Higher magnification as well as resolving power can be obtained by utilizing waves of shorter λ. Electron microscope uses electron waves whose wavelength is given by the formula $\lambda = 12{\cdot}25/\sqrt{V}$. For $V = 10{,}000$ V, $\lambda = 0{\cdot}1225$ Å which is extremely short. Electron microscopes giving magnifications more than 200,000 X are common in medical research laboratories.

An electron microscope consists of the following essential parts :

(*i*) **Electron Gun.** Its function is to provide a narrow beam of electrons of uniform velocity.

(*ii*) **Electrostatic and Magnetic Lenses.** Their function is to refract and properly focus the electron beam.

(*iii*) **Fluorescent Screen or Photographic Plate.** They are used to receive the highly-magnified image of the extremely small object being studied.

Fig. 9.7 gives comparison of an optical and an electron microscope.

9.11. Electron Waves Within the Atom

Recent developments concerning atomic structure have shown that Bohr's picture of the atom with sharply-defined electron orbits is not correct. According to the new theory, the electron in an atom does not behave as though it were a particle. It behaves, instead like a wave (de Broglie's electron wave). In fact, Bohr's quantum conditions which are assumed arbitrarily can be derived in a more natural way by considering the wave-like property of the electron. According to the wave-mechanical model of the atom, an electron behaves as a standing or stationary wave which goes round the nucleus in a circular orbit. The only necessary condition for the establishment of such a stationary wave is that the length of the orbit should be a whole number multiple of the wavelength of the electron wave as shown in Fig. 9.8. If r is the radius of the circular orbit then

$$2\pi r = n\lambda$$

Now,

$$\lambda = h/mv$$

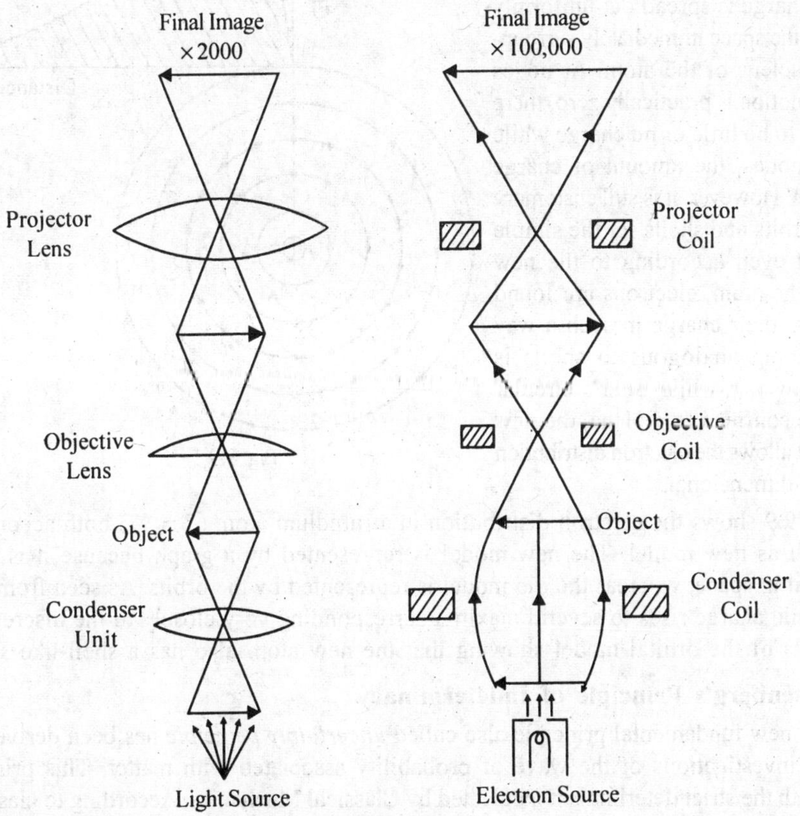

Fig. 9.7

$$\therefore \quad 2\pi r = nh/mv$$

or $$mvr = nh/2\pi$$

where $n = 1, 2, 3$ etc.

Since 'mvr' is the angular momentum of the electron as a particle, we see that wave-mechanical picture leads naturally to Bohr's postulate that the angular momentum is an integral multiple of $h/2\pi$. Other quantum conditions can also be derived similarly.

Moreover, the concept of an electron as a standing wave rather than a particle revolving in an orbit also removes the difficulty met in Bohr's theory regarding non-radiation of energy by the revolving electron.

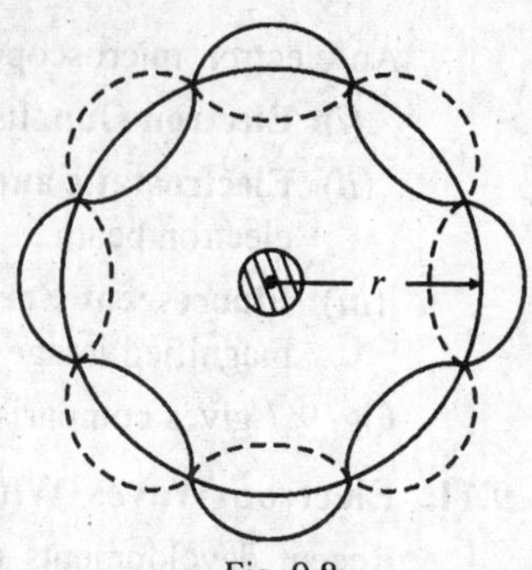

Fig. 9.8

9.12. The New Atomic Picture

In Bohr's atomic model, nucleus of an atom is surrounded by particles known as electrons which revolve in sharply-defined orbits or shells. This model has been replaced by the wave-mechanical model in which electrons as particles have been replaced by de Broglie's electron waves. These waves form stationary waves with their nodes and antinodes. Hence, instead of being localized at a point, the whole of electron mass and charge is spread out uniformly throughout the space immediately surrounding the nucleus of the atom. At nodes where the motion is practically zero, there is assumed to be little or no charge while at the antinodes, the amount of charge is maximum. However, it is still customary to talk of orbits and shells for the simple reason that even according to the new picture of the atom, electrons are found to distribute their charge in such a way that something analogous to shells is formed. However, while Bohr's circular orbits were confined to a plane, the new wave model allows the electron distribution to be three-dimensional.

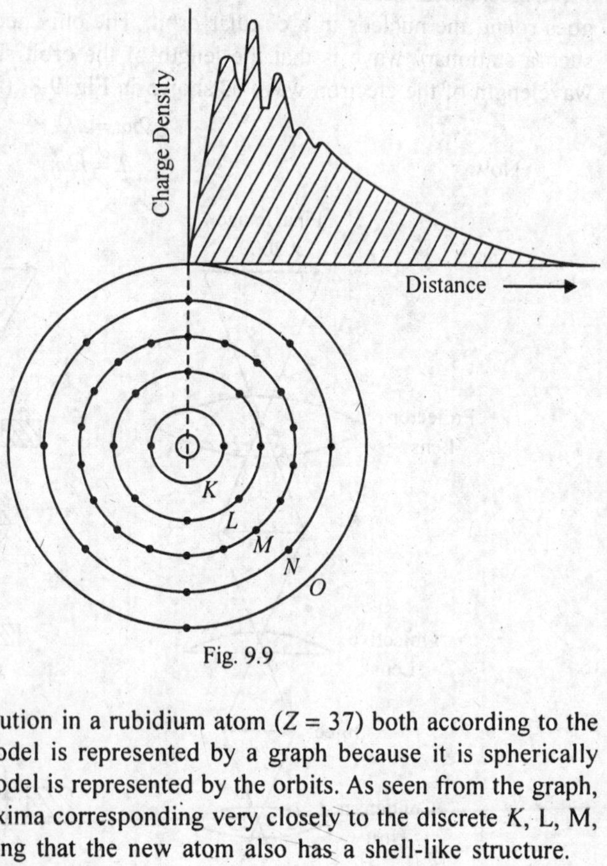

Fig. 9.9

Fig. 9.9 shows the electron distribution in a rubidium atom ($Z = 37$) both according to the odd as well as new model. The new model is represented by a graph because it is spherically symmetrical in space, whereas the old model is represented by the orbits. As seen from the graph, the electronic charge rises to several maxima corresponding very closely to the discrete K, L, M, N etc. shells of the orbital model showing that the new atom also has a shell-like structure.

9.13. Heisenberg's Principle of Indeterminacy

This new fundamental principle also called *uncertainty principle* has been derived from the theoretical investigations of the wave of probability associated with matter. This principle is at variance with the strict determinism inculcated by Classical Mechanics. According to classical ideas, every particle has, at any time, a fixed position in space and has a definite momentum which can be determined if the initial values are known.

However, according to uncertainty principle *it is impossible to determine simultaneously both the position and momentum of a particle with accuracy*. The term 'particle' has a different meaning in Wave Mechanics. It is something made up of a very deep and essential union of both corpuscle and wave characteristics resulting in a certain unavoidable indeterminacy.

The limits of accuracy with which the position and momentum of a particle can be obtained are specified by the uncertainty principle according to which

$$\Delta x \cdot \Delta p \geq h$$

where Δx and Δp are the errors in the determination of the position and momentum. As indicated by this principle, it is impossible to design an experiment which shows the wave and particle aspects of matter at the same time. If position is measured accurately, then measurement of momentum becomes correspondingly inaccurate, and *vice versa*. This principle only gives *probability* of finding a particle in a given space in place of *certainty*.

9.14. Free Electron Theory of Metals

According to the free electron model of a metal, the outermost or valence electrons of an atom in the metal are not bound to that atom but are free to move throughout the whole body of that metal. These valence electrons contributed by all atoms of the metal constitute what is commonly known as 'electron gas' or Fermi gas within the metal. Since atoms lose all their valence electrons, they become positively-charged ions. Hence, one may look upon a metal as consisting of an array of closely-packed ion cores immersed in a sea of free electrons (Fig. 6.8).

One basic assumption of free electron theory is that the potential field due to the ion cores is uniform throughout the metal which leads to the important conclusion that *free electrons have the same potential energy everywhere within the metal*. Strictly speaking, due to electrostatic attraction between a free electron and the ion core, potential energy will have a finite negative value but as we are interested only in *changes* of energy, we can assume this constant potential energy to be zero. In that case, the only energy which an electron possesses is the kinetic energy* $E = mv^2/2$. The de Broglie's wavelength of the electron is given by

$$\lambda = h/mv \qquad \text{—Art. 9.2}$$

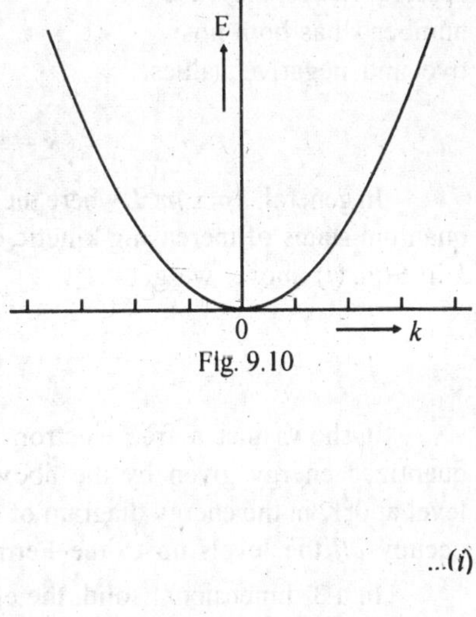

Fig. 9.10

The relation between de Broglie's wavelength and wave number** is given by

$$k = \frac{2\pi}{\lambda}$$

Combining $E = mv^2/2$ and $\lambda = h/mv$, we get

$$\boxed{E = \frac{h^2k^2}{8\pi^2 m}} \qquad \text{...(i)}$$

Obviously, the relation between E and k is parabolic as shown in Fig. 9.10.

---

* Since velocity of the electron is much smaller as compared to velocity of light, we ignore relativistic effect on mass and take the classical equation for kinetic energy.

** This definition of wave-number is commonly used in Quantum Theory. Many-body Theory and Nuclear Physics. In spectroscopy. In spectroscopy, it is, however, customary to define the wave number as = $1/\lambda$ (Art. 3.13).

9.15. Quantization of Free Electron Energy

Consider the motion of a free electron in a linear solid (say, a copper wire) of length L. As shown in Fig. 9.11 (*a*), the wavelength of the free electron can have the following values depending on its kinetic energy

$$L = \lambda/2 \quad \text{— for minimum kinetic energy}$$

$$L = 2 \cdot \lambda/2 \quad \text{— for next higher energy}$$

$$= 3\lambda/2 \text{ and so on}$$

$$\therefore \quad \lambda = 2L, L, 2L/3 \text{ etc.}$$

$$\therefore \quad \lambda = \frac{2\pi}{2L}, \frac{2\pi}{L}, \frac{2\pi}{2L/3} \text{ etc.}$$

$$= \frac{\pi}{L}, \frac{2\pi}{L}, \frac{3\pi}{L} \text{ etc.}$$

However, in a solid that is electrically neutral, a free electron having a certain velocity in one direction is always associated with another having the same speed but moving in the opposite direction. It must be so otherwise there will be a net flow of electrons in one direction even when no external field is applied. Hence, the wave number k has both positive and negative values.

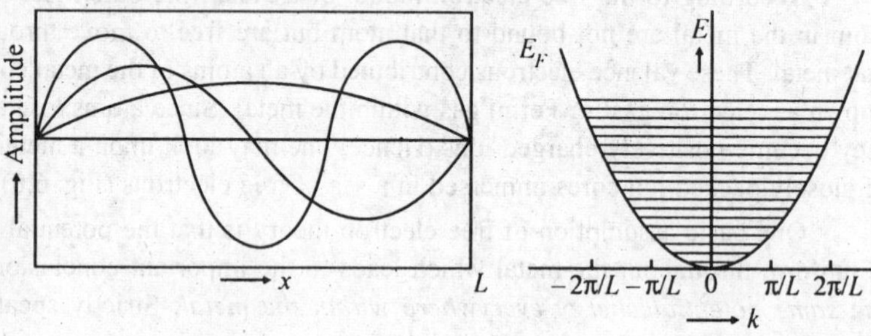

(*a*) (*b*)

Fig. 9.11

$$\therefore \quad k = \pm\frac{\pi}{L}, \pm\frac{2\pi}{L}, \pm\frac{3\pi}{L}, \ldots\ldots$$

In general, $k = \pm n\pi/L$ where the quantum number $n = 1, 2, 3, \ldots\ldots, \infty$. The free electron occupies quantum states of increasing kinetic energy as the value of n increases. Substituting this value of k in Eqn. (*i*) above, we get

$$\boxed{E = \frac{n^2 h^2}{8mL^2}}$$

It shows that a free electron does not possess any arbitrary amount of energy but only quantized energy given by the above relation. The energy corresponding to the highest filled level at 0°K in the energy diagram of Fig. 9.11 (*b*) is called the Fermi energy E_F. The free electrons occupy *all* the levels up to the Fermi level leaving all the levels above it empty.

In a 3-dimensional solid, the electron motion may be in any direction, hence

$$\boxed{E = \frac{h^2 (n_x^2 + n_y^2 + n_z^2)}{8mL^2}}$$

In this case, several combinations of n_x, n_y and n_z can result in the same value of E which means that energy state can be degenerate.

9.16. Energy Distribution Among Free Electrons

The number of free electrons in a metallic solid is nearly 10^{22} per cm^3. These electrons move about in the metal with a random and chaotic motion, exchange energy with each other and with the atoms in the three-dimensional lattice which vibrate about their mean positions. The energies possessed by these electrons vary over a wide range. Since number of electrons is very large and, also, their motion is random, the problem of finding their energy distribution can only be solved by using statistical methods. Such statistical solutions are characterized by distribution functions which make it possible to calculate various distribution such as the distribution of energy or velocity among the free electrons.

(*a*) Energy Distribution Function

Fermi and Dirac and later on, Sommerfeld developed the statistical technique known as Fermi-Dirac Distribution which is based on the following assumptions :

(*i*) that the energies of these electrons are quantized and

(*ii*) that these electrons are subject to Pauli's exclusion principle.

According to Fermi-Dirac statistics, the expression for the distribution of energy among free electrons in the electron gas of a solid is

$$dN = \frac{C\sqrt{E}}{1+e^{(E-E_F)/kT}}\,dE\,*$$

where dN = No. of electrons having energy between E and $(E + dE)$

E = kinetic energy of the electron

E_F = Fermi energy

k = Boltzmann's constant

T = Metal temperature in °K

The distribution has been plotted for T = 0°K and 2500°K in Fig. 9.12. Following facts are worth noting :

(*i*) At T = 0°K, energy distribution is a step function centred around Fermi energy E_F. It means that at this temperature, the number of low-energy electrons is very little but the number of electrons with higher energy increases till a value E_F is reached when it suddenly falls to zero. Put in another way, it means that all states below E_F are filled whereas all those beyond E_F are unfilled.

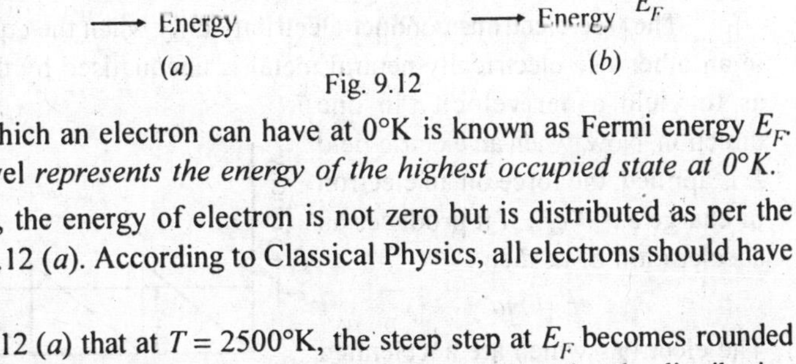

Fig. 9.12

(*ii*) The highest energy which an electron can have at 0°K is known as Fermi energy E_F. Alternately, Fermi level *represents the energy of the highest occupied state at 0°K.*

(*iii*) As seen, even at 0°K, the energy of electron is not zero but is distributed as per the curve shown in Fig. 9.12 (*a*). According to Classical Physics, all electrons should have zero energy at 0°K.

It will be noted from Fig. 9.12 (*a*) that at T = 2500°K, the steep step at E_F becomes rounded with experimental 'tails' extending below and above E_F. However, it is surprising that the distribution

*The value of the constant C is

$$C = \frac{4\pi(2m_e)^{3/2}}{h^3}$$

at as high a temperature as 2500°K is not much different from that as 0°K. In fact, the low energy electrons are hardly disturbed whereas high-energy electrons are given still greater energy. This point can be easily understood from the fact that the distribution curve just to the left of E_F is lowered (indicating that there are now fewer electrons having energies in this neighbourhood) whereas the same curve extends to the right of E_F (indicating that some electrons have energies greater than E_F).

(b) Probability Distribution Function

The probability of finding an electron having any particular value of energy E or (putting it in another way) of finding an electron at an energy level E is given by Fermi-Dirac probability distribution function.

$$P(E) = \frac{1}{1+e^{(E-E_F)/kT}}$$

This function, as plotted for different temperatures, is shown in Fig. 9.13. Following facts can be deduced from the above figure :

(i) At $T = 0°K$, for $E < E_F$, $e^{(E-E_F)/kT} = 0$, so that $P(E) = 1$. This shows that at $T = 0°K$, the probability of finding an electron with energy less than Fermi energy is unity. In simple words, it means that at $T = 0°K$, all energy levels between zero and Fermi level E_F are occupied. Alternately, it can be stated that at 0°K, all energy states below E_F have a probability of occupancy of unity.

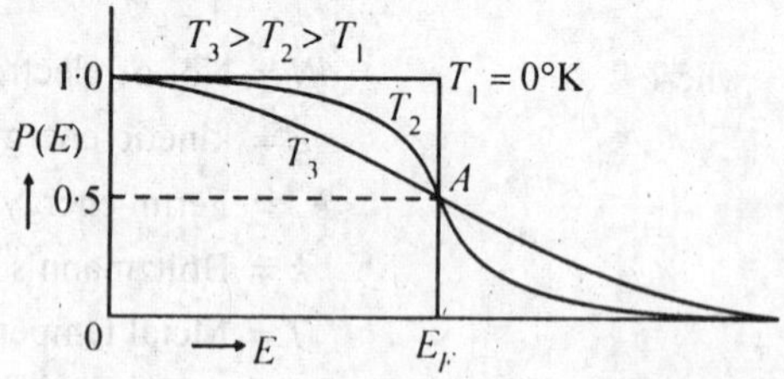

Fig. 9.13

(ii) At $T = 0°K$, for $E > E_F$, $e^{(E-E_F)/kT} = \infty$ so that $P(E) = 0$. It means that there are no electrons having energy greater than E_F or occupying an energy level higher than Fermi level. Put in a slightly different way, it means that energy states above E_F have zero probability of occupancy and are therefore empty.

(iii) At $T > 0°K$ and $E = E_F$, $e^{(E-E_F)/kT} = 1$ so that $P(E) = \frac{1}{2}$.

The above result may be used for defining Fermi level as *that level at any finite temperature at which the probability of finding an electron is 1/2 or 50 per cent as shown in Fig. 9.13.*

9.17. Conduction Process in a Metal Under Electric Field

The free electrons conduct electricity only when the equal and opposite velocity distribution in an otherwise electrically neutral metal is unequalised by the application of an electric field, so as to yield a net velocity in one direction. Now, when an electric field E is applied, the force on an electron of charge e is $= eE$. If it produces an acceleration of a, then

$$eE = ma$$

The electrons which are accelerated in the direction of the applied field do not continue accelerating indefinitely because of their collision with impurity atoms, crystal imperfections and photons. Depending on the time interval between two successive collisions, they acquire an average velocity $\bar{v}$ as shown in Fig. 9.14 (a). The electron is

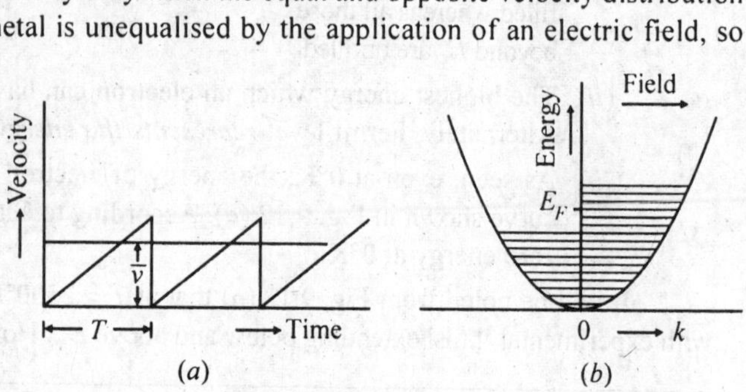

Fig. 9.14

supposed to lose all the extra velocity it has acquired prior to the collision. The *E-k* diagram becomes modified as shown in Fig. 9.14 (*b*). The velocity (and hence energy) of the fastest electron in the direction of the applied field *E* has a larger magnitude than that of the fastest electron in the opposite direction, *i.e.*, against the applied field. That is why the energy levels on the right-hand side of Fig. 9.14 (*b*) are higher than those on the left-hand side. Such distribution is possible only when empty energy levels are available immediately above the Fermi level. This availability is a *basic characteristic of conductors as opposed to that of semiconductors and insulators.*

As seen from Fig. 9.14 (*a*), acceleration $a = \bar{v}/T$

$\therefore \quad eE = m/\bar{v}\,T \qquad$ or $\qquad \bar{v} = eET/m$

The electronic current density J_e is given by

$$J_e = ne\bar{v} = ne^2ET/m \text{ A/m}^2$$

where n represents the number of free electrons which are near the Fermi level and which can increase their kinetic energy by moving into higher quantum states. At ordinary temperatures, n ¾ 1 per cent of total free electron in the solid.

Now, conductivity σ is defined as current density per unit applied electric field

$$\sigma = ne^2T/m$$

$$\therefore \qquad \text{resistivity, } \rho = m/ne^2T$$

Example 9.7. *Compute the energy difference between the first and second quantum states for a free electron in a solid 1 metre cube.*

Solution. (*a*) $$E = \frac{h^2\,(n_x^2 + n_y^2 + n_z^2)}{8mL^2}$$

For the first quantum state, $n_x = n_y = n_z = 1$. Also, $L = 1$

$$\therefore \qquad E_1 = \frac{(6{\cdot}63\times10^{-31})^2\,(1^2+1^2+1^2)}{8\times9{\cdot}1\times10^{-31}\times1^2} = 1{\cdot}81\times10^{-37}\text{ J}$$

There are many equal energy states above the first state having n_x, n_y and n_z as (1, 1, 2), $(1, \bar{2}, 1)$, $(2, \bar{1}, \bar{1})$. For all these states

$$(n_x^2 + n_y^2 + n_z^2) = 6$$

$$E = \frac{(6{\cdot}63\times10^{-34})^2\times6}{8\times9{\cdot}1\times10^{-31}\times1^2} = 3{\cdot}62\times10^{-37}\text{ J}$$

It is seen that energy difference between the first and next higher energy state is only $\mathbf{1{\cdot}81\times10^{-37}}$ **J.**

9.18. Schrödinger Equation

It is a differential equation capable of describing the motion of an electron. It was developed by Schrödinger who extended the application of wave equation to the vast realm of atomic Physics.

In an ordinary material wave, the displacement of whatever is vibrating about its mean position is given by

$$y = a\sin 2\pi\left(ft - \frac{x}{\lambda}\right) \qquad ...(i)$$

where $\qquad y$ = displacement at time t and at distance x from origin

a = maximum displacement from mean position

λ = wavelength

f = frequency of vibration

When differentiated twice with respect to x, it becomes

$$\frac{d^2y}{dx^2}+\frac{4\pi^2}{\lambda^2}y=0 \qquad ...(ii)$$

or

$$\frac{d^2y}{dx^2}+k^2y=0 \qquad ...(iii)$$

The above equation involves only distance as the independent variable. Its solution is $y = a \sin 2\pi x/\lambda$ which defines a standing wave. At each point along the wave in space, y varies periodically with frequency f.

Let us now see how the equation can be applied to specify an electron in motion. As we know, the total energy E of an electron is partly kinetic and partly potential.

$$E = \frac{1}{2}mv^2 + V$$

$$\therefore \quad mv = \sqrt{2m(E-V)}$$

Now, de Broglie's wavelength is given by

$$\lambda = \frac{h}{mv} = \frac{h}{\sqrt{2m(E-V)}}$$

Substituting this value in Eqn. (*i*) above and replacing y by ψ as is customary, we get

$$\frac{d^2\psi}{dx^2}+\frac{8\pi^2m}{h^2}(E-V)\,\psi=0 \qquad ...(iv)$$

It is known as time-independent Schrödinger equation in one dimension. For three-dimensional motion of an electron, this equation becomes

$$\frac{d^2\psi}{dx^2}+\frac{d^2\psi}{dy^2}+\frac{d^2\psi}{dz^2}+\frac{8\pi^2m}{h^2}(E-V)=0$$

or

$$\boxed{\nabla^2\psi+\frac{8\pi^2m}{h^2}(E-V)=0} \qquad ...(v)$$

This Schrödinger equation is the starting point of *wave mechanics* just as $F = ma$ is the starting point of classical mechanics.

It is seen from Eqns. (*iii*) and (*iv*) that

$$k^2 = \frac{8\pi^2m}{h^2}(E-V) \qquad ...(vi)$$

Schrödinger equation can be applied for determining the total energy of an electron when it is moving in an electric field so that its potential energy V is specified in terms of space co-ordinates and not of time. Like other differential equations, its solutions are also governed by boundary conditions. Generally, solutions are only obtained for certain energy values called *characteristic* or *eigen-values*. The corresponding wavefunctions ψ (which are generally complex) are called characteristic or eigen-functions. If we write $\psi = a + jb$ and its conjugate $\psi^* = a - jb$, then their product $\psi\psi^* = a^2 + b^2$ and is real. For a one-dimensional system, $|\psi|^2\,dx$ is supposed to represent the *probability* of finding the electron within the range x to $(x + dx)$. But in three dimensions, $|\psi|^2dV$ represents the probability of finding an electron within the volume range V to $(V + dV)$.

9.19. Applications of Schrödinger Equation

Solution of Schrödinger equation may be found for the following cases :

(*i*) particles in a box (like electrons in metals) (*ii*) particles in a potential well (radioactive disintegration) (*iii*) hydrogen atom and (*iv*) rotating molecule.

(*a*) Particle in a Box

Consider a particle of mass m which is free to move in a one-dimensional box of length L as shown in Fig. 9.15. The potential energy V of the electron at the bottom of the box is constant and may be taken, for convenience, as zero. Hence, *inside* the box, $V = 0$.

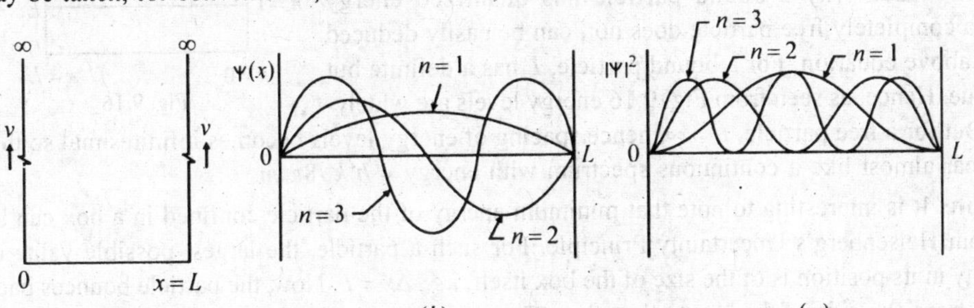

Fig. 9.15

Let potential energy be infinite for $x < 0$ and $x > L$. Obviously, ψ function has to be zero at $x = 0$ and for all negative values of x since the particle is not allowed over the walls of the box. Similarly, ψ function must be zero for all values of $x > L$. Put in another way, it simply means that the particle is confined to the box and does not exist outside.

A suitable general solution of Schrödinger equation is

$$\psi(x) = A \sin kx + B \cos kx$$

From boundary conditions, we have that

$$\psi(x) = 0 \text{ at } x = 0 \quad \text{or} \quad \psi(0) = 0$$

$$0 = A \sin 0 + B \cos 0 \quad \text{or} \quad B = 0$$

Also, $\psi(x) = 0$ at $x = L$ or $\psi(L) = 0$

$\therefore \quad A \sin kL + B \cos kL = 0$

It would be possible only when

$$kL = n\pi \text{ or } k = n\pi/L$$

where n is called quantum number and equals 1, 2, 3, ..., ∞.

Substituting this value in Eqn. (*iv*) above and also putting $V = 0$, we get

$$\frac{n^2\pi^2}{L^2} = \frac{8\pi^2 m}{h^2}E$$

Here, E — the total energy of the electron — becomes equal to its kinetic energy.

$$\therefore \quad E_n = \frac{n^2h^2}{8\pi^2 m} \quad \text{where } n = 1, 2, 3, ..., \infty$$

The meaning of the above equation is that the particle in the box cannot possess any arbitrary amount of energy*. Rather, it can have energy values in a discrete set only. In other words, its energy is quantized. A few of the energy levels are shown in Fig. 9.16.

---

*By putting $h = \frac{2T}{\lambda} = \frac{n\pi}{L}$, the expression for kinetic energy of the particle becoms

$$E = \frac{h^2k^2}{8\pi^2 m} \quad \text{— same as in Art. 9.14}$$

$$E_1 = \frac{h^2}{8mL^2}$$

$$E_2 = \frac{4h^2}{8mL^2}$$

$$E_3 = \frac{9h^2}{8mL^2} \text{ and so on}$$

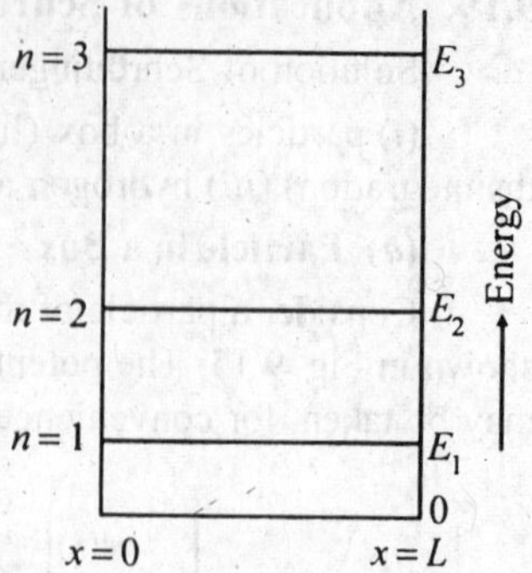

Fig. 9.16

The reason why a bound particle has quantized energy whereas a completely free particle does not, can be easily deduced from the above equation. For a bound particle, L has a definite but small value. Hence, as seen from Fig. 9.16 energy levels are widely-spaced. But for a free particle, $L = \infty$, hence spacing of energy levels becomes infinitesimal so that they appear almost like a continuous spectrum with energy $= h^2k^2/8\pi^2 m$.

Note. It is interesting to note that minimum energy of the particle confined in a box can be found from Heisenberg's Uncertainty Principle. For such a particle, the largest possible value of uncertainty in its position is of the size of the box itself, *i.e.*, $\Delta x = L$. Now, the particle bounces back and forth from one side of the box to the other. The uncertainty in the momentum of the particle is $\Delta p = 2p$ because the momentum continuously changes from $+p$ to $-p$ as collisions with the walls occur. Hence, for a maximally non-localized particle, the uncertainty principle becomes

$$\Delta x \times \Delta p = L\,(2p) = h \quad \text{or} \quad p = h/2L$$

Now,
$$E = \frac{1}{2}\cdot mv^2 = \frac{(mv)^2}{2m} = \frac{p^2}{2m}$$

$\therefore$
$$E = \frac{h^2}{8mL^2}$$

(*b*) Free Electrons in a Metal

The simplest, but sufficiently true, picture of conduction electrons in a solid metal is that they move freely in a region of constant potential. This view is based on the assumption that the electric field of the atoms is completely smoothed out so that the solid is at constant potential throughout its volume. If this constant-potential region is treated as a 3-dimensional box of very large dimensions, then we have to solve the one-dimensional Schrödinger equation three times. The energy levels can be calculated from the following equation :

$$E_n = \frac{h^2}{8mL^2}\,(n_x^{\,2} + n_y^{\,2} + n_z^{\,2})$$

where $n = 1, 2, 3, \ldots\ldots, \infty$.

The energy levels corresponding to each set of quantum numbers n_x, n_y and n_z can now be filled up by putting two electrons of opposite spin in each level in accordance with Pauli's Exclusion principle. The energy of the highest filled state is known as Fermi energy E_F as shown in Fig. 9.17 (*a*) which represents only one-dimensional view. The energy level diagram for three dimensions is given in Fig. 9.17 (*b*).

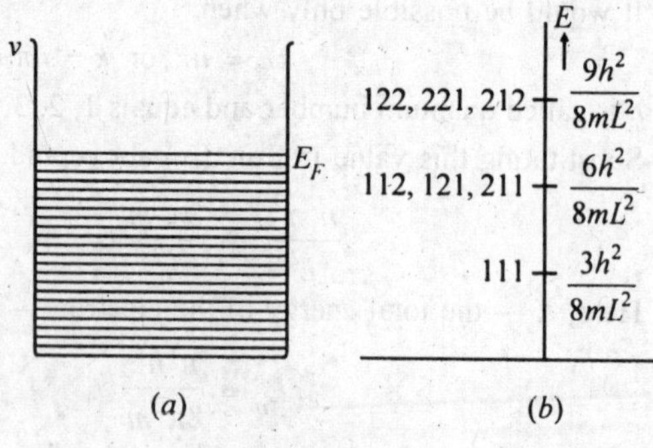

Fig. 9.17

Example 9.8. *An electron is bound by a potential which closely approaches an infinite square well of width 2·5 × 10⁻¹⁰ m. Calculate the lowest three permissible quantum energies the electron can have.*

Solution. As seen from Art. 9.15, the energy of such an electron is given by the equation

$$E_n = \frac{n^2h^2}{8mL^2} \qquad \text{where } n = 1, 2, 3, \ldots, \infty$$

Here,

$$h = 6{\cdot}62 \times 10^{-34}\ \text{J-s};\ m = 9{\cdot}1 \times 10^{-31}\ \text{kg}$$

$$L = 2{\cdot}5 \times 10^{-10}\ \text{m}$$

$$E_1 = \frac{1^2 \times (6{\cdot}62 \times 10^{-34})^2}{8 \times 9{\cdot}1 \times 10^{-31}\ (2{\cdot}5 \times 10^{-10})^2} = 9{\cdot}63 \times 10^{-19}\ \text{J}$$

$$= 9{\cdot}63 \times 10^{-19}/1{\cdot}6 \times 10^{-19} = \mathbf{6\ eV}$$

$$E_2 = 4 \times 6 = \mathbf{24\ eV};\ E_3 = 9 \times 6 = \mathbf{54\ eV.}$$

TUTORIAL PROBLEMS

1. What would be the de Broglie's wavelength associated with (*i*) a 2000 kg car having a constant speed of 25 m/s (*ii*) 80 kg scooter having a speed of 10 m/s. **[(*i*) $1{\cdot}32 \times 10^{-38}$ m (*ii*) $8{\cdot}1 \times 10^{-37}$ m]**
2. Find the wavelength associated with an electron moving with a velocity of 0·5 c. **[4·85 pm]**
3. Compute the wavelength of the de Broglie's wave associated with a proton moving with 5 per cent of velocity of light. Proton has 1836 times the mass of an electron. **[$2{\cdot}64 \times 10^{-19}$ m]**
4. In a TV set, electrons are accelerated by a p.d. of 10 kV. What is the wavelength associated with these electrons ? **[12·3 pm]**
5. What would be the de Broglie's wavelength of a proton after it is accelerated by a p.d. of 1 kV ? **[0·9 m]**
6. What is the momentum of each photon in a beam of violet light λ = 4000 Å ? **[$1{\cdot}66 \times 10^{-27}$ kg-m/s]**
7. A beam of 10 kV electrons is passed through a thin metallic sheet whose interplanar spacing is 55 pm (55×10^{-12} m). Calculate the angle of deviation of the first order diffraction maximum. **[12°44']**
8. Find the de Broglie's wavelength for an electron travelling along the first Bohr orbit in a hydrogen atom. **[3·3 Å]**
9. Calculate the wavelength of the wave associated with an electron having a kinetic energy of 100 eV. **[12·3 Å]**
10. What is the wavelength of a proton moving with a speed of 1000 m/s ? Variation of mass with velocity may be neglected.
11. Find the velocity and kinetic energy of an electron which is associated with a wavelength equal to 1 Å. **[7,270 km/s; 150 eV]**
12. An incident photon of wavelength 0·03 Å recoils at an angle of 60° after collision with a free electron. What is the energy of the recoiling electron ? **[120 eV]**
13. During a photon-electron collision, Compton shift is found to be 0·012 Å. If the wavelength of the incident photon is 0·05 Å, find out the angle between the direction of motion of the scattered photon and of the recoil electron. **[109·5°]**
14. What is de Broglie's wavelength of an electron travelling at a speed of 10^7 m/s. **[0.728 A.U.]**

HIGHLIGHTS

1. The de Broglie's wavelength associated with a matter particle is :

$$\lambda = \frac{h}{mv} = \frac{h}{p}\ \text{metre}$$

$$= \frac{h}{\sqrt{2mV}} = \frac{12{\cdot}25}{\sqrt{V}}\ \text{A}$$

2. Wave number ν of a wave is given by the reciprocal of its wavelength.

$$\nu = \frac{1}{\lambda} = \sqrt{\frac{2emV}{h^2}} \text{ m}^{-1}$$

$$= 8{\cdot}15 \times 10^8 \sqrt{V} \text{ per metre}$$

3. Velocity of matter wave is

$$u = \frac{h}{2m\lambda} \text{ m/s}$$

4. Compton shift is given by

$$\Delta\lambda = \frac{h}{m_0 c}(1 - \cos\phi) \text{ metre}$$

$$= \frac{2h}{m_0 c} \cdot \sin^2 \phi/2 \text{ metre}$$

or

$$\Delta\lambda = 0{\cdot}0243\ (1 - \cos\phi)\ \text{Å}$$

$$= 0{\cdot}0486 \sin^2 \phi/2\ \text{Å}$$

5. Principle of Complementary states that the wave and particle aspects are complementary, *i.e.*, both cannot be shown in a single experiment. If one aspect is there, the other cannot be there.
6. According to Heisenberg's Uncertainty Principle, it is impossible to determine instantaneously both the position and momentum of a particle with accuracy. The limits of accuracy are specified by the relation.

$$\Delta x \cdot \Delta p \geq h$$

7. According to free-electron theory of a metal, the energy of an electron is quantized and is given by

$$E = \frac{n^2 h^2}{8mL^2} \qquad \text{where } n = 1, 2, 3, \ldots.., \infty$$

8. Electronic conductivity of a metal is given by

$$\sigma = ne^2 T/\text{m}$$

9. The time-independent Schrödinger one-dimensional equation is given by

$$\frac{d^2\psi}{dx^2} + \frac{8\pi^2 m}{h^2}(E - V)\,\psi = 0$$

Can You Answer This ?

1. Give in detail with neat diagram an experimental procedure for determining the wavelength of electrons.
2. (*a*) Discuss briefly the phenomenon to show that radiation sometime behaves as particle and at others as wave. What are de Broglie's matter waves ? Mention the factors on which the wavelength of the particle depends.

   (*b*) Describe and explain Davisson and Germer experiment for the determination of wavelength of electrons. Mention one important application of electron waves.
3. Write note on Heisenberg's uncertainty principle.
4. What is uncertainty principle ? Is this principle an outcome of the wave description of particle ? Give at least one experiment to prove its validity.
5. What are the de Broglie's matter waves ? Describe their properties. How are these different from electromagnetic waves ?
6. Explain Heisenberg's Uncertainty Principle. Describe two experiments to illustrate it.

10 THE ATOMIC NUCLEUS

Theories of Nuclear Composition — Proton-Neutron Theory of Nuclear Composition — General Properties of a Nucleus — Packing Fraction — Mass Defect and Atomic Binding Energy — Accurate Expression for Mass Defect — Variation of Binding Energy — Significance of Average Binding Energy — Nuclear Forces — Meson Theory of Nuclear Forces — Nuclear Models — Liquid-Drop Model — Nuclear Shell Model — Highlights — Can You Answer This ?

10.1. Theories of Nuclear Composition

Rutherford's theory of nuclear atom based on large-angle α-scattering shows that nucleus is a very compact and hard structure. Natural radioactivity indicates that nuclear structure though compact must be very complex since it is capable of ejecting different particles like α-particles, β-particles and γ-rays etc. The study of α-ray and γ-rays spectra process that there exists great order in the internal structure of the nucleus in spite of its complexity. Interpretation of the continuous spectra of β-particles has resulted in the introduction of a new elementary particle called *neutrino*. This particle has negligible mass although it plays the important role as carrier of energy. Researchers on artificial radioactivity have shown that many particles like α-particles, protons, neutrons, β-particles and positrons etc. enter into the constitution of the nuclei in one way or the other. Cosmic ray studies have proved the existence of another fundamental particle called *meson* which occurs in many forms as regards charge and mass. At present, main problem is to determine which of these elementary particles enters into the actual constitution of a nucleus in the most complex and compact but orderly fashion. Several theories regarding nuclear composition have been suggested; more important of which is the proton-neutron theory discussed below.

10.2. Proton-Neutron Theory of Nuclear Composition

The discovery of neutron by Chadwick in 1932 led to the assumption that every atomic nucleus consists essentially of protons and neutrons. This hypothesis was used for the first time as the basis of a detailed theory of nucleus by Heisenberg in 1932. According to this proton-neutron hypothesis, the total number of protons and neutrons (collectively called *nucleus*) in a nucleus gives its (atomic) mass number A. The number of protons gives the total positive charge on the nucleus as well as its atomic number (or charge number) Z. In order to neutralize the positive charge of the nucleus and hence to make the atom electrically neutral, the number of extranuclear (or orbital) electrons in Z. Obviously, the number of neutrons is $(A - Z)$.

Pictorial representation of a few nuclei is shown in Fig. 10.1. In ordinary hydrogen, the nucleus consists of one proton only. In helium, the nucleus contains two protons and two neutrons. Carbon nucleus is made up of a 6 protons and 6 neutrons. For elements of low mass number, the atomic number Z is nearly half the mass number, *i.e.*, number of protons

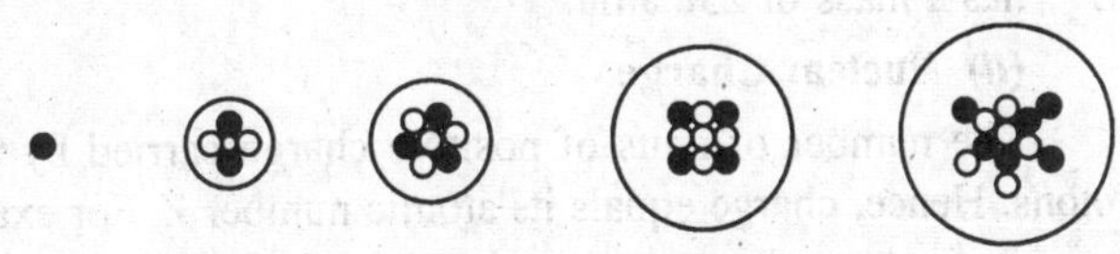

H	He	Li	Be	C
$A=1$	$A=4$	$A=7$	$A=9$	$A=12$
$Z=1$	$Z=2$	$Z=3$	$Z=4$	$Z=6$
$A-Z=0$	$A-Z=2$	$A-Z=4$	$A-Z=5$	$A-Z=6$

Fig. 10.1

is nearly equal to the number of neutrons. With increasing mass number, the value of Z becomes less than half of A which means that number of neutrons exceeds that of protons. In the case of $_{92}U^{238}$ atom, $Z = 92$, $A = 238$ and $(A - Z) = 146$.

Hence, uranium nucleus consists of 92 protons and 146 neutrons. These facts about neutron/proton ratio have an important bearing on the stability of atomic nuclei. For elements of low mass numbers, the neutron-to-proton ratio is close to unity. But when the number of protons in a nucleus is greater than 20, the neutron/proton ratio in stable nuclides is always greater than unity. It seems that in order to maintain nuclear stability, the number of neutrons must exceed the number of protons, the neutron excess increasing with the increasing atomic number. For the heaviest stable nuclides such as $_{82}Pb^{208}$ and $_{83}Bi^{209}$, the neutron/proton ratio exceeds 1·5.

The reason for the above fact is not far to seek. As the number of protons increases, the electrostatic repulsive force (as given by given by Coulomb's Law) between them also increases.* In order to overcome this increasing repulsive force and thus maintain stability in the heaviest elements, the number of neutrons must be increased. The additional short-range $(n - n)$ and $(n - p)$ attractive nuclear forces (Art. 10.9) compensate, to a great extent, the growing proton-proton repulsion.

The β-particle emission by radioactive elements can be explained on the proton-neutron theory as follows :

The electron does not pre-exist in a nucleus but is just formed at the instant of emission. Its production is due to the conversion of a neutron into a proton according to the following equation :

$$_0n^1 \longrightarrow {}_1H^1 + {}_{-1}e^0$$

The positron emission can be likewise explained as due to the conversion of a proton into a neutron as shown below :

$$_1H^1 \longrightarrow {}_0n^1 + {}_{-1}e^0$$

Since a proton and a neutron are mutually convertible, they are regarded as two alternative states of a single nuclear particle.

10.3. General Properties of a Nucleus

Some of the important properties of an atomic nucleus are discussed below.

(i) Nuclear Mass

Strictly speaking, nuclear mass means the mass or weight of the nucleus proper. But it is often called the atomic mass number (or atomic weight) which would rather imply the mass or weight of the whole atom. The mass of the nucleus is equal to the sum of the masses of neutrons and protons each of which has a mass nearly unity or 1 atomic mass unit** (amu). Accordingly, the $_6C^{12}$ nucleus has a mass of 12 amu so that its mass number $A = 12$. Similarly, the uranium nucleus $_{92}U^{238}$ has a mass of 238 amu.

(ii) Nuclear Charge

The number of units of positive charge carried by a nucleus is equal to the number of its protons. Hence, charge equals its atomic number Z. For example, a hydrogen nucleus (*i.e.*, proton) carries a single positive charge whereas nobelium nucleus carries 102 units of positive charge.

---

* This force F is nearly proportional to Z^2/R where R is nuclear radius which itself varies as $A^{1/3}$. Hence,

$$F \propto Z^2/R \propto Z^2/A^{1/3}$$

** Atomic mass unit is equal to one-sixteenth the mass of the neutral O^{16} isotope of oxygen which is taken to have a mass of exactly 16 units.

$$1 \text{ amu} = \frac{1}{16}\cdot\frac{16}{N} = \frac{1}{N} = \frac{1}{6{\cdot}02 \times 10^{23}} = 1{\cdot}66 \times 10^{-24} \text{ gram} = 1{\cdot}66 \times 10^{-27} \text{ kg}$$

(iii) Nuclear Radius

Nuclear radii have been calculated from observations on the maximum scattering angles of α-particles as well as scattering on other particles like protons and neutrons by various target nuclei. According to Rutherford's calculations, greater the angle through which an α-particle has been deflected, the closer has it approached the atomic nucleus before being turned away. By measuring the maximum scattering angle, the distance of the closet approach between the centre of an atomic nucleus and an a-particle can be calculated. This distance, of the order of 10^{-14} m, represents the maximum value for the sum of the radii of an atomic nucleus and an α-particle. The radius of an atomic nucleus (assumed spherical) is given by the following empirical relation :

$$\boxed{R = R_0 A^{1/3}}$$

where R_0 has an average value of $1{\cdot}4 \times 10^{-15}$ m,

$$\therefore \quad R = 1{\cdot}4 \times 10^{-15} A^{1/3}$$

For carbon $(A = 12)$

$$R = 1{\cdot}4 \times 10^{-15} \times 12^{1/3} = 3{\cdot}21 \times 10^{-15} \text{ m}$$

For uranium $(A = 238)$

$$R = 1{\cdot}4 \times 10^{-15} \times 238^{1/3} = 8{\cdot}68 \times 10^{-15} \text{ m}$$

It is seen that radii of all nuclei vary between 10^{-14} to 10^{-15} m.

In the case of α-scattering, an approximate calculation of the radius of the target nucleus can be made in the following manner.

Suppose that an α-particle of mass m, moving with a velocity of v and hence, having a kinetic energy $\frac{1}{2}mv^2$ is approaching a target nucleus along the line joining their centres. At the distance of closet approach d_0 to the target nucleus, the α-particle comes to rest momentarily so that all its kinetic energy is converted into electrical energy.

$$\therefore \quad \frac{1}{2}mv^2 = \frac{1}{4\pi\varepsilon_0}\cdot\frac{Ze}{d_0}\cdot 2e$$

$$\therefore \quad d_0 = \frac{Ze^2}{\pi\varepsilon_0 mv^2}$$

Now, $m = 4/6{\cdot}02 \times 10^{23} = 0{\cdot}664 \times 10^{-23}$ gram $= 0{\cdot}664 \times 10^{-26}$ kg and average velocity of α-particles is about $1{\cdot}5 \times 10^7$ m/s and if $Z = 20$, then

$$d_0 = \frac{20 \times (1{\cdot}602 \times 10^{-19})^2}{\pi \times 8{\cdot}854 \times 10^{-12} \times 0{\cdot}664 \times 10^{-26} \times (1{\cdot}5 \times 10^7)^2}$$

$$= 1{\cdot}2 \times 10^{-14} \text{ m.}$$

(iv) Nuclear Density

Since size of the nucleus is extremely small and its mass very large, its density is fantastically high. The density of all nuclei is found to be about the same and equal to almost 10^7 kg/m^3. Obviously, nuclear density is much higher than that of the matter in bulk which must be comparatively very porous. On the basis of equivalence of mass and energy, it is seen that the concentration of energy is much greater in nuclei than in bulk material.

(v) Nuclear Quantum States

The study of α- and γ-rays spectra as well as artificial radioactivity shows that every nucleus possesses a set of quantum states and a corresponding number of discrete energy levels. Transitions between different nuclear states are accompanied by emission of γ-rays. There is great resemblance between quantum states inside the nucleus and those in the extranuclear or orbital electrons. But there is a fundamental difference as regards the nature of the forces acting in their respective regions. In the extranuclear region, the Coulomb's Law of inverse square reigns supreme whereas inside the nucleus, some other law of short-range nuclear forces comes into play (Art. 10.9).

(*vi*) Nuclear Spin and Magnetic Moment

Since a nucleus has finite size, it possesses, just like orbital electrons, spin motion and the consequent magnetic moment. The earliest evidence of nuclear spin was found in atomic spectra. Many spectral lines were found to have a fine structure which was explained due to electron spin. Many of these fine-structure lines show a still finer structure called *hyperfine structure* which has been attributed to the nuclear spin.

Experimental evidence shows that the constituent particles inside the nucleus are in continuous motion in discrete quantized orbits. This orbital motion endows the nucleus with mechanical angular momentum because of which protons additionally possess magnetic moment. Over and above these orbital motions, the nuclear particles spin about their own axes and so possess spin angular momentum and an associated magnetic moment. Hence, the nucleus as a whole will possess the following :

(*i*) resultant angular momentum which is the resultant of the orbital and spin momenta of different constituent particles. It is referred to as nuclear spin.

(*ii*) resultant angular momentum called nuclear magnetic moment which is the resultant of the moments due to the orbital and spin motions of protons* inside the nucleus.

The nuclear spin or angular momenta of nuclei, in general, is given by

$$p_1 = I \cdot \frac{h}{2\pi}$$

where I is the nuclear spin quantum number. This quantum number I has half-integral values for all isotopes of odd mass numbers and whole-number or integral values for all those with even mass numbers.

It may be noted that magnetic moments of nuclei, in general, are very small as compared to those associated with extranuclear electrons.

For all nuclei, the magnetic moment is given by

$$\mu_1 = g \cdot \frac{h}{2\pi} \cdot \frac{e}{2M}$$

where g-factor varies from nucleus to nucleus and M is the mass of a proton. The product $\frac{h}{2\pi} \cdot \frac{e}{2M}$ is known as nuclear magneton**.

$$\therefore \qquad \mu_n = \frac{h}{2\pi} \frac{e}{2M} = 5{\cdot}05 \times 10^{-27} \text{ A-m.}$$

10.4. Packing Fraction

Although atomic masses are close to whole numbers, they almost invariably differ from integers by small amounts. This divergence of the mass of a nuclide from whole number is expressed in the form of a quantity known as *packing fraction.*

---

* Surprisingly, unchanged particle like a neutron has been experimentally found to possess a negative magnetic moment.

** It will be seen that one nuclear magnetron equals 1/1836 of one Bohr magnetron (Art. 3.25).

Packing fraction, $f = \dfrac{M - A}{A}$

where, M = actual or isotopic mass of the nuclide in amu

A = mass number (*i.e.*, isotopic mass rounded off to the nearest integer)

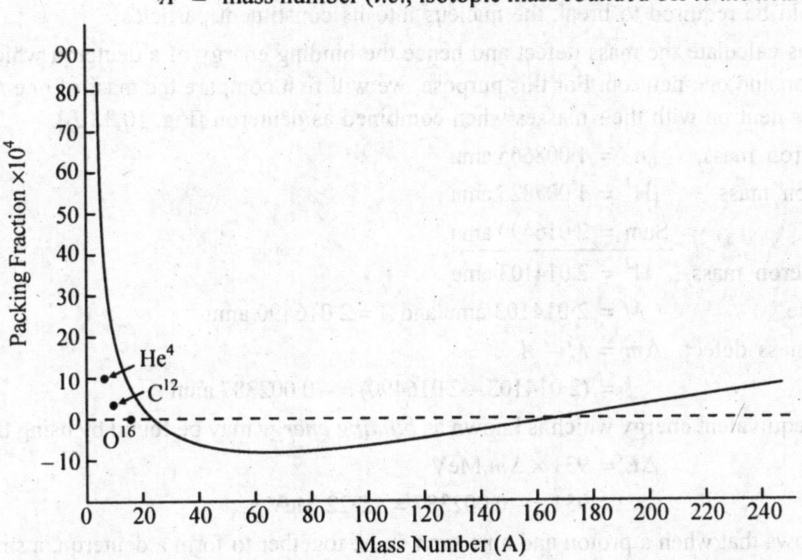

Fig. 10.2

The difference between atomic mass (M) and mass number (A) is called *mass defect*. Obviously, mass defect.

$$\Delta m = (M - A)$$

$$\therefore \quad \boxed{f = \frac{\Delta m}{A} = \frac{\text{mass defect}}{\text{mass number}}}$$

It means that packing fraction may also be defined simply *as mass defect per elementary particle in the nucleus.*

Variation of packing fraction with mass number for large number of nuclides is shown in Fig. 10.2. It is seen that packing fraction is high for elements of low mass numbers (except O^{16}, C^{12}, He^4) but then it decreases rapidly with increasing mass number. After passing through a comparatively flat minimum, the packing fraction starts to increase slowly but steadily.

It has been found *that packing fraction is a fundamental property of a nucleus and is directly related to the availability of nuclear energy and stability.* A negative packing fraction (and hence mass defect) means that the atomic mass of a nucleus is less than the nearest whole number and this suggests that there has been a conversion of mass into energy during the formation of that particular nucleus. Since same amount of energy would have to be supplied in order to break up the nucleus, it means that negative *packing fraction implies exceptional nuclear stability*. On the other hand, a positive packing fraction (and hence mass excess) indicates that the nucleus is somewhat unstable.

10.5. Mass Defect and Atomic Binding

The difference between the measured mass M and the mass number A of a nuclide is called the *mass defect.*

$$\therefore \quad \text{mass defect,} \quad \boxed{\Delta m = (M - A)}$$

Measurement of nuclear mass shows that the mass of a nucleus is always different from the sum of the masses of its constituents, *i.e.*, nucleons. Accordingly, we have either mass defect or excess. When nucleons are bound together into a nucleus, the mass which disappears is converted into energy and is known as the *binding energy* of the nucleus. Conversely, the same amount of energy would be required to break the nucleus into its constituent particles.

Let us calculate the mass defect and hence the binding energy of a deuteron which consists of one proton and one neutron. For this purpose, we will first compare the mass of one *free* proton and one *free* neutron with their masses when combined as deuteron [Fig. 10.3 (*a*)].

$$\begin{aligned} \text{Neutron mass,} \quad {}_0n^1 &= 1{\cdot}008665 \text{ amu} \\ \text{Proton mass,} \quad {}_1\text{H}^1 &= 1{\cdot}007825 \text{ amu} \\ \text{Sum} &= 2{\cdot}016490 \text{ amu} \\ \text{Deuteron mass,} \quad {}_1\text{H}^2 &= 2{\cdot}014103 \text{ amu} \\ \text{Hence,} \quad M &= 2{\cdot}014103 \text{ amu and } A = 2{\cdot}016490 \text{ amu} \\ \therefore \text{ mass defect, } \Delta m &= M - A \\ &= (2{\cdot}014103 - 2{\cdot}016490) = -\,0{\cdot}002387 \text{ amu} \end{aligned}$$

The equivalent energy which is known as *binding energy* may be found by using the relation

$$\begin{aligned} \Delta E &= 931 \times \Delta m \text{ MeV} \\ &= 931 \times -\,0{\cdot}002387 = -\,2{\cdot}22 \text{ MeV.} \end{aligned}$$

It shows that when a proton and a neutron come together to form a deuteron, a small part of their mass is annihilated and is radiation out in the form of energy from the newly-formed nucleus. At close approach, the two particles attract each other (by short-range nuclear forces) so strongly that it takes 2·22 MeV of energy to pull them apart — a fact confirmed experimentally.

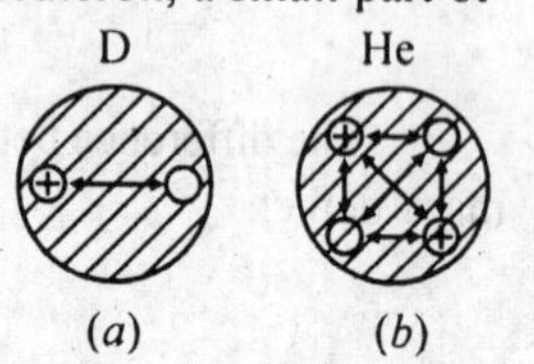

Fig. 10.3

Since, only two nucleus are involved, the binding energy per nucleon is $2{\cdot}22/2 = 1{\cdot}11$ MeV.

Next, consider the nucleus which contains two protons and two neutrons. Comparing the mass of four *free* particles with the mass of the helium nucleus, we get,

$$\begin{aligned} 2\,{}_0n^1 + 2\,{}_0\text{H}^1 &= A = 4{\cdot}032980 \text{ amu} \\ {}_2\text{He}^4 &= M = 4{\cdot}002604 \text{ amu} \\ \Delta m &= M - A = -\,0{\cdot}030376 \text{ amu} \\ \Delta E &= -\,931 \times 0{\cdot}030376 = -\,28{\cdot}3 \text{ MeV.} \end{aligned}$$

This gives the energy required for breaking up the attractive bonds shown in Fig. 10.3 (*b*). The binding energy per nucleon = $28{\cdot}3/4 \simeq 7$ MeV.

It would be of interest to calculate the amount of energy required to form one gram-atom of helium, *i.e.*, 4 grams of helium from its elementary constituents. Since the number of nuclides (or atoms) of helium in one gram-atom is equal to Avogadro's number, the total energy is

$$\begin{aligned} &= 28{\cdot}3 \times 6{\cdot}02 \times 10^{23} = 1{\cdot}69 \times 10^{25} \text{ MeV} \\ &= 2{\cdot}71 \times 10^{12} \text{ joules} = 750{,}000 \text{ kWh.} \end{aligned}$$

Similarly, for ${}_{10}\text{Ne}^{20}$ isotope, whose atomic mass is 19·99244 amu, ΔE = 160 MeV so that binding energy per nucleon = 160/20 = 8 MeV. For ${}_{83}\text{Bi}^{209}$ atom whose mass is 208·9804 amu, ΔE = 1640 MeV so that binding energy per nucleon works out to be $1640/83 \simeq 7{\cdot}8$ MeV. The two values come out to be almost equal despite the fact that the two nuclides lie at the opposite ends of the mass number scale.

If, the above procedure is carried out for the entire periodic table, a graph like the one shown in Fig. 10.4 is obtained. It is seen that binding energies of some of the light nuclei are very small.

The binding energy per nucleon rises sharply and reaches a maximum value of 8·8 MeV in the neighbourhood of $A = 50$. The maximum is quite flat and the binding energy per nucleon is still 8·4 MeV at $A = 140$. For higher mass numbers, the value decreases to 7·6 MeV for uranium. *It is this decrease in binding energy which is the fundamental cause of release of energy during fission of nuclei of high mass numbers.* However, over a considerable range of mass numbers, the binding energy per nucleon is 8 MeV which is usually taken as the average value for all nucleons.* Since

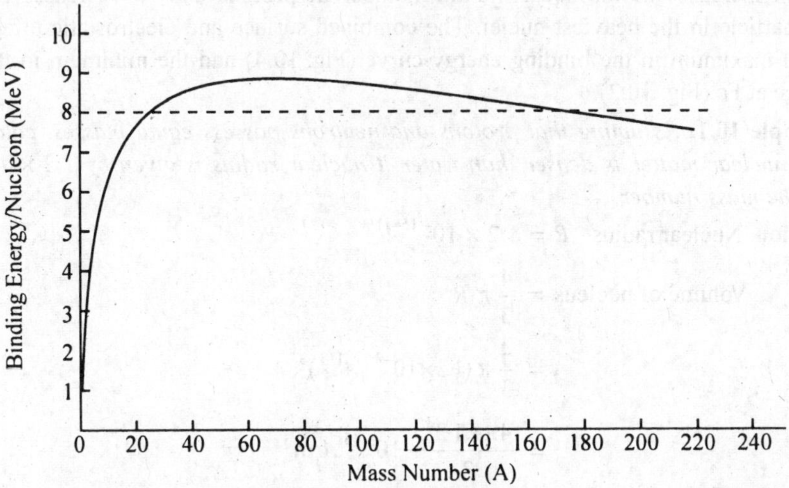

Fig. 10.4

this binding energy arises from the interaction of a nucleon with its immediate neighbours, it means that 8 MeV will have to be introduced to free it from its neighbours.

10.6. Accurate Expression for Mass Defect

A more accurate expression for mass defect can be found as follows :

If the atomic number of an element is Z and its mass number is A, then its nucleus contains Z protons and $(A - Z)$ neutrons. In addition, there are Z extranuclear electrons for neutralizing the positive charge of the nucleus. Obviously, the mass of the atom is equal to the sum of the masses of

(*i*) Z protons (*ii*) $(A - Z)$ neutrons and (*iii*) Z electrons.

Now, the proton and electron masses can be combined into the mass of Z hydrogen atoms *provided we neglect the minute change in mass which accompanies the formation of a hydrogen atom from a proton and an electron.* If m_H and m_n are the masses of the hydrogen atom and a neutron respectively, then

$$\text{true mass defect} = Z \cdot m_H + (A - Z)\, m_n - M$$

where M is the experimentally determined atomic mass. If m_H, m_n and M are expressed in amu, then

$$\text{binding energy} = 931\,[Z \cdot m_H + (A - Z)\, m_n - M]\ \text{MeV}$$

10.7. Variation of Binding Energy

The observed variations in the binding energy per nucleon may be explained as under :

The binding energy of a nucleon is due to its nuclear interaction with its immediate neighbours in a nucleus.

---

* It is seen that nuclear binding energies are in the range of millions of electron-volts in contrast to the few electron-volts for the binding energy of electrons involved in chemical reactions.

Since nucleons on the surface of the nucleus have fewer neighbours than those in the interior, their binding energy is comparatively less. Light nuclei with their small radii have greater surface-to-volume ratio than the heavy nuclei. Hence, they contain relatively more surface particles and consequently have smaller binding energy per particle. This explains the observed increase in binding energy from light nuclei up to Fe. At this stage, another factor comes into play, *i.e.*, the electrostatic repulsive force between protons. This repulsive force gives a negative binding energy which increases as the square of the number of protons and so decreases the binding energy per particle in the heaviest nuclei. The combined surface and electrostatic effects explain the observed maximum in the binding energy curve (Fig. 10.4) and the minimum in the packing fraction curve at Fe (Fig. 10.2).

Example 10.1. *Assuming that protons and neutrons possess equal masses, calculate how many times nuclear matter is denser than water it nuclear radius is given by* $1{\cdot}2 \times 10^{-15}A^{1/3}$ *m where A is the mass number.*

Solution. Nuclear radius $R = 1{\cdot}2 \times 10^{-15}A^{1/3}$

$$\text{Volume of nucleus} = \frac{4}{3}\pi R^3$$

$$= \frac{4}{3}\pi (1{\cdot}2\times10^{-15} A^{1/3})^3$$

$$= \frac{4\pi\times1{\cdot}2^3}{3}\times10^{-45} A \text{ m}^3$$

Let, mass of proton = mass of neutron

$$= 1{\cdot}008 \text{ amu}$$

$$= 1{\cdot}008 \times 1{\cdot}66 \times 10^{-27} \text{ kg}$$

$$= 1{\cdot}67 \times 10^{-27} \text{ kg}$$

Since the combined number of protons and neutrons is A,

$$\text{Mass of nucleus} = A \times 1{\cdot}67 \times 10^{-27} \text{ kg}$$

$$\text{Density of nucleus} = \frac{\text{mass}}{\text{volume}}$$

$$= \frac{A\times1{\cdot}67\times10^{-27}}{4\pi\times1{\cdot}2^3\times10^{-45} A/3}$$

$$= 2{\cdot}3 \times 10^{17} \text{ kg/m}^3$$

$$\text{Density of water} = 1 \text{ gram/cm}^3 = 10^3 \text{ kg/m}^3$$

$$\therefore \quad \frac{\text{density of nucleus}}{\text{density of water}} = \frac{2{\cdot}3\times10^{17}}{10^3} = \mathbf{2{\cdot}3 \times 10^{14}}$$

Hence, nucleus is $2{\cdot}3 \times 10^{14}$ times denser than water.

Example 10.2. *Compute the energy of one atomic mass unit in MeV.*

Solution. By using Einstein's mass-energy equation, the energy equivalence of 1 amu can be found easily.

Now, $\quad 1 \text{ amu} = 1{\cdot}66 \times 10^{-27} \text{ kg}; \; c = 3 \times 10^8 \text{ m/s}$

$$\therefore \quad E = mc^2 = 1{\cdot}66 \times 10^{-27} \times (3 \times 10^8)^2 = 14{\cdot}9 \times 10^{-11} \text{ J}$$

Also, $\quad 1 \text{ kWh} = 1{\cdot}6 \times 10^{-13} \text{ J}$

$$E = 14{\cdot}9 \times 10^{-11}/1{\cdot}6 \times 10^{-13} \cong \mathbf{931 \text{ MeV.}}$$

Example 10.3. *Compute the equivalent energy per gram in kWh.*

Solution. According to Einstein's equation

$$E = mc^2$$

Now, 1 gram $= 10^{-3}$ kg; $c = 3 \times 10^8$ m/s

$\therefore$ $E = 10^{-3} \times (3 \times 10^8)^2 = 9 \times 10^{13}$ J

Also, 1 kWh $= 36 \times 10^5$ J

$\therefore$ equivalent energy per gram $= 9 \times 10^{13}/36 \times 10^5 =$ **2·5 × 10⁷ kWh.**

Example 10.4. *Calculate the binding energy when*

(a) one neutron and one proton combine to form a deuteron

(b) two neutrons and two protons combine to form an α-particle.

Given : mass of neutron = 1·00893 amu; mass of proton = 1·00813 amu; mass of deuteron = 2·01473 amu; mass of α-particle = 4·00389 amu.

Solution. (*a*) $\Delta m = 2{\cdot}01473 - (1{\cdot}00893 + 1{\cdot}00813)$

$= -0{\cdot}00233$ amu $= 0{\cdot}00233 \times 931 =$ **2·17 MeV**

(*b*) $\Delta m = 4{\cdot}00389 - (2 \times 1{\cdot}00893 + 2 \times 1{\cdot}00813)$

$= -0{\cdot}03023$ amu $= 0{\cdot}03023 \times 931 =$ **28·1 MeV.**

Example 10.5. *The mass of $_{17}Cl^{35}$ is 34·9800 amu. Calculate its binding energy. What is the binding energy per nucleons ?*

Mass of $_0n^1$ = 1·008665 amu and $_1H^1$ = 1·007825 amu.

Solution. The given chlorine atom has 17 protons and (35 – 17) = 18 neutrons.

Mass of 17 protons $= 17 \times 1{\cdot}007825 = 17{\cdot}133025$ amu

Mass of 18 neutrons $= 18 \times 1{\cdot}008665 = 18{\cdot}155970$ amu

Total mass of separate constituents $= 35{\cdot}288995$ amu

Mass defect $\Delta m = 35{\cdot}288995 - 34{\cdot}9800$

$= 0{\cdot}308995$ amu

Now, 1 amu = 931 MeV

$\therefore$ Binding energy of the nucleus $= 931 \times 0{\cdot}308995 =$ **288 MeV**

No. of nucleons (*i.e.*, protons and neutrons) = 35

$\therefore$ Binding energy per nucleon = 288/35 = **8·22 MeV.**

Example 10.6. *Calculate the binding energy of lithium nucleus, given that*

mass of proton = 1·00814 amu

mass of neutron = 1·00893 amu

mass of lithium nucleus = 7·01822 amu

Solution. Obviously, the lithium isotope considered is $_3Li^7$. It contains 3 protons and (7 – 3) = 4 neutrons.

Mass of 3 protons $= 3 \times 1{\cdot}00814 = 3{\cdot}02442$ amu

Mass of 4 neutrons $= 4 \times 1{\cdot}00893 = 4{\cdot}03572$ amu

Total mass of 7 nucleons $= 7{\cdot}06014$ amu

Mass defect $\Delta m = 7{\cdot}06014 - 7{\cdot}01822 = 0{\cdot}04192$ amu

Binding energy $= 931 \times 0{\cdot}04192 =$ **39 MeV.**

10.8. Significance of Average Binding Energy

A remarkable fact about nuclei is that each of their protons and neutrons has an average binding energy of 8 MeV. It shows that a given nucleon does not interact with all other nucleons in a nucleus. Because, if it did, the binding energy per particle would have increased linearly with the number of nucleons in a nucleus on the basis of attractive forces between each pair. It is apparent that nucleons interact only with their *immediate* neighbours (about three of them) just as molecules in a liquid or gas interact only with their neighbouring molecules. This fact is known as *saturation* of nuclear forces and is consistent with their observed short range (Art. 10.9).

10.9. Nuclear Forces

According to the proton-neutron theory, a nucleus consists of protons and neutrons. Because of their positive charge, there are repulsive electrostatic forces between the protons tending to push the nucleus apart. Hence, if stable complex nuclei are to exist, there must be some sort of attractive forces in the nuclei strong enough to overcome the repulsive forces. These attractive forces are the specific *nuclear forces* which have the following properties :

(*i*) they are fundamentally different from the gravitational and electrostatic forces.

(*ii*) they are charge independent, *i.e.*, these nuclear forces are the same for proton-proton, proton-neutron and neutron-neutron interactions.

(*iii*) they are spin dependent.

(*iv*) they are short-range forces, *i.e.*, they operate over very short distances of about 2×10^{-15} m or 2 fermi. Outside the nucleus they decrease rapidly whereas coulombic force of repulsion which is a relatively long-range force becomes predominant.

(*v*) they possess saturation character, *i.e.*, each nucleon attracts those which are its immediate neighbours (about three of them) and is relatively indifferent to others just as the molecules in a liquid or gas interact only with neighbouring molecules.

(*vi*) their magnitude is so high that the work required to divide a nucleus into its constituents, *i.e.*, the binding energy is about 8 MeV in contrast to a few electron-volts required to separate the extranuclear electrons from its atom.

Imagine a proton or an α-particle approaching a positively-charged nucleus which is stationary. As the two come closer and closer, they repel each other with greater and greater force as given by Coulomb's Law.

However, this repulsion cannot continue to increase all the way to zero separation. Actually, as the two charged particles come very close to each other, they start attracting because at that separation, short-range nuclear forces become predominant. Graph of the Coulomb's Law of repulsion and short-range forces of attraction is shown in Fig. 10.5. As can be seen, when the particles are at distances greater than 2×10^{-15} m, Coulomb's force of repulsion is predominant. But at lesser distances, nuclear forces of attraction predominate.

Since both these forces exist over a certain range, their combined curve F can be drawn as shown

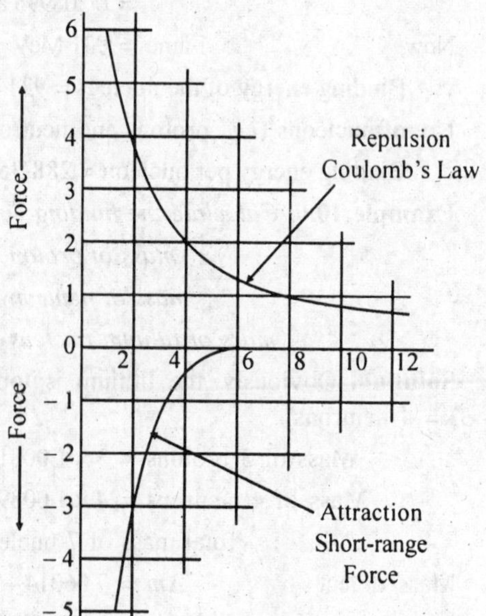

Fig. 10.5

in Fig. 10.6. The potential energy stored between the two particles is given by the dotted potential-energy curve.

Such a curve represents what is known as the *potential barrier* of the nucleus. It acts as barrier in the sense that it prevents the entry of the α-particles unless they possess certain minimum amount of energy.

A very good model of the nucleus, having the form of a well or a orater, can be made by rotating the potential energy line around the vertical axis. When this is done, we get the cross-section diagram of Fig. 10.7. The electric potential barrier between two charged particles is analogous to the potential energy of a rolling ball at any point on the side of the hill because electrostatic force of repulsion is analogous to the force of gravity. The quantity V (Fig. 10.7) gives the height of the potential barrier and R is the radius of this potential barrier or nuclear radius.

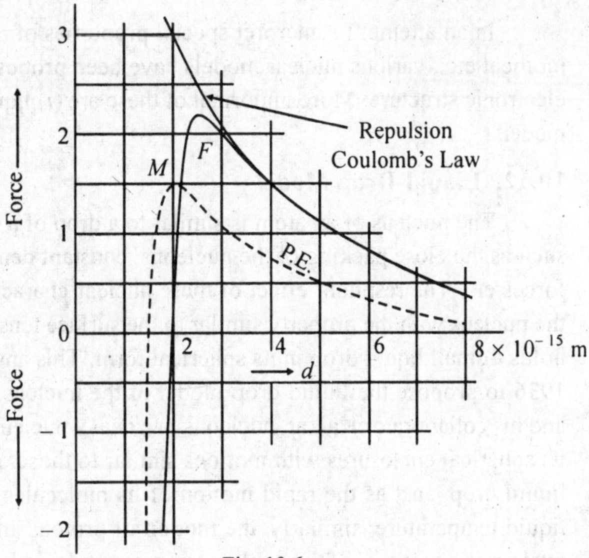

Fig. 10.6

For elements of high mass number, the maximum of the potential-energy curve corresponds to an energy of 25 MeV. It means that an α-particles with lesser energy would be repelled back if it tries to approach the nucleus from outside. For example, an α-particle, having an energy of 10 MeV would be repelled back when it is at a distance of OD from the centre of the nucleus. Obviously, the α-particle must at least possess 25 MeV if it wants to enter the nucleus.

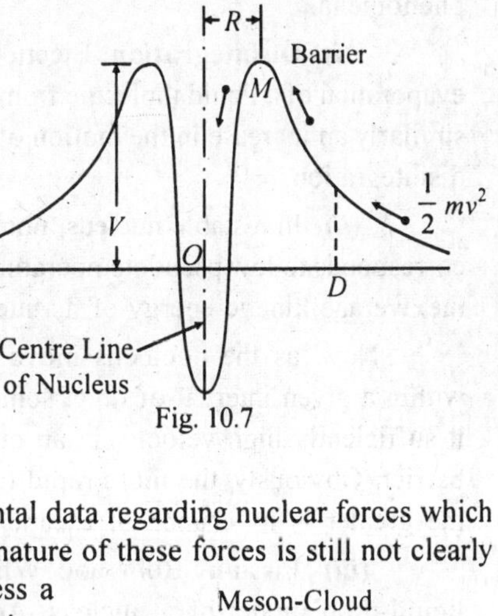

Fig. 10.7

10.10. Meson Theory of Nuclear Forces

Despite the existence of a large amount of experimental data regarding nuclear forces which hold protons and neutrons together in a nucleus, the exact nature of these forces is still not clearly understood. It has been discovered that these forces possess a strong property of being sometime attractive and sometime repulsive depending on the state of the nucleons with respect to each other. The origin of these nuclear exchange forces is not yet understood.

The discovery of a new particles called meson* in the cosmic radiation has led to the *meson theory of nuclear forces*. Several different mesons with positive, zero, and negative charges and different masses have been found by now.

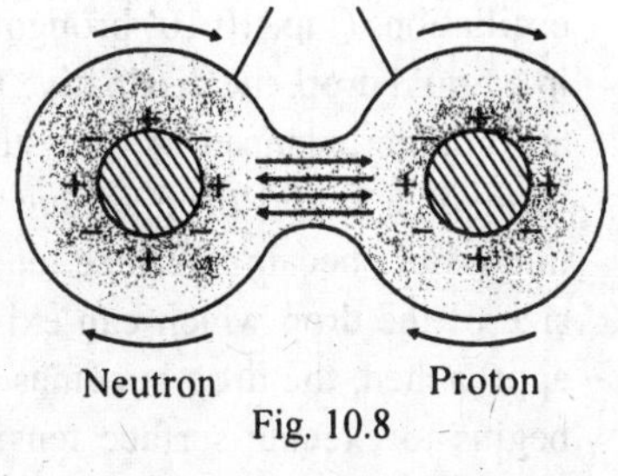

Fig. 10.8

According to the above theory, nucleons consist of some sort of common core surrounded by a pulsating cloud of π mesons called *pions* as shown in Fig. 10.8. These pions are supposed to rapidly jump back and forth between the nucleons thereby changing their identity equally fast and at the same time keeping them bound together.

* This particle was theoretically predicted by Japanese mathematical physicist Yukawa in 1935.

10.11. Nuclear Models

In an attempt to interpret special properties of a nucleus such as stability, spin and magnetic moment etc., various nuclear models have been proposed just as was done in the case or peripheral electronic structure. More important of these are (*i*) liquid or water-drop model and (*ii*) nuclear shell model.

10.12. Liquid-Drop Model

The nucleus of an atom is similar to a drop of a liquid in many ways such as the close packing of the nucleons, constant density and short-range forces etc. The resultant effect of these nuclear characteristics is to endow the nucleus with the property similar to the surface tension of liquids which holds a small liquid drop in its spherical form. This analogy led N. Bohr in 1936 to propose the liquid-drop model of the nucleus. According to Bohr and his collaborator Kalkar, nucleons in a heavy nucleus move about within its spherical enclosures with motions similar to those of the molecules in a liquid drop. Just as the rapid motion of its molecules is a measure of the liquid temperature, similarly, the motion of protons and neutrons within a nucleus is a measure of the nuclear temperature to which Bohr gave the name of pseudotemperature.

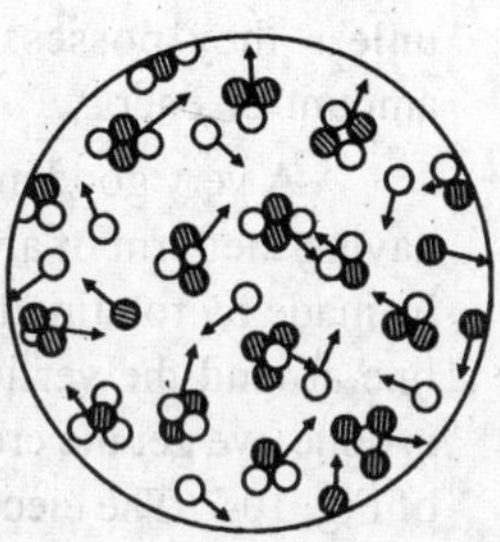
Fig. 10.9

Liquid-drop model of a nucleus helps to give acceptable explanations of the following nuclear phenomena.

(*i*) **Disintegration.** Ejection of a particle from a nucleus is considered analogous to the evaporation of a liquid molecule from a liquid drop. As temperature rise makes the evaporation rapid, similarly an increase in the motion of nucleons within a nucleus gives rise to a higher probability of disintegration.

(*ii*) In a stable nucleus, nucleons move about slowly with very little kinetic energy which corresponds to low pseudotemperature. Injection of a high-energy particle into the nucleus increases the average kinetic energy of its nucleons with a consequent rise in its temperature.

Now, as the nucleons move about inside, there is a certain probability or chance that, within a given interval of time, some particular nucleon would be hit by several nucleons giving it sufficiently high velocity in an outward direction resulting in its escape through the potential barrier. Obviously, the more rapid the internal motion and hence higher the nuclear temperature, the greater is its chance of escape.

(*iii*) **Fission.** Bohr and Wheeler were the first to explain the fission process on the liquid-drop model of a nucleus. According to their explanation, nuclear fission is due to the oscillations (Fig. 10.10) brought about by the impinging neutrons. As is known, in a liquid drop, the dissipative forces acting between the constituent molecules are counterbalanced by the surface tension forces. Moreover, the spherical form of a liquid drop is due to the surface tension. Very large drops are impossible because surface tension forces are feeble and there is a maximum size of the drop which can exist in a stable form permanently. As the size is approached, the drop becomes less stable and gets deformed. At this stage, it begins to execute surface tension oscillations with the slightest provocation from outside which ultimately lead to the disruption of the drop into two or more smaller droplets as shown in Fig. 10.11.

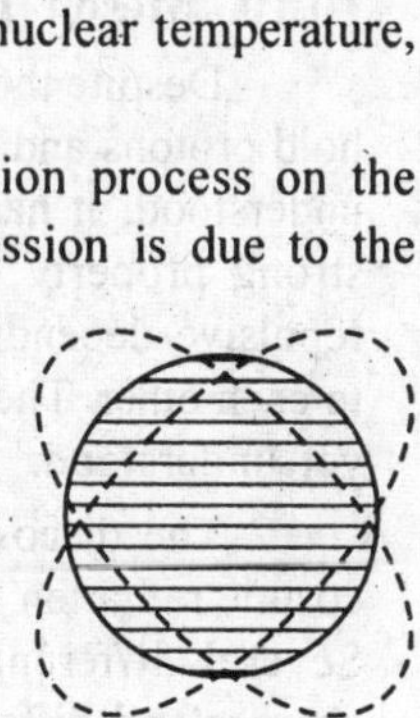
Fig. 10.10

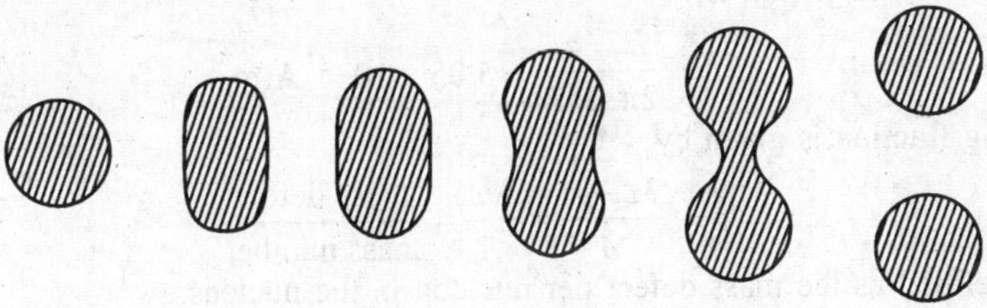

Fig. 10.11

In atomic nuclei, short-range nuclear forces play the part of surface tension and keep the different nucleons together in a stable state. The repulsive forces between protons play the part of disruptive forces tending to destroy nuclear stability. If surface tension forces prevail, the nucleus will not break up by itself. On the other hand, any two nuclei coming together will show a tendency to fuse together just as two ordinary liquid droplets do. If, on the contrary, the forces of repulsion have the upper hand, the nucleus will have a tendency to break up at slight external provocation, into two or more parts which fly apart at high speeds — a process known as *nuclear fission*.

10.13. Nuclear Shell Model

It is also known as *independent particle model* and is one of the most important and useful models of nuclear structure. According to this model, a nucleus is considered analogous to a gas in which the nucleons occupy quantum shells just as do the extranuclear electrons in an atom. This concept is base on the realization that there are specific numbers of protons and neutrons generally referred to as *magic numbers* or *shell numbers* for which the nuclei exhibit exceptional stability just as do the atoms containing certain number of closed shells of electrons. For example, all nuclei containing 2, 8, 20, 50, 82 and 126 nucleons are exceptionally stable because these numbers represent closed shells in the concerned nuclei. The stability of such nuclides is further proved by their small cross-section for neutron capture. In addition the concept of nuclear shells can be fairly well related in a simple manner with the known magnetic moments of nuclei.

HIGHLIGHTS

1. According to proton-neutron theory of nuclear constitution
   (*a*) the number of protons gives the atomic number Z.
   (*b*) the number of protons and neutrons together gives atomic mass number A.
   (*c*) the number of neutrons in $(A - Z)$.
   (*d*) the neutron/proton ratio increases with increasing mass number of atoms in order to ensure nuclear stability.

2. The nuclear radius R is given by the expression

$$R = R_0 A^{1/2}$$

where R_0 has an average value of $1{\cdot}4 \times 10^{-15}$ for all nuclei.

3. The nuclear spin of different nuclei is given by

$$p_I = I \cdot \frac{h}{2\pi}$$

where the spin quantum number I has half-integral values for isotopes of odd mass numbers and whole number values for those having even mass numbers.

4. The nuclear magnetic moment is

$$\mu_I = g \cdot \frac{h}{2\pi} \frac{e}{2M} = g \cdot \mu_n$$

5. The nuclear magneton is

$$\mu_n = \frac{h}{2\pi}\frac{e}{2M} = 5{\cdot}05 \times 10^{-27} \text{ A-m}$$

6. Packing fraction is given by

$$f = \frac{M-A}{A} = \frac{\Delta m}{A} = \frac{\text{mass defect}}{\text{mass number}}$$

It is defined as the mass defect per nucleon in the nucleus.

7. Mass defect (or excess) is given by the expression

$$\Delta m = (M - A)$$

where M is the measured mass and A the mass number of an atom.

8. Nuclear binding energy is given by

$$\Delta E = 931 \times \Delta m \text{ MeV}$$

9. The binding energy per nucleon is

$$= \Delta E/A = 931 \times \Delta m/A \text{ MeV}$$

Its average value is 8 MeV/nucleon.

10. A more accurate expression for true mass defect is

$$\Delta m = Z{\cdot}m_H + (A-Z)\,m_n - M$$

where m_H = mass of one hydrogen atom

m_n = mass of one neutron

$$\Delta E = 931\,[Z{\cdot}m_H + (A-Z)\,m_n - M] \text{ MeV}$$

11. Apart from long-range coulombic forces between nuclear protons, there exist short-range attractive nuclear forces which keep various nucleons bound together in a nucleus. These forces are considered to arise from a continuous and rapid jumping back and forth pions between the two nucleons.

12. According to liquid-drop model, a nucleus resembles a liquid drop in many respects. The short-range nuclear forces play the part of surface tension and keep the nucleus in a spherical shape.

    The electrostatic repulsive forces play the role of disruptive forces which tend to push the nucleons apart. The nuclear stability is determined by the balance of these two forces.

13. Nuclear shell model considers the various nucleons to lead independent existence and move about in quantum shells just as do the extranuclear electrons. Nuclear stability is due to complete and filled nuclear shells.

Can You Answer This ?

1. What do you understand by 'mass defect' and 'binding energy' of a nucleus ? Draw a curve showing the variation of binding energy for nucleon against the mass number. Use this curve to explain the release of energy in fusion of light nuclei and fission of heavy nuclei.
2. Write short note on the structure of the nucleus.
3. What is the binding energy of the nucleus ? Show how this concept of binding energy is related to the stability of an atomic nucleus ?
4. Write a short note on liquid-drop model.
5. Explain the term 'binding energy' of the nucleus.
6. State clearly what you understand by the terms (*i*) binding energy and (*ii*) mass defect, with respect to a nucleus.
7. Explain briefly the liquid-drop model of the nucleus.

11 NATURAL RADIOACTIVITY

Radioactivity — Natural and Artificial Radioactivity — General Properties of Radioactive Radiations — Identification of Radioactive Elements — Identification of α-particles — Properties of α-rays — Properties of β-rays — Beta Disintegration — Conservation of Nuclear Energy During Beta Disintegration — —Properties of γ-rays — Electron Positron Pair Production — Gamma Decay — Radioactive Disintegration or Decay — Laws of Radioactive Disintegration — Radioactive Decay Constant (λ) — Average Life — Methods of Finding Radioactive Constant — Determination of Half-life — Units of Radioactivity — Radioactive Series — Uranium Series — Actinium Series — Thorium Series — Similarities Between Radioactive Series — Isotopes, Isomers, Isobars, Isotones and Isodiapheres — Radioactive Growth and Decay — Radioactive Equilibrium — Cherenkov Radiations — Secondary Radiations — Radiation Detectors — Electrometer — Ionization Chamber — Non-integrating Type Ionization Chamber — Integrating-type Ionization Chamber — Geiger-Mueller Counter — Scintillation Counters — Semiconductor — Detectors — Cloud Chamber — Diffusion Cloud Chamber — Bubble Chamber — Spark Chamber — Photographic Emulsions — Radiation Damage — Applications of Radioactive Radiations — Highlights — Can You Answer This?

11·1. Radioactivity

Radioactivity was originally discovered by Becquerel in 1896 almost one year after the discovery of X-rays by Roentgen. This phenomenon is confined almost entirely to the heaviest element from 83 to 106 in the periodic table. What Becquerel discovered was that uranium element-92 gave out some kind of radiations which were highly penetrating, could affect photographic plates, ionize gases and cause scintillations on fluorescent screen. It was also discovered that radioactive behaviour of uranium was unaffected by temperature, pressure, electric and magnetic fields etc. Since this behaviour was absolutely independent of the environment of uranium atom, the radioactive properties of uranium were attributed to its nucleus.

In 1898, it was found by French husband-and-wife team of Pierre Curie (1859—1906) and Marie Curie (1867—1934) that polonium and radium (found in very minute quantities in a uranium ore called pitchblende) also emitted radioactive radiations. Weight for weight, polonium is about 10 million times more active than uranium and radium is twice as active as polonium.

The radioactive radiations emitted by these elements are found to consist of the following :

(*i*) Alpha (α) rays or α-particles

(*ii*) β-rays or β-particles

(*iii*) γ-rays or photons.

The α-particles are, in fact, doubly-ionized helium atoms *i.e.* helium atoms with both of their electrons removed. Obviously, an α-particle has double the charge of a hydrogen nucleus (or proton) and a mass number or atomic weight four times as great. It is represented by the symbol ${}_2He^4$.

The β-rays consist of ordinary electrons with a mass equal to 1/1836 of the mass of a proton. Since it carries a unit negative charge and its mass is negligible, symbol for electron is ${}_{-1}e^0$.

γ-rays are electromagnetic waves having frequency higher than that of the X-rays. They have no charge.

α-particles are ejected from the nucleus of the radioactive atom with a velocity varying from 1/10 to 1/100 of the velocity of light. The β-particles move faster than α-particles, some of them travelling with 99% of the velocity of light.

This phenomenon of spontaneous emission of powerful radiations exhibited by heavy elements is called radioactivity.

Radioactivity is essentially a nuclear phenomenon and is a drastic process because the element changes in kind. It is spontaneous and an irreversible self-disintegrating activity because the element breaks itself up for good.

Those elements which exhibit this activity are called radioactive elements. Examples are : uranium, polonium, radium, radon, ionium, thorium, actinium and mesothorium.

11·2. Natural and Artificial Radioactivity

Natural radioactivity is that *which is exhibited by elements as found in Nature*. It is always found in heavier elements in the periodic table.

However modern techniques of artificial transmutation of elements have made it possible to produce radioactivity in many other elements much lighter than those that occur in Nature. Such type of radioactivity is known as *artificial* or *induced radioactivity*. But whatever its origin, the activity itself is always spontaneous.

11·3. General Properties of Radioactive Radiations

(*i*) These radiations are highly penetrating, they affect photographic plates, ionize gases, cause scintillations on fluorescent screen, develop heat and produce chemical changes.

(*ii*) As radiations are given out, new elements are formed in an irreversible process—the new elements themselves being usually radioactive.

(*iii*) The emission of radiations is spontaneous and is not affected by external agents.

(*iv*) The emission is not instantaneous but is prolonged *i.e.* it is extended over a period of time otherwise it would not have been discovered at all.

(*v*) Except for radioactivity, there is nothing abnormal about the radioactive elements as regards their physical and chemical properties.

11·4. Identification of Radioactive Elements

It was discovered by Rutherford and his co-workers that radioactive radiations are of *three* quite different kinds. A simple experiment to demonstrate the actual existence of three radiations is illustrated in Fig. 11.1.

A small amount of a radioactive element, say radium, is dropped to the bottom of a small drill hole made in a lead block. This helps in obtaining a narrow beam of radioactive radiations which emerge from the mouth of the hole because other rays entering the walls of the lead block are absorbed before reaching the surface. A photographic plate *P* is placed at a short distance above the lead block to receive the radiations. The whole apparatus is placed in an air-tight chamber which is kept evacuated by a vacuum pump in order to avoid absorption of the radiations. A strong magnetic field is applied in the direction indicated in Fig.11.1

After a long exposure, when *P* is developed, three blackenings or traces are found on it. The one lying to the left is produced by the positively-charged α-particles*. The one to the right is produced by negatively-charged β-particles. The third trace which lies in between the other two is registered by γ-rays which, being unaffected by the magnetic field, go undeviated.

*The direction of bending can be found by using Fleming's left-hand rule.

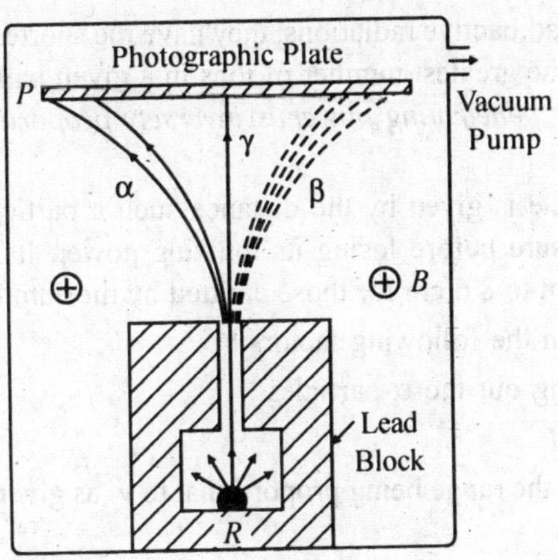

Fig. 11.1

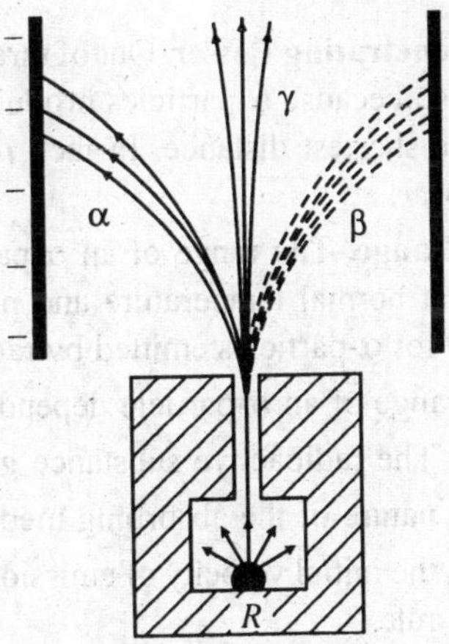

Fig. 11.2

Similar results can be obtained by the arrangement shown in Fig. 11.2 where an electric field has been applied instead of the magnetic field. The negatively-charged β-particles are deflected towards the positive plate whereas the positive α-particles bend towards the negative plate. As expected, the γ-rays do not bend at all.

It may be noted in passing that most *pure* radioactive substances emit γ-rays accompanied by α or β-particles but since samples are seldom pure, one usually finds all three types of radiations present together.

11·5. Identification of α-particles

It was conclusively proved for the first time by Rutherford and Royds in England in 1909 that α-particles are nothing else *but doubly-ionized helium nuclei.* A special discharge tube made of glass (Fig. 11.3) was used for this purpose. A thin-walled glass tube *A* containing radon gas was sealed inside a thick-walled tube *B* containing mercury. At the top was a glass capillary tube *C* with two metal electrodes sealed through glass as shown. Tubes *B* and *C* were evacuated through the side tube *D* after which mercury was raised to the level shown in the figure.

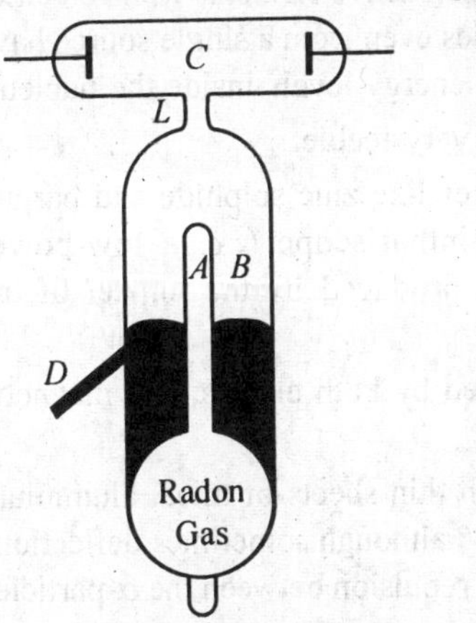

Fig. 11.3

After some time, a few α-particles from radon gas pass through thin walls of *A* and enter *B* where they acquire electrons and become neutral atoms of helium. After standing for a few days, mercury level is raised up to *L* thereby forcing the helium gas into the discharge tube *C*. A high-voltage is applied to the electrodes of *C* and the light from the discharge is observed with a spectroscope. The line spectrum so obtained is found to be identical with that obtained by passing a discharge through another tube containing regular helium gas.

11·6. Properties of α-rays

(*i*) **Ionizing Power.** They produce intense ionization in the gases through which they pass. Their ionizing power is 100 times greater than that of β-rays and 10,000 times greater than that of γ-rays. Hence, they produce very dense cloud-chamber tracks that resemble the vapour trails of a high flying aircraft. These traces are short and thick.

(*ii*) **Penetrating Power.** Out of three radioactive radiations, they have the shortest penetrating power. It is so because α-particles produce the greatest number of ions in a given path, hence they penetrate the shortest distance. In fact, *their penetrating power is inversely proportional to their ionizing power*.

(*iii*) **Range.** The range of an α-particle is given by the distance such a particle will travel through air at normal temperature and pressure before losing its ionizing power. Its value varies from 3·4 cm for α-particles emitted by radium to 8·6 cm for those emitted by thorium *O*.

The range of an α-particle depends on the following factors :

(*a*) The radioactive substance giving out the α-particles

(*b*) nature of the absorbing medium

(*c*) the initial velocity of emission—the range being proportional to v^2 as given by Geiger's rule.

The range of an α-particle can be measured (1) with the help of Wilson cloud chamber, (2) by counting the number of ions produced along the path and (3) by scintillations produced on a fluorescent screen.

When such an α-particle strikes a fluorescent screen, a tiny flash of light called *scintillation* is produced. Each individual scintillation can be observed with the help of a low-power microscope. When the radioactive sample, say, polonium is moved farther and farther away by pulling the rod *R* back, a point is reached when scintillations are no longer observed. The distance *d* where the α-particles just fail to reach the screen and hence produce no scintillation gives the range of the α-particles.

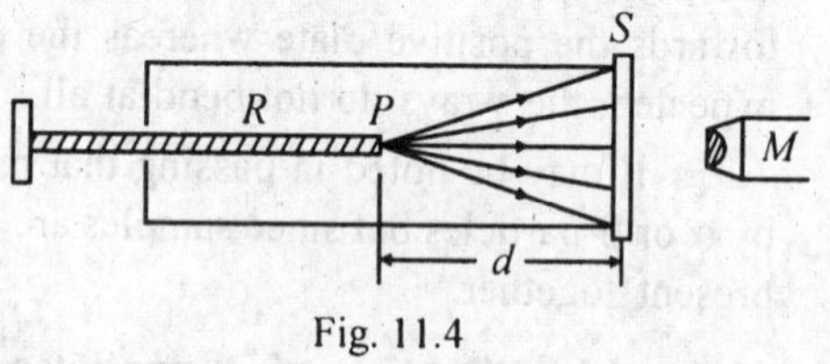

Fig. 11.4

(*iv*) **Velocity.** The α-particles are ejected from the nuclei of radioactive elements with velocities ranging from 1/10 to 1/100 of the velocity of light. Since α-particles even from a single source have different velocities, it means that they occupy discrete quantized energy levels inside the nucleus.

(*v*) They affect photographic plates although the effect is very feeble.

(*vi*) **Fluorescence.** They produce fluorescence in substances like zinc sulphide and barium platinocyanide. On observing the fluorescence through a spinthariscope (*i.e.* a low-power microscope), it is found to consist of successive scintillations produced by the impact of the individual particles.

(*vii*) **Deflection.** Being charged particles, they are deflected by both electric and magnetic fields.

(*viii*) **Scattering.** They are scattered while passing through thin sheets of mica, aluminium and gold foil etc. Generally, the deflection is not more than 2° to 3° although sometimes deflections of more than 90° are also observed. Such deflections are due to the repulsion between the α-particles and the nuclei of the atoms scattering them.

(*ix*) **Alpha Disintegration.** Before their emission, α-particles probably pre-exist in some form or the other (although not as individual units) in the nucleus. Moreover, they do not lie in any haphazard manner but are arranged in discrete energy levels.

11·7. Properties of β-rays

Some of their properties are as follows :

(*i*) β-rays consist of ordinary electrons which are of *nuclear origin* although they do not revolve in orbits as do the extra-nuclear electrons.

(*ii*) β-particles do not emerge out of the nucleus with the same range and velocity ; rather they have a wide band of velocity ranging from 1 per cent to 99 per cent of the velocity of light.

(*iii*) They produce ionization in air although their ionizing power is 1/100 of that of α-particles. That is why the β-tracks, as seen in a cloud chamber, are thin beady lines.

(*iv*) They are readily scattered when passing through matter because of their extremely small mass. Generally, they undergo multiple scattering although single scattering also takes place by impact both with electrons and with nuclei.

(*v*) They affect photographic plates more strongly than α-particles.

(*vi*) They produce fluorescence in barium platinocyanide, calcium tungstate,, Willemite etc.

(*vii*) They are deflected by electric and magnetic fields.

11·8. Beta Disintegration

The velocity of β-particles from a radioactive source can be measured with the help of a β-ray magnetic spectrograph shown in Fig. 11.5. Here *A* is a small radioactive source and consists of a fine wire with a deposit of the radioactive material on its surface. *S* is a slit which allows β-particles to pass through. When a uniform magnetic field is applied to the β-particles, they follow circular paths and are brought to focus on a photographic plate or at the open slit of an ionization chamber. Sometimes, Geiger-Mueller counter tube is used instead. The lead block protects detector from parasitical radiations. The whole apparatus is placed in a highly evacuated brass box in order to avoid scattering and the consequent energy loss of the deflected β-particles.

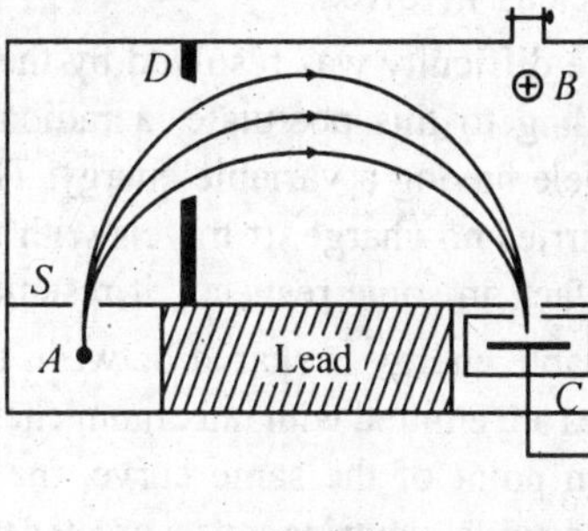

Fig. 11.5

Suppose the value of flux density *B* is kept constant and a photographic plate is used to receive the β-particles. On developing the plate after suitable exposure, a photograph like the one shown in Fig. 11.6 (*b*) is obtained. In addition to a darkened background all along the plate, one finds several black lines parallel to the slit. The background indicates the presence of β-particles having a wide range of velocities while the lines signify groups with discrete and definite velocities and hence energies.

However, if a G-M counter tube is used instead of the photographic plate and the number of electrons with different velocities is counted, then on plotting the experimental data, a curve like the one shown in Fig. 11.6 (*a*) is obtained. The sharp peaks indicate the presence of groups of electrons having definite energy.

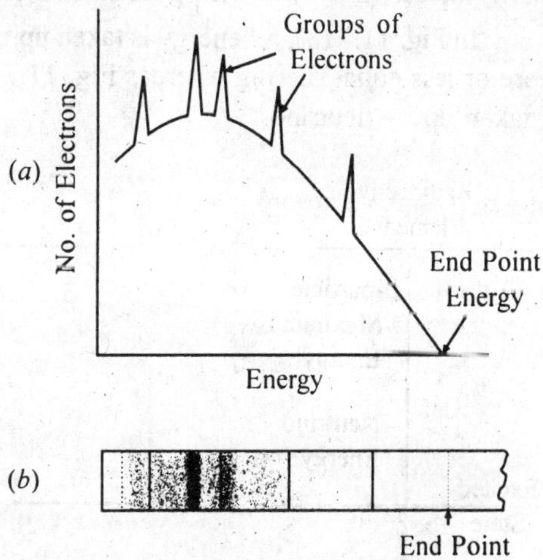

Fig. 11.6

The β-particles from a radioactive source have been found to be an admixture of the following two categories of electrons :

(*i*) *primary* or *disintegration* electrons which are emitted by the nucleus itself with varying velocities. These electrons give rise to the continuous spectrum of Fig. 11.6 (*a*)

(*ii*) *secondary* or *internal conversion* (*IC*) *electrons*. These are the orbital or extra-nuclear electrons which are knocked off by the γ-rays during their escape from the nucleus.

These *IC* electrons are emitted simultaneously with the disintegration electrons. The *IC* electrons always consist of discrete groups which register on the photographic plate as dark lines [Fig. 11.6 (*b*)] or as sharp peaks when plotted [Fig. 11.6 (*a*)].

The continuous β-spectrum which is due to the intra-nuclear electrons has an *end-point* or a *cut-off point* at the high-energy end and signifies an upper limit to the velocity or energy of these electrons.

The fact that primary electrons are emitted with varying energy having a definite end-point limit was difficult to explain particularly when nuclear spectroscopy had shown that there exist discrete quantum energy levels in the nuclei. According to this view, all the emitted β-particles must have a constant and definite value of energy equal to the difference of the energies of the two states of the nucleus involved.

The above difficulty was resolved by the neutrino postulate proposed by Pauli of Germany in 1931. According to this postulate, a radioactive nucleus emits β-particle accompanied by a companion particle having a variable energy. This particle is called *neutrino.* It has practically no rest mass and carries no charge. It travels with the velocity of light and acts merely as a carrier of energy only. In fact, in some respects, it resembles a photon.

The available energy is shared between the β-particle and neutrino. When neutrino gets no energy β-particles are emitted with maximum energy as shown by the end-point energy of Fig. 11.6(*a*). At the maximum point of the same curve, the two share the energy equally whereas in the low energy part of the curve, neutrino gets a greater part of the energy and the β-particle correspondingly less.

This variable distribution of energy between β-particle and neutrino can be easily understood with the help of Fig. 11.7 which shows the energy-level diagram for a typical β-particle emitter. As seen, a nucleus having an atomic number Z emits a β-particle and thus changes to excited state of a new nucleus $(Z + 1)$. It then goes to stable state by emitting two γ-ray photons one after another.

In Fig. 11.7 (*a*), all energy is taken up by β-particle and none by neutrino. Fig. 11.7 (*b*) shows more or less equal sharing whereas Fig. 11.7 (*c*) represents the condition when most of the energy is taken up by neutrino.

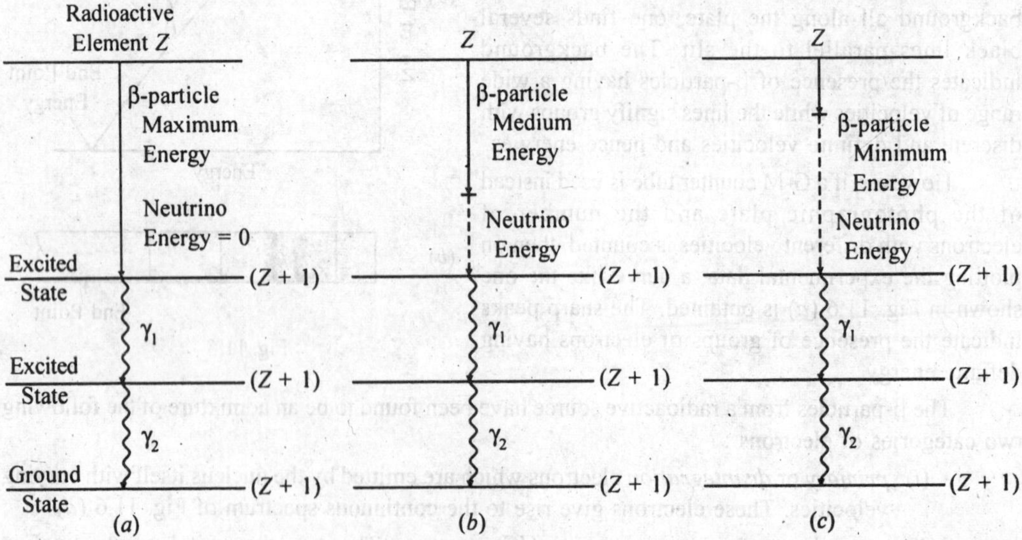

Fig. 11.7

Consider the case of RaE which is the radioactive bismuth isotope ${}_{83}Bi^{210}$. The reaction for β-emission by a RaE nucleus can be written as

$${}_{83}Bi^{210} \longrightarrow {}_{84}Po^{210} + \beta^{-1} + \nu$$

where ν represents neutrino.

In the end, it may be remarked that existence of neutrino was experimentally observed in 1956 by Reines and Cowan.

11·9. Conservation of Nuclear Energy During Beta Emission

The simultaneous emission of a β-particle and neutrino from the nucleus of a radioactive atom requires energy. The disintegrating nucleus gives up not only the mass of the two particles but some additional mass which it converts into kinetic energy.

Consider the emission of β-particle as represented by the above given equation. The masses of the atoms are

$${}_{83}Bi^{210} = 209{\cdot}984110 \text{ amu}$$
$${}_{84}Po^{210} = 209{\cdot}982866 \text{ amu}$$
$$\Delta m = 0{\cdot}001244 \text{ amu*}$$

Since annihilation energy of 1 amu = 931 MeV, the energy released by the annihilation of this mass is

$$\Delta E = 931 \times 0{\cdot}001244 = 1{\cdot}17 \text{ MeV}$$

This is just the end-point energy of the β-spectrum given by RaE nucleus.

11·10. Properties of γ-rays

Some of the important properties are as under :

(*i*) Gamma rays are electromagnetic radiations consisting of very energetic photons. They are emitted by an excited nucleus and have extremely short wavelength ranging from 0·5 Å to 0·005 Å.

(*ii*) They ionize the gas through which they pass although the ionization produced by them is comparatively small. Consequently, their penetrating power is very high as compared to that of β-rays.

(*iii*) They are capable of producing fluorescence and affecting photographic plates more intensely than β-rays.

(*iv*) Their absorption by a thin slab of matter follows the exponential law $I = I_0 e^{-\mu x}$ where μ is the absorption coefficient.

There are three processes responsible for the absorption of γ-rays by matter.

(*a*) **Photoelectric absorption.** As shown in Fig. 11.8, in this process, all the energy of the incident γ-ray photon is given to the bound electron. This energy is utilized partly for removing the electron from the atom and partly for giving it kinetic energy *T*.

$$\therefore \; hf = T + I$$

where *I* is the ionizing potential of the electron. It is also known as ***internal conversion electron process***. This process of absorption takes place when γ-rays have low-photon energy.

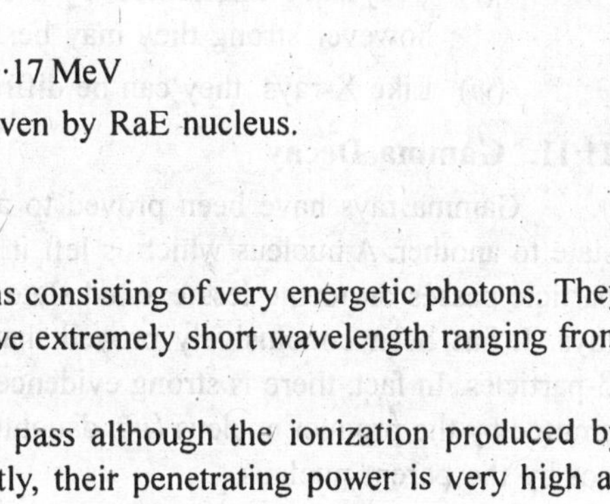

Fig. 11.8

*One a.m.u. = $1{\cdot}66 \times 10^{-27}$ kg.

If by internal conversion, an electron is ejected from the *K*-shell, then the vacancy so created may be filled by an electron from the *L*-shell or *M*-shell. During this jumping of the electron from either *L*- or *M*-shell to the *K*-shell, energy is radiated out in the form of X-rays which are found to be identical with those emitted by the same element in an X-ray tube.

(*b*) **Compton Scattering.** In the process which takes place when γ-rays have medium photon energies, the incident photon is scattered by one of the atomic electrons and the latter is separated from its atom. The photon moves off at an angle with its original direction and with less energy than it has initially.

(*c*) **Electron-positron* Pair Production.** This pair production takes place with very high-energy γ-ray photons. In this process of absorption, a γ-ray photon disappears in the Coulomb field of an atomic nucleus and an electron-positron pair is created (Fig. 11.9).

$$\therefore \qquad hf = T + 2m_0c^2$$

where T is the kinetic energy of the *pair* and m_0 is the rest mass of either the electron or the positron.

Obviously, pair production cannot take place if photon energy $hf < 2m_0c^2$ or $hf < 1{\cdot}02$ MeV.**

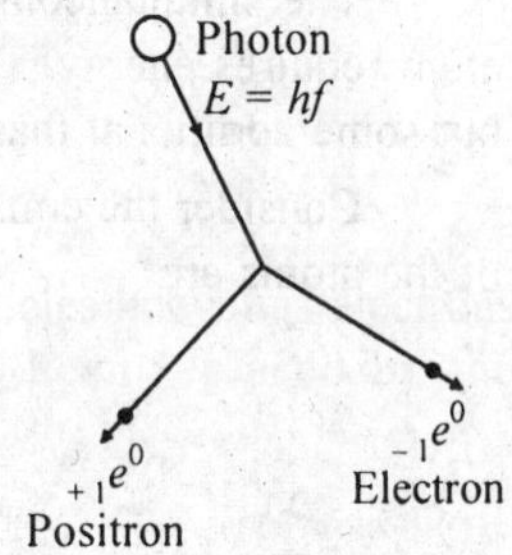

Fig. 11.9

(*v*) γ-rays are unaffected by electric and magnetic fields, however strong they may be.

(*vi*) Like X-rays, they can be diffracted by crystals.

11·11. Gamma Decay

Gamma rays have been proved to arise from the transition of a nucleon from one energy state to another. A nucleus which is left in an excited state following the emission of an α- or β-particle passes on to the less excited state or sometimes to the ground state itself, by emitting γ-rays. It has been conclusively proved that emission of γ-rays always follows that of the α- or β-particles. In fact, there is strong evidence that γ-rays are emitted by the product nucleus (*i.e.* daughter product) and not by the parent nucleus.

Consider the case of RaD (a lead isotope ${}_{82}Pb^{210}$). Its reaction of beta and gamma emission can be written as

$${}_{82}Pb^{210} \longrightarrow {}_{83}Bi^{210} + \beta^{-1} + \nu + \gamma\text{-ray}$$

There emissions are shown in the nuclear energy level diagram of Fig. 11.10. When a ${}_{82}Pb^{210}$ nucleus emits a β-particle and neutrino, the end-point energy is found to be 18 keV. This leaves a ${}_{83}Bi^{210}$ nucleus in an excited state. A transition down to ground level is accompanied by the emission of γ-rays with an energy equal to 47 keV. Although not shown in the diagram, this nucleus too is radioactive and emits another β-particle to become ${}_{84}Po^{210}$.

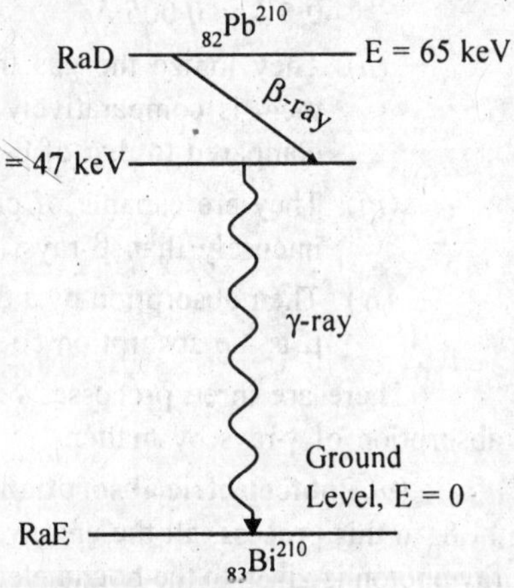

Fig. 11.10

*Positron particle is identical with electron in every respect except that its charge is positive.

** $$2\,m_0c^2 = 2 \times 9{\cdot}1 \times 10^{-31} \times (3 \times 10^8)^2 \text{ J}$$

Now $$1 \text{ eV} = 1{\cdot}6 \times 10^{-19} \text{ J} \text{ or } 1 \text{ MeV} = 1{\cdot}6 \times 10^{-13} \text{ J}$$

$$\therefore \qquad 2m_0c^2 = \frac{2 \times 9{\cdot}1 \times 10^{-31} \times 9 \times 10^{16}}{1{\cdot}6 \times 10^{-13}} = 1{\cdot}02 \text{ MeV}$$

Particles like neutrons and protons etc. which reside inside the nucleus are collectively known as nucleons.

11·12. Radioactive Disintegration or Decay

Atoms of heavy elements like uranium, thorium, polonium and radium etc. are constantly breaking up into fresh radioactive atoms with the emission of α-, β- and γ-rays from their nuclei. In the process, the original (or parent) atom disappears and gives rise to new (daughter) atom. These new atoms are also, in general, radioactive and hence spontaneously break up, in their turn, thereby leading to a long chain of different radioactive elements in the form of a series until an inactive element (usually lead) is reached. This *spontaneous breaking up of the nucleus is known as radioactive disintegration.*

For example, when a radium (Ra) atom disintegrates by ejecting an α-particle, its nucleus loses a net positive charge of 2 units and mass of 4 units. Hence, it changes into a new atom called radon (Rn). It means that there occurs transmutation of radium into radon.

It is common practice to designate all atomic nuclei in an abbreviated form. The radon nucleus is written as ${}_{88}Rn^{226}$. The subscript to the left gives the *atomic number i.e.* the number of protons or positive charge on the nucleus and the superscript on the right gives the *atomic weight* (*i.e.* weight of the protons and neutrons put together) or *atomic mass number*.

The disintegration of radium nucleus can be represented by the following equation :—

$${}_{88}Ra^{226} \rightarrow {}_{86}Rn^{222} + \alpha$$

or

$${}_{88}Ra^{226} \rightarrow {}_{86}Rn^{222} + {}_{2}He^{4}$$

Similarly, the equation for polonium can be written as

$${}_{84}Po^{210} \rightarrow {}_{82}Pb^{206} + {}_{2}He^{4}$$

When a radium-B (RaB) nucleus disintegrates by ejecting a β-particle to become radium-C (RaC), the nuclear positive charge increases by one unit. However, mass of the nucleus remains practically unchanged because electronic mass is negligible. Accordingly, disintegration of RaB atom may be represented by the following equation

$${}_{82}Ra^{214} \rightarrow {}_{83}Ra^{214} + {}_{-1}e^{0} + \gamma\text{-ray}$$

It is worth noting that atomic symbol for a β-particle (which is just an electron) is ${}_{-1}e^{0}$. Since ejection of β-particle is followed by the emission of γ-ray, it is also included in the equation.

It should be noted that in all equations representing nuclear disintegration, the *sum of the subscripts as well as superscripts on the right side of the equation is equal to the sum of the same quantities on the left side.*

11·13. Laws of Radioactive Disintegration

Radioactive disintegration is found to obey the following two laws :

(*i*) *atoms of all radioactive elements undergo spontaneous disintegration to form fresh radioactive products with the emission of* α-, β– and γ-*rays.*

(*ii*) *The rate of radioactive disintegration i.e. the number of disintegrations per second is not affected by environmental factors* (*like temperature, pressure and chemical combination etc.*) *but depends on the number of the atoms of the original kind present at any time.*

Suppose at the beginning of disintegration *i.e.* at $t = 0$, the number of radioactive atoms present in a sample is N_0. As time passes, the number of the original atoms decreases due to continuous disintegration. At any time *t*, let the number of atoms of the original kind be *N*. At this instant, the rate of disintegration ($-dN/dt$) which is also called *activity* is proportional to *N*

$$\frac{dN}{dt} \propto N \quad \text{or} \quad \frac{-dN}{dt} = \lambda N \qquad \ldots (i)$$

where λ is the constant of proportionality and is known as the *disintegration or decay constant or radioactive constant.*

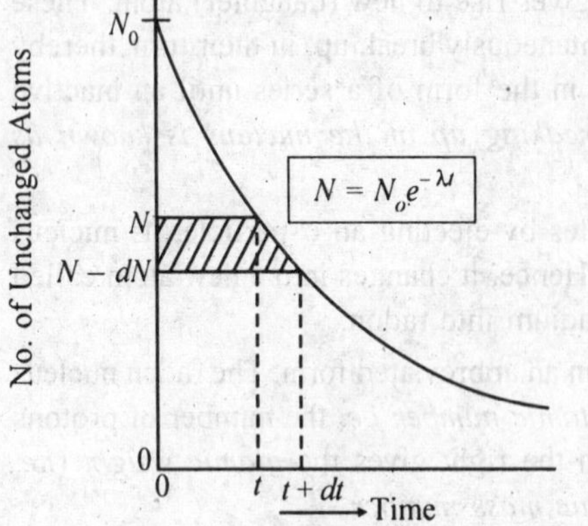

Fig. 11.11

Rearranging the above equation, we get

$$\frac{dN}{dt} = -\lambda\, dt$$

$$\therefore \quad \int_{N_0}^{N} \frac{dN}{dt} = -\lambda \int_0^t dt$$

$$|\log_e N|_{N_0}^{N} = -\lambda |t|_0^t$$

$$\therefore \quad \log_e N / N_0 = -\lambda t$$

$$\text{or} \quad \frac{N}{N_0} = e^{-\lambda t}$$

$$\therefore \quad \boxed{N = N_0\, e^{-\lambda t}} \quad ...(ii)$$

This decay equation may be used to find the number of the atoms of the original kind still present at any time *t*.

If M_0 is the original mass of the material and *M* its present mass, then

$$\boxed{M = M_0\, e^{-\lambda t}}$$

11·14. Radioactive Decay Constant (λ)

(*a*) Equation (*i*) above may be put as

$$\lambda = \frac{-dN/dt}{N}$$

Hence, radioactive constant may be defined as *the ratio of the amount of the substance which disintegrates in a unit time to the amount of the substance present.*

(*b*) Suppose, we put $t = 1/\lambda$ in Eq. (*ii*) above. Then the number of the original atoms present after this time is

$$N = N_0\, e^{-\lambda \times 1/\lambda} = N_0\, e^{-1}$$

$$= \frac{N_0}{e} = \frac{N_0}{2{\cdot}718}$$

$$N = 0{\cdot}368\, N_0 \simeq 0{\cdot}37\, N_0$$

Hence, radioactive constant may also be defined as *the reciprocal of the time during which the number of atoms of a radioactive substance falls to 37 per cent of its original value.*

11·15. Half-life Period

Half-life period of a radioactive element may be defined as *the time during which a given amount of that element is reduced by disintegration to half its initial amount.*

It is written as $t_{1/2}$ or *T*. Its value may be found by putting $N = N_0/2$ and $t = T$ in Eq. (*ii*) of Art. 11.13.

$$\therefore \quad N_0/2 = N_0 \cdot e^{-\lambda T} \quad \text{or} \quad 1/2 = e^{-\lambda T} \quad \text{or} \quad e^{\lambda T} = 2$$

$$\therefore \quad \lambda T \log_e e = \log_e 2 \quad \text{or} \quad \lambda T = 2{\cdot}3 \times \log_{10} 2 = 0{\cdot}693$$

$$\boxed{T = \frac{0{\cdot}693}{\lambda}}$$

As seen, *T* is inversely proportional to the radioactive constant and hence is a characteristic constant which can be employed to differentiate radioelements. For example, it takes nearly 1620 years for one-half of a given quantity of radium to change into radon. In another 1620 years, 1/2 of the remainder would have disintegrated leaving 1/4 of the original amount behind. Accordingly, we say that half-life period of radium is 1620 years. However, radon has a half-life of only 3.8 days.

In general, it may be said that at the end of time T, only 50% of the radioactive atoms remain unchanged, at the end of $2T$, only 25%, after $3T$, only 12·5%, after $4T$, only 6·25% and at the end of $10T$, only 0·1%. Hence, the life time of a radioactive element, though theoretically infinite (Fig. 11.12), is finite for all practical purposes.

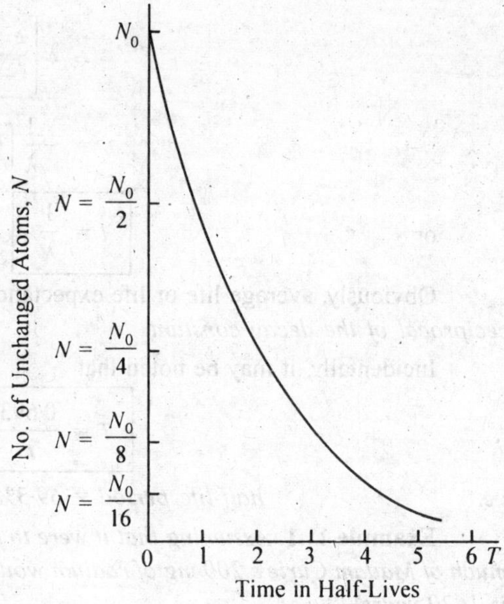

Fig. 11.12

As seen from Art. 11.13, λN represents the rate at which N atoms disintegrate.

∴ rate of disintegration

$$\text{or activity} = \lambda N = \frac{0{\cdot}693N}{T}$$

If T is expressed in seconds, then the above equation gives the emission rate in terms of the number of particles expelled per second from N atoms of the given nuclide.

11·16. Average Life

Since atoms of a radioactive element are constantly disintegrating one after another, the life of every atom is different. The atoms which disintegrate earlier have very short life whereas others which disintegrate at the end have a long life. Hence, the average or mean life of a radioactive atom is given as

$$l = \frac{\text{sum of lives of all atoms}}{\text{total number of atoms}}$$

From the decay curve of Fig. 11.11, it is seen that $-dN$ atoms disintegrate between time t and $(t + dt)$ so that these atoms have a life of t seconds.

∴ total life of $-dN$ atoms $= -dN \cdot t$

Since the possible life of any of the given number N_0 of atoms varies from 0 to ∞, total life of all the N_0 atoms is given by

$$\int_0^{\infty} -t \cdot dN.$$

$$\therefore \qquad \text{average life, } l = \frac{1}{N_0}\int_0^{\infty} -t \cdot dN$$

Now, $N = N_0\, e^{-\lambda t}$. Hence, $-dN = \lambda N_0\, e^{-\lambda t}\, dt$. Substituting this value of $-dN$ in the above equation, we get

$$l = \frac{1}{N_0}\int_0^{\infty} \lambda N_0\, te^{-\lambda t} \cdot dt = \lambda \int_0^{\infty} t \cdot e^{-\lambda t}\, dt$$

Using double integration, we get

$$l = \lambda \left[\frac{e^{-\lambda t}}{-\lambda} \cdot t - \int \frac{e^{-\lambda t}}{-\lambda} \cdot dt \right]_0^{\infty}$$

$$= \lambda\left[\frac{e^{-\lambda t}}{-\lambda}\cdot t - \frac{e^{-\lambda t}}{-\lambda^2}\cdot dt\right]_0^{\infty}$$

$$= -\frac{1}{\lambda}\left[(\lambda t+1)\,e^{-\lambda t}\right]_0^{\infty} = \frac{1}{\lambda}$$

or
$$\boxed{l = \frac{1}{\lambda}}$$

Obviously, average life or life expectancy of an individual radioactive atom *is equal to the reciprocal of the decay constant.*

Incidentally, it may be noted that

$$\boxed{T = \frac{0{\cdot}693}{\lambda} = 0{\cdot}693\,l}$$

i,e. *half-life period = 69·3% of average life.*

Example 11.1. *Assuming that it were to remain undisturbed since 1898 A.D., calculate how much of Madam Curie's 200 mg of radium would be left in the year 8378 A.D. (Half-life of radium is 1620 years).*

Solution. Total time gap = 8378 – 1898 = 6480 yrs

No. of half-periods = 6480/1620 = 4

Amount of radium which would be left un-disintegrated

$$= M_0\left(\frac{1}{2}\right)^n = 200\left(\frac{1}{2}\right)^4 = \mathbf{12{\cdot}5\ mg.}$$

Example 11.2. *A radioactive substance has a half-life period of 30 days. Calculate (i) radioactive disintegration constant, (ii) time taken for 3/4 of the original number of atoms to disintegrate and (iii) time taken for 1/8 of the original number of atoms to remain unchanged.*

Solution. (*i*) $\lambda = 0{\cdot}693/T = 0{\cdot}693/30\ \text{day}^{-1}$

(*ii*) Here, remaining mass is $M = M_0/4$

$\therefore\ M_0/4 = M_0\,e^{\lambda t}$ $\qquad \therefore\ \mathbf{t = 60\ days}$

(*iii*) Here, $M = M_0/8$ $\qquad \therefore\ M_0/8 = M_0\,e^{\lambda t}$ $\qquad \therefore\ \mathbf{t = 90\ days}$

Example 11.3. *A counter rate meter is used to measure the activity of a radioactive sample. At a certain instant, the count rate was recorded as 4750 counts per minute. Five minutes later, the count rate recorded was 2700 counts per minute. Compute (i) the decay constant and (ii) the half-life of the sample.*

Solution. $\qquad N = N_0\,e^{-\lambda t}$

or $\qquad N_0/N = e^{\lambda t}$ or $\log_e{}^{N_0/N} = \lambda t$

$$\therefore \qquad \lambda = \frac{\log_e{}^{N_0/N}}{t} = \frac{2{\cdot}3\log_{10}{}^{N_0/N}}{t} \qquad ...(i)$$

Now, $\qquad dN_1/dt = \lambda N_0$ and $dN_2/dt = \lambda N$

$$\therefore \qquad \frac{N_0}{N} = \frac{dN_1/dt}{dN_2/dt} = \frac{4750}{2700} = 1{\cdot}76$$

Substituting this value in the Eq. (*i*) above, we get

(*a*) $\lambda = 2{\cdot}3 \times \log_{10}^{1{\cdot}76/5} = \mathbf{0{\cdot}113\ per\ minute}$

(*b*) $T = 0{\cdot}693/0{\cdot}113 = \mathbf{6{\cdot}1\ minutes}$

Example 11.4. *A radioactive source contains one microgram (μg) of Pu-239. This source is estimated to emit 2200 α-particles per second in all directions. Estimate half-life of plutonium from this data.*

Solution. Mass of one α-particle = 4·00387 amu

$$= 4{\cdot}00387 \times 1{\cdot}66 \times 10^{-24}\ \text{g} = 6{\cdot}646 \times 10^{-24}\ \text{g}$$

Mass of 2300 α-particles = $2300 \times 6{\cdot}646 \times 10^{-24}\ \text{g} = 14{\cdot}867 \times 10^{-21}\ \text{g}$

Now,
$$\lambda = \frac{-dM/dt}{M} = \frac{-14{\cdot}867 \times 10^{-21}/1}{1 \times 10^{-6}} = -14{\cdot}867 \times 10^{-15}\ \text{s}^{-1}$$

$$\text{Half-period life } T = \frac{0{\cdot}693}{\lambda}$$

$$= \frac{0{\cdot}693}{14{\cdot}865 \times 10^{-15}} \text{ seconds}$$

$$= \mathbf{1{\cdot}478 \times 10^6\ years}$$

Example 11.5. *A radioactive sample initially contains 4·00 mg of $_{92}U^{234}$. (a) How much of it will remain unchanged after 62,000 years? (b) What will be its $_{92}U^{234}$ activity at the end of that time ?*

Given. $T = 2{\cdot}48 \times 10^5$ *years and* $\lambda = 8{\cdot}88 \times 10^{-14}\ s^{-1}$

Solution. (*a*) Let M_0 and M be the masses of the radioactive material at the beginning and the end of 62,000 years.

Now,
$$M = M_0\, e^{-\lambda t}$$

Also,
$$\lambda t = \frac{0{\cdot}693}{T} \times t = \frac{0{\cdot}693 \times 62{,}000}{2{\cdot}48 \times 10^5} = 0{\cdot}173^*$$

$\therefore$
$$M = M_0\, e^{-0{\cdot}173}$$

or
$$M_0 = M\, e^{0{\cdot}173} = 1{\cdot}189\, M \text{ or } M = M_0/1{\cdot}189$$

$\therefore$ mass left unchanged $= 4/1{\cdot}189 = \mathbf{3{\cdot}364\ mg.}$

(*b*) Activity is given by $A = dN/dt = \lambda N$

where N is the number of atoms still left intact. Now, number of atoms in a mass of 3·364 mg

$$= \frac{3{\cdot}364 \times 10^{-3} \times 6{\cdot}02 \times 10^{23}}{234}$$

$\therefore$
$$\text{activity} = \lambda N = 8{\cdot}88 \times 10^{-14} \times \frac{3{\cdot}364 \times 10^{-3} \times 6{\cdot}02 \times 10^{23}}{234}$$

$$= \mathbf{7{\cdot}69 \times 10^5\ disintegrations/second.}$$

Example 11.6. *The half-life period of radium is 1620 years. In how many years will one gram of pure element (a) lose one centigram and (b) be reduced to one centigram?*

Solution. (*a*) Radioactive constant is given by

$$\lambda = \frac{0{\cdot}693}{T} = \frac{0{\cdot}693}{1620} = 4{\cdot}28 \times 10^{-4}\ \text{years}^{-1}$$

Let t be the time during which one centigram of radium is lost due to disintegration. Out of one gram of radium (1 – 1/100) = 0·99 gram would be left behind.

---

*Also $\lambda t = 8{\cdot}88 \times 10^{-14} \times 62{,}000 \times 365 \times 24 \times 3600 = 0{\cdot}173$

Now, $N = N_0 e^{-\lambda t}$ or $N_0/N = e^{\lambda t}$

or $\log_{10} N_0/N = \lambda t \times 0.4343$

$N_0/N = 1/0.99 = 100/99$ —given

$\therefore$ $\log_{10} {}^{100/99} = 4.28 \times 10^{-4} \times t \times 0.4343$ $\therefore$ $t =$ **23·68 years**

(*b*) Here, $N_0/N = 1/0.01 = 100$

$\therefore$ $\log_{10} {}^{100} = 4.28 \times 10^{-4} \times t \times 0.4343$ $\therefore$ $t =$ **10,760 years**

11·17. Units of Radioactivity

The *activity* of a radioactive substance is measured in terms of disintegrations/second (or *dN/dt*). The unit for measuring *activity* is Curie (Ci). One curie is defined as *that quantity of a radioactive material which gives* $3{\cdot}7 \times 10^{10}$ disintegrations per second.

Obviously, 1 millicurie (mCi) $= 3.7 \times 10^7$ dis/s

1 microcurie (μCi) $= 3.7 \times 10^4$ dis/s

A new absolute unit of activity or radioactive disintegration rate has been recently suggested. Its name is, rutherford (rd). One rutherford is defined *as that quantity of a radioactive substance which gives* 10^6 *dis/s.*

Smaller units like millirad (mrd) and microrad (μrd) are also used.

Let us find the mass in gram of 1 Ci and 1 rd of radium-B *i.e.* RaB ($_{82}Pb^{214}$) from its half-life period of 26·8 minutes Its $\lambda = 0.693/26.8 \times 60 = 4.31 \times 10^{-4}$ per second. If m is the mass of RaB which gives 3.7×10^{10} dis/s, then number of atoms contained in this mass is

$$= \frac{m \times 6.02 \times 10^{23}}{214}$$

Now, $$-\frac{dN}{dt} = \lambda N$$

$\therefore$ $$3.7 \times 10^{10} = 4.31 \times 10^{-4} \times \frac{m \times 6.02 \times 10^{23}}{214}$$

$\therefore$ $$m = 3.1 \times 10^{-8} \text{ gram}$$

Similarly, for 1 rd, $$\frac{dN}{dt} = 10^6 \text{ dis/s}$$

$\therefore$ $$10^6 = 4.31 \times 10^{-4} \times \frac{m \times 6.02 \times 10^{23}}{214}$$

$\therefore$ $$m = 8.3 \times 10^{-13} \text{ gram}$$

It is obvious that for a substance of short half-life, a very small amount of material is required to provide one curie or one rutherford of activity.

Let us now consider $_{92}U^{238}$ which has very long half-life period of 4.5×10^9 years. Its $\lambda = 0.693/4.5 \times 10^9 \times 365 \times 24 \times 3600 = 4.9 \times 10^{-18}\ s^{-1}$. Using, $-dN/dt = \lambda N$, we get

for one curie

$$3.7 \times 10^{10} = 4.9 \times 10^{-18} \times \frac{m \times 6.02 \times 10^{23}}{238}$$

$\therefore$ $$m = 3.2 \times 10^6 \text{ gram}$$

for one rutherford

$$10^6 = 4.9 \times 10^{-18} \times \frac{m \times 6.02 \times 10^{23}}{238}$$

$\therefore$ $$m = 80.9 \text{ gram}$$

11·18. Radioactive Series

When radioactive disintegration occurs with the emission of α- or β-particles, the original atom (called parent atom) changes into something else called *daughter* atom which is also radioactive. The daughter element, sooner or later, ejects another particle to become still different element. This process of parent-daughter formation continues through a series of elements ending up finally with a type of element that is stable and not radioactive. It is known that all natural disintegration processes end with stable lead atoms (atomic number 82).

In 1903 Rutherford and Soddy proposed that nature of daughter could be inferred from the nature of the parent and the particle emitted. The Rutherford-Soddy rules for balancing nuclear reactions are as follows :

(*a*) *the total electric charge (atomic number) or the algebraic sum of the charges before disintegration must equal the total electric charge after disintegration.*

(*b*) *the sum of the mass numbers of initial particles must equal the sum of the mass numbers of the final particles.*

The following points should also be kept in mind :

(*i*) When a parent atom ejects an α-particle, its atomic number is reduced by 2 units and atomic mass number by 4 units.

(*ii*) If the parent atom is β-emitter, then atomic number of the decay-product *i.e.* daughter is *increased* by one unit but its atomic mass number remains unaffected because electronic mass is negligible.

There are three series of naturally-occurring radioactive elements which form a sequence of parent-daughter relationship. These series are named (*i*) uranium, (*ii*) actinium and (*iii*) thorium series respectively after the element at or near the head of the series. There is a fourth series which has been obtained from an *artificially-produced* radioactive material. The first element in this series is $_{94}Pu^{241}$ and the stable end product is the ordinary bismuth $_{83}Bi^{209}$ rather than an isotope of lead as in the uranium, actinium and thorium series. This series is also known as $(4n + 1)$ series.*

As stated earlier, all natural radioactive series end with stable isotope of lead. Some elements emit α-particles whereas some others emit β-particles. Although no one atom can go both ways, some atoms go either of the two ways and cause branches in the series. No matter which way the parent goes, the daughter goes the other way so that even though the series branch, they always come together again.

11·19. Uranium Series

It is also known as $(4n + 2)$ series. Here, uranium-238 ($_{92}UI^{238}$) emits α-particle and is transformed into uranium-X_1 ($_{90}UX_1{}^{234}$) which is quite different from its parent element UI, being 4 units less in mass and two units less in atomic number. This daughter element is radioactive and emits a β-particle by a *branching* process forming $_{91}UZ^{234}$ and $_{91}UX_2{}^{234}$. Both these elements emit a β-particle each and form a common product $_{92}UII^{234}$. The UII atom emits an α-particle and becomes ionium ($_{90}I_0{}^{230}$) which is an isotope of UX_1. Ionium, in turn, emits an α-particle and is transformed into radium ($_{88}Ra^{226}$). Radium ejects an α-particle and gives rise to radon ($_{86}Rn^{232}$) which is known as *radium emanation.* Two successive α-particle emissions cause the formation of RaA and RaB respectively.

---

*The uranium series is also called $(4n + 2)$ series because the mass numbers of the atoms in it are given by this expression. Here, n is an integer whose value decreases by unity when we go from one radioelement to the next one below it. The actinium series can be represented by $(4n + 3)$ and thorium series by $4n$. The value of n is not the same for the first element in each series.

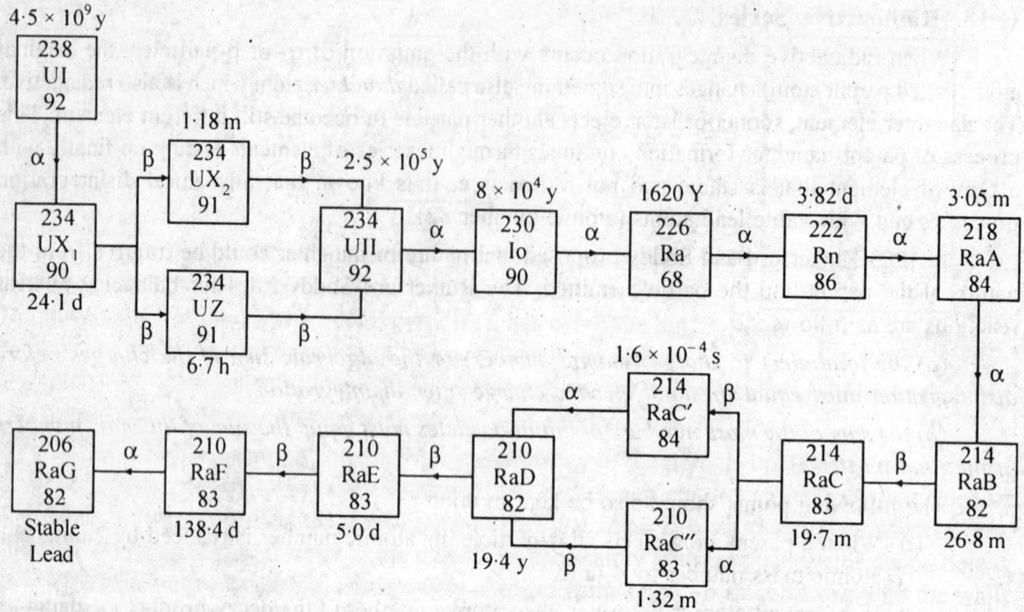

Fig. 11.13

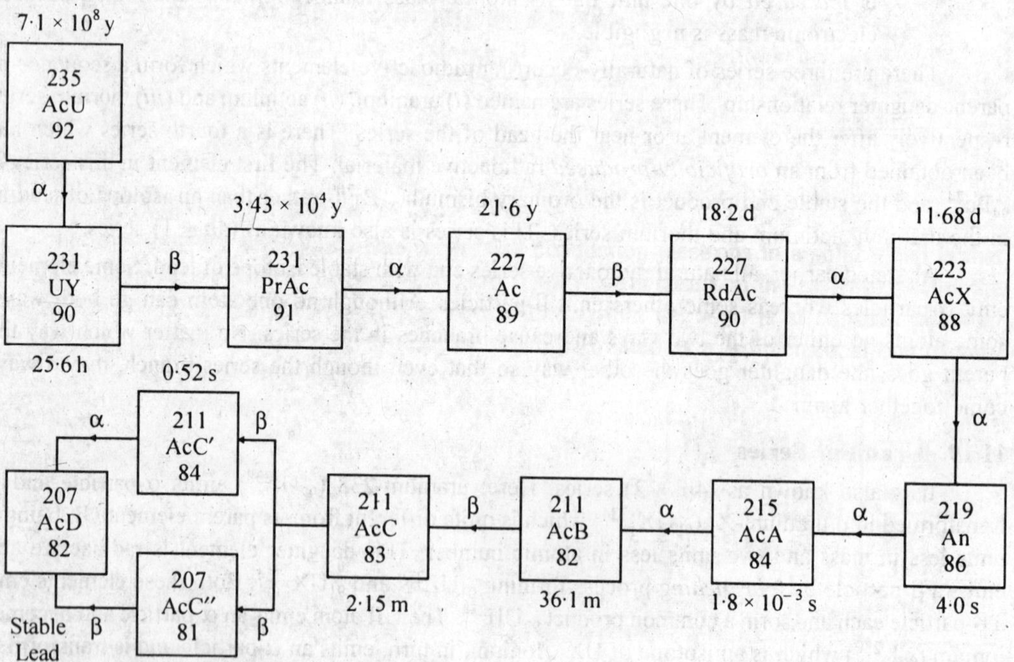

Fig. 11.14

The latter loses a β-particle and is converted into RaC which breaks by a branching process is two ways ; most of it emits β-particles and becomes RaC′ and the rest emits α-particles and becomes RaC″. RaC′ and RaC″ emit an α-particle and a β-particle respectively and produce a common product RaD. This emits β-particle and is transformed into RaE which after emitting a β-particle becomes RaF (polonium). RaF emits an α-particle and is transformed into a stable end product RaG which is an isotope of lead (Fig. 11.14).

11·20. Actinium Series

It is also known as $(4n + 3)$ series and starts with actinouranium ($_{92}AcU^{235}$) and by successive transformations ends up in a stable lead isotope of actinium-D (AcD) which is $_{82}Pb^{207}$. The whole chain of elements is shown in Fig. 11.14.

11·21. Thorium Series

It is also known as '$4n$' series and starts with thorium ($_{90}Th^{232}$). As shown in Fig. 11.15, by a successive transmutation by disintegration it ends up in thorium-D which is a stable lead isotope $_{82}Pb^{208}$.

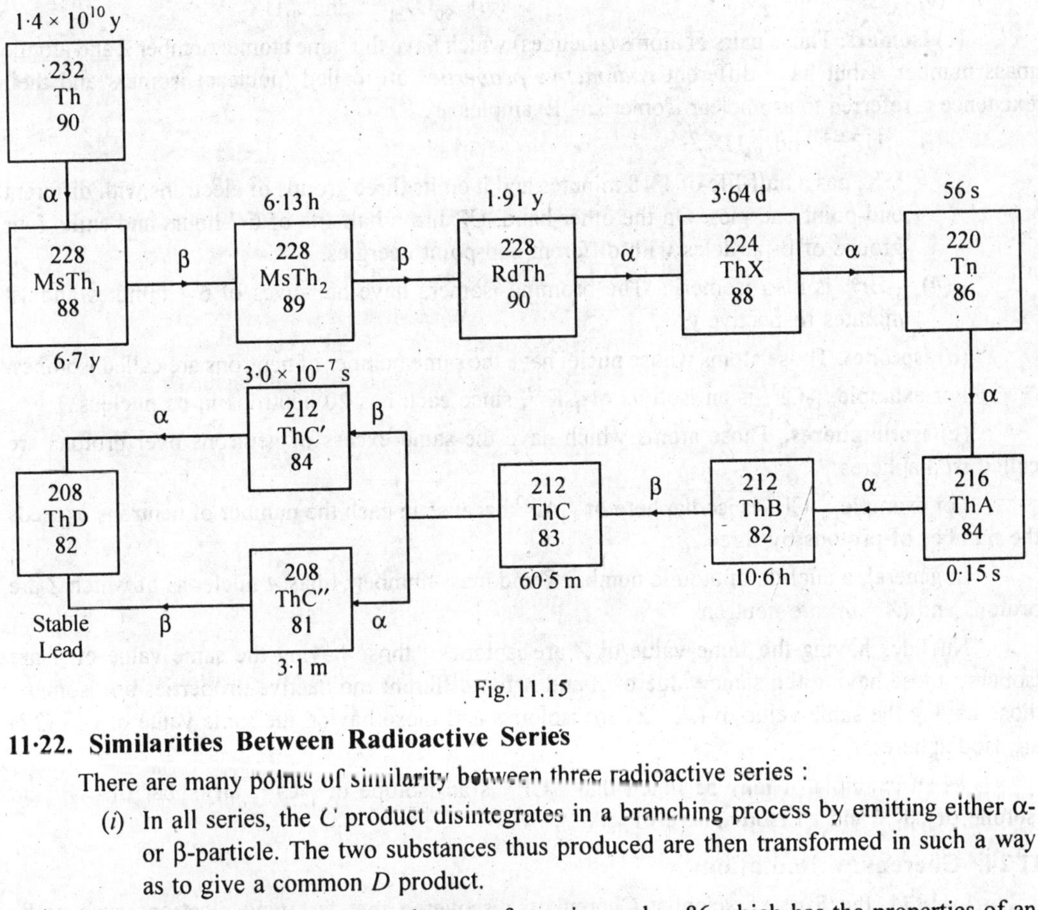

Fig. 11.15

11·22. Similarities Between Radioactive Series

There are many points of similarity between three radioactive series :

(*i*) In all series, the *C* product disintegrates in a branching process by emitting either α- or β-particle. The two substances thus produced are then transformed in such a way as to give a common *D* product.

(*ii*) In all series, there is an element of atomic number 86 which has the properties of an inert gas and is called *emanation*.

(*iii*) The stable end-product in all the three series having an atomic number of 82 is an isotope of lead *i.e.* Pb^{206}, Pb^{207} and Pb^{208}.

11·23. Isotopes, Isobars, Isomers, Isotones and Isodiapheres

(*a*) **Isotopes.** Those pairs of atoms which have the same *atomic number Z* and hence similar chemical properties but different *atomic mass number A* (*i.e.* atomic weights) are called *isotopes* (meaning the same place in the periodic table). Examples are :

(*i*) $_{92}UI^{238}$ and $_{92}UII^{224}$

(*ii*) RaB, $_{82}Pb^{214}$ and RaD, $_{82}Pb^{210}$

(*iii*) Actinium-A (AcA) $_{84}Po^{215}$ and Actinium-C′ (AcC′) $_{84}Po^{211}$

(*iv*) $_{90}Th^{232}$ and radiothorium (RdTh), $_{90}Th^{228}$.

Obviously, the different isotopes of an element contain different number of neutrons in their nuclei.

(*b*) **Isobars.** Those atoms which have the same mass number A but different atomic numbers Z are called isobars. Obviously, these atoms belong to different chemical elements. Examples are :

(*i*) $_{18}Ar^{40}$ and $_{20}Ca^{40}$ (*ii*) $_{24}Cr^{54}$ and $_{26}Fe^{54}$

(*iii*) $_{32}Ge^{76}$ and $_{34}Se^{76}$ (*iv*) $_{37}Rb^{87}$ and $_{38}Sr^{87}$

(*v*) $_{38}Sr^{92}$ and $_{42}Mo^{92}$ (*vi*) $_{90}UX_1{}^{234}$ and $_{91}UX_2{}^{234}$

(*c*) **Isomers.** Those pairs of atoms (nuclides) which have the same atomic number Z and atomic mass number A but have different *radioactive properties* are called (nuclear) isomers and their existence is referred to as nuclear isomerism. Examples are :

(*i*) $_{91}UZ^{234}$ and $_{91}UX_2{}^{234}$

UX_2 has a half-life of 1·18 minutes and it emits three groups of electrons with different end-point energies. On the other hand, UZ has a half-life of 6·7 hours and emits four groups of β-particles with different end-point energies.

(*ii*) $_{35}Br^{80}$ is also isomeric. The bromine isomers have half-lives of 6·4 minutes and 18 minutes respectively.

(*d*) **Isotones.** Those atoms whose nuclei have the same number of neutrons are called isotones. For example, $_{17}Cl^{37}$ is an isotone of $_{19}K^{39}$, since each has 20 neutrons in its nucleus.

(*e*) **Isodiapheres.** Those atoms which have the same excess of neutrons over protons are called isodiapheres.

For example, $_{17}Cl^{37}$ is isodiaphere of $_{18}Ar^{39}$ because in each the number of neutrons exceeds the number of protons by three.

In general, a nuclide of atomic number Z and mass number A has A nucleons of which Z are protons and $(A - Z)$ are neutrons.

Nuclides having the same value of Z are isotopes ; those having the same value of A are isobars : those having the same value of Z and A but different radioactive properties are isomers, those having the same value of $(A - Z)$ are isotones and those having the same value of $(A - 2Z)$ are isodiapheres.

As an example, it may be noted that $_{17}Cl^{37}$ is an isotope of $_{17}Cl^{35}$, an isobar of $_{16}S^{37}$, an isotone of $_{19}K^{39}$ and an isodiaphere of $_{18}Ar^{39}$.

11·24. Cherenkov Radiations

In 1934, the Russian scientist Cherenkov discovered that fastening electrons such as β-particles from a radioactive source produce ghostly bluish light while passing through the transparent medium such as plastic, water and glass etc. whenever their velocity is greater than the speed of light *in that medium.* For example, when radioelement $_{27}Co^{40}$ is immersed in water, β-particles emitted by it have a relativistic velocity of 0·8 c where c is the velocity of light in vacuum. The velocity of light in water is $\frac{c}{\mu} = \frac{c}{4/3} = 0{\cdot}75\,c$. Obviously, the electrons travel faster than the velocity of light *in water.* This situation is similar to the production of (*i*) V-shaped bow-waves from a ship when it travels through water at a speed greater than the wave velocity, (*ii*) the shock waves from a missile travelling through the air at a speed greater than that of sound.

As shown in Fig. 11.16, let an electron move with a velocity of v in water from point A to B in time t. It causes a sequence of disturbances in water which combine constructively to form the wavefronts represented by the lines from B to C and from B to D. These wavefronts move with a phase velocity v_p and while the electron moves from A to B, the Huygens' wavelets from A go to C, a distance = $v_p\, t$. The angle θ between the direction of travel of the electron and the conical wavefront is given by

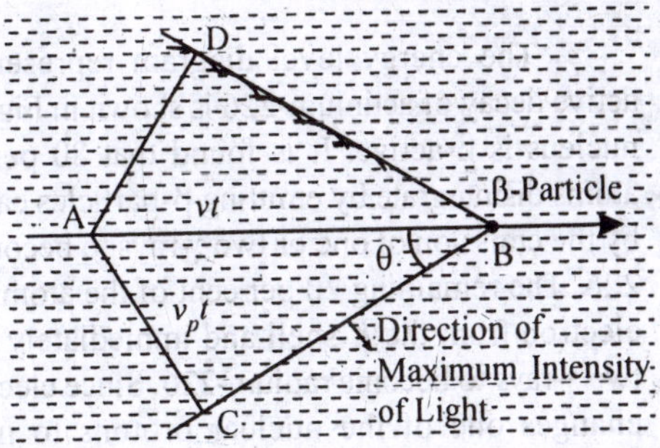

Fig. 11.16

$$\sin\theta = \frac{v_p t}{vt} = \frac{v_p}{v} = \frac{c/\mu}{v} = \frac{c}{\mu v}$$

The direction of maximum intensity of light produced is at right angles to the conical wavefront as shown in Fig. 11.16.

It can be shown that the number of photons emitted in a wavelength interval $\Delta\lambda$ is proportional to $[1 - (c/\mu v)^2]/\lambda^2$. Obviously, short wavelength will be predominant and hence colour would be bluish white.

Cherenkov radiations have been used for measuring electron velocities by using the above relation *i.e.* $\sin\theta = c/\mu v$. If the electron beam is collimated so that its direction is known and if the direction of maximum radiation intensity is measured, then θ can be determined. Since c and μ for the particular colour are known, the electron velocity v can be found.

11·25. Secondary Radiations

Extensive studies of radiations from disintegrating nuclei have shown that there are two additional processes which result in the emission of electromagnetic waves. The two processes are as follows :

(*i*) **K-electron capture.** Many artificial radioactive nuclides are found unstable to the extent of one extra positive charge. Although most of these nuclides disintegrate and become stable by the emission of a positron, some of them capture an extranuclear or orbital electron usually from the nearest *K*-shell of the same atom.* This decay by orbital electron capture is usually referred to as *K*-electron capture or just *K*-capture. Since electron-capture changes one of the nuclear protons into a neutron, the atomic number of the decay product decreases by unity.

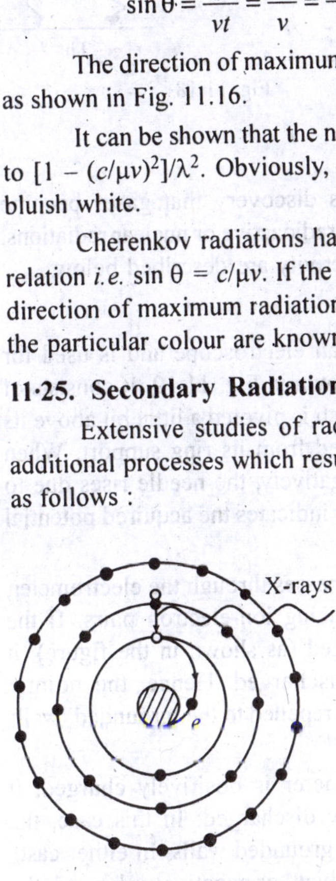

Fig. 11.17

The possibility of electron-capture was predicted by the Japanese physicists H. Yukawa and S. Sakata in 1936 and the proof of its reality was obtained in USA by L.W. Alvarez in 1938.

Electron-capture leaves a vacancy in the *K*-shell which is filled up by the jumping of another electron from the *L*- or *M*-shell. The energy so released is radiated out in the form of X-rays. Beryllium-7 and gallium-65 are specific examples of unstable nuclei in which *K*-capture is known to take place. The reaction can be represented as under :

$${}_{4}Be^{7} + {}_{-1}e^{0} = {}_{3}Li^{7}$$

$${}_{31}Ga^{65} + {}_{-1}e^{0} = {}_{30}Zn^{65}$$

*Instances of an electron being captured from *L* shell are known although these are not much common.

The energy level diagram representing radioactive decay of actinium-226 is shown in Fig. 11.18. The nucleus is β-active. It is found that 80 per cent of its atoms disintegrate by emitting β-particles each followed by the emission of one or two γ-rays to become thorium-226. The remaining 20 per cent of the atoms capture an electron from the *K*-shell and immediately emit one or two γ-rays to become radium-226. Since electron capture changes one of the nuclear protons to neutron, the atomic number decreases from 89 to 88. On the other hand, emission of a β-particle changes one of the neutrons into a proton, hence atomic number rises by unity from 89 to 90 as shown in Fig. 11.18.

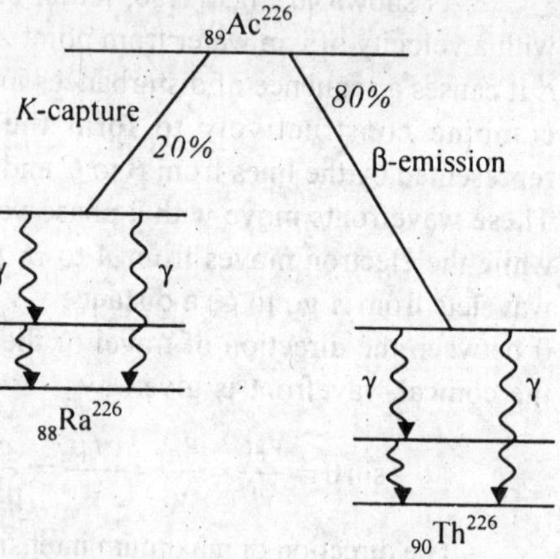

Fig. 11.18

(*ii*) **Internal Conversion Process.** It has already been discussed in Art. 11.10.

11·26. Radiation Detectors

Most of the radiation detectors are based on Becquerel's discovery that gases become electrical conductors due to ionization produced by their exposure to radioactive or nuclear radiations. Some of the instruments used for detecting and measuring radioactivity are described below.

11·27. Electrometer

This simple instrument operates, in principle, exactly like an electroscope and is used for detecting radioactive radiations. The Braun type electrometer is shown in Fig. 11.19. It consists of a light weight metal pointer which is pivoted a little bit above its centre of gravity and is insulated from its ring support. When charged either positively or negatively, the needle rises due to mutual repulsion and the pointer indicates the acquired potential on a pre-calibrated scale.

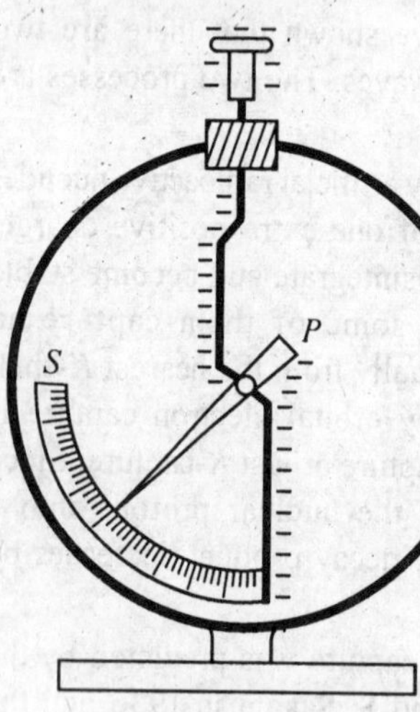

Fig. 11.19

When a charged particle passes through the electrometer, it ionizes the air thereby producing ion-electron pairs. If the electrometer is negatively-charged (as shown in the figure), it attracts positive ions and is discharged. Hence, the pointer returns to zero. The electrons are repelled to the 'grounded' walls where they are neutralized.

However, if the electrometer is positively charged, it attracts electrons and is thereby discharged. In this case, the positive ions are repelled to the grounded walls. In either case, whether the electrometer is positively or negatively charged, the pointer falls to zero on being discharged thereby indicating the presence of radioactive radiations around it.

11·28. Ionization Chamber

It is used for the detection of *moving charged* particles. It is based on the principle that charged particles in motion produce ionization in a gas which varies with their nature and velocity. By estimating the number of ions produced, one can calculate the nature and energy of the ionizing particle.

Ionization chambers are of the following two types :

(*i*) *non-integrating* (*or pulse*) *type* in which each ionizing particle is recorded separately and (*ii*) *integrating* (*or current measuring*) *type* in which current pulses are not separated ; rather they are allowed to combine into a steady flow of current.

Basically, the difference between the two lies in the value of the time constant (or response time) of the system relative to the frequency of arrival of the ionizing particles. The pulse-counting ionization chamber has a very short time constant whereas the integrating type has a comparatively long time constant.

11·29. Non-Integrating Type Ionization Chamber

One form of a non-integrating type ionization chamber is shown in Fig. 11.20. It consists of a cylindrical conducting chamber containing a central metallic electrode *AB* located on the axis of the chamber and insulated from it. The chamber is filled with a gas at a pressure of one atmosphere or less. A suitable voltage *V* is applied between the walls of the chamber and the central electrode.

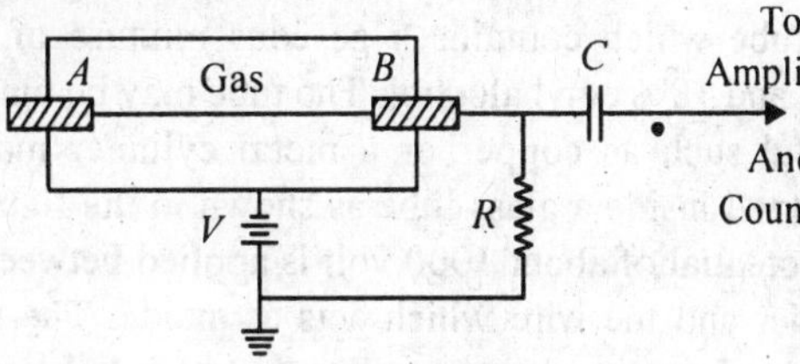

Fig. 11.20

As the charged particle passes through the gas chamber, ion-electron pairs are produced. Due to the presence of voltage *V*, positive ions go to the chamber walls and electrons to the central electrode. This gives rise to an extremely short electric flow or pulse which develops a drop across *R* and hence some charge on *C*. This pulse can be amplified by a vacuum-tube or transistor amplifier and then passed on to a suitable electronic counter. If the arrival rate of pulses exceeds the maximum rate of response of the counter (which is about 5-10 seconds), an electronic device called *scaler* is used. This scaler selects precisely *m*th pulse and passes it on to the recorder. The scaling ratio is usually a power of 2 or 10 and very high counting rates are achieved with the help of the scaler. The number of ionizing particles entering the chamber per unit time can be found by multiplying the number of pulses indicated on the counter per unit time with the appropriate factor of the scaler. In this way, ionization chambers of the pulse-counting type can be made to respond to particles arriving at the rate of 100,000 or more per second.

Ionization chambers were the first to be used for the measurement of the intensity of nuclear radiations, in the analysis of transmutation and for the measurement of neutron-induced radioactivity in various elements.

11·30. Integrating Type Ionization Chamber

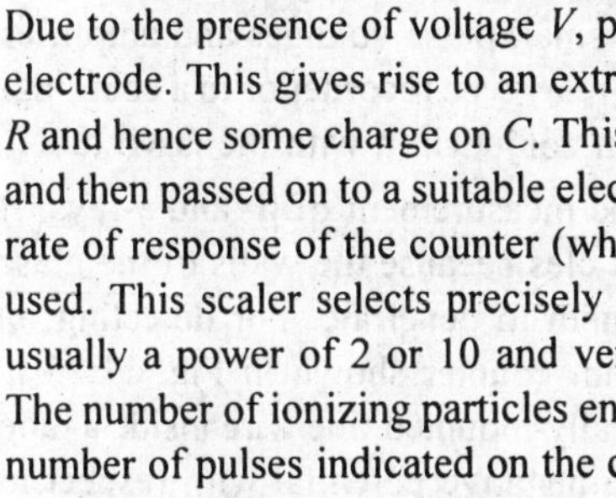

Fig. 11.21

As shown in Fig. 11.21, the integrating (*i.e.* current-measuring) type ionization chamber consists of a cylindrical box containing two plates *M* and *N* which are kept at some distance from each other. The battery *B* maintains suitable potential difference between the two plates thereby setting up an electric field in the chamber. When charged particles pass through the chamber, there is a steady production of ion-pairs which results in a very weak and continuous current through *R*. The magnitude of this current may be measured with the help of a voltmeter connected across *R* or it may be directly measured by means of a sensitive instrument. The strength of the current is found to be directly proportional to the rate at which the ionization radiations (α-, β- or γ-rays etc.) are entering the ionization chamber.

11·31. Geiger-Mueller Counter

The Geiger-Mueller counter, named after its inventors, is evolved from the ionization chamber and is a very efficient and accurate device for counting individual particles. Because the detecting portion is generally in the form of a tube, the names Geiger-Mueller tube or G-M tube are frequently used.

Construction

As shown in Fig. 11.22, a G-M counter usually consists of a fine tungsten wire mounted along the axis of a tube which contains a gaseous mixture of 90% argon and 10% ethyl alcohol. The tube may be made of a metal such as copper or a metal cylinder may be supported inside a glass tube as shown in the figure. A d.c. potential of about 1000 volt is applied between the cylinder and the wire which acts as anode. The value of this voltage is adjusted to be somewhat below the breakdown potential of the gaseous mixture. When an ionizing particle passes through the tube, it ionizes the gaseous mixture, thereby releasing a number of ion-pairs. The positive and negative ions so produced are drawn towards the appropriate electrodes and the resulting pulse of a current through the tube produces a voltage pulse (as small as 10 μV) across R. An electronic pulse amplifier accepts these small pulse voltages and amplifies them to about 5 to 50 V. This amplified output is then applied to a relay type recorder or to a counting-rate meter. For very high pulse rates, a suitable scaler is used in conjunction with the G.M. tube.

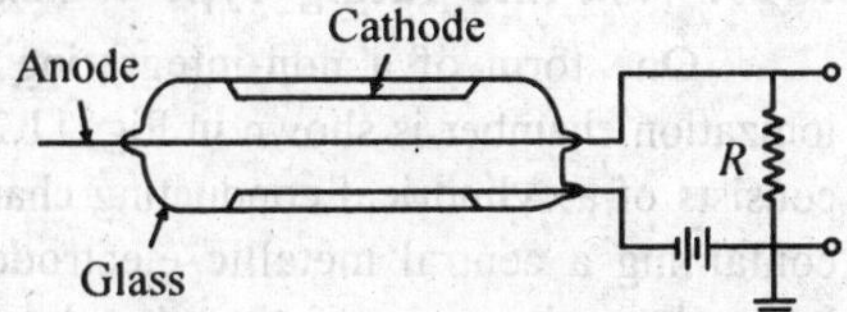

Fig. 11.22

The G.M. counter tube discussed above is suitable for the measurement of β- and γ-rays. It is not suitable for α-particles because the walls of the glass tube are too thick for them to penetrate. For detecting α-particles, the Geiger point counter shown in Fig. 11.23 is used. It consists of an axially-mounted fine wire inside a tube which is maintained at a negative potential with respect to the wire. The tube contains dry air at atmospheric pressure and has an extremely thin foil window at one end. The α-particles to be detected enter the tube through this window.

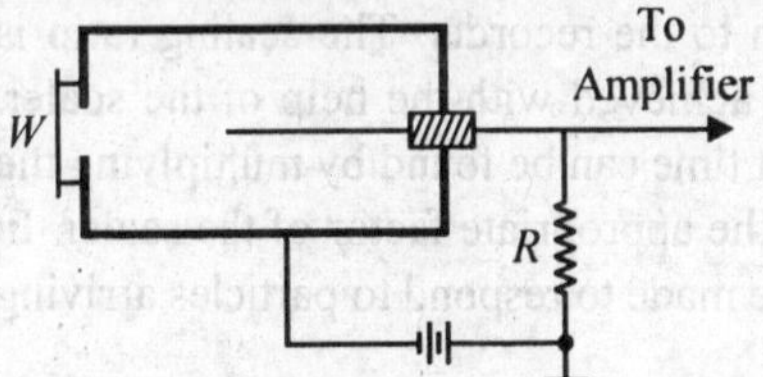

Fig. 11.23

Example 11.7. *A G-M counter with a dead time of 2×10^{-4} second registers 500 pulses per second. Find the intensity of the incoming beam of particles n_0 and the corresponding relative error of counting.*

Solution. Dead time or rest time is the time for which the counter is inoperative after registering one pulse. If T is the dead time and n the number of pulses *actually* registered in one second, then the total time for which the counter cannot register is nT. Now if n_0 is the number of incoming particles per second *i.e.* intensity of the incoming beam, then the number of particles which pass through the counter/second *but are not registered by it is* $= n_0 nT$. Also, the number of particles/second not registered by the counter is $(n_0 - n)$.

$$\therefore \quad n_0\, n\, T = n_0 - n$$

$$\therefore \quad n_0 = \frac{n}{1 - nT} = \frac{500}{1 - 500 \times 2 \times 10^{-4}}$$

$$= \textbf{536 particles/second}$$

Relative error

$$= \frac{\Delta n}{n_0} = \frac{n_0 - n}{n_0} = nT$$

$$= .500 \times 2 \times 10^{-4} = 10^{-1} \text{ or } 10^{-1} \times 100 = \textbf{10\%}$$

11·32. Scintillation Counters

A scintillation counter is a very sensitive device used in nuclear physics study for the detection and measurement of high-energy atomic radiations. Of late, it has assumed great importance because of the discovery of phosphors which are sensitive to β- and γ-rays and the development of highly efficient photo-multiplier tube.

Working Principle

The working of a scintillation counter is based on the fact that when high-energy charged particles pass through certain transparent materials, fluorescent light is produced all along the path. As the fast-moving particle collides with the atoms and molecules of the material, electrons are raised to excited energy levels and in returning to their ground states, they emit light (called scintillation). For many crystals and particles, this fluorescent light is blue or violet in colour whereas for many others, it is ultraviolet or infra-red. This tiny flash of light is made to fall on the photocathode of a very sensitive photo-electric cell called *photomultiplier tube* where it ejects photoelectrons. These photoelectrons are accelerated by the successive dynodes which are kept at progressively higher voltages. These primary photoelectrons knock off secondary electrons as they travel from one anode to another. Ultimately, the number of electrons is no much increased in successive stages that a measurable current pulse is obtained. This current pulse is recorded by a counter-rate meter.

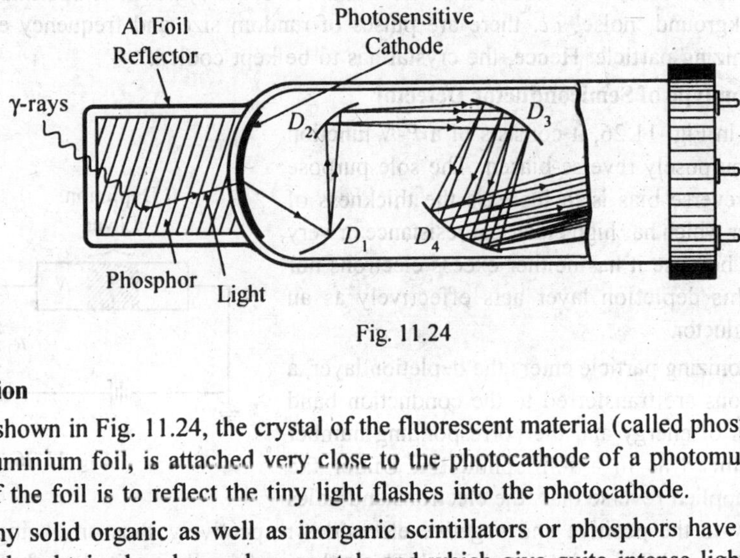

Fig. 11.24

Construction

As shown in Fig. 11.24, the crystal of the fluorescent material (called phosphor) surrounded by thin aluminium foil, is attached very close to the photocathode of a photomultiplier tube. The purpose of the foil is to reflect the tiny light flashes into the photocathode.

Many solid organic as well as inorganic scintillators or phosphors have been discovered which can be obtained as large clear crystals and which give quite intense light flashes. Among such solid organic crystals are included anthracene and stilbene. Inorganic salts containing small amounts of impurities as activators for luminescence are sodium or potassium iodide activated with thallium. Liquid scintillators include solutions of organic compounds such as terphenyl dissolved in toluene or xylene.

Modern scintillators can detect and record many millions of flashes per second and can be used with intense radiations. Because of their simplicity, great flexibility and versatility, high sensitivity to all forms of nuclear radiations, rapid response, very short resolving time and ability to discriminate between and measure photon energy, scintillation counters have found many uses. Their extremely rapid response makes them specially suited for accurate timing of nuclear particles moving with very high speeds. For example, scintillation counters were employed in the detection of antiproton where it was necessary to count the particles traversing a distance of 12 metres in 52×10^{-9} seconds.

11·33. Semiconductor Detectors

Such detectors can be subdivided into the following two types :

(*i*) Crystal detectors and

(*ii*) Junction type detectors.

(*a*) Crystal Detector

As shown in Fig. 11.25, it consists of a thin wafer of an intrinsic semiconductor, say, pure silicon or germanium whose opposite faces are covered with very thin layers of gold metal. The two conducting surfaces are connected to a d.c. source of 1–20 volts and an amplifier.

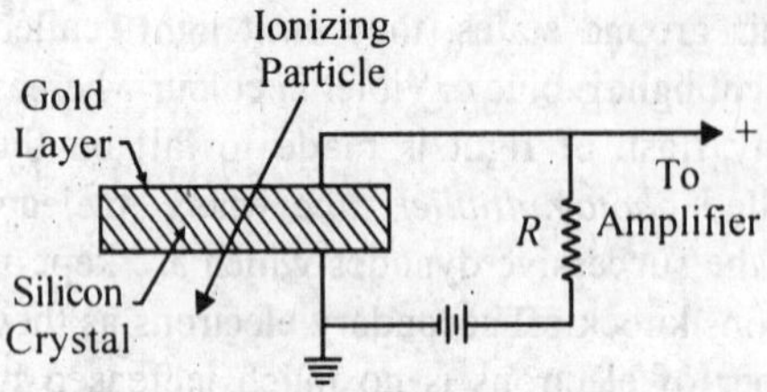

Fig. 11.25

Under ordinary conditions, the electrical conductance of an* intrinsic semiconductor crystal is relatively small. But when an ionizing particle enters, the crystal, free electrons and holes develop which increase its conductivity. Due to the movement of electrons and holes in opposite directions under the influence of the electric field, a sudden current is created. The current pulse is amplified and then either measured or counted.

The chief drawback of the above detector is that at ordinary room temperature, there is considerable background 'noise' *i.e.* there are pulses of random size and frequency even in the absence of an ionizing particle. Hence, the crystal has to be kept cooled.

(*b*) Junction type of Semiconductor Detector

As shown in Fig. 11.26, it consists of a *P-N* junction which has been purposely reverse-biased. The sole purpose of applying the reverse bias is to increase the thickness of the depletion layer which has high electrical resistance or very low conductance because it has neither excess electrons nor holes. In fact, this depletion layer acts effectively as an intrinsic semiconductor.

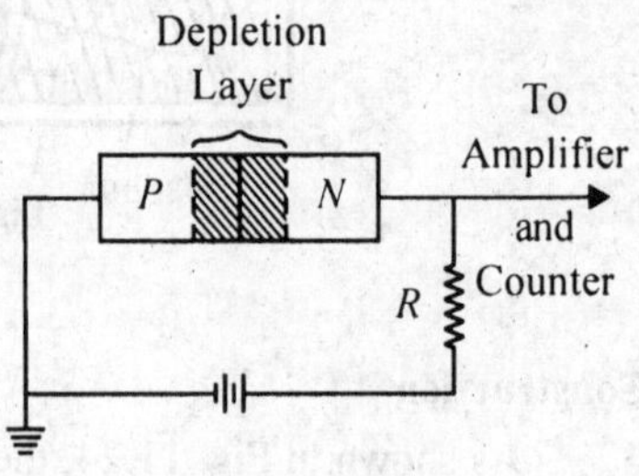

Fig. 11.26

When an ionizing particle enters the depletion layer, a number of electrons are transferred to the conduction band by the absorption of energy and the corresponding number of holes is produced as in a semiconductor. Under the influence of the applied reverse bias, the electrons and holes are swept rapidly to the positive and negative electrodes respectively, thus producing a current pulse which can be amplified and then recorded. This pulse is of small magnitude and of extremely short duration (from 10^{-7} second to 10^{-9} second). When used in conjunction with a linear amplifier and pulse height analyser, a semiconductor detector can be utilized for measuring the energy distribution (*i.e.* spectrum) of the nuclear radiation. Its excellent resolution makes it a very useful tool in nuclear spectroscopy.

11·34. Cloud Chamber

It is one of the most important instruments for basic research on radiations like detecting and measuring moving charged particles, studying the behaviour of individual atoms, photographing the actual paths of ionizing radiations and analyzing the complicated interactions which take place between charged particles and individual atoms.

---

*Other intrinsic semiconductors used are silver chloride, cadmium sulphate and diamond etc.

Basic Principle

The working of a cloud chamber is based on the fact that supercooled vapour needs some discontinuity or nuclei to initiate its condensation. In the absence of such nuclei, a supercooled vapour remains in the vapour phase and does not condense. As an example, imagine a certain volume of air which is saturated with water vapour at room temperature. If this air is subjected to an adiabatic expansion, its temperature is lowered so that its saturation limit is decreased. Hence, air now contains more water vapour than is necessary for saturation at this lower temperature. If any particles of dust or smoke are present in air, they will act as condensation nuclei and the excess of water vapour will separate out as fine droplets of liquid in the form of a cloud or mist. If, on the other hand, there are no dust or smoke particles, no condensation is possible so that air will become supersaturated with water vapours. In Wilson cloud chamber, ions act as nuclei for condensation of supersaturated water vapours. These ions are produced due to the passage of a high-energy particle like α- or β-particle etc. through the chamber which leads to the ionization of the air or gas contained therein.

Construction

There are two types of cloud or condensation chambers differing mainly in the method employed for obtaining supersaturation. The two types are :

(*a*) *expansion-type cloud chamber and*

(*b*) *diffusion-type cloud chamber.*

(*a*) **Expansion-type Cloud Chamber**

A typical chamber of this type, designed by C.T.R. Wilson, is shown in Fig. 11.27. It consists of an air-tight cylinder *C* whose one end is closed by a glass plate *P* and the other by a movable metal piston. The air chamber formed between the two contains a saturated vapour from an excess amount of some liquid, usually a mixture of alcohol and water. A small amount of this mixture kept in the trough *T* inside the chamber keeps the air fully saturated. The chamber is strongly illuminated by a mercury vapour lamp *L* whose light passes through a window in the side of the cylinder.

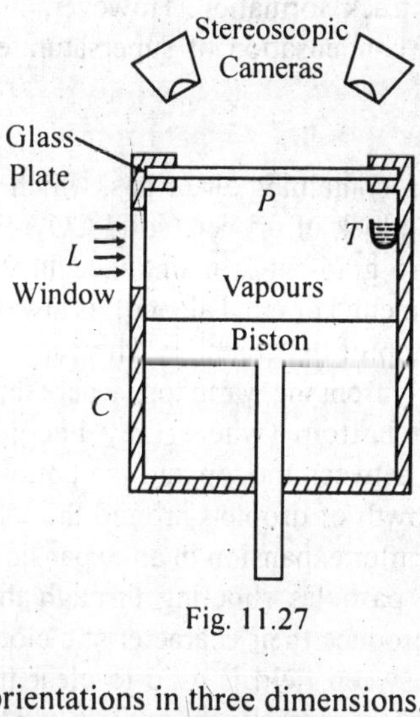

Fig. 11.27

Working

If the volume of chamber is increased by suddenly moving the piston downwards, the adiabatic expansion causes cooling which renders the vapours supersaturated, unstable and likely to condense. If ions are present within the chamber, condensation occurs preferentially on them.

If an ionizing particle enters the chamber either immediately before, during or immediately after the expansion, the ions left in its path will act as condensation nuclei so that a close array of fine droplets *i.e.* a kind of linear cloud called a *cloud track* will be formed. When illuminated, the track appears like a white line on a black background. The tracks can be photographed by a stereoscopic camera whose pictures can be used for studying track orientations in three dimensions. The chamber is then cleared of the ions by means of a sweep field applied across the chamber and the piston is returned to its original position so that the chamber is ready for another burst of radiations.

Instead of operating the chamber regularly but intermittently, many methods have been devised to make the chamber function automatically at the critical moment. Geiger counters are placed at the top and bottom of the chamber and when an ionizing particle passes through *both* of them and hence through the chamber, a relay is operated which causes expansion of the gas and condensation of droplets of water on the ions left by the particle. At the same time, cameras are triggered to photograph the track immediately. When the cloud chamber is placed between the poles of a strong electromagnet, it is possible to distinguish between the positively and negatively charged particles from the direction of the curvature. The curvature gives not only the sign of the charge but momentum of the ionizing particle as well.

Different particles produce different types of tracks. Heavy, slow and highly-ionizing particles like α-particles produce short, broad, densely-packed straight line tracks whereas slow β-particles which are light and less ionizing produce thin, beaded and tortuous tracks. Fast particles, both light and heavy, produce narrow beaded straight tracks. It may be noted that γ-rays are never observed in a cloud chamber because they produce so few ions.

Cloud chamber tracks have been used for the following purposes :

(*i*) for measuring the range and hence the energy carried by a particle.

(*ii*) for determining the nature of the particle by finding out its specific ionization *i.e.* number of droplets per unit length of the cloud track. This has helped to distinguish between an α-particle, proton, meson and electron etc.

(*iii*) for determining the momentum of a charged moving particle by measuring the radius of curvature of the cloud tack in a magnetic field.

(*iv*) for determining the sign of the charge by noting the direction of the curvature of the track in a magnetic field. This fact helped in the discovery of positron.

(*v*) for studying the phenomenon of artificial transmutation.

(*b*) Diffusion-type Cloud Chamber

One of the major limitations of an expansion cloud chamber is that it has a relatively long recovery time (10 to 60 seconds) after expansion so that it may often miss an interesting nuclear event. Moreover, it is sensitive only for a short interval after the expansion. The diffusion chamber proposed by A. Langsdroff in 1939 is *continuously* sensitive to track formation. However, like expansion-type cloud chamber, it also works on the principle of condensation of supersaturated water vapours on the ions left behind by an ionizing particle.

Construction and Working

Essentially, the diffusion cloud chamber consists of a vessel, containing air or gas, which is kept warm at the top and cold at the bottom. The vessel rests on a block of dry ice (solid CO_2). In this way, a vertical temperature gradient is maintained in the chamber. A volatile liquid (like methyl or ethyl alcohol) is always available in a trough near the top of the chamber. The vapours so produced diffuse continuously from the warm top (where they are unsaturated) to the cold bottom (where they become saturated). Hence, somewhere between the top and the bottom, conditions are right for the growth of droplets around the ions just as is the case immediately after expansion in an expansion-type cloud chamber. Charged particles shooting through this sensitive part of the chamber produce their characteristic cloud tracks. As is usual, a suitable sweep field is used to clear the chamber of ions so that newly-formed tracks are not masked by the dense fog of droplets.

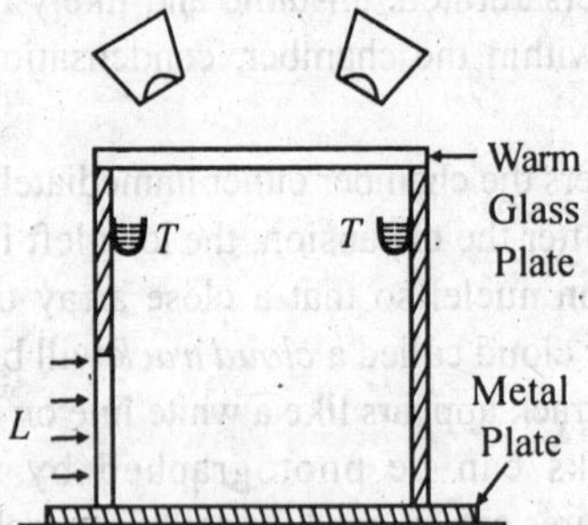

Fig. 11.28

The diffusion cloud chamber has the special advantage of being continuously sensitive and mechanically simple.

11·35. Bubble Chamber

Invented in 1952 by D.M. Glaser, this instrument has assumed great importance in the study of minute details of high-energy nuclear events. It contains superheated *liquid* instead of supersaturated *vapours* as in condensation chambers of both types. In this instrument, the tracks consist of a series of closely-spaced bubbles which develop on the ions produced by the passage of a charged particle through the superheated liquid contained in the chamber.

Basic Principle

It is well known that boiling point of a liquid can be raised by increasing pressure on its surface. For example, water boils at 100°C at standard atmospheric pressure. If pressure on its surface is increased, as in a pressure cooker, boiling does not start till a higher temperature is reached. If pressure is then suddenly decreased, boiling begins with sudden formation of tiny bubbles that grow quickly in size. If at that exact moment, a charged particle passes through the liquid, bubbles grow on the ions produced by it and form a bubble track which can be photographed.

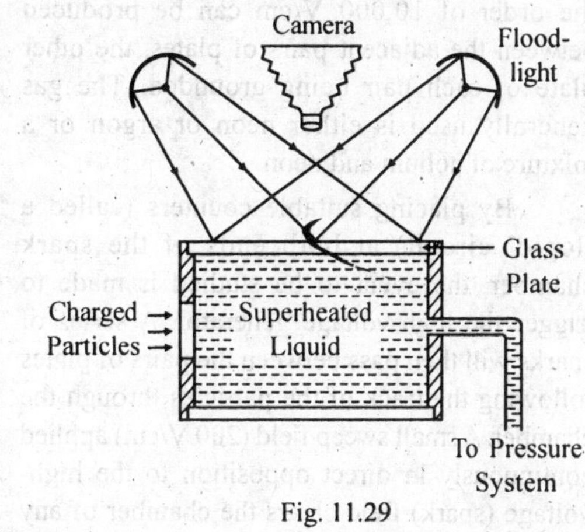

Fig. 11.29

Construction

Fig. 11.29 shows a bubble chamber in a very simplified form. It consists of a box having thick glass walls which is filled with a liquid and is also connected to a pressure system. The liquid is heated to a predetermined temperature. High-energy particles enter the liquid through a thin window.

Working

The valve in the pressure system is suddenly released and is quickly followed by the flash of the floodlights and the snap of the camera shutter. In this way, sharply-defined trails of bubbles formed on the ions produced by the traversing particle can be photographed.

Bubble chambers are nearly always located in strong magnetic field so that the charge and momentum of the particle may be obtained from track curvature.

Some of the drawbacks of a bubble chamber are that

(*i*) it is not continuously sensitive and

(*ii*) it has a relatively large recovery time (from 1 to 5 seconds).

However, these drawbacks are overcome by using the bubble chamber in conjunction with a high-energy particle accelerator which functions in a pulse *i.e.* intermittent manner. The chamber can then be timed to operate in synchronism with each pulse of the particle from the accelerator. Use of a bubble chamber has grown to such an extent that a laboratory associated with an accelerator is faced with the immense task of analysing millions of track photographs each year. Various mechanical and electronic devices including computers are being utilized to expedite the process of interpreting these numerous photographs.

11·36. Spark Chamber

Introduced in 1959, this chamber is used for the study of cosmic rays and high-energy particles produced by accelerators.

Basic Principle

Its working depends on the passage of a spark through a gas that has been ionized by nuclear radiations. Usually, a high voltage is maintained between two plates or electrodes situated in a gas such that the electric field is not strong enough to produce a spark in the gas under normal conditions. If an ionizing particle enters the gas between the plates, a spark will pass which would tend to follow the path of the ion-pairs produced by the particle. In this way, the track of the ionizing particle can be delineated by the spark and recorded photographically by means of stereoscopic cameras for three-dimensional representation.

Construction

A multiplate spark chamber (Fig. 11.30) consists of a number of parallel plates made of a conducting material and spaced 2 to 3 mm apart. Alternate plates are connected to a high-voltage d.c. pulse generator so that an electric field of the order of 10,000 V/cm can be produced between the adjacent pairs of plates, the other plate of each pair being grounded. The gas generally used is either neon or argon or a mixture of helium and neon.

By placing suitable counters (called a 'logic' circuit) at both ends of the spark chamber, the event to be studied is made to trigger the high-voltage generator. A series of sparks will then pass between the pairs of plates following the track of the particles through the chamber. A small sweep field (200 V/cm) applied continuously in direct opposition to the high-voltage (spark) field clears the chamber of any ions present in about 1 μs. The chamber can be operated in a magnetic field so that charge and momentum of the ionizing particle can be measured from the track curvature.

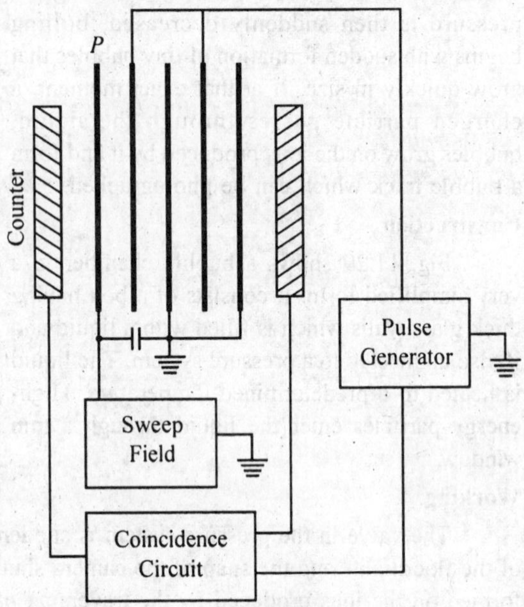

Fig. 11.30

Since 1963, many filmless spark chambers have been developed in which photography of the spark discharge is avoided. The coordinates of the sparks are obtained directly in a form suitable for a computer to determine the geometry of the nuclear event.

11·37. Photographic Emulsions

This technique is employed for detecting high-speed particles. When such an ionizing particle travels through the sensitive emulsion of a photographic plate, it leaves a track consisting of a number of silver bromide grains which turn into black silver upon being developed by a regular film-developing process. Special nuclear emulsions have been manufactured for this purpose. Their silver bromide content, grain size and emulsion thickness is much greater as compared to the optical film. The study of detailed structure of tracks under a microscope yields information about the mass, charge and energy of the ionizing particle.

As compared to a cloud chamber, the photographic emulsion has the advantage that it is permanently sensitive and the tracks are short.

11·38. Radiation Damage

Energy absorbed by the passage of radiations through human body gives rise to structural changes called radiation damage. The nature and extent of damage depends on the type of radiation, on the depth to which radiation has penetrated, on the extent of the body exposed, on the amount of radiation absorbed and also upon whether the exposure was *chronic i.e.* repeated and prolonged or *acute i.e.* was received in one large dose.

The changes produced by radiation in the body are of two types : (*i*) somatic and (*ii*) genetic. Somatic effects are those which are experienced directly by the exposed individual. The genetic effects, on the other hand, are not evident in the irradiated person but appear in subsequent generations.

Since biological effects vary widely between body tissues and depend on the type of radiation, it is usual to measure radiation absorption by the number of ions produced in air by the radiation. The exposure is measured in terms of a unit called *roentgen*. One roentgen (R) is that quantity of X- or γ-radiations which under standard pressure and temperature produces $2{\cdot}08 \times 10^{15}$ ion pairs per m^3 of dry air.

Since on the average it takes 34 eV to produce one ion-pair, hence

$$1\ R = 2{\cdot}08 \times 10^{15} \times 34 = 70{\cdot}7 \times 10^{15}\ eV$$

or

$$1\ R = 70{\cdot}7 \times 10^{15} \times 1{\cdot}602 \times 10^{-19} = 113{\cdot}3 \times 10^{-4}\ J$$

Now, density of air at N.T.P. is $1{\cdot}293\ kg/m^3$, hence in one kg of air, one roentgen will produce an absorption of

$$113{\cdot}3 \times 10^{-4}/1{\cdot}293 = 88 \times 10^{-4}\ J.$$

Roentgen Equivalent Physical (rep). This unit (now obsolete) was used for measuring the dose of radiation not covered by definition of roentgen.

One rep has been defined as that dose of any nuclear radiation that produces an energy absorption of 0·01 joule per kg of soft tissue. The alternate name of the unit is *rad*.

Roentgen Equivalent, Man (rem). It is defined as that quantity of radiation which produces the same biological damage in man as that produced by one roentgen (R) of X- or γ-rays.

Relative Biological Effectiveness (RBE). The RBE factor of a given radiation is defined as the ratio of the dose in roentgen of X- or γ- rays (of a specified energy, usually 200 keV) to the dose of given radiation producing the same biological effect.

11·39. Applications of Radioactive Radiations

(*i*) Many radiation sources such as cobalt-60 (a γ-ray emitter) have been used for industrial radiography *i.e.* for investigating the interiors of metallic castings for detecting any flaws or defects.

(*ii*) Radioactive nuclides emitting α- or β-particles have been used for the production of electric power by thermo-electric conversion. When α- or β-particles are absorbed in matter, the energy of the radiation is converted into heat which can be utilized for production of electric power. Such radionuclide power generators are particularly suitable for space vehicles because of their light weight, long life and absolute reliability.

(*iii*) Radionuclide have been used as compact sources of heat energy because of which they find many space-related applications. Radionuclides of promethium-147, polonium-210 and plutonium-238 have been used to heat the propellant gas (hydrogen) in low-thrust rockets.

(*iv*) Chemical changes initiated by radiations from radioactive substances is the basis of a newly-developed subject called *radiation chemistry* which has been applied in the production and modification of plastics and in the production of irradiated wood-plastic combinations.

(*v*) Nuclear radiations like γ-rays have been utilized for the preservation of food. The irradiation of foodstuff, mainly meat, poultry, fish and fruits, is achieved by exposure to γ-rays from cobalt-60 or caesium-137. A dose of 2 to 5 million rads is sufficient to destroy almost all bacteria in food. This increases the shelf-life of these articles without refrigeration. Radiations have also been employed for insect disinfestation of wheat and flour.

(*vi*) Gammas radiations from cobalt-60 are used in hospitals for sterilization of materials like dressings, hypodermic syringes and surgical sutures.

(*vii*) Population of insects which causes considerable damage to both plant crops and livestock can be controlled by irradiating the male members of these insects so that they become sterile.

(*viii*) Radiation mutations in plants have been practised to produce new varieties of these plants.

(*ix*) Self-luminous paints for use on instruments and watch dials have been made by adding a natural α-emitting radioactive substance to phosphor.

(*x*) The ionization produced by β-particles has been widely used for the elimination of static electricity which constitutes a serious fire and explosion hazard in the paper, textile, rubber and plastic industries.

TUTORIAL PROBLEMS

1. The half-life of a radioactive element is 10 days. Calculate how long it takes for 90 per cent of a given mass of this element to disintegrate? **[33·24 days]**
2. Given that the half-life of radioactive cobalt (cobalt-60) is 5 years, calculate how long will it take for the activity of a specimen of this radioactive isotope to decrease to 20 per cent of its initial value? **[11·6 y]**
3. The half-life of radiosodium (Na^{24}) is 15 hours. Calculate the percentage loss in activity which occurs in a sample of radiosodium in 10 minutes. **[0·76%]**
4. Compute the time required by 10% of a sample of thorium to disintegrate. Given that half-life of thorium is $1{\cdot}4 \times 10^{19}$ years. **[$2{\cdot}1 \times 10^9$ y]**
5. What is the weight of one rd of radium (Ra^{226})? Its half-life is = 1620 years ; 1 rd = 10^6 dis/s.
6. If the half-life of a radioactive element is 100 days, find in what time its initial amount gets reduced to 1/10th amount. **[332 days]**
7. The half-life of Io^{230} is $8{\cdot}0 \times 10^4$ years. Calculate its disintegration constant and mean life. Calculate the amount of ionium in one rutherford. Avogadro's number = $6{\cdot}02 \times 10^{23}$ atoms per gram-atom.
8. A source containing iodine-131 has an initial activity of 2·0 curies. If iodine-131 has a half-life of 8 days, calculate (*i*) activity of the source after 24 days and (*ii*) the time required for the activity to fall to 1·0 μCi. **[(*i*) 0·25 Ci; (*ii*) 167 days]**
9. If caesium-137 has a half-life of 30 years, calculate its mass which has an activity of 5 μCi. One Ci is equal to $3{\cdot}7 \times 10^{10}$ disintegrations/second. **[$5{\cdot}74 \times 10^{-11}$ kg]**

HIGHLIGHTS

1. The phenomenon of spontaneous emission of powerful radiations exhibited by heavy elements is called radioactivity.
2. Radioactive radiations consist of α-rays, β-rays and γ-rays.
3. The two laws of radioactive disintegration are :

   (*i*) Atoms of all radioactive elements undergo spontaneous disintegration to form fresh radioactive products with the emission of α-, β- and γ-rays.

   (*ii*) The rate of radioactive disintegration is not affected by environmental factors but depends on the number of the atoms of the original kind present at any time.

   The decay is given by $N = N_0 e^{-\lambda t}$

   Also, $M = M_0 e^{-\lambda t}$
4. The decay constant λ may be defined as the reciprocal of the time during which the number of atoms of a radioactive substance falls to 37 per cent of its original value.

5. Half-life (T) of a radioactive element is equal to the time during which a given amount of that element is reduced by disintegration to half its initial amount

$$T = 0{\cdot}693/\lambda.$$

6. The average life expectancy of a radioactive atom is equal to the reciprocal of the decay constant.

$$l = 1/\lambda.$$

7. The rate of disintegration or activity of a radioactive material is given by

$$-dN/dt = \lambda t$$

8. Curie (Ci) is the unit of activity *i.e.* rate of disintegration. One curie is that amount of radioactive material which gives $3{\cdot}7 \times 10^{10}$ disintegrations per second.

9. Rutherford (rd) is also the unit of activity and is equal to that quantity of radioactive material which gives 10^6 disintegrations per second.

10. For radioactive equilibrium

$$(dN/dt)_{parent} = (dN/dt)_{daughter}$$

11. Radioactive exposure is measured in terms of roentgen (R).

One roentgen is that quantity of X- or γ-rays which under standard pressure and temperature produces $2{\cdot}08 \times 10^{15}$ ion-pairs per m^3 of dry air.

Can You Answer This ?

1. Explain the terms : half and mean life.
2. Derive an expression for half-life of radioactive elements.
3. (*a*) What are the basic laws of radioactive disintegration? Derive the expression for the mean life time for a radioactive element.

   (*b*) Give one decay series of any one of the following :—

   (*i*) uranium (*ii*) radium (*iii*) thorium.
4. (*a*) Distinguish between natural and artificial radioactivity. Explain the terms : radioactive decay constant, half-life time and mean life time.

   (*b*) Obtain an expression of mean life time of a radioactive element in terms of its half-life time.
5. Write short note on artificial and natural radioactivity.
6. (*a*) Explain the term radioactivity equilibrium. Derive the necessary condition for the radioactive secular equilibrium.

   (*b*) Describe how you would determine the half-life of a given radioactive sample.
7. (*a*) Explain the following terms :

   (*i*) half-life (*ii*) mean-life (*iii*) decay constant of a radioactive substance. Obtain the relation between them.

   (*b*) What percentage of a radioactive element will decay during the time equal to the mean-life of this element?
8. (*a*) Distinguish between natural and artificial radioactivity.

   (*b*) Explain the terms radioactive decay constant and half-life.

   (*c*) Derive the expression for half-life period.

ARTIFICIAL RADIOACTIVITY 12

Discovery of Artificial Radioactivity — Discovery of Radiosodium — Researches in Artificial Radioactivity — Radioisotopes of Transuranic Elements — Isotope Tables and Nuclide Charts — Uses of Isotopes — Diagnostic Applications of Radioisotopes — Industrial Applications of Radioisotopes — Isotopic Dating in Geology — Highlights —Can You Answer This?

12·1. Discovery of Artificial Radioactivity

It was discovered by Mme. Irene Curie-Joliot and her husband M. Frederick Joliot that products of some artificial or induced nuclear reactions were radioactive. These radionuclides were found to decay spontaneously according to the same laws that apply to the disintegration of naturally-occurring radioactive elements. However, since this radioactivity is produced by artificial means, it is called *induced* or *artificial* radioactivity.

This phenomenon of artificial radioactivity came to light during the experiments conducted by Curie-Joliots for studying effects of α-particle bombardment on the light elements like boron, aluminium and magnesium etc.

As shown in Fig. 12.1, Curie-Joliots bombarded aluminium with α-particles from polonium and measured the energies of the ejected neutrons by the recoiling of protons from paraffin. They found, to their great surprise, that even after the polonium source was taken away, the detector continued to respond to some kind of penetrating radiations. These were later on found to be positrons.*

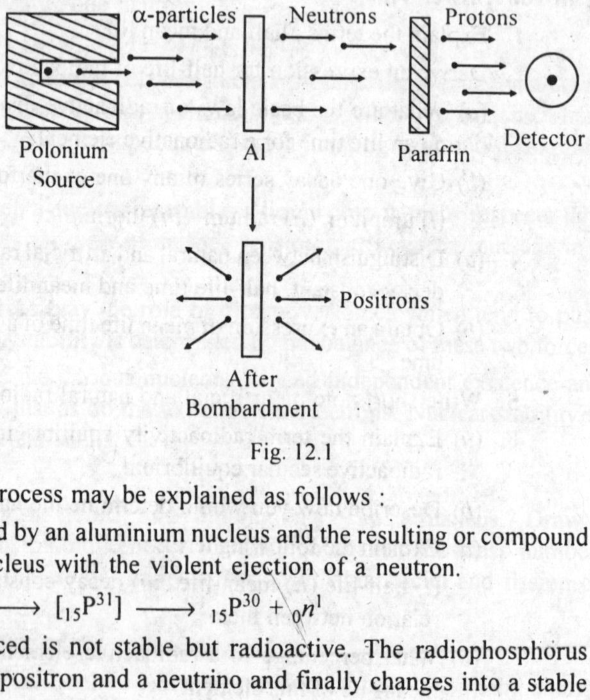

Fig. 12.1

By repeated experiments it was confirmed that under the bombardment of α-particles, aluminium had become radioactive by its own right. The whole process may be explained as follows :

The striking α-particles are captured by an aluminium nucleus and the resulting or compound nucleus is converted into phosphorus nucleus with the violent ejection of a neutron.

$$_2\text{He}^4 + {}_{13}\text{Al}^{27} \longrightarrow [{}_{15}\text{P}^{31}] \longrightarrow {}_{15}\text{P}^{30} + {}_0n^1$$

The phosphorus nucleus so produced is not stable but radioactive. The radiophosphorus disintegrates spontaneously by emitting a positron and a neutrino and finally changes into a stable silicon atom as shown below :

$$_{15}\text{P}^{30} \longrightarrow {}_{14}\text{Si}^{30} + {}_1e^0 + \nu^{**}$$

*Positron is a fundamental particle and was discovered by Anderson in 1932. Its rest mass is the same as that of the electron. However, its charge has the same magnitude but opposite sign from that of the electron.

**It is generally omitted from such equations since it carries no charge and has no mass.

Radiophosphorus is found to have a half-life period of 3·25 minutes.

Similar reaction with a boron nucleus is as follows :

$$_5B^{10} + {}_2He^4 \longrightarrow [{}_7N^{14}] \longrightarrow {}_7N^{13} + {}_0n^1$$

$$_7N^{13} \longrightarrow {}_6C^{13} + {}_1e^0$$

The radionitrogen has a half-life period of 14 minutes and decays according to the laws of natural radioactivity.

At present, nearly 1300 artificial radionuclides have been identified and their properties determined by combined chemical and physical techniques. The study of the radioactivity of the artificially-made atomic nuclei has rendered much help in clarifying the concepts regarding the constitution and stability of the nuclei.

12·2. Discovery of Radiosodium

Soon after the discovery of artificial radioactivity by Curie-Joliots, Lawrence bombarded sodium (in the form of NaCl) with 2 MeV deuterons from a cyclotron as shown in Fig. 12.2. He found that sodium also became radioactive. During bombardment, Lawrence found that protons with energies of 7 MeV were ejected by Na nuclei. When the sodium target was removed from the deuteron beam and tested for radioactivity, it was found to emit β-particles (*i.e* electrons) and γ-rays.

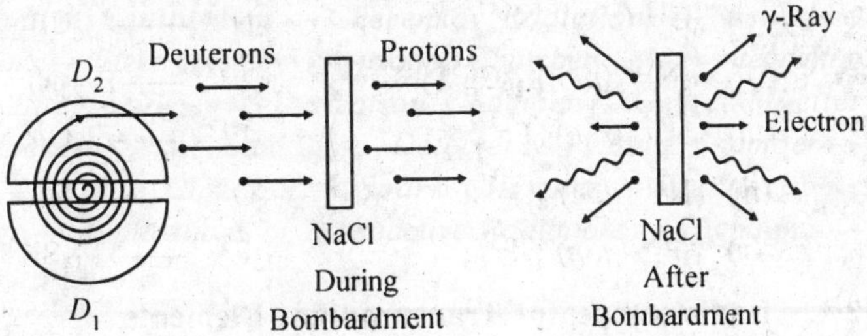

Fig. 12.2

The bombardment reaction is as under :

$$_{11}Na^{23} + {}_1H^2 \longrightarrow [{}_{12}Mg^{25}] \longrightarrow {}_{11}Na^{24} + {}_1H^1$$

The unstable radiosodium Na^{24} decays immediately according to the following equation :

$$_{11}Na^{24} \longrightarrow {}_{12}Mg^{24} + {}_{-1}e^0 + \gamma$$

These processes are shown in Fig. 12.3 (*a*). Radiosodium has a half-life period of 15 hours and is not found in nature. The energy level diagram for decay of radiosodium is shown in Fig. 12.3 (*b*). As seen radiosodium nucleus emits a β-particle and then becomes a magnesium nucleus in an excited state. With two successive transitions in which γ-rays arc emitted, it becomes normal and stable isotope of magnesium $_{12}Mg^{24}$.

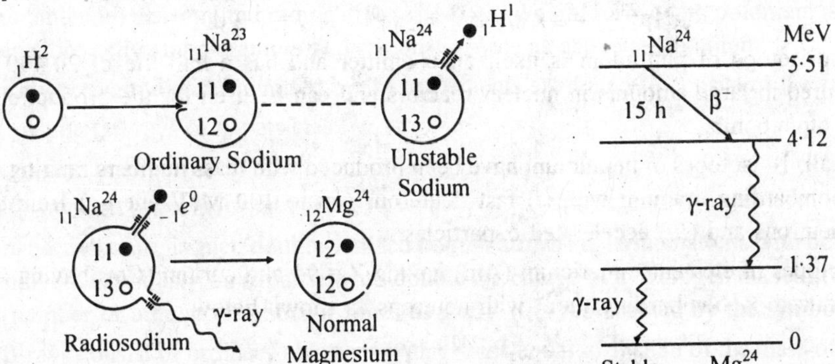

Fig. 12.3

12·3. Researches in Artificial Radioactivity

It is well known that nuclear reactions induced by protons, deuterons, neutrons and photons can also result in radioactive products. It has also been discovered that there is some correlation between the activity of a radionuclide and the transformation by means of which it is produced.

(*a*) Electron emission is common for activities produced by (n, γ), (n, p), (n, α) and (d, p) reactions since they *decrease* the charge-to-mass ratio. Some of the examples are :

(*i*) ${}_{49}In^{115}\,(n, \gamma)\,{}_{49}In^{116}$; $\quad {}_{49}In^{116} \longrightarrow {}_{50}Sn^{116} + {}_{-1}e^0 \quad T = 13\text{ s}$

(*ii*) ${}_{7}N^{14}\,(n, p)\,{}_{6}C^{14}$; $\quad {}_{6}C^{14} \longrightarrow {}_{7}N^{14} + {}_{-1}e^0 \quad T = 5568\text{ y}$

(*iii*) ${}_{13}Al^{27}\,(n, \alpha)\,{}_{11}Na^{24}$; $\quad {}_{11}N^{24} \longrightarrow {}_{12}Mg^{24} + {}_{-1}e^0 \quad T = 15\text{ h}$

(*iv*) ${}_{15}P^{31}\,(d, p)\,{}_{15}P^{32}$; $\quad {}_{15}P^{32} \longrightarrow {}_{16}S^{32} + {}_{-1}e^0 \quad T = 14{\cdot}3\text{ d}$

(*b*) Positron emission is common for those activities produced by (p, λ), (p, n), (α, n), (d, n), (γ, n) reactions which *increase* the charge-to-mass ratio. Examples are :

(*i*) ${}_{6}C^{12}\,(p, \gamma)\,{}_{7}N^{13}$; $\quad {}_{7}N^{13} \longrightarrow {}_{6}C^{13} + {}_{1}e^0 \quad T = 10\text{ min}$

(*ii*) ${}_{28}Ni^{58}\,(p, n)\,{}_{29}Cu^{58}$; $\quad {}_{29}Cu^{58} \longrightarrow {}_{28}Ni^{58} + {}_{1}e^0 \quad T = 2{\cdot}6\text{ s}$

(*iii*) ${}_{7}N^{14}\,(\alpha, n)\,{}_{9}F^{17}$; $\quad {}_{9}F^{17} \longrightarrow {}_{8}O^{17} + {}_{1}e^0 \quad T = 70\text{ s}$

(*iv*) ${}_{7}N^{14}\,(d, n)\,{}_{8}O^{15}$; $\quad {}_{8}O^{15} \longrightarrow {}_{7}N^{15} + {}_{1}e^0 \quad T = 2{\cdot}1\text{ min}$

(*v*) ${}_{15}P^{31}\,(\gamma, n)\,{}_{15}P^{30}$; $\quad {}_{15}P^{30} \longrightarrow {}_{14}Si^{30} + {}_{1}e^0 \quad T = 2{\cdot}5\text{ min}$

12·4. Radioisotopes of Transuranium Elements

All elements which have values of Z greater than 92 are called transuranium elements because they lie beyond uranium in the periodic table.

In 1934, Fermi suggested the possibility that the bombardment of uranium with neutrons might result in the production of transuranium elements. By 1940, it was proved that bombardment of uranium with *slow* neutrons produces a new isotope U^{239} which has a half-life of 23 minutes and is an electron emitter. It decays into an isotope of a new element with $Z = 93$ (neptunium, Np) which is also an electron emitter and has a half-life of 2·3 days. The decay of Np^{239} yields an isotope of another new element with $Z = 94$ (plutonium, Pu). The whole process can be represented by the following equations :

$${}_{92}U^{238} + {}_{0}n^1 \longrightarrow {}_{92}U^{239} + \gamma$$

$${}_{92}U^{239} \longrightarrow {}_{93}Np^{239} + {}_{-1}e^0 \qquad T = 23\text{ min}$$

$${}_{93}Np^{239} \longrightarrow {}_{94}Pu^{239} + {}_{-1}e^0 \qquad T = 2{\cdot}3\text{ d}$$

The isotope of plutonium is itself an α-emitter and has a half-life of 24,400 years. It is manufactured in large amounts in nuclear reactors and can be used for the production of nuclear power or atom bomb.

In all, 10 isotopes of neptunium have been produced with mass numbers ranging from 231 to 240 by bombarding uranium with (*i*) fast deuterons (up to 100 MeV energy) from a cyclotron, (*ii*) fast neutrons and (*iii*) accelerated α-particles.

Isotopes of elements americium (Am) having $Z = 95$ and curium (Cm) having $Z = 96$ have been produced by bombarding Pu^{239} with neutrons as shown below.

$${}_{94}Pu^{239} + {}_{0}n^1 \longrightarrow {}_{94}Pu^{240} + \gamma$$

$${}_{94}Pu^{240} + {}_{0}n^1 \longrightarrow {}_{94}Pu^{241} + \gamma$$

$${}_{94}Pu^{241} \longrightarrow {}_{95}Am^{241} + {}_{-1}e^0$$

At the same time, $_{96}Cm^{242}$ is also formed thus :

$$_{95}Am^{241} + _{0}n^{1} \longrightarrow _{95}Am^{242} + \gamma$$

$$_{95}Am^{242} \longrightarrow _{96}Cm^{242} + _{-1}e^{0}$$

The nuclide Am^{241} when bombarded with α-particles gives an isotope of a new element with $Z = 97$ (berkelium, Bk)

$$_{95}Am^{241} + _{2}He^{4} \longrightarrow _{97}Bk^{243} + 2\ _{0}n^{1}$$

Similar reactions were used in the first preparation and discovery of an isotope of a new element californium, $_{98}Cf^{243}$. Starting with Pu^{238} and going through a sequence of neutron capture and β-decay, we reach nuclides as heavy as fermium $_{100}Fm^{254}$.

It is also possible to produce isotopes of transuranium elements by bombarding uranium with highly ionized nuclei such as C^{12}, N^{14} and O^{16}. Isotopes of californium (Cf), einsteinium (E) and fermium (Fm) have been produced by the bombardment of uranium with high-energy (100 MeV or more) ions of carbon, nitrogen and oxygen respectively as shown below :

$$_{92}U^{238} + _{6}C^{12} \longrightarrow _{98}Cf^{244} + 6\ _{0}n^{1}$$

$$_{92}U^{238} + _{7}N^{14} \longrightarrow _{99}E^{246} + 6\ _{0}n^{1}$$

$$_{92}U^{238} + _{8}O^{16} \longrightarrow _{100}Fm^{250} + 4\ _{0}n^{1}$$

An isotope of element mendelevium, $_{101}Md^{256}$, has been prepared by bombarding a sample of $_{99}E^{253}$ with α-particles, the element being discovered by means of its decay products. Similarly, an isotope of nobelium having $Z = 102$ has been made by bombarding a mixture of Cm^{244} and Cm^{246} with C^{12} ions at energies of 60-100 MeV.

12·5. Isotope Tables and Nuclide Charts

Isotope tables which are published periodically are compilations of vast information about different isotopes and other latest nuclear data.

However, the most convenient method for quick reference is provided by a nuclide chart in which atomic number Z is plotted against the number of neutrons *i.e.* $(A - Z)$ in a nucleus. Each nuclide occupies a square but stable, natural radioactive and artificial nuclides are differentiated by colour or shading of the square. Its symbol, mass number, half-life, type of decay and the energies of the emitted particles and γ-rays are indicated in the square. Such charts can be used to obtain quick information about the products of nuclear disintegration.

12·6. Uses of Radioisotopes

Radioisotopes have provided a powerful tool for the solution of numerous problems in biology, physiology, chemistry, physics and other sciences in addition to industry. Some of the problems, particularly those connected with life processes, could not have been solved without the use of isotopes. The more important applications of isotopes depend on the fact that the chemical properties of the isotope of a given element are essentially identical. A radioisotope can be detected by its radioactivity and a stable isotope by a mass spectrometer. Consequently, the radioactivity or mass of an isotope can act as a 'tag' or 'label' which permits the fate of the element or of a compound containing this element to be traced through a series of chemical and physical changes. The element which has been labelled or tagged is called a *tracer element*.

Following simple illustrations will prove how isotopic tracers can be used to get an unequivocal solution of some types of important problems.

(*i*) When a plant grows, it absorbs phosphorus both from the soil and from the added fertilizer. In order to check the usefulness of the fertilizer, it is important to know what proportion of phosphorus comes from each source. Ordinary analytical methods cannot answer this question because phosphorus from both sources is exactly the same. However, if radioisotope is used in the fertilizer, the exact proportion can be determined accurately.

(*ii*) In cases of blood transfusions, it is important to know as to what happens to the red corpuscles in the body of the patient. Once transfusion has taken place it is impossible to distinguish between the red corpuscles already present in the body of the patient and those which have been added. But by labelling the iron in the haemoglobin—the red colouring material of the transfused blood—it is possible to determine its behaviour in the body of the recipient.

(*iii*) When exposed to sunlight, green plants can take up carbon dioxide and water which are converted into sugars and starches. However, at the same time plants give up oxygen. Now, both CO_2 and H_2O contain oxygen and the question which arises is : does the oxygen produced by a plant originate from CO_2 or H_2O molecules or from both? By labelling oxygen atoms either in CO_2 or H_2O, it was found that oxygen came only from H_2O.

Isotopic tracers have been widely used in cellular biology. Behaviour of cells during mitosis has been studied with the help of tagged DNA (deoxyribonucleic acid) molecules.

12·7. Diagnostic Applications of Radioisotopes

Radioisotopes have become a very valuable tool in the diagnosis and understanding of many diseases which has eluded conventional diagnostic procedures so far. Some of the examples are given below :

(*i*) Radiosodium has been used to study cases of restricted circulation of blood. A small quantity of NaCl solution which contains labelled sodium is injected into the vein of a patient's forearm and a γ-ray detector is then placed in contact with one of his feet. If blood circulation is normal, the presence of radioactivity is soon discovered in the foot. It increases rapidly and becomes maximum within an hour. However, if there is any circulatory impairment, radioactivity will increase slowly showing that blood has some difficulty in reaching the foot. By moving the detector to different parts of the body, the position of restriction can be located and the necessary treatment applied.

(*ii*) Pumping action of heart has been studied by using radiosodium or radioiodine. A small quantity of labelled NaCl solution is injected into the blood stream and a detector, attached to an automatic pen recorder, is placed on the heart. As radiosodium enters the right side of the heart (*i.e.* atrium) the pen of the recorder rises and then drops as the venous blood enters the lungs for oxygenation. After some time, the radiosodium appears with the arterial blood on left side of the heart (*i.e.* ventricle) and there is another rise and fall of the recording pen. By examining the resulting curve, the pumping action of the two sides of the heart can be compared and abnormalities, if any, can be discovered.

(*iii*) By using iron-59, many diseases caused by deficiency of red blood cells in the human body have been distinguished from each other.

(*iv*) Radioisotopes have been used to located and detect the presence of tumours particularly brain tumours which are difficult to detect and especially to locate.

(*v*) The functioning of various organs like liver, kidney and spleen etc. under normal and diseased conditions has been studied with the help of labelled iodine-131.

12·8. Industrial Applications of Radioisotopes

Radioisotopes have found many applications in industry both in research and in process control. Some of the examples are as under :

(*i*) The wear of piston rings and of gears in engines and its prevention by means of suitable lubricants have been studied by radioactivity. A steel piston ring is made radioactive by exposing it to neutrons in a nuclear reactor. The piston ring is then fitted into the cylinder of an internal combustion engine which is operated normally with a particular lubricating oil. By determining the radioactivity removed by the oil, the amount of piston wear can be determined.

(*ii*) Alloys are frequently subjected to different treatments such as age-hardening, annealing, quenching and cold-rolling. Radioisotopes have been used to find the effect of such treatments.

(*iii*) The phenomenon of self-diffusion in metals *i.e.* the movement of the atoms of a metal within the crystal lattice has been studied with the help of radioisotopes.

(*iv*) Uniformity of mixing during the blending of petrols, lubricating oils and greases etc. has been achieved by labelling one of the constituents with a radioactive tracer.

(*v*) The functioning of various organs like liver, kidney and spleen etc. under normal and diseased conditions has been studied with the help of labelled iodine-131.

12·9. Isotopic Dating in Geology

The age of a rock or mineral containing uranium can be determined by its radioactive decay. The radioactive decay is given by the equation $N = N_0 e^{-\lambda t}$ where N_0 is the amount of U-238 originally present in a small quantity at the time it was formed and N is the amount still left after the lapse of time t which represents the age of the mineral. Since all U-238 which has decayed has been converted into lead-206, the quantity of the latter present is equal to $(N_0 - N)$. Hence,

$$U^{238} = (U^{238} + Pb^{206})\, e^{-\lambda t}$$

where the symbols U^{238} and Pb^{206} refer to the *amounts* of thc respective nuclides existing at the present time in a given piece of the mineral. Since these quantities can be determined experimentally and the decay constant λ of the uranium is known, the age t of the mineral can be found.

A similar relation holds good for the uranium-235 (the parent of the actinium series) and its end product lead 207.

$$U^{235} = (U^{235} + Pb^{207})\, e^{-\lambda' t}$$

where λ' is the decay constant of uranium-235.

However, if the above-given two equations are divided by each other, we have,

$$\frac{Pb^{206}}{Pb^{207}} = \frac{U^{238}\,(e^{\lambda t} - 1)}{U^{235}\,(e^{\lambda' t} - 1)}$$

The present ratio of U^{238} to U^{235} is known to be 139 and λ and λ' can be determined from the known half-lives of two isotopes of uranium, hence the age t of the rock or mineral can be calculated if the ratio Pb^{206}/Pb^{207} is determined by a mass spectrometer.

The above method is useful for dating minerals and rocks of considerable age. For recent objects, particularly those of animal or vegetable origin, radiocarbon dating method is commonly employed.

Example 12.1. *The ratio of mass of Pb^{206} to the mass of U^{238} in a certain rock specimen is found to be 0·5. Assuming that the rock originally contained no lead, estimate its age. Half-life of uranium is $4{\cdot}5 \times 10^9$ years.*

Solution. As seen from Art. 12.9, the relation between the existing masses of U^{238} and Pb^{206} in any rock specimen is given by

$$U^{238} = (U^{238} + Pb^{206})\, e^{-\lambda t}$$

or
$$1 = \left(1+\frac{Pb^{206}}{U^{238}}\right)e^{-\lambda t}$$

or
$$1 = (1+0\cdot5)\, e^{-\lambda t} = 1\cdot5\, e^{-\lambda t}$$

$\therefore$
$$e^{\lambda t} = 1\cdot5$$

or
$$\lambda t = 2\cdot3 \log_{10}{}^{1\cdot5} = 2\cdot3 \times 0\cdot1761$$

$\therefore$
$$\frac{0\cdot693}{4\cdot5\times10^9}\times t = 0\cdot23 \times 0\cdot1761\ ;\ t = \mathbf{2\cdot63 \times 10^9\ years}$$

HIGHLIGHTS

1. Artificial radioactivity was discovered by Curie-Joliots when they bombarded aluminium with α-particles from polonium.
2. Radiosodium was produced by bombarding Na with deuterons obtained from a cyclotron.
3. Transuranium elements have been produced by bombarding uranium with fast deuterons and neutrons and accelerated α-particles.
4. Radioisotopes have found the following uses :
   (*i*) as tracer elements
   (*ii*) in diagnosis and understanding of many life processes and diseases which cannot be handled by conventional methods.
   (*ii*) in research and industry
   (*iv*) for dating purposes in geology.

Can You Answer This ?

1. Explain the terms : natural and artificial radioactivity.
2. Distinguish between natural and artificial radioactivity.
3. What is artificial radioactivity? Describe in short how this phenomenon was discovered.
4. (*a*) What is artificial radioactivity? How does it differ from natural radioactivity? Give two examples of each.
   (*b*) Describe Rutherford's experiment on artificial disintegration of nitrogen by bombardment with particles. What conclusion do you derive therefrom?
5. What do you mean by artificial radioactivity? What is a radioisotope? State two uses of radioisotopes.
6. What is artificial radioactivity? How does it differ from natural radioactivity? Give two examples of each.
7. Write a short note on artificial disintegration.
8. (*a*) Describe how the phenomenon of artificial radioactivity was discovered.
   (*b*) Describe how radioisotopes are produced artificially. Give suitable examples.
   (*c*) Outline any two uses of radioisotopes in industry.

NUCLEAR REACTIONS 13

Transmutation or Disintegration — Rutherford's Experiment for Artificial Disintegration — Disintegration of Lithium with Protons — Q-Value of a Nuclear Reaction — Transmutation Equations — Importance of Transmutation Equations — Nuclear Projectiles — Discovery of Neutron — Classification of Artificial Nuclear Reactions — Neutron Sources — Transmutation by Different Particles — Nuclear Reaction Cross-section — Highlights —Can You Answer This?

13·1. Transmutation or disintegration

Transmutation, in general, means the conversion of one element into another *i.e.* of one type of an atom into another. Artificial transmutation is such a conversion achieved by artificial means as distinct from the spontaneous transmutations witnessed in natural radioactive elements. Transmutation is also called *disintegration* because the transmuted element disintegrates, forming a different and a new element. The earliest experiments on artificial nuclear disintegration were performed by Sir Ernest Rutherford. He bombarded the atoms of different stable elements of low atomic number such as nitrogen, aluminium and potassium etc. by using high-velocity α-particles (as projectiles) obtained from natural radioactive substances such as RaC′. Now-a-days, other bombarding particles like protons ($_1H^1$) and deuterons ($_1H^2$) etc. are also used for the same purpose.

13·2. Rutherford's Experiment for Artificial Disintegration

The simple but sensitive apparatus used by Rutherford is shown schematically in Fig. 13.1. At one end of the box or chamber *B* was cut an opening which was covered by a silver foil *F* for observing α-particles. A zinc sulfide screen was placed at *S* just outside the opening and the scintillations produced on the screen observed by means of a microscope *M*. The sources of α-particles was RaC′ which was placed at *R* on a small disc whose distance from screen could be varied at will by means of a long rod.

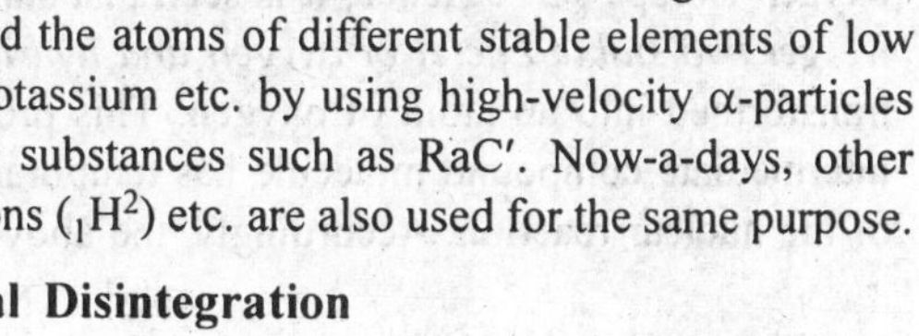

Fig. 13.1

The range of these α-particles was known to be as 7 cm in dry air at atmospheric pressure. Different gases could be introduced into the chamber and removed through the side tubes *T*.

When the box was filled with O_2 or CO_2 at standard atmospheric pressure, no scintillations were seen on the screen so long as the radioactive source was 7 cm or more away from *S*. It was so because this thickness of the gas was sufficient to absorb all α-particles from the source even without the silver foil. However, with nitrogen in the box, scintillations were observed even when *R* was as far away as 40 cm. Since it was known that α-particles from RaC′ could not penetrate 40 cm of air, Rutherford concluded that these scintillations were caused by long-range, high-energy particles ejected from the N_2 nucleus by the impact of α-particles. Measurement of the magnetic deflection showed that they were protons ($_1H^1$) which conclusion was later confirmed by more precise work.

The exact nature of the nuclear reaction which takes place when an α-particle makes a head-on collision with a N_2 molecule was not clearly understood at first. However, soon a satisfactory explanation backed by experimental verification was achieved. It was proved that α-particle is captured by N_2 nucleus thus forming an unstable *compound molecule* which immediately disintegrates by ejecting a proton. The whole process described above is shown in Fig. 13.2.

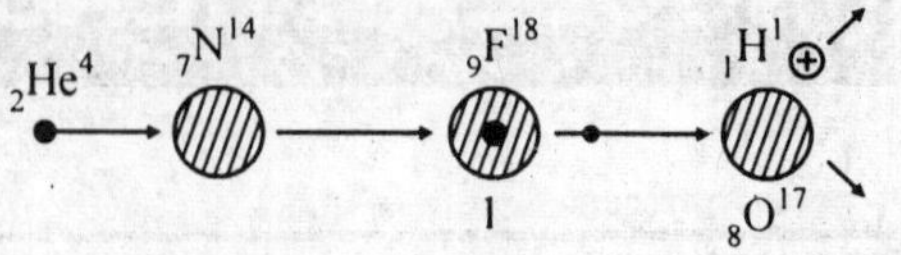

Fig. 13.2

This disintegration or nuclear rearrangement may be represented by an equation similar to that used for chemical reactions.

$$_7N^{14} + {}_2He^4 \longrightarrow [{}_9Fe^{18}] \longrightarrow {}_8O^{17} + {}_1H^1$$

Since, like chemical equations, such equations must balance, the charge and mass numbers have to be the same on the two sides of these equations. The intermediate unstable compound is shown in brackets.

It is seen that when the N_2 molecule captures an α-particle, it is converted into an unstable fluorine atom (not found in nature) which, then, ejects one proton ($_1H^1$) to become a stable but rare oxygen isotope $_8O^{17}$. Hence, it is seen that starting *with two stable nuclei of helium and nitrogen, we get two stable nuclei of oxygen and hydrogen.* In other words, an atom of nitrogen has been transformed into an atom of oxygen. This process is called *transmutation of elements.* Since the intermediate compound molecule has temporary existence, it is often omitted from any discussion of the nuclear reaction. Accordingly, the above disintegration reaction can be simply written as

$$_2He^4 + {}_7N^{14} = {}_8O^{17} + {}_1H^1$$

The above reaction may be represented by the abbreviated notation N^{14} (α, p) O^{17} which should be interpreted as follows : a nitrogen (N^{14}) nucleus, called *target nucleus*, interacts with an alpha particle (α), referred to as *incident* or *projectile*, a proton (p) is ejected and an oxygen (O^{17}) nucleus called *recoil* nucleus remains behind.

Rutherford and Chadwick afterwards bombarded all the light elements from boron to potassium with the exception of oxygen and carbon with α-particles and found that in all such cases, protons were emitted. Some of these nuclear reactions are given below :

$$_5B^{10} + {}_2He^4 \longrightarrow [{}_7N^{14}] \longrightarrow {}_6C^{12} + {}_1H^1$$
$$_{11}Na^{23} + {}_2He^4 \longrightarrow [{}_{13}Al^{27}] \longrightarrow {}_{12}Mg^{26} + {}_1H^1$$
$$_{13}Al^{27} + {}_2He^4 \longrightarrow [{}_{15}P^{31}] \longrightarrow {}_{14}Si^{30} + {}_1H^1$$
$$_{16}S^{32} + {}_2He^4 \longrightarrow [{}_{18}A^{36}] \longrightarrow {}_{17}Cl^{35} + {}_1H^1$$
$$_{19}K^{39} + {}_2He^4 \longrightarrow [{}_{21}Sc^{43}] \longrightarrow {}_{20}Ca^{42} + {}_1H^1$$

It will be seen that in each case, charge of the product nucleus is increased by one unit and the mass is increased by three units. Such reactions are also referred to as (α, p) reactions.

13·3. Disintegration of Lithium with Protons

By 1924, Rutherford and Chadwick had transmuted almost all elements up to potassium with α-particles obtained from natural sources. However, they found that with heavier atoms, there was only scattering of these particles indicating that they did not have sufficient energy to penetrate the target nucleus. Hence, attention was directed towards exploring the means to obtain high-velocity charged particles from suitable high-voltage accelerators.

In 1932, Cockcroft and Walton achieved successful transmutation of lithium atoms by using artificially-accelerated protons. This represented the first case of a nuclear disintegration brought about by purely artificial means. This apparatus is schematically shown in Fig. 13.3.

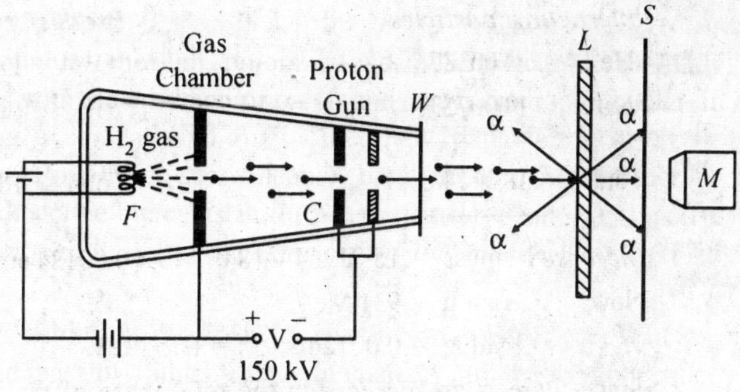

Fig. 13.3

Electrons from the hot filament *F* ionize the hydrogen gas contained in the gas chamber. The hydrogen nuclei or protons are accelerated towards the other end of the accelerating chamber by a voltage of 150 kV. After passing through the opening *C* and window *W*, they emerge from the chamber as a narrow beam of protons.

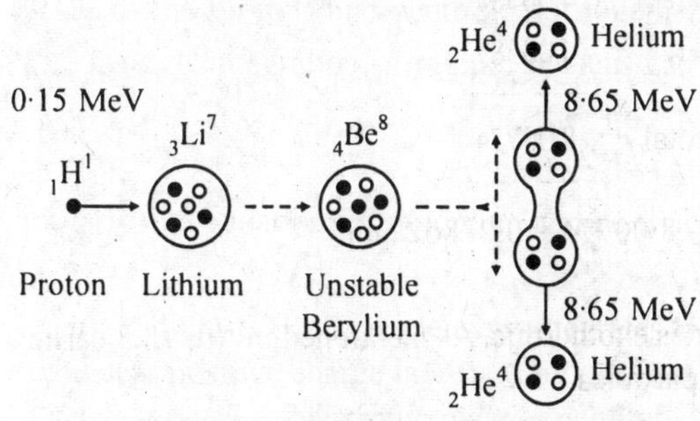

Fig. 13.4

The accelerating chamber acting as a proton gun is aimed at a target consisting of lithium metal in the form of a thin layer of lithium oxide. Bright scintillations due to ejected particles were immediately seen on the zinc sulfide screen placed a short distance away.

These ejected particles were found to be α-particles. The nuclear reaction thus achieved may be written as

$$_1H^1 + {}_3Li^7 \longrightarrow [{}_4Be^8] \longrightarrow {}_2He^4 + {}_2He^4$$

The above process could be abbreviated as Li^7 (p, α) He^4.

As shown in Fig. 13.4, a proton of energy 0·15 MeV enters a lithium nucleus to form a new but unstable beryllium nucleus. This nucleus splits up into two α-particles which are driven apart with great violence. Since each particle is found to have an energy of 8·65 MeV, it means that each disintegration involves liberation of 2 × 8·65 = 17.3 MeV. The source of this energy is found to be in the annihilation of a part of the total atomic mass.

13·4. Q-Value of Nuclear Reaction

Nuclear reactions, involving rearrangement of nucleons, resemble ordinary chemical reactions in the sense that they are accompanied by energy changes. The overall energy liberated or absorbed in a nuclear disintegration is called *nuclear reaction energy*. This energy, which is generally represented by the symbol *Q*, is also called the *energy balance* of the reaction or more commonly, its *Q*-value. The *Q*-value of a reaction can be positive or negative depending on the nature of the reaction. As per Einstein's theory of energy-mass equivalence, *Q*-value must be balanced by the changes in mass associated with the nuclear reaction. If *Q*-value is positive *i.e.* if the reaction is accompanied by *liberation* of energy, there must be a *decrease* in mass. In that case, total mass of products will be *less* than that of the interacting nuclei by an amount equivalent of the energy. Such a reaction is said to be *exothermic* or *exoergic*. On the other hand, if *Q*-value is negative *i.e.* there is *absorption* of energy, then there is a *gain* of mass in the reaction which implies that the total mass of the products is *greater* than that of the reactants. Such reactions are called *endothermic* or *endoergic*. As an example of calculating *Q*-value of a nuclear reaction, consider the transmutation of N_2 into O_2 discussed in Art. 13.2. The change in mass in atomic-mass-units (amu) can be found as follows :

Interacting particles		*Product particles*	
$_2He^4$	4·00387	$_8O^{17}$	17·00450
$_7N^{14}$	14·00753	$_1H^1$	1·00814
Total	18·01140	Total	18·01264

Increase in mass = 18·01264 – 18·01140 = 0·00124 amu

Now, 1 amu = 931 MeV

∴ Q-value = 0·00124 × 931 = 1·15 MeV

Since there is an *increase* in the total mass of the products, there must be a *decrease* in energy during this reaction. Accordingly, this reaction is *endothermic*.

Similarly, consider the Li^7 (p, α) He^4 reaction. Here, the mass values are as under :

Interacting particles		*Product particles*	
$_1H^1$	1·00814	$_2He^4$	4·00387
$_3Li^7$	7·01822	$_2He^4$	4·00387
Total	8·02636	Total	8·00774

The *decrease* in mass is given as = 8·02636 – 8·00774 = 0·01862 amu

∴ Q-value = 0·01862 × 931 = 17·3 MeV

Since there is *increase* in energy, this reaction is exothermic. As mentioned in Art. 13.3 earlier, this energy is divided equally between the two α-particles.

If we incorporate the Q-value, then the above two reactions may be written as

$$_2He^4 + {_7N^{14}} = {_8O^{17}} + {_1H^1} + Q$$

or $$4·00387 + 14·00753 = 17·00450 + 1·00814 + Q$$

∴ $$Q = -0·00124 \text{ amu}$$

$$= -0·00124 \times 931 = -1·15 \text{ MeV}$$

Similarly, $$_1H^1 + {_3Li^7} = {_2He^4} + {_2He^4} + Q$$

or $$1·00814 + 7·01822 = 4·00387 + 4·00387 + Q$$

∴ $$Q = 0·01862 \text{ amu}$$

$$= 0·01862 \times 931 = +17·3 \text{ MeV}$$

13·5. Transmutation Equations

As pointed out earlier, nuclear disintegrations are expressed in the form of equations similar to chemical equations. These equations are based on the universal law of conservation of energy in which masses are regarded as equivalent to energy according to the famous mass-energy equation $E = mc^2$. Such equations must balance as regards (*i*) charge and (*ii*) mass and energy.*

(*a*) *Total charge* (*or atomic number*) *of the reactants must be equal to the total charge of the product particles.*

In the case of N^{14} (α, p) O^{17} reaction, total charge or sum of atomic numbers of reactants is = 7 + 2 = 9 and that of the product particles is = 8 + 1 = 9.

(*b*) *Total sum of atomic mass numbers on both sides of the equation must be equal.*

Considering the above reaction, we find that the sum of atomic mass numbers on the left side is = 14 + 4 = 18 and that on the right side is 17 + 1 = 18.

*In fact, in any nuclear reaction there is conservation of momentum, charge and energy.

A nuclear reaction equation may be written in the following general form :

$$M_0 + (M_1 + E_1) = (M_2 + E_2) + (M_3 + E_3)$$

where M_0 = mass of target nucleus assumed stationary

M_1, E_1 = mass and energy of the projectile.

M_2, E_2 = mass and energy of the product nucleus.

M_3, E_3 = mass and energy of the ejected or outgoing particle.

As defined in Art. 13.4, the Q-value is equal to the difference in mass of the reactants and the products (*i.e.* the mass defect).

$$\therefore \quad Q = (M_0 + M_1) - (M_2 + M_2)$$

$$= (E_2 + E_2) - E_1$$

Accordingly, the Q-value of a reaction may be found in the following two ways :

(*i*) from the known masses of the interacting nuclei (reactants) and the product nuclei involved in the reaction.

(*ii*) from the known kinetic energies of the particles involved.

The above equation may be rearranged as below :

$$M_0 + M_1 = M_2 + M_3 + (E_2 + E_3 - E_1)$$

or

$$M_0 + M_1 = M_2 + M_3 + Q$$

13·6. Importance of Transmutation Equations

Such equations have many practical applications and utility in nuclear physics :

(*i*) they not only help us to check up the actual disintegration represented by them but also aid us in predicting other possible reactions.

(*ii*) they offer a reliable means of accurate determination of atomic masses of different elements.

(*iii*) they help us to discover and identify new isotopes.

(*iv*) they provide an experimental verification of Einstein's equation $E = mc^2$.

13·7. Nuclear Projectiles

Although, historically, α-particles were the first to be used as projectiles for nuclear reactions, many other particles both charged and uncharged (like neutrons and γ-rays etc.) have been often used for the same purpose.

(*i*) **α-particle ($_2He^4$).** It is helium nucleus and was used by Rutherford in 1919 for transmutation of nitrogen nucleus. The nuclear reaction he discovered was of the (α, p) type which is common with elements of low atomic number. As the latter increases, the electrostatic (or Coulomb) barrier increases thereby making the entry of an α-particle into the target nucleus difficult. However, by using more energetic α-particle, reactions of the (α, n) type have been achieved, one of which is given below :

$$_4Be^9 + {_2He^4} \longrightarrow {_6C^{12}} + {_0n^1}$$

When the energy of α-particles is very large, two or more nucleons may be expelled in a nuclear reaction. Processes of the (α, 2n), (α, np), (α, 3n), (α, 4n), (α, 3np) etc. have been detected especially with target nuclei of medium and high mass numbers.

(*ii*) **Protons ($_1H^1$).** It is a hydrogen nucleus and was first used for nuclear disintegration by Cockcroft and Walton as described in Art. 13.2. It has less atomic number and atomic mass as compared to an α-particle and hence experiences much less repulsive Coulomb force in the field of a target nucleus.

With proton as the projectile, *radiative capture* processes of the (p, γ) type have been observed for a number of lighter elements. One example is

$$_{13}Al^{27} + {}_1H^1 \longrightarrow {}_{14}Si^{28} + \gamma$$

Other instances of the same kind are Li^7 (p, γ) Be^8, N^{14} (p, γ) O^{15} and Cr^{50} (p, γ) Mn^{81}.

If proton has sufficient energy, probability of (p, n) type of reaction becomes high. One such reaction is

$$_{11}Na^{23} + {}_1H^1 = {}_{12}Mg^{23} + {}_0n^1$$

Protons having energies in excess of 20 MeV are found to lead to the expulsion of two or more nucleons from the compound molecules. Examples of such reactions are : Cu^{63} (p, pn) Cu^{62}, Cu^{63} (p, 2n) Zn^{62} and Sr^{88} (p, 3n) Y^{86}. Processes of the (p, d) type are more common with heavier elements and protons of fairly high energy.

(*iii*) **Deuteron ($_1H^2$ or $_1D^2$).** It is the nucleus of a heavy isotope of hydrogen called *deuterium*. Although predicted theoretically in 1920 by Harkins and Rutherford, it was actually obtained in 1931 by Urey, Brickwedde and Murphy. Deuteron has been found capable of initiating several nuclear transformations through the formation of compound nucleus at relatively low energies.

(*iv*) **Neutrons ($_0n^1$).** It is a particle of unit mass and zero charge. This uncharged particle was discovered by Chadwick in 1932 for which he was awarded the Nobel Prize for Physics in 1935. He obtained fast-moving neutrons by bombarding beryllium with α-particles. The nuclear reaction is as under :

$$_2He^4 + {}_4Be^9 = {}_6C^{12} + {}_0n^1$$

Before the development of the chain-reacting pile or nuclear reactor, the main sources of neutron were reactions such as H^2 (d, n) H^3, Be^9 (d, n) B^{10} and Be^9 (α, n) C^{12} shown above. The accelerated deuterons for the above reactions were obtained from a cyclotron.

Neutrons have proved very effective in producing nuclear transformations. Since they are uncharged, they are not subject to repulsive electrostatic (Coulomb) force in the neighbourhood of a positively-charged nucleus and are, therefore, more likely to penetrate nuclei than other charged particles like protons, deuterons and α-particles. Not only are high-energy neutrons capable of causing nuclear reactions, but even slow neutrons are also extremely effective. Because of these properties, neutrons have been used to produce many more disintegrations than any other particle. Neutrons have been successfully employed for producing *nuclear fission i.e.* breaking up of the heavy unstable elements with the liberation of large amounts of energy.

Similarly, they have been extensively employed for inducing radioactivity in many elements—a process known as artificial radioactivity (Chapter 12).

Neutrons are found to produce intense biological effects (even greater than X-rays or γ-rays) on normal and tumour tissues. Since neutrons are more destructive to neoplastic than normal tissue, they are being extensively employed in cancer therapy.

(*v*) **Triton ($_1H^3$ or $_1T^3$).** It is the nucleus of isotope of hydrogen of mass 3 called *tritium* (symbol T). It is produced by the interaction of an accelerated deuteron ($_1D^2$ or $_1H^2$) with a stationary one as given by the following reaction :

$$_1H^2 + {}_1H^2 = {}_1H^3 + {}_1H^1$$

or

$$_1D^2 + {}_1D^2 = {}_1T^3 + {}_1H^1$$

(*vi*) **Photons.** High-energy photons have also been used for nuclear disintegration which is usually referred to as *photodisintegration*. Since photons have neither mass nor energy, they can only give their kinetic energy to the nuclear reaction. This energy must be at least equal to the binding energy of the nuclear particle before such a particle can be ejected out of the nucleus. Photodisintegration processes are, therefore, endothermic and usually have threshold energies of about 10 MeV. Usually, (γ, p) and (γ, n) reactions are obtained depending on the energy of the incident photons.

13·8. Discovery of Neutron

As early as 1920, Rutherford had hinted at the possibility of a particle consisting of a proton and electron in close combination. A number of experiments were performed in 1924 to check this proposition but without success. In 1930, Bothe and Becker bombarded light nuclei like lithium and beryllium with α-particles and observed a penetrating radiation which they assumed to be hard γ-rays. In the same year, Curie and Joliot found that these penetrating radiations could knock energetic protons out from paraffin wax. In their experiment (Fig. 13.5) α-particles from a polonium source were used to bombard beryllium beyond which was placed a piece of paraffin wax W that provided a high concentration of hydrogen atoms. They supposed that the energy of γ-rays was transferred to the proton by Compton effect. However, they calculated that for producing protons with the observed energy (4·7 MeV), these radiations must have an energy of 55 MeV which is much in excess of that emitted by radioactive substances. Later in 1932, it was showed by Chadwick that these radiations could also knock out energetic nitrogen nuclei (1·2 MeV) from nitrogen gas. Again calculations showed that for this to happen, these radiations must possess an energy of 90 MeV which is inconsistent with hydrogen result.

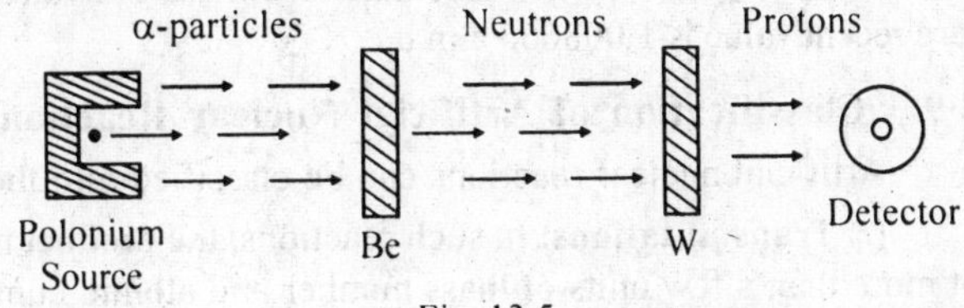

Fig. 13.5

At this stage, Chadwick revived the idea first suggested by Rutherford of a particle having no charge but of mass comparable with proton—a particle he named *neutron*. The difficulties experienced by earlier workers in accounting for the energies of protons disappeared when it was assumed that the unknown so-called γ-radiations were nothing else but neutrons.

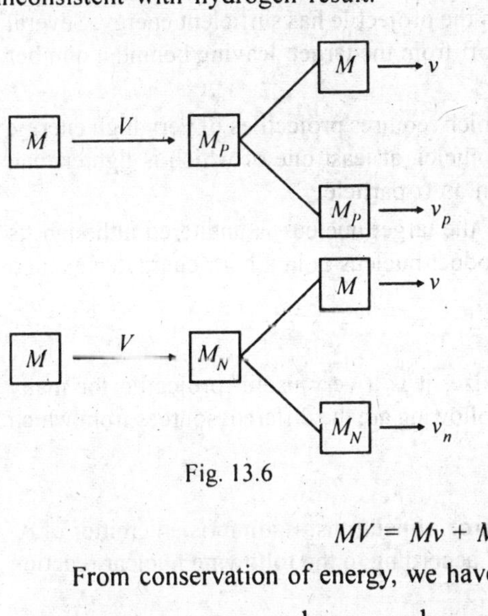

Fig. 13.6

The calculations on which Chadwick based his conclusion are as under :

Suppose a neutron of mass M and velocity V makes a head-on elastic collision with a stationary proton of mass M_p (Fig. 13.6). For conservation of momentum,

$$MV = Mv + M_p v_p \qquad \dots (i)$$

From conservation of energy, we have

$$\frac{1}{2} MV^2 = \frac{1}{2} Mv^2 + \frac{1}{2} M_p v_p^2 \qquad \dots (ii)$$

By eliminating v from Eqs. (*i*) and (*ii*) above, we get

$$v_p = 2MV / (M_p + M)$$

Similarly, the velocity of nitrogen nuclei is given by

$$v_N = 2MV / (M_N + M)$$

Dividing one equation by the other, we get

$$\frac{v_p}{v_N} = \frac{M_N + M}{M_p + M}$$

Now, observed values are :

$$v_p = 3{\cdot}3 \times 10^7 \text{ m/s} ; v_N = 0{\cdot}47 \times 10^7 \text{ m/s}$$
$$M_p = 1 \text{ a.m.u.} ; M_N = 14 \text{ a.m.u.}$$

Substituting the values, we get $M = 1{\cdot}16$ a.m.u.

In this way, Chadwick calculated the mass of neutron as 1·16 times that of proton. However, more recent value is 1·008665 a.m.u.

13·9. Classification of Artificial Nuclear Reactions

Artificial nuclear reactions can be classified into the following six general categories :

(*i*) **Transmutations.** In such reactions, the product nucleus differs from the target nucleus by not more than a few units of mass number and atomic number. The reaction is accompanied by the emission of one or two nucleons or a photon.

(*ii*) **Stripping and Pickup.** Stripping reactions resemble transmutations in the respect that product nucleus does not differ greatly from the target nucleus. However, the mechanism of the process is quite different. In a stripping reaction, one or more nucleons from the projectile are captured by the target while the remainder continues on its way.

Pickup reactions are related to stripping reactions although in their case the projectile captures a nucleon (or nucleons) from the target nucleus.

(*iii*) **Fission.** In fission reaction, the target nucleus splits mainly into two lighter nuclei of comparable mass while, at the same time, emitting a few other particles, usually neutrons.

(*iv*) **Spallation.** Such a reaction takes place when the projectile has sufficient energy. Several nucleons up to 20 or 30 and some α-particles are split off from the target, leaving behind a number of nuclei of *significantly smaller mass.*

(*v*) **Fragmentation.** In fragmentation reaction which requires projectiles of very high energy, the target nucleus is disintegrated into several smaller nuclei, at least one of which is lighter than a fission-fragment nucleus but appreciably heavier than an α-particle.

(*vi*) **Scattering.** In such processes, the nature of the target nucleus is unaltered although its energy is changed. However, in some instances, the product nucleus is in a high-energy or excited state.

13·10. Neutron Sources

Because neutron has no charge and has small size, it is a very useful projectile for many experimental purposes, particularly for nuclear fission. Following are the different sources from which neutrons are obtained for experimentation purposes :

(*i*) Radioactive Source

A mixture of radium and beryllium is a good source of neutrons. Radium is an emitter of α-particles which knock out neutrons from beryllium $_4Be^9$ according to the following nuclear reaction equation

$$_2He^4 + {}_4Be^9 \longrightarrow {}_6C^{12} + {}_0n^1$$

However, the neutron beam from Ra-Be mixture is not mono-energetic.

(*ii*) Accelerated Charged Particles

In this case, deuterons are accelerated and then made to hit a tritium target. Neutrons are produced according to the following nuclear reaction :

$$_1H^2 + {}_1H^3 \longrightarrow {}_2He^4 + {}_0n^1$$

(*iii*) **Photodisintegration**

Neutrons are also produced when energetic γ-ray photons react with certain nuclei as shown below :

$$\gamma + {}_4Be^9 \longrightarrow {}_4Be^8 + {}_0n^1$$

The threshold energy of γ-ray photons for this reaction is found to be 1·67 MeV.

(*iv*) **Stripping Reactions**

Very energetic neutrons are produced when deuterons of energy of several hundreds of MeV strike a target. The neutrons so separated from the projectile have high energy.

13·11. Transmutations by Different Particles

Some important nuclear reactions produced by different particles dealt with in Art. 13.7 will be discussed.

(*a*) **Transmutations by Alpha-particles.** Some of the (α, p) reactions have already been given in Art. 13.2. Some of the (α, n) reactions are given below :

$${}_{18}Ar^{40} + {}_2He^4 \longrightarrow [{}_{20}Ca^{44}] \longrightarrow {}_{20}Ca^{43} + {}_0n^1$$

$${}_3Li^7 + {}_2He^4 \longrightarrow [{}_5B^{11}] \longrightarrow {}_5B^{10} + {}_0n^1$$

$${}_{11}Na^{23} + {}_2He^4 \longrightarrow [{}_{13}Al^{27}] \longrightarrow {}_{13}Al^{26} + {}_0n^1$$

$${}_7N^{14} + {}_2He^4 \longrightarrow [{}_9F^{18}] \longrightarrow {}_9F^{17} + {}_0n^1$$

When these reactions are compared with those mentioned in Art. 13.2, it is seen that some atoms such as N^{14} and Na^{23} emit either a proton or a neutron when bombarded with α-particles.

(*b*) **Transmutations by Protons**

(*i*) (p, α) *reactions*

$${}_3Li^7 + {}_1H^1 \longrightarrow [{}_4Be^8] \longrightarrow {}_2He^4 + {}_2He^4$$

$${}_9F^{19} + {}_1H^1 \longrightarrow [{}_{10}Ne^{20}] \longrightarrow {}_8O^{16} + {}_2He^4$$

An interesting reaction occurs when the target nucleus is B^{11}.

$${}_5B^{11} + {}_1H^1 \longrightarrow [{}_6C^{12}] \longrightarrow {}_4Be^8 + {}_2He^4$$

The Be^8 is very unstable and splits apart into two more α-particles

$${}_4Be^8 \longrightarrow {}_2He^4 + {}_2He^4$$

The above nuclear reaction is shown in Fig. 13.7. The proton is captured by ${}_5B^{11}$ nucleus to form an unstable carbon nucleus ${}_6C^{12}$ which disintegrates by emitting an α-particle with an energy of several MeV, leaving behind a beryllium nucleus ${}_4Be^8$. Being unstable, ${}_4Be^8$ splits to produce two more α-particles.

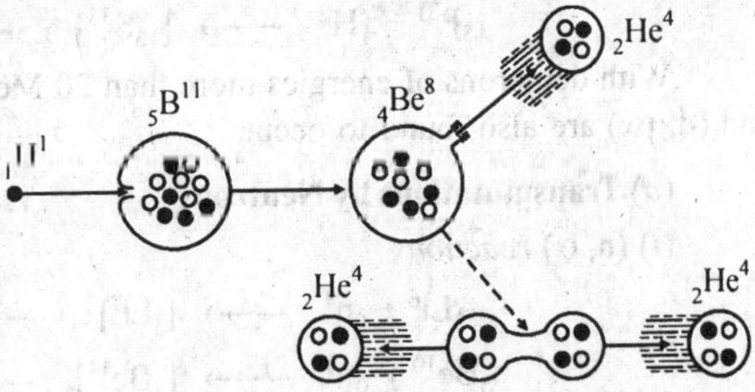

Fig. 13.7

(*ii*) (p, n) *reactions*

$${}_8O^{18} + {}_1H^1 \longrightarrow [{}_9F^{19}] \longrightarrow {}_9F^{18} + {}_0n^1$$

$${}_{18}Ni^{58} + {}_1H^1 \longrightarrow [{}_{29}Cu^{59}] \longrightarrow {}_{29}Cu^{58} + {}_0n^1$$

$${}_{29}Cu^{65} + {}_1H^1 \longrightarrow [{}_{30}Zn^{66}] \longrightarrow {}_{30}Zn^{65} + {}_0n^1$$

$${}_{11}Na^{23} + {}_1H^1 \longrightarrow [{}_{12}Mg^{24}] \longrightarrow {}_{12}Mg^{23} + {}_0n^1$$

(*iii*) (p, γ) *reactions*

$$_3Li^7 + {}_1H^1 \longrightarrow [{}_4Be^8] \longrightarrow {}_4Be^8 + \gamma$$

$$_{13}Al^{27} + {}_1H^1 \longrightarrow [{}_{14}Si^{28}] \longrightarrow {}_{14}Si^{28} + \gamma$$

$$_6C^{12} + {}_1H^1 \longrightarrow [{}_7N^{13}] \longrightarrow {}_7N^{13} + \gamma$$

(*iv*) (p, d) *reactions*

$$_4Be^9 + {}_1H^1 \longrightarrow [{}_5B^{10}] \longrightarrow {}_4Be^8 + {}_1H^2$$

$$_3Li^7 + {}_1H^1 \longrightarrow [{}_4Be^8] \longrightarrow {}_3Li^6 + {}_1H^2$$

(*c*) **Transmutations by Deuterons**

(*i*) (d, α) *reactions*

$$_3Li^6 + {}_1H^2 \longrightarrow [{}_4Be^8] \longrightarrow {}_2He^4 + {}_2He^4$$

$$_8O^{16} + {}_1H^2 \longrightarrow [{}_9F^{18}] \longrightarrow {}_7N^{14} + {}_2He^4$$

$$_{20}Ca^{40} + {}_1H^2 \longrightarrow [{}_{21}Sc^{42}] \longrightarrow {}_{19}K^{38} + {}_2He^4$$

(*ii*) (d, p) *reactions*

$$_{15}P^{31} + {}_1H^2 \longrightarrow [{}_{16}S^{33}] \longrightarrow {}_{15}P^{32} + {}_1H^1$$

$$_{11}Na^{23} + {}_1H^2 \longrightarrow [{}_{12}Mg^{25}] \longrightarrow {}_{11}Na^{24} + {}_1H^1$$

(*iii*) (d, n) *reactions*

$$_6C^{12} + {}_1H^2 \longrightarrow [{}_7N^{14}] \longrightarrow {}_7Na^{13} + {}_0n^1$$

$$_{52}Te^{130} + {}_1H^2 \longrightarrow [{}_{53}I^{120}] \longrightarrow {}_{53}I^{131} + {}_0n^1$$

$$_{83}Bi^{209} + {}_1H^2 \longrightarrow [{}_{84}Po^{211}] \longrightarrow {}_{84}Po^{210} + {}_0n^1$$

$$_1H^3 + {}_1He^2 \longrightarrow [{}_2He^5] \longrightarrow {}_2He^4 + {}_0n^1$$

(*iv*) (d, t) *reactions*

$$_3Li^7 + {}_1H^2 \longrightarrow [{}_4Be^9] \longrightarrow {}_3Li^6 + {}_1T^3\,({}_1H^3)$$

$$_{47}Ag^{109} + {}_1H^2 \longrightarrow [{}_{48}Cd^{111}] \longrightarrow {}_{47}Ag^{108} + {}_1T^3$$

$$_{15}P^{31} + {}_1H^2 \longrightarrow [{}_{16}S^{33}] \longrightarrow {}_{15}P^{30} + {}_1T^3$$

With deuterons of energies more than 20 MeV, reactions of the type (d, 2n), (d, 3n), (d, 2p) and (d, p*x*) are also found to occur.

(*d*) **Transmutations by Neutrons**

(*i*) (n, α) *reactions*

$$_3Li^6 + {}_0n^1 \longrightarrow [{}_3Li^7] \longrightarrow {}_1H^3 + {}_2He^4$$

$$_5Be^{10} + {}_0n^1 \longrightarrow [{}_5Be^{11}] \longrightarrow {}_3Li^7 + {}_2He^4$$

(*ii*) (n, p) *reactions*

$$_7N^{14} + {}_0n^1 \longrightarrow [{}_7N^{15}] \longrightarrow {}_6C^{14} + {}_1H^1$$

$$_{30}Zn^{64} + {}_0n^1 \longrightarrow [{}_{30}Zn^{65}] \longrightarrow {}_{29}Cu^{64} + {}_1H^1$$

(*iii*) (n, γ) *reactions*

$$_{49}In^{115} + {}_0n^1 \longrightarrow [{}_{49}In^{116}] \longrightarrow {}_{49}In^{116} + \gamma$$

$$_{92}U^{238} + {}_0n^1 \longrightarrow [{}_{92}U^{239}] \longrightarrow {}_{92}U^{239} + \gamma$$

(*iv*) (n, 2n) *reactions*

$$_6C^{12} + {}_0n^1 \longrightarrow [_6C^{13}] \longrightarrow {}_6C^{11} + 2\,{}_0n^1$$

$$_{92}U^{238} + {}_0n^1 \longrightarrow [_{92}U^{239}] \longrightarrow {}_{92}U^{237} + 2\,{}_0n^1$$

(*e*) **Transmutations by Tritons**

(*i*) (t, p) *reaction*

$$_{27}Co^{59} + {}_1H^3 \longrightarrow [_{28}Ni^{62}] \longrightarrow {}_{27}Co^{61} + {}_1H^1$$

(*ii*) (t, d) *reactions*

$$_3Li^6 + {}_1H^3 \longrightarrow [_4Be^9] \longrightarrow {}_3Li^7 + {}_1H^2$$

$$_{29}Cu^{63} + {}_1H^3 \longrightarrow [_{30}Zn^{66}] \longrightarrow {}_{29}Cu^{64} + {}_1H^2$$

(*iii*) (t, n) *reaction*

$$_{16}S^{32} + {}_1H^3 \longrightarrow [_{17}Cl^{35}] \longrightarrow {}_{17}Cl^{34} + {}_0n^1$$

(*iv*) (t, α) *reaction*

$$_3Li^7 + {}_1H^3 \longrightarrow [_4Be^{10}] \longrightarrow {}_2He^6 + {}_2He^4$$

(*f*) **Transmutations by Photons**

(*i*) (γ, n) *reactions*

$$_1H^2 + \gamma \longrightarrow [_1H^2] \longrightarrow {}_1H^1 + {}_0n^1$$

$$_4Be^9 + \gamma \longrightarrow [_4Be^9] \longrightarrow {}_4Be^8 + {}_0n^1$$

$$_{15}P^{31} + \gamma \longrightarrow [_{15}P^{31}] \longrightarrow {}_{15}P^{30} + {}_0n^1$$

(*ii*) (γ, p) *reaction*

$$_{12}Mg^{25} + \gamma \longrightarrow [_{12}Mg^{25}] \longrightarrow {}_{11}Na^{24} + {}_1H^1$$

(*iii*) (γ, 2pn) *reaction*

$$_{13}Al^{27} + \gamma \longrightarrow [_{13}Al^{27}] \longrightarrow {}_{11}Na^{24} + 2\,{}_1H^1 + {}_0n^1$$

Example 13.1. *Calculate the energy available by disintegration of 100 grams of $_3Li^7$ into $_2He^4$ by bombardment of protons.*

Given, mass of $_3Li^7$ = 7·0183 amu ; mass of $_2He^4$ = 4·0040 amu

mass of $_1H^1$ = 1·0082 amu ; Avogadro number, N = 6·025 × 10^{26}/kg-atom

Solution. The nuclear reaction can be represented by

$$_3Li^7 + {}_1H^1 \longrightarrow [_4Be^8] \longrightarrow {}_2He^4 + {}_2He^4$$

$\Delta m = (7{\cdot}0183 + 1{\cdot}0082) - 2 \times 4{\cdot}0040 = 0{\cdot}0185$ amu

Energy available/disintegration = 0·0185 × 931 = **17·22 MeV**

No. of atoms in 100 grams or 0·1 kg of lithium = 0·1 × 6·025 × 10^{26}/7 = 8·61 × 10^{24}

∴ Total energy available = 8·61 × 10^{24} × 17·22 = **14·83 × 10^{25} MeV**

Example 13.2. *Calculate the energy* (*Q*) released in the following nuclear reaction :

$$_3Li^7 + {}_0n^1 \longrightarrow {}_2He^4 + {}_1H^3 + Q$$

Explain the mechanism of this reaction.

Given, $_3Li^7$ = 6·015126 amu $_2He^4$ = 4·002604 amu

$_0n^1$ = 1·00865 amu $_1H^3$ = 3·016049 amu

Solution. It is a (*n*, α) reaction in which neutron is captured by lithium nucleus thus forming an unstable *compound nucleus* which immediately disintegrates by ejecting α-particle and a triton.

Substituting the given values of different masses, we have

$$6{\cdot}015126 + 1{\cdot}00865 = 4{\cdot}002604 + 3{\cdot}016049 + Q$$

$$\therefore \quad 7{\cdot}023791 = 7{\cdot}018653 + Q$$

or $$Q = 0{\cdot}0051380 \text{ amu}$$

Now, $$1 \text{ amu} = 931 \text{ MeV}$$

$$Q = 0{\cdot}0051380 \times 931 = \mathbf{4{\cdot}7835\ MeV}$$

Example 13.3. *What is the Q-value of a nuclear reaction ? Calculate it for the reaction taking place in Rutherford's experiment on artificial disintegration of nitrogen by bombardment with α-particles.*

Relevant masses in a.m.u. are :

$${}_7N^{14} = 14{\cdot}007515 \qquad {}_2He^4 = 4{\cdot}003837$$

$${}_8O^{17} = 17{\cdot}004533 \quad \text{and} \qquad {}_1H^1 = 1{\cdot}008142$$

Solution. The nuclear reaction is given by

$${}_7N^{14} + {}_2He^4 \longrightarrow {}_8O^{17} + {}_1H^1$$

Mass of reactants $= 14{\cdot}007515 + 4{\cdot}003837$

$= 18{\cdot}011352$ a.m.u.

Mass of products $= 17{\cdot}004533 + 1{\cdot}008142$

$= 18{\cdot}012675$ amu

∴ increase in mass, $\Delta m = 18{\cdot}012675 - 18{\cdot}011352$

$= 0{\cdot}001323$ amu

Q-value $= 0{\cdot}001323 \times 931 = \mathbf{1{\cdot}232\ MeV}$

Example 13.4. *A nuclear reaction is given by*

$${}_7N^{14} + {}_0n^1 = {}_6C^{14} + {}_1H^1 + 0{\cdot}55 \text{ MeV}$$

Find the mass of ${}_6C^{14}$ in amu given that

$${}_7N^{14} = 14{\cdot}007520 \text{ amu} \qquad {}_1H^1 = 1{\cdot}008145 \text{ amu}$$

$${}_0n^1 = 1{\cdot}008986 \text{ amu}$$

Solution. Let us convert 0·55 MeV into equivalent mass in amu. Its value is 0·55/931 = 0·000591 amu.

$$\therefore {}_6C^{14} = ({}_7M^{14} + {}_0n^1) - ({}_1H^1 + 0{\cdot}000599) = \mathbf{14{\cdot}00777\ amu}$$

Example 13.5. *In a cloud chamber photograph, a proton is seen to have undergone an elastic collision, its track being deviated by 60°. The struck particle makes a track at 30° with the incident proton direction. What mass does the particle possess ?*

Solution. The collision is shown in Fig. 13.8.

The struck particle of assumed mass m is supposed to be stationary before collision. Applying the law of conservation of *lineur* momentum to X-axis and Y-axis, we get

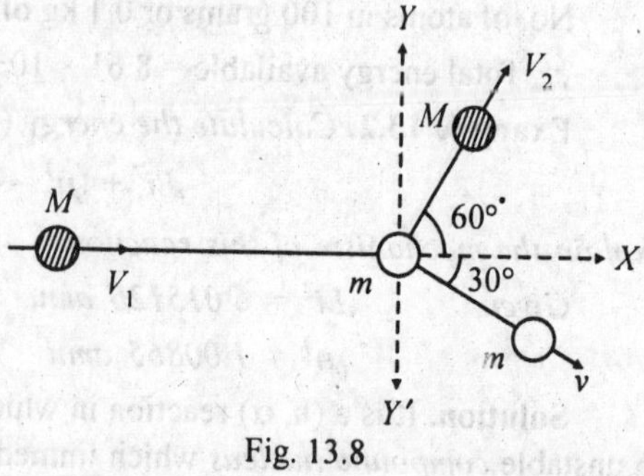

Fig. 13.8

(*i*) *Along X-axis*

$$MV_1 = MV_2 \cos 60° + mv \cos 30°$$

$$= \frac{1}{2} MV_2 + \frac{\sqrt{3}}{2} mv$$

or $$2MV_1 = MV_2 + \sqrt{3}\, mv \quad ...(i)$$

(*ii*) *Along Y-axis*

$$MV_2 \sin 60° = mv \sin 30°$$

or $$\frac{\sqrt{3}}{2} MV_2 = \frac{1}{2} mv \quad \text{or} \quad mv = \sqrt{3}\, MV_2$$

$$\therefore \quad \sqrt{3}\, mv_2 = 3MV_2$$

Substituting this value in Eq. (*i*) above, we get

$$2MV_1 = MV_2 + 3\, MV_2 = 4\, MV_2$$

$$\therefore \quad V_1 = 2V_2 \quad \text{or} \quad V_2 = V_1/2$$

Since the velocity of the proton after collision is reduced by half, it means that struck particle must be of equal mass.

Example 13.6. *Explain mass defect, packing fraction and binding energy.*

In the reaction

$$_5B^{11} + {}_2He^4 \longrightarrow {}_7N^{14} + {}_0n^1$$

the incident α-particle has 5·250 MeV of kinetic energy and the resultant neutron and nitrogen atoms had energies of 3·260 and 2·139 MeV respectively. If the mass of boron and nitrogen atoms and α-particle are 11·01280, 14·00752 and 4·00387 amu respectively, find the mass of neutrons.

Solution. As shown in Art 13.5, a nuclear reaction may be written in the following form :

$$M_0 + (M_1 + E_1) = (M_2 + E_2) + (M_3 + E_3)$$

where M_0 = mass of target nucleus assumed stationary

M_1, E_1 = mass and kinetic energy of projectile

M_2, E_2 = mass and kinetic energy of product nucleus

M_3, E_3 = mass and kinetic energy of ejected or outgoing particle

As explained in Art. 13.5, in such transmutation equations, we have to take into account not only the masses but also kinetic energies of both reactants and product nuclei.

In the present case, target nucleus is boron, projectile is α-particle, the product nucleus is nitrogen and the outgoing particle is neutron. For uniformity of units, let us convert given masses into equivalent energies in MeV.

It should be kept in mind that

$$1 \text{ amu} = 931 \text{ MeV}$$

Mass of boron, M_0 = 11·01280 amu

= 11·01280 × 931

= 10,253 MeV

Mass of α-particle, M_1 = 4·00387 amu

= 4·00387 × 931

= 3,727·6 MeV

Mass of nitrogen, M_2 = 14·00752 amu

= 14·00752 × 931

= 13,041 MeV

Mass of neutron = M_3

Substituting the values of masses and kinetic energies of different nuclides in the above equation, we get

$$10{,}253 + (3{,}727{\cdot}6 + 5{\cdot}250) = (13{,}041 + 2{\cdot}139) + (M_3 + 3{\cdot}260)$$

$$\therefore \quad M_3 = 13{,}985{\cdot}85 - 13{,}046{\cdot}40$$
$$= 939{\cdot}45 \text{ MeV}$$

This mass of the neutron can be converted back into amu.

$$\therefore \quad M_3 = 939{\cdot}45/931$$
$$= \mathbf{1{\cdot}009 \text{ amu.}}$$

It is seen that it is (α, n) nuclear reaction. Here, a stationary boron atom is hit by a fast-moving α-particle. The particle is captured by boron nucleus which emits one neutron and consequently gets converted into nitrogen isotope ${}_7N^{14}$.

Example 13.7. *Compute the following disintegration reactions*

(i) ${}_1H^2 + {}_8O^{16} \longrightarrow ? + {}_2He^4$

(ii) ${}_1H^2 + {}_5B^{10} \longrightarrow ? + {}_0n^1$

(iii) ${}_2He^4 + {}_{13}Al^{27} \longrightarrow ? + {}_1H^1$

(iv) ${}_1H^1 + {}_3Li^6 \longrightarrow ? + {}_2He^3$

Solution. For finding the missing particle, the rule is to equalize the sum of the total charge and the total mass on both sides of the nuclear reaction equations. In other words, sum of A-values and Z-values on both sides of the equation must be equal.

(i) L.H.S. charge $= 1 + 8 = 9$, R.H.S. charge $= X + 2$

$$\therefore \quad X + 2 = 9 \text{ or } X = 7$$

Hence, unknown element on RHS must have a nuclear charge of 7.

Obviously, the element with $Z = 7$ is nitrogen. Its A-value $= (16 + 2) - 4 = 14$.

Hence, missing isotope is ${}_7N^{14}$ and the reaction is

$${}_1H^2 + {}_8O^{16} \longrightarrow {}_7N^{14} + {}_2He^4$$

(ii) Here, A-value of RHS atom $= (10 + 2) - 1 = 11$

$$\text{Z-value} = (4 + 2) - 0 = 6$$

Obviously, missing element is carbon isotope ${}_8C^{11}$

$$\therefore \quad {}_1H^2 + {}_5B^{10} \longrightarrow {}_8C^{11} + {}_0n^1$$

(iii) Here, missing nuclide has $A = (27 + 4) - 1 = 30$ and $Z\,(13 + 2) - 1 = 14$. It is silicon isotope ${}_{14}Si^{30}$.

(iv) Here, Z-value of missing element is $(3 + 1) - 2 = 2$ and A-value $= (6 + 1) - 3 = 4$.

It is α-particle, ${}_2He^4$.

$$\therefore \quad {}_1H^1 + {}_3Li^6 \longrightarrow {}_2He^4 + {}_2He^3$$

Example 13.8. *Complete the following nuclear reactions :*

(i) ${}_{17}Cl^{35}$ + ? $\longrightarrow$ ${}_{16}S^{32}$ + *alpha particle*

(ii) ${}_5B^{10}$ + ? $\longrightarrow$ ${}_6C^{13}$ + *proton*

(iii) ${}_3Li^7$ + ${}_2He^4$ $\longrightarrow$ ${}_5B^{10}$ + ?

(iv) ${}_7N^{14}$ + *alpha particle* $\longrightarrow$? + *proton*

Solution. (i) ${}_{17}Cl^{35} + ? \longrightarrow {}_{16}S^{32} + {}_2He^4$

Here, Z-value of unknown particle is $(18 - 17) = 1$. Hence, it is hydrogen nucleus or proton. A-value $= (36 - 35) = 1$. Obviously, the unknown particle is confirmed as **Proton**.

(*ii*) $_5B^{10} + ? \longrightarrow {_6C^{13}} + {_1H^1}$

Z-value difference = (6 + 1) – 5 = 2

A-value difference = (13 + 1) – 10 = 4

Obviously, the unknown particle is α-particle ($_2He^4$)

(*iii*) $_3Li^7 + {_2He^4} \longrightarrow {_5B^{10}} + ?$

Z-value difference = (3 + 2) – 5 = 0

A-value difference = (7 + 4) – 10 = 1

The unknown particle is **neutron** ($_0n^1$)

(*iv*) $_7N^{14} + {_2He^4} \longrightarrow ? + {_1H^1}$

Z-value difference = (7 + 2) – 1 = 8

A-value difference = (14 + 4) – 1 = 17

The unknown nucleus is **oxygen** isotope $_8O^{17}$.

13·12. Nuclear Reaction Cross-section

It is quantitative measure of the *probability* of a given nuclear reaction. It represents the effective area of cross-section of a single nucleus of a given species for a particular nuclear reaction. Thus when the probability of the reaction is high, the nuclear cross-section will be large and when the probability is low, the cross-section will be small.

If *n* is the number of incident particles striking per second per m² of the target material containing N_t target nuclei per m² and *N* is the number of these nuclei that undergo interaction per second, then the nuclear cross-section σ expressed as m²/nucleus is defined by

$$\sigma = \frac{N}{n \cdot N_t} \text{ m}^2/\text{nucleus}$$

If $N_t = 1$, then $\sigma = N/n$.

Hence, nuclear cross-section may be defined as *the fraction of the particles falling on the target that interact with a single nucleus.* Obviously, larger the fraction of the incident particles reacting, greater is the probability that the process will occur under the given conditions.

The unit of measuring nuclear reaction cross-section is *barn.*

1 barn = 10^{-28} m²/nucleus.

TUTORIAL PROBLEMS

1. Americium ($_{95}Am^{241}$) is obtained by the decay of Pu-241 as a result of β-decay. The Pu-241 is made by bombarding U^{238} with α-particles. Write down the respective equations for the above reactions. **[33·24 days]**
2. Complete the following nuclear reactions :

   (*i*) $_{11}Na^{23} + {_2He^4} \longrightarrow {_{13}Al^{26}} + \ldots\ldots$

   (*ii*) $_4Be^9 + {_1H^1} \longrightarrow {_4Be^8} + \ldots\ldots$

   (*iii*) $_3Li^7 + {_1H^2} \longrightarrow {_3Li^6} + \ldots\ldots$

   (*iv*) $_6C^{12} + {_0n^1} \longrightarrow {_6C^{11}} + \ldots\ldots$

   [(*i*) $_0n^1$ (*ii*) $_1H^2$ (*iii*) $_1H^3$ (*iv*) 2 $_0n^1$]
3. Complete the following disintegration reactions :

   (*i*) $_1H^2 + {_8O^{16}} \longrightarrow {_7N^{14}} + ?$

   (*ii*) $_2He^4 + {_{13}Al^{17}} \longrightarrow {_{14}Si^{30}} + ?$

(*iii*) $_1H^2 + {_5B^{10}} \longrightarrow {_6C^{11}} + ?$

(*iv*) $_1H^1 + {_3Li^6} \longrightarrow {_2He^4} + ?$

[(*i*) $_2He^4$ (*ii*) $_1H^1$ (*iii*) $_0n^1$ (*iv*) $_2He^3$]

4. Complete the following nuclear reactions :

(*i*) $_{18}A^{40} + {_2He^4} = {_{20}Ca^{43}} + ?$

(*ii*) $_{11}Na^{13} + {_2He^4} = ? + {_0n^1}$

(*iii*) $_{29}Cu^{65} + {_1H^1} = ? + {_0n^1}$

(*iv*) $_1H^2 + {_3Li^6} = {_2He^4} + ?$

(*v*) $_{15}P^{31} + {_1H^2} = {_{15}P^{30}} + ?$

HIGHLIGHTS

1. The overall energy liberated or absorbed in a nuclear disintegration is called nuclear reaction energy Q.

2. Those nuclear reactions in which energy is liberated are called exothermic or exoergic. They have positive Q-value.

3. Those reactions in which energy is absorbed are called endothermic or endoergic. They have a negative Q-value.

4. In general, the Q-value of a nuclear reaction is given by

$$Q = (M_0 + M_1) - (M_2 + M_3)$$
$$= (E_2 + E_3) - E_1$$

Can You Answer This ?

1. What is nuclear reaction? How is it different from a chemical reaction?

2. What is a nuclear reaction?

Give one example of each of the following nuclear reactions :

(*i*) (α, p) reaction (*ii*) (α, *n*) reaction (*iii*) disintegration of nuclei by deuterons and (*iv*) disintegration of nuclei by protons.

3. (*a*) What is a nuclear reaction? How is it different from a chemical reaction?

(*b*) Why neutron is considered an important particle in artificial transmutation? What are the neutron sources?

4. Discuss how nuclear reactions by α-rays lead to the discovery of neutron.

5. Why are neutrons chosen as missiles in nuclear reactions?

NUCLEAR FISSION AND FUSION 14

Introduction — Discovery of Fission — Types of Fission Reactions — Mass Distribution of Emission Products — Energy Distribution of Fission Products — Neutron Emission in Nuclear Fission — Fissile and Fissionable Nuclides — Fission Energy — Theory of Fission Process — Nuclear Fusion — Stellar Thermonuclear Fusion Reactions — Proton-Proton Chain — Carbon-Nitrogen Cycle — Controlled Thermonuclear Reactions — Conditions for Controlled Fusion — Hydrogen Bomb — Cobalt Bomb — Fission and Fusion — Highlights — Can You Answer This ?

14·1. Introduction

Continuous research on artificial transmutation and especially the study of induced radioactivity culminated in the discovery of nuclear fission which is accompanied by the release of enormous amounts of energy. In ordinary nuclear disintegrations, both natural and artificial, the nucleus is only chipped off rather than broken and accordingly, the amount of energy released is comparatively less *i.e.* from about 10 to 23 MeV.

It was discovered in 1939 that the heavy unstable uranium nucleus when bombarded by neutrons splits into two almost equal fragments which fly apart with great speed and the amount of energy released per fission is about 200 MeV.

This division of a nucleus into two approximately equal parts is called nuclear fission.

14·2. Discovery of Fission

The starting point in the discovery of nuclear fission can be traced to the attempts of Fermi, in 1934, to produce transurani elements by bombarding uranium* with neutrons. However, the fission process itself was discovered in 1939 by German radiochemists Otto Hahn and his two associates Meitner and Strassmann. After bombarding uranium with neutrons, they performed a series of chemical separations to identify the products. To their great surprise, they found that the atoms produced by the bombardment of uranium belonged to elements which lie near the centre of the periodic table. Obviously, a uranium nucleus after capturing neutron and become so unstable that instead of disintegrating by ejecting one or two particles, it had split up into two parts.

The actual fission process can be understood with the help of Fig. 14.1 which shows a uranium nucleus capturing a neutron. The newly-formed nucleus of Fig. 14.1 (*b*) is unstable and starts breaking up into two parts. In breaking up, the uranium nucleus, behaving like a liquid drop, splashes out small droplets *i.e.* neutrons and γ-rays. So great is the release of energy that the two fission fragments fly apart in opposite directions with tremendous speeds. It may, however, be noted that not all uranium nuclei break into Sb and Nb as shown in Fig. 14.1. There are at least 30 different ways in which a fissile nuclide can divide itself. The experimental evidence seems to favour pairs of fission fragments of unequal masses (asymmetrical fission) accompanied by one to five or sometimes more neutrons. In general, fission fragments are unstable nuclei containing an excess number of neutrons. After a series of β-emissions in which neutrons are converted into protons in the nucleus, a stable nuclide results (Fig. 14.3).

Out of all the neutrons ejected during the fission of uranium, about 99 per cent are ejected in an extremely short interval of time and are called *prompt* neutrons. The remaining one per cent of

*In nature, it occurs as a very dense black metal.

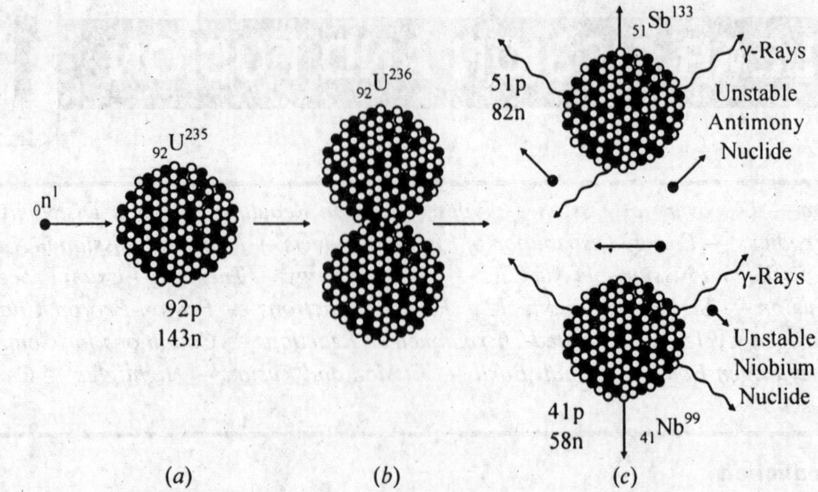

Fig. 14.1

neutrons are emitted a little later and are called *delayed* neutrons. The delayed neutrons originate from unstable fragments that decay by neutron emission on their way to becoming stable nuclei.

It may be noted that division of a fissile nucleus into *three* fragments of comparable sizes (ternary fission) has been observed although it is a rare event, occurring about 5 times per million binary fissions.

14·3. Types of Fission Reactions

Historically speaking, uranium was the first element to undergo fission. However, soon after it was found that other elements of high atomic weight could also be made to undergo fission and that particles other than neutron could be equally effective in this respect.

Natural uranium contains three principal isotopes with the following relative abundance :

U^{238}	99·28%	$4·51 \times 10^9$ Y
U^{235}	0·714%	$7·1 \times 10^8$ Y
U^{234}	0·006%	$2·48 \times 10^5$ Y

It is found that *slow* neutrons cause fission of U^{235} but not of more abundant isotope U^{238} which requires fast neutrons with energies exceeding 1 MeV. Similarly, Th^{232} and Pa^{231} undergo fission when bombarded with fast neutrons. Fission can also be produced in uranium and thorium by high-energy α-particles, protons, deuterons and γ-rays etc. Two other nuclides which do not occur in nature but have proved to be fissionable by neutrons of all energies are $_{92}U^{233}$ and $_{94}Pu^{239}$. In 1947, successful fission of bismuth, lead, thallium, mercury, gold, platinum and tantalum was achieved in USA by means of α-particles, deuterons, and neutrons of 100 MeV and more. With bismuth ($Z = 83$) fission was detected with 50 MeV deuterons whereas tantalum ($Z = 73$) required α-particles of 400 MeV energy.

It is worth noting that only three fissile materials U^{233}, U^{235} and Pu^{239} are important in the large-scale application of nuclear energy.

Finally, some heavy nuclei have been found to undergo *spontaneous* fission. In this process nucleus divides in the ground state without bombardment by particles from outside.

14·4. Mass Distribution of Fission Products

During uranium fission, a large number of nuclides of intermediate charge and mass are found. Their study is a promising source of information about the mechanism of the fission process itself and also offers the possibility of discovering hitherto unknown nuclides. Investigations of the fission products of U^{235} have shown that the range of their mass numbers is from 72 to 158. About 97% of the U^{235} nuclei undergoing fission give fragments which fall into two groups as shown in the *fission yield curve* of Fig. 14.2.

(*i*) *light* group with mass numbers from 85 to 104 and

(*ii*) *heavy* group with mass numbers from 130 to 149.

The most probable type of fission, which occurs in about 7% of the total cases, gives fission products with mass numbers 95 and 139.

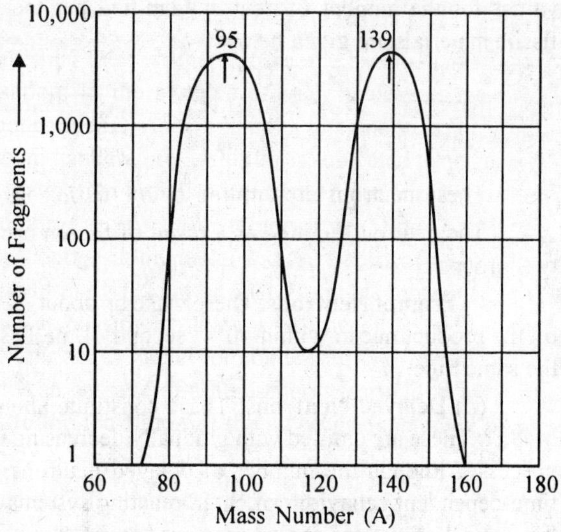

Fig. 14.2

As mentioned earlier, fission fragments have too many neutrons in their nuclei for stability. Consequently, most of them decay by electron emission. Each fragment starts a short radioactive series involving many emissions of β-particles. These series are called *fission decay series* and each chain has three members on the average although longer and shorter chains occur frequently. One such fission decay chain is shown in Fig. 14.3 which starts with one of the unstable fragments of the fission of U^{235} nucleus.

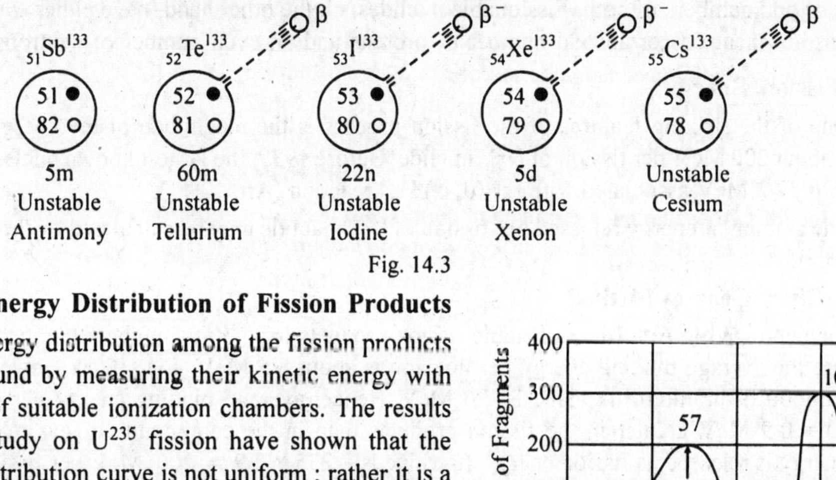

Fig. 14.3

14·5. Energy Distribution of Fission Products

Energy distribution among the fission products can be found by measuring their kinetic energy with the help of suitable ionization chambers. The results of such study on U^{235} fission have shown that the energy distribution curve is not uniform ; rather it is a double-peaked curve with maxima at 67 MeV and 100 MeV. It is seen that while the greatest probability is for a fragment of 100 MeV, the areas under the two peaks which represent the total number of particles in the two groups are approximately equal.

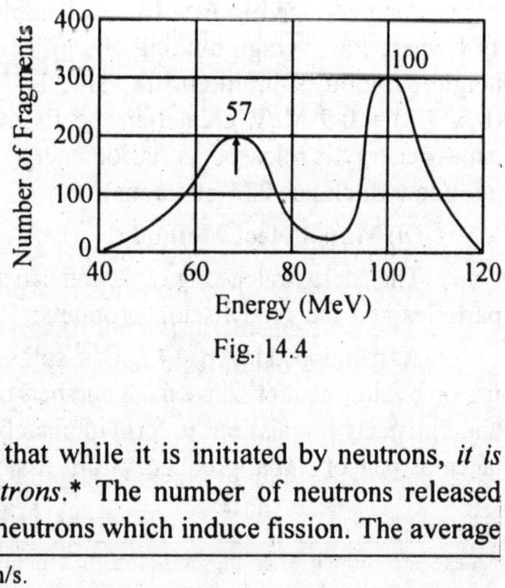

Fig. 14.4

14·6. Neutron Emission in Nuclear Fission

One of the notable features of nuclear fission is that while it is initiated by neutrons, *it is also accompanied by the emission of fast-moving neutrons*.* The number of neutrons released depends on the mode of fission and on the energy of the neutrons which induce fission. The average

*The velocity of these fast fission neutrons is about 16,000 km/s.

values for the number of neutrons emitted per thermal neutron absorbed by the three important fissile materials are given below :

U^{235}	2·43
U^{233}	2·50
Pu^{239}	2·89

These neutrons are emitted *by the fission fragments and not by the compound nucleus.*

The neutrons emitted as a result of fission process (*i.e.* fission neutrons) can be divided into two groups :

(*i*) **Prompt Neutrons.** These make up about 99·36% of the total fission neutrons and are ejected by the product nuclei within 10^{-14} second of the fission process. Prompt γ-rays are also emitted at the same time.

(*ii*) **Delayed Neutrons.** These constitute about 0·64% of the total neutrons from the fission of U^{235}. These are emitted with gradually decreasing intensity for several minutes after actual fission process. Although the number of delayed neutrons is small, they have a strong influence on the time-dependent behaviour of chain-reacting systems based on fission and play an important role in the control of nuclear-fission reactors (Art.15.8).

14·7. Fissile and Fissionable Nuclides

Elements like U^{235}, U^{233} and Pu^{239} undergo fission by neutrons of energy from almost zero upwards. Such nuclei are referred to as *fissile* nuclides. On the other hand, U^{238} and Th^{232} nuclei which have a fission threshold at 1 MeV are said to be *fissionable* nuclides.

In general, fissile nuclides have either an even number of protons and an odd number of neutrons or odd numbers of both. Fissionable nuclides, on the other hand, have either even number of protons and neutrons or an odd number of protons and an even number of neutrons.

14·8. Fission Energy

One of the striking features of the fission process is the magnitude of the energy released which is about 200 MeV per fission of U^{235} nuclide. Before 1939, the largest known nuclear reaction energy was 22·2 MeV associated with Li^6 (d, α) He^4 reaction (Art. 13.11).

The amount of energy released per fission of U^{235} nuclide may be calculated by the following three methods :

(*i*) **Binding-energy Method**

As mentioned in Art. 14.4, all stable fission products have mass numbers in the range 72 to 158 where the average binding energy per nucleon is about 8·5 MeV (Fig. 10.4). However, in the neighbourhood of uranium, its value is 7·6 MeV. Hence, average binding energy per nucleon is (8·5–7·6) = 0·9 MeV greater in the fission products than in the compound nucleus of U^{235}. The excess energy is released as fission energy. Its value is = 235 × 0·9 = 200 MeV per fission of U^{235} nuclide (which has 235 nucleons).

(*ii*) **Mass Defect Method**

The energy released per fission can also be estimated by comparing the mass of the interacting particles and the final fission products.

As mentioned in Art. 14.2, U^{235} splits in many ways and the nuclei obtained in the greatest yield in fission by slow neutrons have mass numbers of 95 and 139. The fission products being initially radioactive, undergo many β-emissions to form ultimately stable nuclides. If molybdenum-95 and lanthanum-139 are taken as pair of *stable* products* from fission of U^{235}, the fission reaction can be written as**

$$_{92}U^{235} + {_0n^1} \longrightarrow [_{92}U^{236}] \longrightarrow {_{42}Mo^{95}} + {_{57}La^{139}} + 2\,{_0n^1}$$

*These are formed after the emission of a number of β-particles.

**β-particles emitted during fission have been neglected.

Comparing masses on both sides of the above equation, we get

mass of U^{235} nuclide = 235·124 amu

mass of one neutron = 1·009 amu

Total = 236·133 amu

mass of Mo^{95} nuclide = 94·946 amu

mass of La^{139} nuclide = 138·955 amu

mass of two neutrons = 2·018 amu

Total = 235·919 amu

Mass defect = 236·133 – 235·919 = 0·214 amu

Therefore, energy released per fission of U^{235} nucleus

= 0·214 × 931 = 200 MeV

(*iii*) **Kinetic Energy Measurement Method**

The total amount of energy released per fission is equal to the sum of the following kinetic energies :

(*a*) the kinetic energy of the fission fragments. As seen from Fig. 14.4, the average value of this energy for U^{235} is 167 MeV.

(*b*) the kinetic energy of fission neutrons. Since the average number of neutrons emitted per fission of U^{235} is 2·43 or say 2·5 and the average kinetic energy of these neutrons is 2 MeV, total kinetic energy of fission neutrons is 2·5 × 2 = 5 MeV.

(*c*) the kinetic energy of prompt γ-rays. Its value is about 7 MeV.

(*d*) total energy of the decay processes in the fission decay chains.

This includes the energy carried away by radiations like β-rays, γ-rays and neutrons. Its value is nearly 21 MeV.

The total of all the above energies is = 167 + 5 + 7 + 21 = 200 MeV

Example 14.1. *A U-235 nucleus is fissioned by a thermal neutron and two fission fragments and two neutrons are produced. Compute the fission energy released if the average binding energy per nucleon is 7·8 MeV in the fissioned U-235 nucleus and 8·6 MeV in the fission fragments.*

Solution. Greater binding energy of the fission fragments indicates that there has been release of energy during fission of low-binding energy nucleus U-235.

Fission energy released is = (234 × 8·6 – 236 × 7·8) = **171·6 MeV**

14·9. Theory of Fission Process

The first attempt to explain the mechanism of fission process was made by Bohr and Wheeler who accounted for many of the properties of fission on the basis of the liquid-drop model of the nucleus.

As discussed earlier in Art. 10.12, in the spherical liquid-drop nucleus, the shape of the drop depends on a balance between the surface tension forces and the Coulombic repulsive forces. The excitation energy given to the drop during the capture of the slow or thermal neutron sets up oscillations within the drop. These oscillations

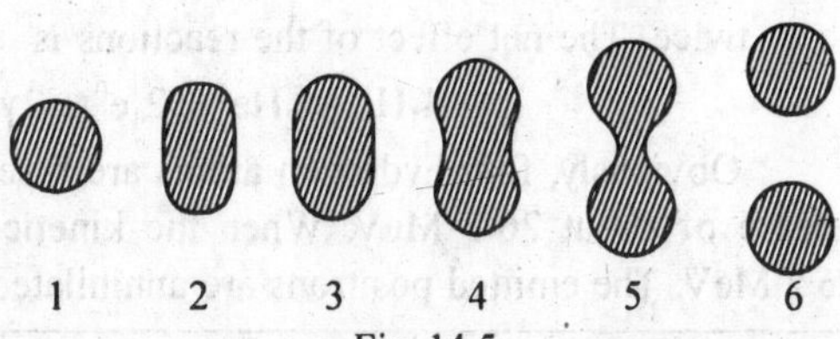

Fig. 14.5

tend to distort the spherical shape so that the drop becomes ellipsoid in shape. The surface tension forces try to make the drop return to its original spherical shape while the excitation energy tends to distort the shape still further. If the excitation energy and hence oscillations are sufficiently large, the drop attains the 'dumbell' shape as shown in Fig. 14.5. The Coulombic repulsive forces then push the two 'bells' further apart until the dumbell splits into two similar drops each of which assumes a spherical shape. The sequence of steps leading to fission is shown in Fig.14.5.

However, if the excitation energy is not large enough, the ellipsoid will return to the spherical shape. In that case, the excitation energy is given out in the form of γ-rays and the process becomes a *radioactive capture process* rather than fission process.

14·10. Nuclear Fusion

It is the process of combining or fusing two lighter nuclei into a stable and heavier nuclide*. In this case also, large amount of energy is released because mass of the product nucleus is *less* than the masses of the two nuclei which are fused.

Many reactions between nuclei of low mass numbers have been brought about by accelerating one or the other nucleus in a suitable manner. These are often fusion processes accompanied by release of energy. However, reactions involving artificially-accelerated particles cannot be regarded as of much significance for the utilization of nuclear energy. To have practical value, fusion reactions must occur in such a manner as to make them self-sustaining, *i.e.* more energy must be released than is consumed in initiating the reaction. It is thought that the energy liberated in the Sun and other stars of the main sequence type is due to the nuclear fusion reactions occurring at the very high stellar temperature of 30 million °K. Such processes are called *thermonuclear reactions* because they are temperature-dependent.

14·11. Stellar Thermonuclear Reactions

Following two sets of thermonuclear reactions have been proposed as sources of energy in the Sun and other stars of the main sequence :

(*i*) proton-proton (p-p) chain and (*ii*) carbon-nitrogen (C-N) cycle.

At low temperatures corresponding to those in the Sun when it was first formed, the proton-proton chain was predominant. In the present state of the Sun with its higher central temperature and larger He^4 concentration, the C-N cycle is supposed to be the main source of its energy.

14·12. Proton-Proton Chain

It is so called because the first step involves the combination of two protons. When two protons fuse, they produce a deuteron nucleus, a positron and a neutrino thus :

$$_1H^1 + {}_1H^1 = {}_1D^2 + {}_1e^0 + \nu \qquad \times 2$$

The deuteron then combines with another proton to yield helium-3.

$$_1D^2 + {}_1H^1 = {}_2He^3 + \gamma \qquad \times 2$$

The two helium-3 nuclei fuse to produce helium-4 thus

$$_2He^3 + {}_2He^3 = {}_2He^4 + 2\,{}_1H^1 + {}_1e^0$$

It should be noted that for the third reaction to occur, each of the first two reactions must occur twice. The net effect of the reactions is

$$4\,{}_1H^1 = {}_2He^4 + 2\,{}_1e^0 + 2\gamma + 2\nu$$

Obviously, four hydrogen atoms are fused to produce one helium atom with a total energy release of about 26·7 MeV. When the kinetic energy of neutrinos is subtracted, the energy is 26·2 MeV. The emitted positrons are annihilated by free electrons with the production of γ-rays.

*In this respect it is just the opposite of nuclear fission process in which one heavier nucleus is split into two lighter nuclei.

14·13. Carbon-Nitrogen Cycle

It was proposed by H.A. Bethe to account for the energy production in the Sun and other stars of main sequence. In this cycle, carbon acts as a nuclear catalyst. The cycle starts when a proton (hydrogen atom) first interacts with carbon-12 nucleus with the releases of fusion energy thus

$$_6C^{12} + {_1H^1} = {_7N^{13}} + \gamma$$

The product N^{13} is known to be radioactive, emitting a positron with a half-life of 10 minutes. Hence, it decays in a very short time according to the relation

$$_7N^{13} = {_6C^{12}} + {_1e^0} + \nu$$

The stable C^{13} nucleus reacts with another proton, thereby liberating more energy.

$$_6C^{13} + {_1H^1} = {_7N^{14}} + \gamma$$

The stable product N^{14} combines with third proton thus

$$_7N^{14} + {_1H^1} = {_8O^{15}} + \gamma$$

The O^{15} nucleus is a positron emitter with a half-life of 2·06 minutes which decays by the process

$$_8O^{15} = {_7N^{15}} + {_1e^0} + \nu$$

Finally, the resulting N^{15} nucleus interacts with the fourth proton thus :

$$_7N^{15} + {_1H^1} = {_6C^{12}} + {_2He^4}$$

By adding up the above six equations and cancelling out those nuclei which appear on both sides, it is seen that four hydrogen atoms are consumed and, in return, 2 positrons, 3 γ-rays and one helium nucleus are created. In other words, hydrogen is burned and helium is created.

The overall process may be written as

$$4\,{_1H^1} = {_2He^4} + 2\,{_1e^0} + 24{\cdot}7 \text{ MeV}$$

The annihilation of positrons supplies an additional energy of 2 MeV so that total energy released is 26·7 MeV.

The fusion energy released may also be found by the loss of mass during the above reaction :

$$4\,{_1H^1} = 4 \times 1{\cdot}008144 = 4{\cdot}032576 \text{ amu}$$

$$_2He^4 = 4{\cdot}003873 \text{ amu}$$

$$2\,{_1e^0} = 2 \times 0{\cdot}000558 = 0{\cdot}001115 \text{ amu}$$

$$\therefore \text{ mass loss} = 4{\cdot}032576 - (4{\cdot}003873 + 0{\cdot}001115)$$

or $$\Delta m = 0{\cdot}028857 \text{ amu}$$

$$\therefore \text{ energy released} = 931 \times 0{\cdot}028857 = 26{\cdot}7 \text{ MeV}$$

It is worth noting that the above energy release is less than that in a nuclear fission. However, its value is 26·7/4 = 6·7 MeV per nucleon as compared to less than 1 MeV per nucleon in fission process.*

14·14. Controlled Thermonuclear Reactions

The fact that nuclear fusion reactions release large amounts of energy, as in stars, has attracted much attention and continuous search is being made for finding practical means for controlled release of such energy. It has, however, been found that reactions of the C-N cycle and proton-proton chain occur two slowly to be of any practical use. Other thermonuclear reactions which occur in nature rapidly and depend on abundant hydrogen isotope like deuterium ($_1D^2$ or $_1H^2$) and tritium ($_1T^3$ or $_1H^3$) and hence seem more stable composition, are as under :

(*i*) $$_1H^2 + {_1H^2} = {_2He^3} + {_0n^1} + 3{\cdot}3 \text{ MeV}$$

or $$_1D^2 + {_1D^2} = {_2He^3} + {_0n^1} + 3{\cdot}3 \text{ MeV}$$

*As seen from Art. 14.8, its value is 200/235 = 0·85 MeV per nucleon.

(*ii*) $_1H^2 + {}_1H^2 = {}_1H^3 + {}_1H^1 + 4\text{ MeV}$

or $_1D^2 + {}_1D^2 = {}_1T^3 + {}_1H^1 + 4\text{ MeV}$

(*iii*) $_1H^2 + {}_1H^3 = {}_2He^4 + {}_0n^1 + 17{\cdot}6\text{ MeV}$

or $_1D^2 + {}_2T^3 = {}_2He^4 + {}_0n^1 + 17{\cdot}6\text{ MeV}$

The most important aspect of the above nuclear fusion reactions is that deuterium is available easily and abundantly. It occurs in nature with an abundance of one part in six thousands of hydrogen and can be separated from the lighter isotope quite cheaply. Five litres of water contain about 1/8 gram of deuterium but its energy content if it could be used as a fuel in a thermonuclear reactor, would be equivalent to 130 litres of petrol! The more than 5×10^{19} kg of water present in the oceans could thus supply world's power requirement for several million years at negligible cost if the deuterium could be utilized to provide energy by fusion reactions. However, as discussed below, there are some difficult problems to be solved before man-made controlled fusion reactors can become a reality.

14·15. Condition for Controlled Fusion

In order to provide useful energy, the fusion process must be self-sustaining. Once the temperature of deuterium (or a mixture of D^2 and T^3) has been raised to the point at which fusion occurs at an appreciable rate, the energy released must be sufficient, at least, to maintain that temperature. The minimum temperature is known as *critical ignition temperature* and may be defined as *that temperature above which the rate of energy production by fusion exceeds the rate of energy loss*. Its value is about 5 keV (*i.e.* 50 million °K) for a D-T reaction and 40 keV (*i.e.* 400 million °K) for a D-D reaction.* At these temperatures, the atoms are entirely stripped of their electrons. The result is a completely ionized gas or plasma consisting of atomic nuclei (like deuterons, tritons and protons) and electrons in rapid random motion. It is practically impossible to contain such plasma in walls of ordinary materials. In the Sun, the fusion reactions are contained by a tremendous gravitational pressure. Such a high pressure is yet not available for controlled thermonuclear reactions on earth although the plasma can be contained in a magnetic field. Hence, main problem is to devise an apparatus in which plasma can be contained by means of a magnetic field at the kinetic temperatures required for the fusion reactions to proceed.

Another necessary condition for a self-sustaining thermonuclear system is known as Lawson criterion. It is based on the requirement that in the operation of a fusion reactor, the total useful *recoverable* energy should be at least sufficient to maintain the temperature of the reacting nuclei. Lawson criterion can be expressed in terms of the product 'nt' where n is the number of reacting nuclei per m^3 and t is the time in seconds in which the thermonuclear reaction takes place. The minimum value of 'nt' for D-T system is 7×10^{19} and D-D system is 2×10^{21}. In controlled thermonuclear reactions, t is taken as the time during which the high-temperature plasma can be confined.

It will be seen from above that both the critical ignition temperature and Lawson criterion are much more favourable for D-T system than for D-D system. But the former system has the drawback that it requires tritium which has to be obtained by nuclear reactions (because it does not occur in nature). Tritium can, however, be made by bombarding lithium with slow neutrons in a reactor thus :

$$_3Li^6 + {}_0n^1 = {}_1H^3 + {}_2He^4 + 4{\cdot}8\text{ MeV}$$

This reaction can be made to serve two useful purposes. In the thermonuclear reactor using D-T reaction, the escaping neutrons carry off much of the energy (about 14 MeV per neutron). This energy can be converted into heat by slowing down these fast neutrons in a blanket of beryllium surrounding the reactor in which plasma is produced. The slow neutrons are captured by lithium which produces tritium. The blanket could thus consist of moderator (*i.e.* beryllium), coolant and lithium. The heat generated by the moderation and absorption of neutrons could thus be transferred by coolant to external heat exchangers and then to turbines which could run alternators.

*In thermonuclear work, the temperatures are expressed in keV which is basically the unit of work. It is equal to kT where k is Boltzmann's constant = $8{\cdot}61 \times 10^{-5}$ eV or $8{\cdot}61 \times 10^{-8}$ keV per degree kelvin and T is temperature in °K. Obviously, a temperature of 1 keV = $1/8{\cdot}61 \times 10^{-8} = 1{\cdot}16 \times 10^7$ °K. Roughly, 1 keV = 10^7 °K and 1 eV = 10^4 °K.

14·16. Hydrogen Bomb

This bomb is 1000 times more powerful than the atomic bomb which is based on nuclear *fission*. Hydrogen bomb is based on the *fusion* of hydrogen atoms into heavier ones by the thermonuclear reactions with release of enormous energy. The essential conditions for the operation of the hydrogen bomb are the extremely high temperatures and pressures required for the fusion to start. Once started, the fusion itself maintains the temperature to keep the process going. For this purpose, the atom bomb (fission bomb) is used as a *primer* which, by first exploding, provides the high temperature and pressure necessary for the successful working of the hydrogen bomb (fusion bomb).

A fusion bomb is superior to a fission bomb because of the following reasons :

(*i*) The energy release in a hydrogen bomb is *open-ended i.e.* it has no upper limit. It depends on how much fusible material is present in the bomb.

(*ii*) It has no limitation of a critical size of the fusible material unlike an atomic bomb. If the active material in an atomic bomb exceeds the critical size, *spontaneous* explosion results. Hydrogen bomb cannot explode unless 'ignited' *i.e.* heated to critical ignition temperature and any amount of fusible material is safe until ignited. Thus the amount of fusible material in a hydrogen bomb is not limited.

14·17. Cobalt Bomb

It consists of a hydrogen bomb which is encased in a sheath of metallic cobalt and is more lethal and destructive than a simple uncased hydrogen bomb. When the hydrogen bomb explodes, it gives off neutrons which act on the cobalt cover and render it intensively radioactive due to the formation of Co^{60} that is 300 times more powerful than radium. During explosion, the radioactive cobalt is pulverised and converted into a gigantic radioactive cloud which can spread over thousands of kilometres killing everything living in that area.

14·18. Fission and Fusion

One thing common between the two nuclear processes is that they release very large amounts of energy. But there are many differences in the mechanisms of the two processes.

(*i*) Fission involves *breaking* up of a heavy nucleus into lighter nuclei. Fusion, on the other hand, involves *combining* of two lighter nuclei into one heavy nucleus.

(*ii*) The links of the fission process are neutrons while the links of a fusion process are protons.

(*iii*) Fission proceeds best with *thermal neutrons* where thermal means room temperature. Fusion proceeds best with *thermal particles* where thermal means temperatures of millions of °K.

Example 14.2. *In neutron-induced binary fission of* ${}_{92}U^{235}$ *(235·044) two stable end-products usually found are* ${}_{42}Mo^{98}$ *(97·905) and* ${}_{54}Xe^{136}$ *(135·917). Assuming that these isotopes have come from the original fission process, find* (*i*) *what elementary particles are released,* (*ii*) *mass defect of the reaction* and (*iii*) *the equivalent energy released*

Solution. (*i*) The reaction can be written as

$${}_0n^1 + {}_{92}U^{235} = {}_{42}Mo^{98} + {}_{54}Xe^{136}$$

It is seen that the total *Z*-value of the two *stable* fission products is (42 + 54) = 96. It is 4 units more than that of ${}_{92}U^{235}$. For balance, the original unstable fission products must have got $Z = 92$. Obviously, the original unstable products must have emitted 4 β-particles before becoming stable. Now, mass number on right-hand side is 2 units less than on the left-hand side. It means that two fission neutrons must have been produced. Hence, the fission reaction can be represented by the following equation :

$${}_0n^1 + {}_{92}U^{235} = {}_{42}Mo^{98} + {}_{54}Xe^{136} + 4\ {}_{-1}e^0 + 2\ {}_0n^1$$

(*ii*)
$$\Delta m = \text{LHS mass} - \text{RHS mass}$$
$$\text{LHS mass} = (1{\cdot}009 + 235{\cdot}044) = 236{\cdot}053 \text{ amu}$$
$$\text{RHS mass} = (97{\cdot}905 + 135{\cdot}917 + 4 \times 0{\cdot}00055 + 2 \times 1{\cdot}009)$$
$$= 235{\cdot}842 \text{ amu}$$
$$\therefore \quad \Delta m = 236{\cdot}053 - 235{\cdot}842 = \mathbf{0{\cdot}211 \text{ amu}}$$

(*iii*)
$$\text{Energy released} = 0{\cdot}211 \times 931 = \mathbf{196 \text{ MeV}}$$

Example 14.3. *The masses of $_1H^1$ and $_2He^4$ atoms are 1·00813 amu and 4·00386 amu respectively. How much hydrogen must be converted to helium if solar constant is 1·35 kW/m² and the earth is 1·5 × 10¹¹ m from the Sun.*

Solution. The source of energy in the Sun is the proton-proton chain as discussed in Art. 14.12. This thermonuclear reaction may be written, in its essentials, as

$$4\ {}_1H^1 = {}_2He^4 + 2\ {}_1e^0$$

Neglecting the two positrons, it is seen that 4 hydrogen atoms fuse to produce one atom of helium.

$$\text{Mass of 4 hydrogen atoms} = 4 \times 1{\cdot}00813 = 4{\cdot}03252$$
$$\text{Mass of one helium atom} = 4{\cdot}00386 \text{ amu}$$
$$\text{decrease in mass, } \Delta m = 4{\cdot}03252 - 4{\cdot}00386$$
$$= 0{\cdot}02866 \text{ amu}$$
$$\text{energy produced} = 0{\cdot}02866 \times 931 = 26{\cdot}68 \text{ MeV}$$

This is the energy released when four hydrogen atoms fuse. Hence, energy produced by one hydrogen atom

$$= 26{\cdot}68/4 = 6{\cdot}67 \text{ MeV} = 6{\cdot}67 \times 1{\cdot}6 \times 10^{-13} = 10{\cdot}67 \times 10^{-12} \text{ J}$$

Energy released by 1 kg of hydrogen

$$= 6{\cdot}02 \times 10^{26} \times 10{\cdot}67 \times 10^{-12} = 6{\cdot}42 \times 10^{14} \text{ J}$$

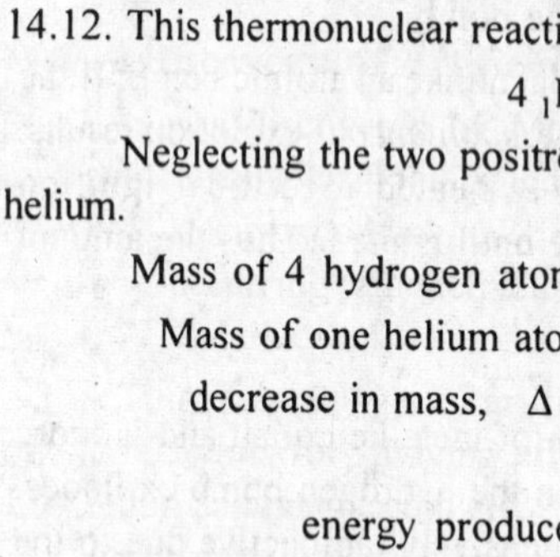

Fig. 14.6

The solar constant represents the amount of energy received per second by 1 m² area held perpendicular to the Sun's rays at a distance equal to the mean distance of the earth from the Sun.

Given value of solar constant

$$= 1{\cdot}35 \text{ kW/m}^2$$
$$= 1{\cdot}35 \times 1000 \text{ W/m}^2$$
$$= 1350 \text{ J/s/m}^2$$

Total energy emitted by the Sun is equal to the energy received by the inner surface of the imaginary sphere drawn with Sun as centre and radius = 1·5 × 10¹¹ m (Fig. 14.6)

Surface area of the sphere

$$= 4\pi R^2 = 4\pi \times (1{\cdot}5 \times 10^{11})^2 = 28{\cdot}28 \times 10^{22} \text{ m}^2$$

∴ energy received by this surface area per second is

$$= 1350 \times 28{\cdot}28 \times 10^{22} = 38{\cdot}18 \times 10^{25} \text{ J/s}$$

It also represents the energy emitted by the Sun per second.

Mass of hydrogen consumed is

$$= \frac{38{\cdot}18 \times 10^{25}}{6{\cdot}42 \times 10^{14}} = 5{\cdot}95 \times 10^{11} \text{ kg/s} = \mathbf{5{\cdot}95 \times 10^8 \text{ tonnes/second}}$$

Example 14.4. *Calculate the energy liberated when a helium nucleus is formed by the fusion of two deuterium nuclei. The mass of* $_1H^2$ *= 2·01478 a.m.u. and mass of* $_2He^4$ *= 4·00388 a.m.u.*

Solution. The reaction may be written as

$$_1H^2 + {}_1H^2 = {}_2He^4 + Q$$

$$\therefore \quad 2{\cdot}01478 + 2{\cdot}01478 = 4{\cdot}00388 + Q$$

$$\therefore \quad Q = 0{\cdot}02568 \text{ a.m.u.} = 0{\cdot}02568 \times 931 = \mathbf{23{\cdot}9\ MeV.}$$

TUTORIAL PROBLEMS

1. What would be the energy liberated if four hydrogen nuclei are fused to form one helium atom. Express it in eV and joules. Atomic mass of helium atom = 4·0039 amu, that of hydrogen molecules = 1·0076 amu. **[24·67 × 10^6 eV ; 39·5 × 10^{-18} J]**
2. When Pu^{239} is fissioned, the products obtained are Ce^{140} and Ru^{98} and two neutrons. The masses of Pu^{239}, Ce^{140} and Ru^{98} are 239·125 amu, 139·958 amu and 97·942 amu respectively. Mass of neutron is 1·009 amu. Calculate the energy released per fission. How much energy is liberated if one gram of Pu^{239} is fissioned?

HIGHLIGHTS

1. The division of a nuclide into two approximately equal parts when hit by neutron is called nuclear fission.
2. The fission fragments are initially radioactive and ultimately become stable after emitting a few β-particles.
3. Fissile nuclides are those which undergo fission with neutrons of energy ranging from almost zero upwards. Examples are : U^{235}, U^{233} and Pu^{239}.

   Fissionable nuclides are those which have fission threshold at 1 MeV. Examples are : U^{238} and Th^{232}.
4. Fission energy per U^{235} nuclide is nearly 200 MeV and per nucleon = 200/235 = 0·85 MeV.
5. Nuclear fusion means fusing of two lighter nuclei into one stable and heavier nuclide. The energy released is called fusion energy and its value is nearly 6·7 MeV/nucleon.

Can You Answer This ?

1. Explain the terms : fission and fusion.
2. Write short notes on :— Nuclear fission and nuclear reactor.
3. Explain the phenomenon of nuclear fission. How is the phenomenon used?
4. Describe nuclear fusion. How does it differ from nuclear fission?
5. Write short notes on nuclear fission and fusion.
6. Distinguish between nuclear fission and nuclear fusion.
7. Write a short note on nuclear fission.
8. Write a short note on nuclear fusion.
9. Describe the proton-proton cycle in nuclear fusion.
10. Describe the important features of the phenomenon of nuclear fission.
11. Write a short note on fusion energy.

NUCLEAR ENERGY SOURCES 15

Nuclear Fission as a Source of Energy — A Chain Reaction — Explosive Chain Reaction and Critical Size — Chain Reaction with Natural Uranium — Power of a Nuclear Reactor — Neutron Balance in a Reactor — Effective Multiplication Factor of Thermal, Natural-Uranium Reactor — Simple Derivation of the Multiplying Factor — Heavy Water — Nuclear Reactors and Their Classifications — Essentials of a Nuclear Reactor — Power Reactors — Pressurised Water Reactors — Essentials of a Pressurised Water Reactor — Fast Breeder Reactor — Highlights — Can You Answer This ?

15·1. Nuclear Fission as a Source of Energy

Nuclear fission can be employed as a source of energy due to the following reasons :

(*i*) Firstly, large amount of energy is released by the fission of each U^{235} nucleus. Its value is about 200 MeV*.

(*ii*) Secondly, the fission process is accompanied by liberation of neutrons which are capable of causing fission of other U^{235} nuclei. The emission, on the average, of 2·5 neutrons in the fission of a U^{235} nucleus permits a self-sustaining chain reaction in which these neutrons produce more fission and more fission produces more neutrons and so on. The energy release can be either very fast resulting in a violent explosion (as in an atomic bomb) or it can be controlled and maintained at a steady rate (as in chain reacting nuclear piles or nuclear reactors).

Before proceeding further, it would be interesting to calculate the amount of energy or power involved in nuclear fission in terms of units more familiar to an engineer.

(*a*) **Nuclear Energy.** Let us calculate the amount of energy released by the complete fission of 1 kg of uranium.

Energy released by the fission of one U^{235} nucleus is 200 MeV $= 200 \times 1{\cdot}6 \times 10^{-13} = 3{\cdot}2 \times 10^{-11}$ joule. Now, the number of individual atoms (or nuclei) is one kg-atom** of uranium (or for that matter, of any element) is equal to Avogadro's number which is $= 6{\cdot}02 \times 10^{26}$/kg-atom. Hence, energy released by the fission of one kg-atom of uranium $= 3{\cdot}2 \times 10^{-11} \times 6{\cdot}02 \times 10^{26} = 1{\cdot}93 \times 10^{16}$ joules. It is the amount of energy which would be released by the fission of all nuclei contained in one kg atom *i.e.* in 235 kg of U^{235}.

Hence, energy released per kg of uranium-235 is $= 1{\cdot}93 \times 10^{16}/235 = 8{\cdot}2 \times 10^{13}$ J.

Since 1 kWh $= 36 \times 10^5$ J, energy released per kg of U^{235}

$= 8{\cdot}2 \times 10^{13}/36 \times 10^5 = 2{\cdot}3 \times 10^7$ kWh

Now, 1 kW-day = 24 kWh. Hence, energy released per kg of $U^{235} = 2{\cdot}3 \times 10^7/24 \simeq 10^6$ kW-day = 1000 MW-day. Hence, one kg of uranium, on complete fission, can supply 1000 MW of power for one day.

(*b*) **Heat Equivalence**

Let us find the heat equivalent of the above energy release :

$$E = 8{\cdot}2 \times 10^{13} \text{ J per kg of } U^{235}$$
$$= 2 \times 10^{10} \text{ kcal per kg of } U^{235}$$

*This energy is about 50 million times as great as that released per atom in the combination of hydrogen which is the primary heat-producing constituent of coal or oil.

**kg-atom is the unit of mass. One kg-atom of an element represents its mass which is equal to its atomic weight (or atomic mass number) expressed in kg. For U^{235}, 1 kg atom = 235 kg ; for Pu^{239}, 1 kg-atom = 239 kg and so on.

This is roughly equal to the energy released by 20 kilotonnes of TNT.

Also, if the above heat could be changed into electricity with a conversion efficiency of 30 per cent, the electric power generated would be = 1000 × 0·3 MW = 300 MW. This is equal to the output of a thermal power plant which consumes about 3000 tonnes of coal per day! Obviously, one kg of fissile material like U^{235} is equivalent to 3000 tonnes = 3×10^6 kg of coal as a source of electric power!

In general, for any nuclear fuel,

$$E = 1{\cdot}93 \times 10^{16}/A \text{ joules/kg}$$
$$= 5{\cdot}36 \times 10^{9}/A \text{ kWh/kg}$$

where A is the atomic mass number (or atomic weight) of the element undergoing fission.

(*c*) **Nuclear Power**

As discussed above, the fission of one U^{235} nucleus releases $3{\cdot}2 \times 10^{-11}$ J or W-s. Hence, fission rate of U^{235} for producing one watt of electric power = $1/3{\cdot}2 \times 10^{-11} = 3{\cdot}1 \times 10^{10}$ fissions/second.

Since large amount of fission energy is released in an extremely short interval of time (of the order of a few microseconds), its production leads to a very powerful explosion unless its rate of production is controlled.

15·2. Chain Reaction

A chain reaction is that process in which the number of neutrons keeps on multiplying rapidly (in geometrical progression) during fission till whole of the fissionable material is disintegrated.

The chain reaction will become self-sustaining or self-propagating only if, for every neutron absorbed, at least one fission neutron becomes available for causing fission of another nucleus. This condition can be conveniently expressed in terms of a *multiplication factor* or *reproduction factor* of the system which may be defined as

$$k = \frac{\text{No. of neutrons in any particular generation}}{\text{No. of neutrons in the preceding generation}}$$

If $k > 1$, chain reaction will continue and if $k < 1$, chain reaction cannot be maintained. For example, if a particular generation starts with 100 neutrons and $k = 1$, there will be 100 neutrons at the beginning of second generation, also 100 at the beginning of third generation and so on. In this way, once started, the fission will continue at the same rate as at the commencement.

However, if any particular generation starts with 100 neutrons and has $k = 0{\cdot}95$, the number of neutrons at the beginning of the second generation would be 95 and will keep on decreasing

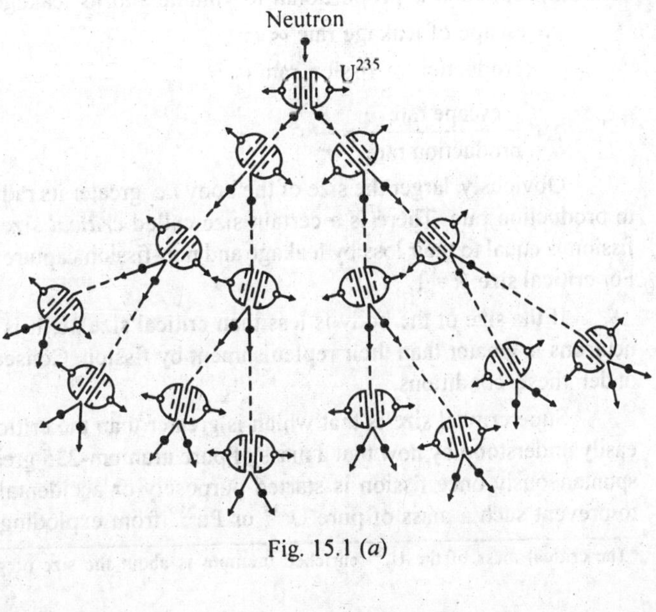

Fig. 15.1 (*a*)

exponentially for other generations till, after 92 generations, the number would be reduced to one neutron ! Obviously, under these circumstances, chain reaction cannot be maintained.

On the other hand, if $k = 1{\cdot}05$ and the initial number of neutrons is again 100, it would grow to 105 at the beginning of second generation and will keep on increasing from one generation to the next till after 100 generations, the number will touch the figure of 15,000 ! In order to prevent such a chain reaction to go out of control, some suitable neutron absorber (like cadmium rods) will have to be introduced into the system.

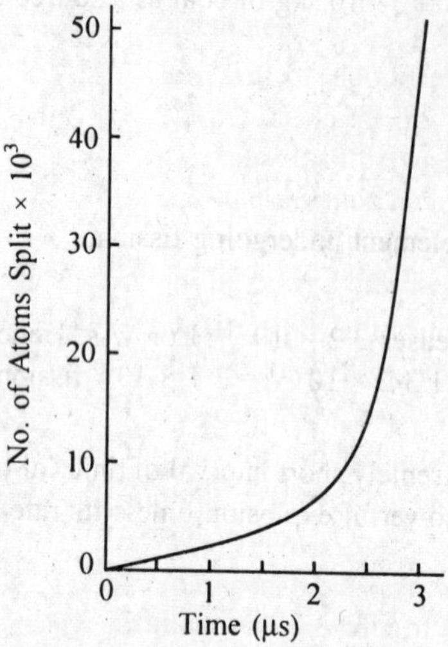

Fig. 15.1 (*b*)

Suppose, we take a certain mass of *pure* U^{235} and initiate fission by one neutron supplied either by cosmic rays or some radioactive source. Further, suppose, for simplicity, that for each U^{235} nucleus suffering fission, two neutrons are liberated. If each of these causes fission with the release of two neutrons in each case, there would become available four neutrons for further fission. These would induce fission in four more U^{235} nuclei accompanied by the emission of 8 neutrons and so on. This chain will continue till no more fissile nuclei remain. In this way, a chain reaction shown schematically in Fig. 15.1 (*a*) is set off which ultimately leads to a rapidly growing avalanche having the characteristic of an explosion. The rate of growth of the chain process is shown in Fig. 15.1 (*b*).

15·3. Explosive Chain Reaction and Critical Size

For a mass of fissile material like pure U^{235} or Pu^{239} to become explosive, two conditions are necessary :

(*i*) the time between fission events must be extremely small and

(*ii*) there should be very small loss of neutrons either by leakage from the surface of the material or by non-fission capture.

The concept of critical size results from the fact that whereas fission occurs throughout the *volume* of the reacting body, neutron leakage or escape takes place from its *surface*. In other words, neutron production is proportional to volume and its leakage to surface area.

$\therefore$ escape of leakage rate $\propto r^2$

production of fission rate $\propto r^3$

$$\therefore \quad \frac{\text{escape rate}}{\text{production rate}} \propto \frac{1}{r}$$

Obviously, larger the size of the body *i.e.* greater its radius, lesser the escape rate as compared to production rate. There is a certain size called *critical* size for which production of neutrons by fission is equal to their loss by leakage and non-fission capture and the chain reaction is *just* possible. For critical size, $k = 1$.

If the size of the body is less than critical size *i.e.* it is *subcritical* so that $k < 1$, then loss of neutrons is greater than their replenishment by fission. Consequently, chain reaction is not possible under these conditions.

Supercritical size is that which is greater than the critical size and for which $k > 1$. It can be easily understood by now that a mass of pure uranium-235 greater than its critical mass will explode spontaneously once fission is started purposely or accidentally by a stray neutron. The only way to prevent such a mass of pure U^{235} or Pu^{239} from exploding is to keep it at *subcritical size**.

*The critical mass of the U^{235} enriched uranium is about the size of grapefruit.

In an atomic bomb, detonation is achieved by bringing together two or more subcritical masses of either U^{235} or Pu^{239} into a supercritical mass. As shown in Fig. 15.2, three or more subcritical masses can be suitably arranged at short distances from each other such that chain reaction can be developed. By suddenly firing explosives like TNT lying behind each block, they can be driven towards each other so as to form one large mass which becomes supercritical. Neutrons entering this supercritical mass from a Ra-Be source *S* initiate a rapid chain reaction which results in a violent explosion with tremendous release of energy.

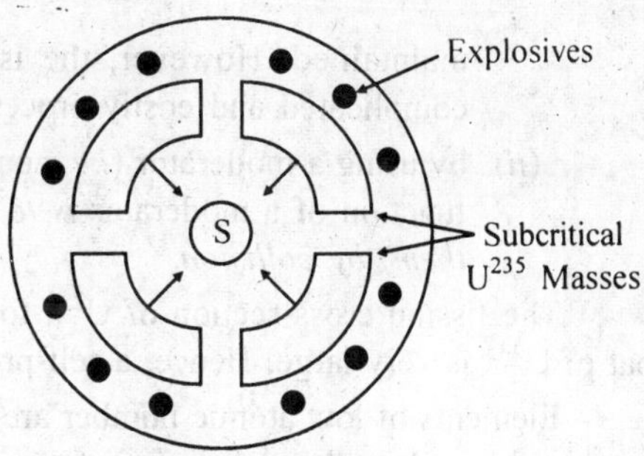

Fig. 15.2

15·4. Chain Reaction with Natural Uranium

Uranium metal, as it is usually refined, has the following isotopic abundance :

U^{238}	99·280%
U^{235}	0·714%
U^{234}	0·006%

Out of these, both U^{238} and U^{235} undergo fission with fast neutrons. Let us consider the possibility of a chain reaction in a mass of *natural* uranium which is so large that leakage of neutrons from its surface is practically zero. If some spontaneous fission occurs, it will release about 2·5 fast neutrons per fission with an average energy of 2 MeV. The fission cross-sections of the two isotopes for these fast neutrons are almost the same. Hence, if fission neutrons cause further fissions, these will chiefly occur in U^{238} (because of its much greater abundance) although there will be a negligible amount of fast fission with U^{235}. Hence, it is only necessary to consider the interaction of the initial fission neutrons with U^{238}. The neutrons can suffer (*i*) fission, (*ii*) radiative capture and (*iii*) elastic or inelastic scattering. Since the two last mentioned processes are more probable, the original fission neutrons are only slightly increased in number by fast fission of U^{238}. Since most of the neutrons are captured before they can reach thermal energies by elastic collisions, chain reaction is not possible in *natural* uranium.

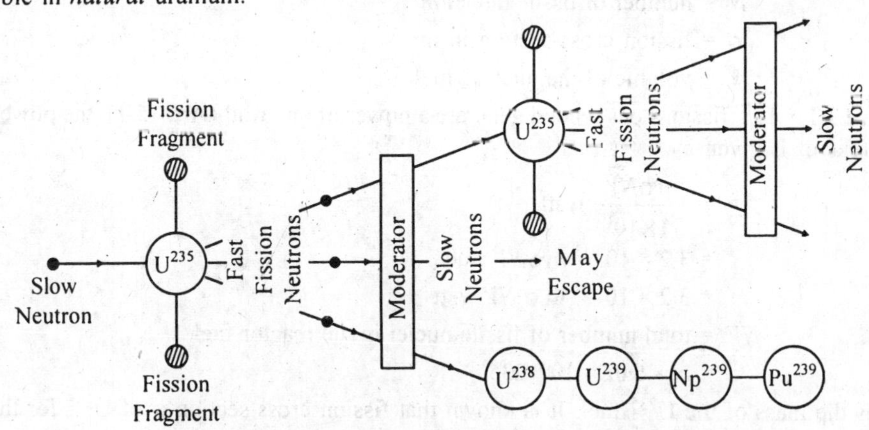

Fig. 15.3

A chain reaction in *natural* uranium can be made possible by the following two methods :

(*i*) by removing U^{238} from natural uranium so that only pure U^{235} is left behind. In that case, neutrons of any energy can cause fission of U^{235} and a chain reaction can be maintained. However, the isotopic separation of U^{238} and U^{235} is an extremely complicated and costly process.

(*ii*) by using a moderator (*i.e.* neutron slowing material) along with natural uranium. The function of a moderator *is to slow down the fast fission neutrons* (*i.e. to thermalise them*) *by collision.*

The fission cross-section of U^{238} for slow or thermal neutrons is practically zero, whereas that of U^{235} is very large. Hence, a self-propagating chain reaction becomes possible.

Elements of low atomic number are particularly suited to serve as moderators because they considerably reduce the speed of neutrons at each collision. Heavy water (D_2O) and graphite have been used successfully as moderators with natural uranium because apart from being light, they have low cross-section for neutron absorption. When D_2O is the moderator, uranium can be dissolved in it. This type of reactor is known as *homogeneous* reactor. When graphite is used as a moderator, uranium is in the form of rods which are distributed in a regular way throughout the graphite thereby forming a lattice. This kind of reactor in which the fuel and moderator are separated is called *heterogeneous.*

Another additional advantage of the second method is that another nuclear fuel Pu^{239} is produced in a *non-fission capture* of neutrons by U^{238} as illustrated by the following equation :

$$_{92}U^{238} + {}_0n^1 \longrightarrow {}_{92}U^{238} \xrightarrow[23\text{ min}]{\beta-1} {}_{93}Np^{239} \xrightarrow{\beta-1} {}_{94}Pu^{239}$$

When pure, Pu^{239} is as good (if not better) a nuclear fuel as pure U^{235}.

15·5. Power of a Nuclear Reactor

The fission rate of a reactor *i.e.* total number of nuclei undergoing fission per second in a reactor is

$$= n\,v\,\sigma\,NV = \Phi\,\sigma\,NV$$

where n = average neutron density *i.e.* number per m^3

v = average speed in m/s

$\Phi = nv$ = average neutron flux

N = number of fissile nuclei/m^2

σ = fission cross-section in m^2

V = volume of the nuclear fuel

Since $3{\cdot}1 \times 10^{10}$ fissions per second generate a power of one watt (Art. 15.1), the power P of a nuclear reactor is given by

$$P = \frac{nv\sigma NV}{3{\cdot}1\times10^{10}} \text{ watt}$$

$$= 3{\cdot}2 \times 10^{-11}\, nv\sigma NV \text{ watt}$$

$$= 3{\cdot}2 \times 10^{-11}\, \Phi\,\sigma\, NV \text{ watt}$$

Now NV = total number of fissile nuclei in the reactor fuel

$$= m \times 6{\cdot}02 \times 10^{26}/235$$

where m is the mass of the U^{235} fuel. It is known that fission cross-section σ of U^{235} for thermal neutrons is 582 barns = 582×10^{-28} m^2

$$\therefore \quad P = \frac{3{\cdot}2\times10^{-11}\times\Phi\times582\times10^{-28}\times m\times6{\cdot}02\times10^{26}}{235}$$

$$= 4{\cdot}77 \times 10^{-12}\, m\,\Phi \text{ watt}$$

$$= 4{\cdot}8 \times 10^{-12}\, mnv \text{ watt}$$

15·6. Neutron Balance in a Reactor

Let us consider a typical neutron balance in a reactor using U-235 as fuel and with multiplication factor $k = 1$ assuming that each fission event produces an average of 2·5 neutrons. Suppose we start the fission process with 100 slow neutrons which are absorbed in U-235 and cause fission. The neutron balance is as under :

100 slow neutrons absorbed in U-235 and cause fission
↓
250 fast fission neutrons produced in the core.
→ 20 fast neutrons escape from the core
→ 96 fast neutrons are absorbed in U-235 during slowing down (without producing fission)
↓
140 neutrons slow down
→ 10 slow neutrons may escape from the core
↓
130 neutrons are absorbed in core
→ 30 neutrons are absorbed in U-235 (without fission) and in moderator, coolant etc.
↓
100 slow neutrons are absorbed in U-235 and cause new fission

It will be seen that at the end of the process, we obtain as many neutrons as we had initially. With such a neutron balance, the reaction is self-sustaining.

15·7. Effective Multiplication Factor of Thermal, Natural-Uranium Reactor

It can be found by considering the neutron balance in such a reactor, in terms of a cycle that starts with the fission of a U^{235} nucleus by a slow (or thermal) neutron as shown in Fig. 15.4. Suppose, ν fast fission neutrons are produced in this process. Since their average energy is above the fission threshold for U^{238}, some of them can cause fission of U^{238} nuclei. However, the chance for such fission will depend on whether fission neutrons can collide with U^{238} nuclei before colliding with moderating nuclei. Some of the neutrons that do produce fast fission in U^{238} produce additional neutrons. Hence, total number of fast neutrons is increased from ν to $\nu\in$ where $\in$ is called *fast fission* factor whose value varies from unity for a homogeneous reactor to 1·1 in others.

$$\in = \frac{\text{total No. of neutrons produced by both thermal and fast fission}}{\text{No. of neutrons produced by thermal fission}}$$

The $\nu\in$ neutrons diffuse through the pile but a fraction n_1 escapes before being slowed down to thermal energy. So far as the reactor is concerned, $\nu\in n_f$ neutrons, are lost. The remaining $\nu\in (1 - n_f)$ neutrons are slowed down by the moderator. But, during this slowing down, some of them may suffer resonance capture by U^{238} nuclei which form U^{239}, then Np^{239} and finally Pu^{239}. If p is the *resonance escape probability*, then $\nu\in (1 - n_f)\, p$ neutrons escape capture while $\nu\in (1 - n_f)(1 - p)$ neutrons are captured and, as shown, are used to form Pu^{239}.

Some of the thermal neutrons diffuse without being captured and ultimately escape from the system. If the fraction escaping is n_t, then the number of thermal neutrons per fission of U^{235} nucleus that escape is = $\nu\in (1 - n_f)\, p n_t$. Of the remaining $\nu\in (1 - n_f)\, p\, (1 - n_t)$ neutrons, a fraction f is absorbed in uranium and the fraction $(1 - f)$ is absorbed in other materials. Obviously, the number of neutrons still available for carrying on the chain is = $\nu\in (1 - n_f)\, p\, (1 - n_t)\, f$. The quantity f is known as *thermal utilization factor.* Now, not all the neutrons absorbed in uranium cause fission of U^{235}. Some are absorbed in U^{238} which eventually form Pu^{239} while some others are absorbed in U^{235} to form U^{236}. The fraction of the thermal neutrons absorved in uranium that causes fission of U^{235} is just $\sigma_f(U)/\sigma_a(U)$. Hence, multiplication factor is given by

$$k = \nu\in (1 - n_f)\, p\, (1 - n_t)\, f \frac{\sigma_f(U)}{\sigma_a(U)}$$

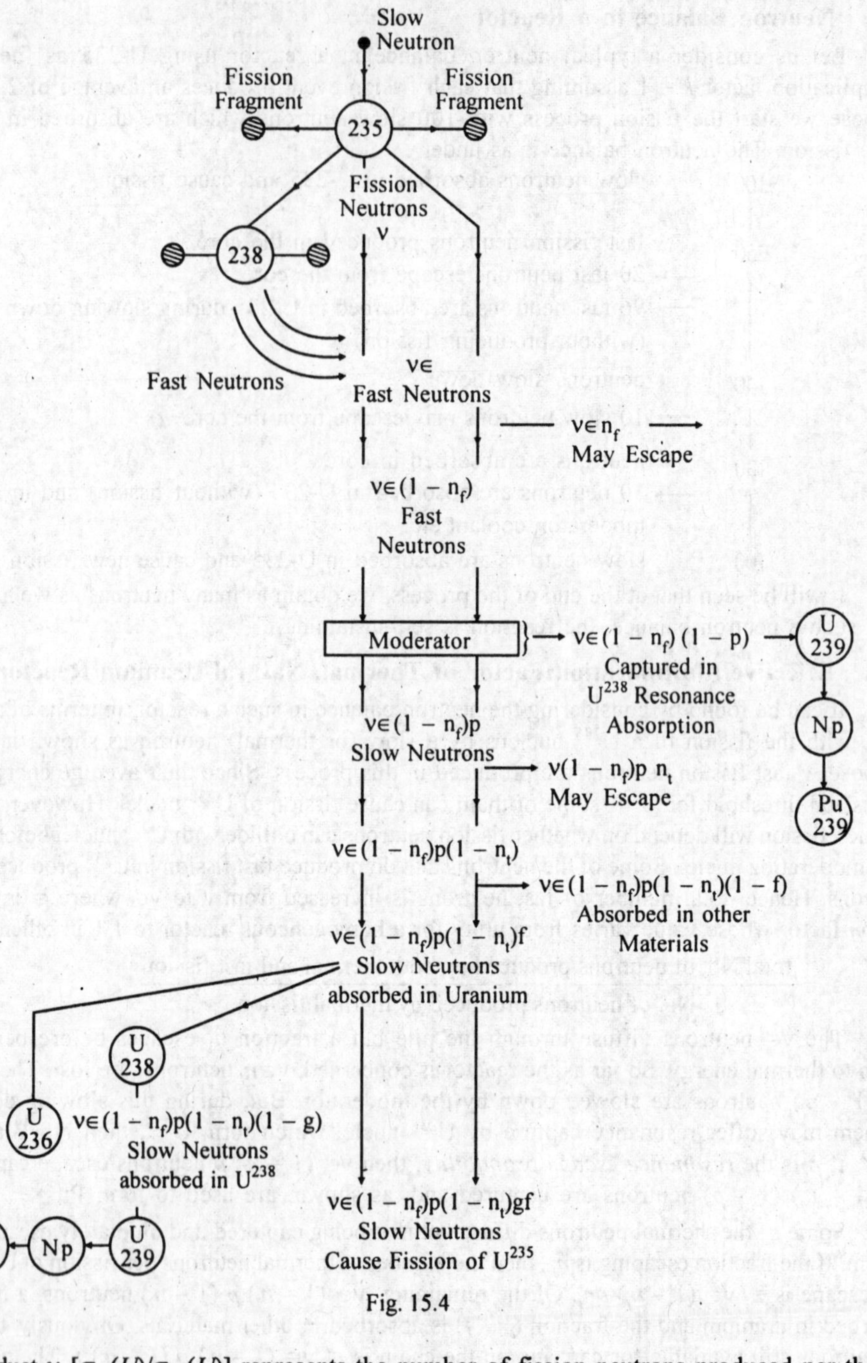

Fig. 15.4

Now, product $\nu\,[\sigma_f\,(U)/\sigma_a\,(U)]$ represents the number of fission neutrons produced per thermal neutron absorbed in uranium and is denoted by η.

$$\therefore \qquad k = \eta \epsilon p f (1 - n_f)(1 - n_t)$$

The factor k as given above is called the *effective* multiplication factor and corresponds to an assembly of finite size. The system is critical for $k = 1$, subcritical for $k < 1$ and supercritical for $k > 1$.

If the body of the material is of infinite size, then there can be no leakage of neutrons from such a system so that both n_f and n_t are zero. Then, multiplication factor is denoted by k_∞ and is called *k-infinity* or infinity (k)

$$\therefore \quad k_\infty = \eta \in p f$$

The above formula is known as *four-factor* formula and is an important factor in the theory and design of a nuclear reactor.

In case the fuel contains only U^{238} nuclei, then the fast fission factor $\in$ and the resonance escape probability p are both nearly unity. In that case, $k_\infty = \eta f$.

15·8. Simple Derivation of the Multiplying Factor

The multiplying factor of an *infinitely large* reactor in which *no leakage of neutrons* takes place can be derived as follows :

Suppose a single neutron initially produces ν neutrons by thermal neutron fission of U^{235}. This number will increase to $\nu\in$ before these neutrons are slowed down in the moderator. The *fast fission factor* ($\in$) is defined as

$$\in = \frac{\nu \in}{\nu} = \frac{\text{total No. of neutrons produced by both thermal \& fast fission}}{\text{No. of neutrons produced by thermal fission}}$$

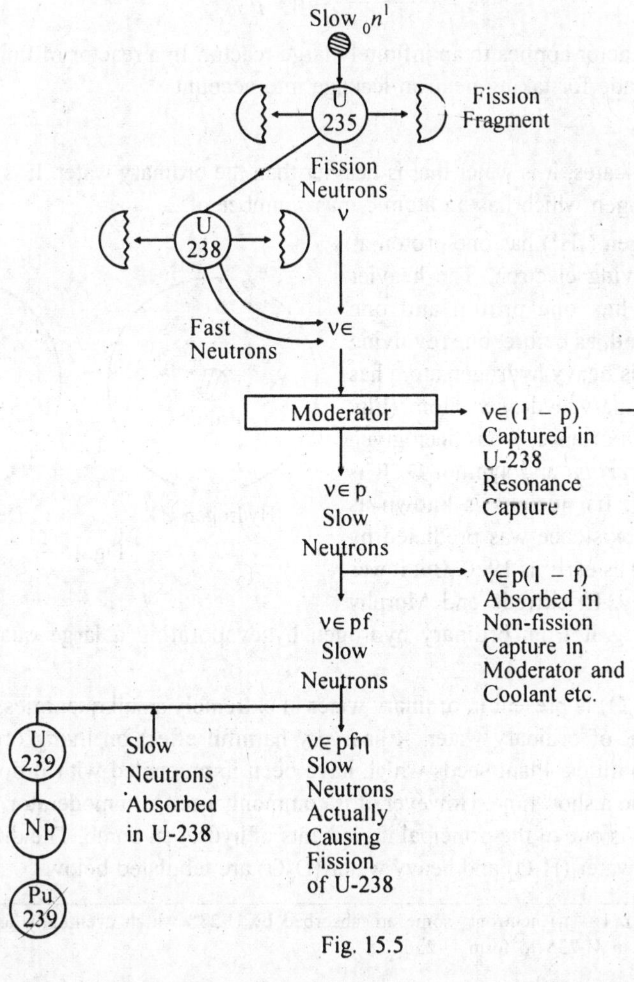

Fig. 15.5

It should be remembered that in reactor using natural or slightly-enriched fuel, there is fast neutron fission of not only the U-235 atoms but also of some of the U-238 atoms.

The *resonance escape probability* (p) represents the probability of a neutron being slowed down to thermal energies without being captured by U-238 at intermediate energies where U-238 has high capture possibility. Taking this factor into account, the number of neutrons reaching thermal energies is $\nu \in p$.

Now, thermal utilization factor (f) is given by

$$f = \frac{\text{No. of thermal neutrons absorbed in } U^{235} \text{ for fission}}{\text{total No. of thermal neutrons}}$$

Of course, some of the neutrons are absorbed by non-fission capture in the moderator, coolant and other materials in the reactor.

Hence, number of thermal neutrons which are capable of causing further fission becomes $\nu \in pf$.

Finally, if η represents the fraction of thermal neutrons absorbed in uranium that actually cause fission of U-235, then the number of available neutrons becomes $\nu \in pf$.*

Hence, multiplication factor is given by

$$k_\infty = \frac{\nu \in p f \eta}{\nu} = \eta \in p f$$

As stated above, this reactor applies to an infinitely large reactor. In a reactor of finite size, a small correction has to be made for taking neutron leakage into account.

15·9. Heavy Water

As the name indicates, it is water that is heavier than the ordinary water. It is formed from a heavy isotope of hydrogen which has an atomic mass number of 2.

Ordinary hydrogen ($_1H^1$) has one proton as nucleus and one revolving electron. The heavier isotope of hydrogen has one proton and one neutron in its nucleus and, as before, one revolving electron. Obviously, this heavy hydrogen atom has twice the mass of ordinary hydrogen atom (Fig. 15.6). Because of its importance, it has been given a special name of *deuterium* and symbol D. It is written as $_1D^2$ or $_1H^2$. Its nucleus is known as *deuteron*. Its possible existence was predicted by Harkins and Rutherford as early as 1920. But it was only in 1931 that Urey, Brickwede and Murphy succeeded in separating it from ordinary hydrogen by evaporating a large quantity of liquid hydrogen.

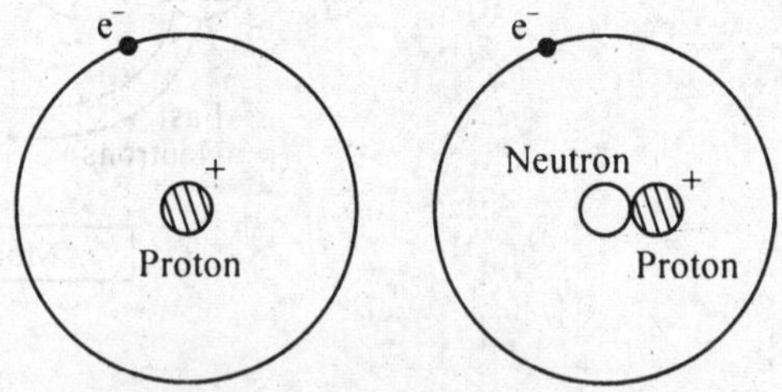

Hydrogen $_1H^1$ Deuterium $_1D^2$

Fig. 15.6

Heavy water (D_2O) is present in ordinary water in extremely small quantities. To be precise, it forms 0·015 per cent of ordinary water. It has very harmful effect on living organisms and is poisonous in large quantities. Plant seeds which have been impregnated with it do not germinate and fish put into it die in a short time. However, it is commonly used as a moderator in certain types of nuclear reactors and is one of the principal ingredients of hydrogen bomb. The differences in the properties of ordinary water (H_2O) and heavy water (D_2O) are tabulated below.

*Out of the remaining $\nu \in p(1 - \eta)$ neutrons, some are absorbed by U-238 which eventually form Pu-239 while some others get absorbed in U-235 to form U-236.

Property	H_2O	D_2O
Freezing point (°C)	0·0	3·82
Boiling temperature (°C)	100	101·4
Maximum density (g/cm^3)	1·0000	1·107
Temperature corresponding to maximum density (°C)	4·0	11·6

15·10. Nuclear Reactors and Their Classification

A nuclear reactor is *an apparatus in which nuclear fission is produced in the form of a controlled self-sustaining chain-reaction.* In other words, it is a controlled chain-reacting system supplying nuclear energy. It may be looked upon as a sort of nuclear furnace which burns fuels like U^{235}, U^{233} or Pu^{239} and, in turn, produces many useful products like heat, neutrons and radioisotopes.

The mechanism of heat production is this : most of the energy is imparted to the two fission fragments into which the nucleus divides causing them to move at high speed. However, because they have taken birth in a dense mass of metal, they are rapidly slowed down and brought to rest by colliding with other atoms of the metal. In so doing, their energy is converted into heat in much the same way as energy given up by a slowing motor car is converted into heat in the brake lining. In this way, the mass of uranium metal gets heated up.

A nuclear reactor can take various forms depending on its neutron spectrum, its construction, composition and the purpose for which it is required. However, different reactors may be differentiated from one another by their following features :—

1. **Neutron Energies**
   (*a*) high-energy —as in fast reactors
   (*b*) intermediate energy
   (*c*) low (thermal) energy —as in thermal reactors
2. **Type of Fuel**
   (*a*) Natural uranium (99·28% U^{238} and 0·741% U^{235})
   (*b*) Enriched uranium (containing higher percentage of U^{235})
   (*c*) U^{233}—obtained from Th^{232}
   (*d*) Pu^{239}—obtained from U^{238}
3. **Type of Fuel-moderator Assembly**
   (*a*) homogeneous—one in which fuel is dissolved in the moderator
   (*b*) heterogeneous—one in which fuel and moderator are separate
4. **Moderator**
   (*a*) heavy water (D_2O)
   (*b*) ordinary water (H_2O)
   (*c*) Graphite
   (*d*) Beryllium or beryllium oxide
5. **Type of Coolant**
   (*a*) gaseous (air, CO_2 etc.)
   (*b*) water or organic liquids
   (*c*) liquid metal

6. The Purpose

(*a*) for research

(*b*) for production of fissile material like Pu^{239}

(*c*) for power production

Depending on which of the above factors are involved, nuclear reactors are generally described as follows :

(*i*) thermal, natural-uranium, graphite-moderated, heterogeneous, air-cooled research reactor

(*ii*) fast, enriched-uranium, heterogeneous, liquid metal-cooled power-breeder reactor etc.

15·11. Essentials of a Nuclear Reactor

The essential compounds of a nuclear reactor are as follows :—

(1) Reactor Core

It is that part of a nuclear power plant where fission chain reaction is made to occur and where fission energy is liberated in the form of heat for operating power conversion equipment.

The *core* of the reactor consists of an assemblage of fuel elements, control rods, coolant and moderator. Reactor cores generally have a shape approximating a right circular cylinder with diameters ranging from 0·5 m to 15 m. The pressure vessel which houses the reactor core is also considered a part of the core (Fig. 15.7).

The fuel elements are made of plates or rods of uranium metal. These plates or rods are usually clad in a thin sheath of stainless steel, zirconium or aluminium to provide corrosion resistance, retention of radioactivity and in some cases, structural support. Enough space is provided between individual plates or rods to allow free passage of the coolant.

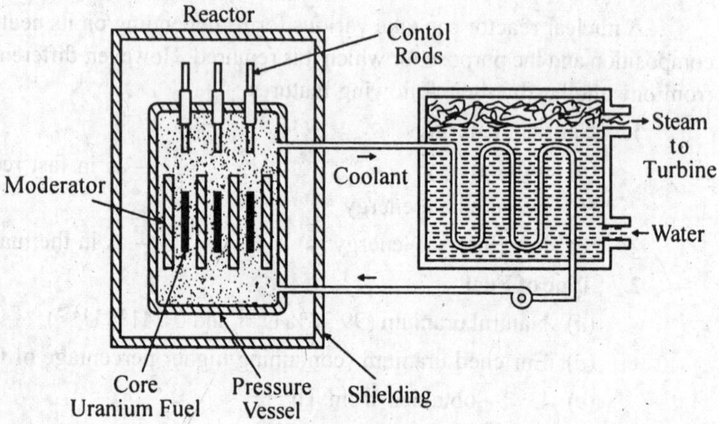

Fig. 15.7

(2) Reflector

A *reflector* is usually placed around the core to reflect back some of the neutrons that leak out from the surface of the core. It is generally made of the same material as the moderator (it is not shown in the figure).

(3) Control Mechanism

It is an essential part of a reactor and serves the following three purposes :—

(*a*) for start up *i.e.* to bring the reactor up to its normal operating level and

(*b*) for maintaining it at that level *i.e.* keep power production at a steady state.

(*c*) for shutting the reactor down under normal or emergency conditions.

The control system is also necessary to prevent the chain reaction from becoming violent and consequently damaging the reactor. The effective multiplication factor of the reactor is always kept greater than unity in order that the number of neutrons keeps on increasing in successive generations. As the number of neutrons and hence the neutron flux density increases, the temperature also increases. Unless the growth is checked at some point, the reactor is likely to be damaged as a result of too rapid liberation of energy.

The control system works on the simple principle of absorbing the excess neutrons with the help of control rods either made of boron steel or cadmium strips. Both these materials have very large cross-section* for thermal neutrons *i.e.* they are very good absorbers of slow neutrons and also have the advantage of not becoming radioactive due to neutron capture. By pushing these rods deeper into the central core, any amount of excess neutrons can be absorbed. Once the reactor has reached predetermined power level, these control rods serve to keep the value of $k = 1$ so that there is no further increase in the number of neutrons from one generation to another. If, at some stage, it is desired to increase the neutron flux density and hence the power level, the rods are partially pulled out thereby allowing k to exceed unity. For shutting down the reactor, the control rods are inserted to a considerable depth so that k becomes less than unity and the chain reaction can no longer be maintained. To start up the reactor, all that is necessary is to carefully withdraw the control rods and then adjust them till required output level is attained.

Due to the production of delayed neutrons in nuclear fission, the neutron density and the power output do not undergo sudden changes either up or down and control by means of neutron absorbers is a relatively simple affair.

Movement of control rods can be manual or made automatic with the help of carefully designed servomechanism.

(4) Moderator

Its function is to thermalise *i.e.* slow down the fast fission neutrons in order to increase their fission cross-section with U^{235}. For a material to act as an efficient moderator, following properties are essential :

(*a*) it should have low mass number in order that maximum neutron energy may be lost per collision with its nuclei. In other words, it should have high slowing-down power.

(*b*) it should have low cross-section for neutron absorption.

The suitability of a moderator is measured in terms of its moderating ratio which is equal to the ratio of its moderating power to its neutron capture coefficient. The higher the moderating ratio, the more suitable is the given substance for slowing down neutrons in a reactor.

The materials such as heavy water (D_2O), graphite, beryllium (or beryllium oxide) are efficient moderators. Hydrogen (in the form of H_2O) has been found useful for some types of reactors in spite of its large neutron absorption coefficient.

As a moderator, D_2O is the best material available, because (*i*) it has excellent neutron slowing properties, (*ii*) it has very small cross-section for neutron capture and (*iii*) it can be used as a coolant as well. The moderating ratio of D_2O is 12,000 as compared to 72 for H_2O and 170 for carbon.

However, it has the following drawbacks :

(*i*) it has low boiling point so that it necessitates pressurisation.

(*ii*) it is very expensive.

But, the advantages of D_2O as moderator (or moderator-coolant) outweigh its high cost.

Beryllium can withstand high temperatures though it is brittle and becomes susceptible to both air and water at those temperatures. Being very expensive, it is used in special cases.

Reactor-grade graphite, though not as good a moderator as Be, is very cheap and can be machined easily into any shape. Unlike other materials, its mechanical strength increases with temperature. However, its disadvantages are that it (*i*) breaks easily when struck and (*ii*) interacts with O_2 (air) and CO_2 at high temperatures.

*The units of cross-section is barn. One barn = 10^{-28} m^2.

(5) **Coolants**

Their function is to remove the intense heat produced in the reactor and to bring it out for being utilized. The materials proposed for coolants can be divided into four general categories :—

(*i*) ordinary and heavy water,

(*ii*) liquid metals,

(*iii*) organic liquids (hydrocarbons), and

(*iv*) gases.

Both ordinary and heavy water are good coolants. In fact, they can serve as both moderator and coolant. However, due to their low boiling point, they need pressurization. Moreover, water becomes very corrosive at high temperatures.

Liquid metals like molten sodium or potassium are excellent coolants for use at high temperatures. However, sodium is highly reactive with O_2 and H_2 and becomes radioactive due to neutron capture. Hence, it needs special shielding. It is usually used in fast reactors.

Organic compounds like polyphenyls have the following advantages :

(*i*) they do not require pressurisation.

(*ii*) they do not become radioactive due to neutron capture

(*iii*) they are not corrosive

(*iv*) they are fairly stable to the action of heat.

However, their heat removing properties are inferior to those of water or liquid metals.

Air cooling with the help of fans, though simple, is suitable only for low-power reactors. At high temperatures, air is not satisfactory as a coolant because it reacts chemically with many materials like graphite, aluminium, beryllium and magnesium etc. used in the construction of the reactor.

The most widely-used gaseous coolant is CO_2 particularly in large power reactors. It is cheap, does not attack metals at reasonable temperatures and has small cross-section for neutron capture.

(6) **Measuring Instruments**

Main instrument required is for the purpose of measuring thermal neutron flux which determines the power developed by the reactor.

(7) **Shielding**

It is an important component of a reactor installation although it plays no role in its operation. Its purpose is to attenuate *i.e.* weaken the γ-rays and neutrons coming out from the reactor to such an extent that they are no longer a hazard to persons in immediate vicinity.

In high-power reactors, there are two shields :

(*i*) *Thermal shield*—It is very close to the reactor core and consists of a thick iron or steel covering. By absorbing most of the γ-rays, it protects the biological shield from overheating.

(*ii*) *Biological shield*—It is a layer of concrete, a few decimetres thick and surrounds the thermal shield and the reactor core. Its function is to absorb the γ-rays and neutrons coming out from the inner shield.

15·12. Power Reactors

Power reactors are those which are primarily intended for the production of usable power. They transform the heat which is produced from the energy released in fission into electricity. The heat is transferred to a fluid passing through the core of the reactor. The hot fluid is then passed through a heat exchanger where energy is transferred and utilized by more or less conventional means.

The basic principle of one type of power reactor is shown in Fig. 15.8. A quantity of enriched uranium in the form of a pure metal forms the centre of the heat energy source. The fission rate and hence the rate of production of heat energy is controlled by the movement of cadmium rods. For decreasing the temperature, the cadmium rods are pushed in a little farther to absorb more neutrons thereby reducing the fission rate. For raising the temperature, they are pulled out a little farther.

Liquid sodium is used as a primary coolant because not only is it a dense material with good thermal conductivity but it also has very low absorption cross-section for thermal neutrons. Sodium is usually alloyed with potassium to lower the melting temperature. The energy absorbed by this primary coolant is transferred to water in a heat exchanger. The steam that is generated drives a turbine which is coupled to an electric generator. The electric power so produced can be used for lighting cities or running factories or for driving ships and submarines or large planes.

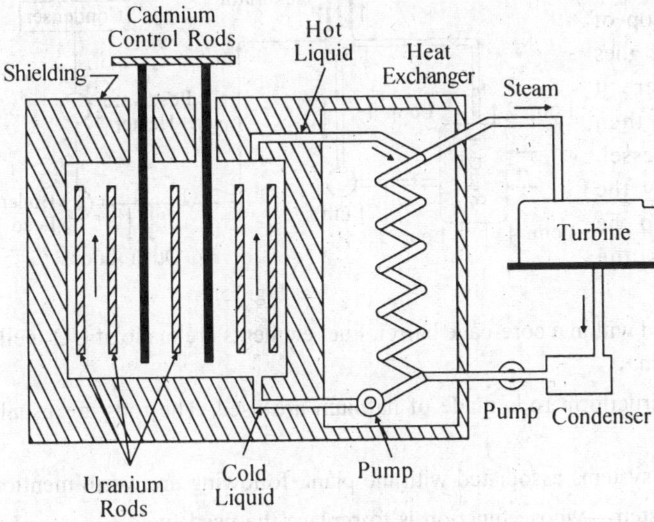

Fig. 15.8

15·13. Pressurised Water Reactors

It may be defined as a reactor in which fission heat is removed from the fuel elements by the coolant without bulk boiling occurring. It implies the use of a two-circuit heat transfer system :

(*i*) a primary loop containing the reactor and one side of a steam generator and

(*ii*) a secondary loop containing the steam side of the steam generator and the turbine-driven alternator.

Pressurised water reactors are of two types :—

(*a*) **Pressure vessel type**—in which the reactor vessel bears the full pressure of the coolant in the primary circuit.

(*b*) **Pressure tube type**—in which the reactor core is pierced by coolant tubes which bear the full coolant pressure while the reactor vessel is at low pressure.

All pressurised water reactors employ a closed primary loop in which water *at high pressure* is circulated from the reactor to a steam boiler and back to the reactor. Since the temperature of the circulating water, if it is to remain as water, cannot be raised above its saturation temperature (about 300°C, depending on pressure) the steam produced in the boiler is at relatively low temperature. Such reactors are very popular because pressurised water technology is well known to engineers and also such reactors have high operating reliability and safety.

15·14. Essentials of a Pressurised Water Reactor

All pressurised water reactors of pressure-vessel type employ high-pressure water both as coolant and moderator with no separation of coolant from moderator. The flow diagram of a typical reactor of this type is shown in Fig. 15.9.

The reactor plant consists of a single reactor with four primary coolant loops. The coolant enters the reactor vessel at its bottom and after being heated in passing through the core, it leaves via outlet nozzles near the top of the vessel. From there, it goes to steam generator where it gives up its heat and then returns to the reactor vessel. The steam produced by the steam generator is fed to the turbines which drives the electric generators.

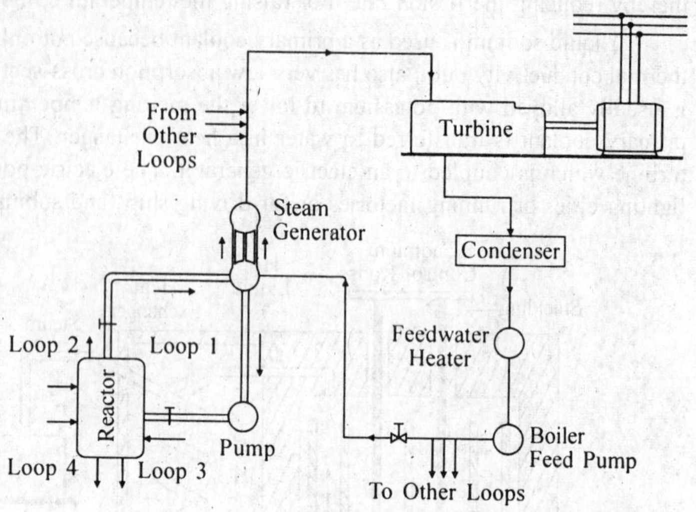

Fig. 15.9

The core is mounted within a core-cage barrel. Fuel elements are made of UO_2 pellets clad in stainless steel or zirconium.

Extended-surface cruciform rods made of hafnium are used. The rods are usually located above the reactor core.

Amongst auxiliary systems associated with the plant, following are worth mentioning :—

(*a*) refuelling system—whose function is to replace the used fuel.

(*b*) water purification system—whose function is to continuously purify the coolant by passing it through suitable ion exchangers.

(*c*) water disposal system—whose function is to remove and dispose of the radioactive wastes from the plant by employing natural decay, concentration, storage and dilution.

15·15. Fast Breeder Reactor

It is a type of nuclear reactor which uses fast fission neutrons to produce more fissile fuel than it consumes. It does not use any moderator for thermalising the neutrons.

This reactor consists of a small core of *natural* uranium in which fast neutrons are allowed to produce a fairly fast chain reaction (but, of course, not as fast as the pure U-235 reaction of the atomic bomb). Neutrons are allowed to escape from the central core into a surrounding blanket of natural uranium where some of them are captured by U-238 nuclei with the consequent formation of plutonium (Art. 15.4). By suitable control of conditions, more Pu^{239} can be produced in the natural uranium blanket than U^{235} that is used up in the reaction. It is for this reason that this type of reactor is called a breeder reactor.

Example 15.1. *Calculate the maximum fraction of the kinetic energy lost by a neutron in elastic collision with the following nuclei at rest :—*

(*a*) H^2 (*b*) C^{12} (*c*) U^{238}

Solution. The maximum fractional loss of neutron energy during elastic collision is

$$\eta = \frac{4\,mM}{(m+M)^2}$$

Percentage loss is

$$\eta = \frac{4\,mM}{(m+M)^2} \times 100$$

where m = mass of neutron

M = mass of the other nucleus

(*a*) Here, $m = 1, \; M = 2$

$$\therefore \quad \eta = \frac{4 \times 1 \times 2}{(1+2)^2} \times 100 = \mathbf{89\%}$$

(*b*) $$\eta = \frac{4 \times 1 \times 12}{(1+12)^2} \times 100 = \mathbf{36\%}$$

(*c*) $$\eta = \frac{4 \times 1 \times 238}{(1+238)^2} \times 100 = \mathbf{1{\cdot}67\%}$$

Example 15.2. *Calculate (i) the fission rate of U^{235} for producing a power of one watt and (ii) the energy released in the complete fissioning of 1 kg of U^{235}.*

Assume 200 MeV are released per fission of the uranium nucleus.

Solution. (*i*) Energy released per fission of U^{235} nuclide

$= 200 \text{ MeV} = 200 \times 1{\cdot}6 \times 10^{-13} \text{ J} = 3{\cdot}2 \times 10^{-11}$ J or W-s

Hence, fission rate for producing one watt of power

$= 1/3{\cdot}2 \times 10^{-11} = \mathbf{3{\cdot}1 \times 10^{10}}$ **fissions/second**

(*ii*) One kg-atom of U^{235} *i.e.* a mass of 235 kg of U^{235} contains $6{\cdot}02 \times 10^{36}$ atoms (or nuclides). Hence, energy released by 1 kg-atom of U^{235} is $= 200 \times 6{\cdot}02 \times 10^{26}$ MeV.

$\therefore$ energy released per kg of U^{235} is

$$= \frac{200 \times 6{\cdot}02 \times 10^{26}}{235} \text{ MeV}$$

$$= \frac{200 \times 6{\cdot}02 \times 10^{26} \times 1{\cdot}6 \times 10^{-13}}{235} \text{ J}$$

$= \mathbf{8{\cdot}2 \times 10^{13}}$ **J**

Example 15.3. *A reactor is developing nuclear energy at the rate of 30,000 kW. How many atoms of U-235 undergo fission per second ? How many kg of U-235 would be used in 1000 hours of operation assuming that 200 MeV of energy is released per fission ?*

Solution. Rate of energy development = 30,000 kW

$= 3 \times 10^7$ J/s

Energy produced/fission = 200 MeV $= 3{\cdot}2 \times 10^{-11}$ J

If n is the number of U-235 atoms which undergo fission per second, then

$$n \times 3{\cdot}2 \times 10^{-11} = 3 \times 10^7$$

$$n = \mathbf{9{\cdot}4 \times 10^{17}}$$

Energy produced in 1000 hours $= 3 \times 10^7 (1000 \times 3600) = 10{\cdot}8 \times 10^{13}$ J

Since energy released per kg of U-235 is $8{\cdot}2 \times 10^3$ J (Art. 15.1), hence mass of U-235 consumed is

$$= 10{\cdot}8 \times 10^3/8{\cdot}2 \times 10^3$$

$$= \mathbf{1{\cdot}32\ kg}$$

Example 15.4. *A railway engine develops an average power of 1200 kW during a 10-hour run from one station to another. If the engine is driven by an atomic power plant of 20% efficiency, how much U-235 would be consumed on the run? Each U-235 atom on fission releases 180 MeV of energy.*

Solution. Energy consumed by the engine for the run

$$= 1200 \text{ kW} \times 10 \text{ h} = 12{,}000 \text{ kWh}$$

Since $\quad 1 \text{ kWh} = 36 \times 10^5 \text{ J}$

$\therefore \quad$ energy consumed $= 12{,}000 \times 36 \times 10^5 = 432 \times 10^8$ J

Since efficiency of atomic power plant is only 20%, energy required to be produced by it

$$= 432 \times 10^8/0{\cdot}2 = 216 \times 10^9 \text{ J}$$

Energy produced per disintegration of U-235 atom is

$$= 180 \text{ MeV} = 180 \times 1{\cdot}6 \times 10^{-13} \text{ J}$$

Hence, number of U-235 atoms which must disintegrate for producing 216×10^9 J is $= 216 \times 10^9/180 \times 1{\cdot}6 \times 10^{-13} = 7{\cdot}5 \times 10^{31}$

Now, $6{\cdot}02 \times 10^{26}$ atoms are contained in 235 kg of U-235.

Hence, the mass which contains $7{\cdot}5 \times 10^{21}$ atoms is

$$m = \frac{235 \times 7{\cdot}5 \times 10^{21}}{6{\cdot}02 \times 10^{26}} = 2{\cdot}93 \times 10^{-3} \text{ kg}$$

$$= \mathbf{2{\cdot}93\ gram}$$

Example 15.5. *A nuclear reactor produces 200 MW of electric power. The thermal efficiency is 30%. Assuming 200 MeV of energy per U-235 fission, calculate the amount of natural uranium fuel (in kg) consumed per day. Natural uranium contains 0·7% of U-235.*

Solution. $\quad$ Reactor input $= 200/0{\cdot}3 = 666{\cdot}7$ MW

$$= 666{\cdot}7 \times 10^6 \text{ W}$$

Fission rate of U-235 for producing a power of 1 Watt $= 3{\cdot}1 \times 10^{10}$ fissions/second. Hence, fission rate for $666{\cdot}7 \times 10^1$ W is

$$= 3{\cdot}1 \times 10^{10} \times 666{\cdot}7 \times 10^6 = 2{\cdot}0667 \times 10^{19} \text{ fissions/second.}$$

No. of U-235 atoms (or nuclides) which would undergo fission in one day $= 2{\cdot}0667 \times 10^{19} \times (24 \times 3600) = 1{\cdot}786 \times 10^{24}$

Now, 1 kg-atom of U-235 *i.e.* 235 kg of U-235 contains $6{\cdot}025 \times 10^{26}$ atoms, hence

$$\text{mass of } U^{235} \text{ required} = \frac{235 \times 1{\cdot}786 \times 10^{24}}{6{\cdot}025 \times 10^{26}} = 0{\cdot}697 \text{ kg}$$

amount of *natural* uranium consumed/day $= 100 \times 0{\cdot}697/0{\cdot}7 =$ **99·57 kg.**

Example 15.6. *A city requires 100 MW of electrical power on an average. If this is to be supplied by a nuclear reactor of efficiency 20%, using U-235 as the nuclear fuel, calculate the amount of fuel required for one day's operation. Given that energy released per fission of U-235 nuclide = 200 MeV.*

Solution. Energy consumed by the city in one day $= 100 \times 10^6 \times 24 \times 3600 = 864 \times 10^{10}$ J

Since efficiency is 20%, energy required to be produced by the nuclear reactor is

$$= 864 \times 10^{10}/0{\cdot}2 = 432 \times 10^{11}\ \text{J}$$

$$\text{Energy released/atom} = 200 \times 1{\cdot}6 \times 10^{-13} = 32 \times 10^{-12}\ \text{J}$$

$$\therefore \quad \text{No. of atoms to be fissioned} = \frac{432 \times 10^{11}}{32 \times 10^{-12}} = 13{\cdot}5 \times 10^{23}$$

Now, $6{\cdot}02 \times 10^{26}$ atoms are contained in 235 kg of U-235, hence $13{\cdot}5 \times 10^{23}$ atoms are contained in a mass

$$m = \frac{235 \times 13{\cdot}5 \times 10^{23}}{6{\cdot}02 \times 10^{26}} = 530 \times 10^{-3}\ \text{kg} = \mathbf{0{\cdot}53\ kg.}$$

Example 15.7. *Bombay requires 3000 MWh of electric energy per day. It is to be supplied by a reactor which converts nuclear energy into electric energy with an efficiency of 20 per cent. If reactor uses nuclear fuel of U^{235}, calculate the mass of U^{235} needed for one day's operation.*

Solution.

$$\text{Efficiency} = \frac{\text{electrical output}}{\text{nuclear energy input}}$$

$$\therefore \quad \text{nuclear energy input} = \frac{3000}{0{\cdot}20} = 15{,}000\ \text{MWh}$$

Now,

$$1\ \text{MWh} = 36 \times 10^{8}\ \text{J}$$

$$\therefore \quad \text{Nuclear energy input per day} = 15{,}000 \times 36 \times 10^{8} = 5{\cdot}4 \times 10^{13}\ \text{J}$$

Now, energy released per fission of U^{235} nuclide

$$= 200\ \text{MeV} = 200 \times 1{\cdot}6 \times 10^{-13}\ \text{J}$$

$$= 3{\cdot}2 \times 10^{-11}\ \text{J}$$

$$\therefore \text{No. of nuclides required per day} = \frac{5{\cdot}4 \times 10^{13}}{3{\cdot}2 \times 10^{-11}} = 1{\cdot}7 \times 10^{24}$$

Since 235 kg of U^{235} contains $6{\cdot}02 \times 10^{26}$ number of atoms, the mass of U^{235} required for $1{\cdot}7 \times 10^{24}$ atoms is

$$= 235 \times 1{\cdot}7 \times 10^{24}/6{\cdot}02 \times 10^{26} = \mathbf{0{\cdot}660\ kg.}$$

Hence, the daily fuel requirement of Bombay is 0·660 kg or **660 gram.**

Example 15.8. *The motors of an atomic ice breaker deliver 32.824 kW. Calculate the fuel consumption of reactor per day if its efficiency is 20 per cent. Average fission energy release of U^{235} nuclide is 200 MeV.*

What would be the daily amount of 7×10^{3} kcal/kg coal needed to obtain the same power if the efficiency now is 80 per cent?

Solution.

$$\text{Power output} = 32{,}824{,}000\ \text{W or J/s}$$

$$\text{Daily output} = 32{,}824{,}000 \times 24 \times 36000 = 28{\cdot}36 \times 10^{11}\ \text{J}$$

Since efficiency is 20 per cent, daily nuclear energy input

$$= 28{\cdot}36 \times 10^{11}/0{\cdot}2 = 14{\cdot}18 \times 10^{12}\ \text{J}$$

Now, as seen from Ex. 15.2, 1 kg of U^{235} provides $8{\cdot}2 \times 10^{13}$ J of energy.

$$\therefore \quad \text{daily fuel consumption} = \frac{14{\cdot}18 \times 10^{12}}{8{\cdot}2 \times 10^{13}} = \mathbf{173\ gram}$$

With an efficiency of 80%, daily energy input is $= 28{\cdot}36 \times 10^{11}/0{\cdot}8$ J

Since, $1 \text{ kcal} = 4{,}186 \text{ J}$

$$\text{Daily input} = 28{\cdot}36 \times 10^{11}/0{\cdot}8 \times 4186 \text{ kcal}$$

$$\text{Coal reqd/day} = \frac{28{\cdot}36 \times 10^{11}}{0{\cdot}8 \times 4186 \times 7 \times 10^{3}}$$

$$= \mathbf{12{,}100 \text{ tonnes}}$$

TUTORIAL PROBLEMS

1. How many kilowatt-hour of energy will be released by the complete fission of 1 gm of U-235? Given that the energy released per fission is 200 MeV.
2. How much energy will be produced on complete fission of 1 g of U^{235} (235·044) assuming that only 0·1 per cent of its mass is completely converted into energy? **[219 MeV]**

HIGHLIGHTS

1. Energy released by complete fission of 1 kg of U^{235} is
$$= 8{\cdot}2 \times 10^{13} \text{ J/kg}$$
$$= 8{\cdot}2 \times 10^{10} \text{ J/g}$$
2. Rate of nuclear fission of U^{235} necessary to produce a power of one watt = $3{\cdot}1 \times 10^{10}$ fissions/second.
3. A chain reaction is that fission process in which number of neutrons keeps on multiplying during fission till whole of the fissile material is disintegrated.
4. The multiplying factor or reproduction factor of a chain reaction is
$$k = \frac{\text{No. of neutrons in any particular generation}}{\text{No. of neutrons in the preceding generation}}$$
5. Chain reaction will continue if $k \geq 1$.
6. Critical size of a fissile material is that for which production of neutrons by fission is equal to their loss by leakage and non-fission capture. For this size, $k = 1$.
7. For supercritical size $k > 1$ and for sub-critical size, $k < 1$.
8. A chain reaction in *natural* uranium is possible only if moderator is used whose purpose is to thermalise the fast fission neutrons.
9. Power of a nuclear reactor is
$$P = 3{\cdot}2 \times 10^{-11} \, \Phi\sigma NV \text{ watt}$$
10. The four-factor formula for multiplication factor is
$$k_{\infty} = \eta \in p f$$
11. Essentials of a nuclear reactor are :
(*i*) reactor core (*ii*) control mechanism (*iii*) moderator
(*iv*) coolant (*v*) measuring instruments (*vi*) shielding.

Can You Answer This ?

1. Explain with illustrations the working of a nuclear reactor bringing out clearly the role of moderator in its working.
2. What is chain reaction in nuclear fission? Discuss the factors on which the chain reaction depends.
3. Write a short note on nuclear reactor.
4. (*a*) What are the essential components of a nuclear reactor? Describe the function of each of these components.

(*b*) What do you understand by the multiplication factor of a nuclear chain-reacting system? Discuss the factors on which it depends.

5. Write a short note on nuclear reactor.
6. Describe the working of a nuclear reactor explaining the function of each of the components.
7. (*a*) What do you understand by the multiplication factor of a thermal reactor? On what factors does it depend? What is its importance in relation to the working of a reactor ?
   (*b*) Explain the function of a moderator and coolant in an atomic reactor with suitable examples.
8. Write a note on atomic reactor.
9. (*a*) For an atomic reactor, explain the terms : fuel, moderator and coolant giving suitable examples.
   (*b*) Describe the design and working of an atomic power reactor. Draw diagrams.
10. (*a*) Outline the technical difficulties in designing a controlled nuclear fusion reactor.
    (*b*) What is the role played by a moderator in a nuclear fission reactor ?
11. Write a short note on atomic reactors in India.
12. Write a short note on atomic power reactor.
13. (*a*) Explain how a controlled chain reaction is achieved in an atomic reactor.
    (*b*) How are radioactive isotopes produced in an atomic reactor ?

PARTICLE ACCELERATORS 16

Introduction — Van de Graaff Generator — Linear Accelerator or Lineac — Cyclotron — Synchrocyclotron or Frequency-modulated Cyclotron — Electron Synchrotron — Proton Synchrotron — Highlights.

16·1. Introduction

The purpose of a particle accelerator is to accelerate and hence to increase the kinetic energy of a charged particle so that it can be used as a powerful projectile or missile (or bullet) to bombard various substances and thereby produce nuclear disintegrations and transmutations of all kinds. In fact, accelerators have become the basic tools in the rapidly growing branch of nuclear science known as high-energy particle Physics. Nuclear reactions carried out with particles of very high energy have not only led to the discovery of new particles (like antiprotons, antineutrons, hyperons and K-mesons etc.) but have also yielded valuable information about the complex problem of nuclear structure and nuclear forces which hold the nucleons together in a nucleus.

The design and construction of accelerators depends not only on electrical engineering, electronics and vacuum techniques but also on precise mechanical engineering before accurately-collimated beams of charged particles can be obtained for nuclear bombardment experiments.

Present-day accelerators may be divided into the following three classes :

(*i*) static, (*ii*) linear and (*iii*) circular.

(*a*) **Static Accelerator or Statitron**

An accelerator of this type still in use is Van de Graaff generator which can give particle energies up to 12 MeV. It is used for accelerating charged particles like electrons, protons, deuterons and α-particles.

(*b*) **Linear Accelerator or Lineac**

It can accelerate electrons and protons up to maximum energies of 40 GeV and 50 MeV respectively.

(*c*) **Circular Accelerators**

These may be further categorised as under :

(*i*) **Lawrence Cyclotron or Fixed-frequency Cyclotron**

It is used for accelerating particles like protons, deuterons and α-particles (up to 40 MeV).

(*ii*) **Synchrocyclotron or Frequency-modulated Cyclotron**

It is a modification of the simple Lawrence cyclotron and gives comparatively much higher particle energies.

(*iii*) **Betatron**

It is used for accelerating *electrons* and gives energies up to 300 MeV.

(*iv*) **Electron Cyclotron**

It is a modulated version of betatron and gives electron energies up to 6 GeV.

(*v*) **Proton Cyclotron**

In principle, it is similar to electron cyclotron and produces protons of 10 GeV energy.

(*vi*) **Alternating Gradient Synchrotron (AGS)**

It is a proton accelerator and provides very high proton energies. The alternating-gradient synchrotron at Serpukhov near Moscow in USSR has been designed to produce 70 GeV protons. Another AGS nearing completion at the small town Weston, near Chicago, Illinois, is expected to produce 200 GeV protons although planning is underway for a 1000 GeV accelerator of this type.

(*vii*) **Omnitron**

It is an accelerator which gives strong beams of accelerated particles of all masses from hydrogen (proton) to uranium and beyond. Such accelerators, still under development in U.S.A., are expected to give peak energies of 500 MeV per nucleon for light elements up to argon decreasing to 300 MeV per nucleon for uranium. The beam intensities are expected to reach as high a value as 10^{12} particles per second.

16·2. Van de Graaff Generator

The purpose of this accelerator is to produce a parallel and homogeneous beam of charged particles like electrons, protons and α-particles etc. It is a combination of an electrostatic generator and an evacuated accelerator tube. It was developed in 1931 by Van de Graaff at Princeton University, USA.

Design Principle

The design of this machine is based on the principle that if a charged conductor is brought into an *internal* contact with a hollow sphere, then all of its charge is transferred to the hollow sphere no matter how high the potential of the latter may be. For this purpose, the advantage is taken of the fact that there is no electric field *inside* a hollow charged conductor. Provided insulation difficulties do not crop up, the charge and hence the potential of a hollow sphere could be raised to any desired value by successively adding charges to it by internal contact. However, the hollow sphere must be supported by non-conducting supports and, moreover, its maximum potential is limited to that at which the rate of leakage of charge from it equals the rate at which charge is delivered to it.

Construction and Working

A schematic diagram of this accelerator is shown in Fig. 16.1. An endless belt *B* made of silk or rayon or some other non-conducting material is given positive charge by a 20 kV d.c. set through a row of needles or spray points located at *C*. Because of the intense electric field around the stray points, ions are formed. The negative ions are attracted towards the needles, whereas the positive ions are repelled towards the belt. Consequently, the positive ions attach themselves to the flexible belt which is nearly 3 dm in width. The positive charge is carried upwards by the belt towards the hollow metallic shell

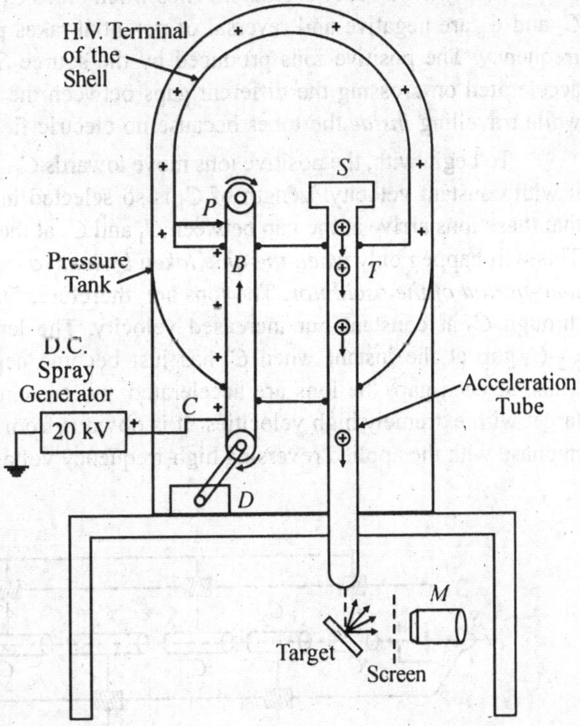

Fig. 16.1

that forms the upper terminal. The charge of the belt is transferred through another set of spray points at P to the internal surface of the shell from where it immediately goes to its outer surface. As the charge transfer process continues, the potential of the shell increases till it reaches a value when the electric charge leaks away as fast as it is collected from the moving belt. Leakage can be minimised and the maximum attainable voltage increased by enclosing the apparatus in an air-tight steel pressure tank containing nitrogen, methane or freon at about 15 atmospheric pressure.

A gas discharge tube which supplies the positive ions for being accelerated and afterwards for being used as projectiles, is located inside the shell at point S. These ions are accelerated down the evacuated tube T through the potential of the shell which may be as high as 20 million volts. These accelerated ions finally hit the target placed below.

The Van de Graaff generator can be used for accelerating electrons by reversing the polarity of the spray voltage and by using a hot filament for thermionic electrons instead of the ion chamber.

16·3. Linear Accelerator or Lineac

As its name indicates, in a linear accelerator, the charged particles move in a straight line through a series of coaxial cylindrical hollow conductors or *drift* tubes. These particles are accelerated by a high-frequency oscillating electric field which is applied to these drift tubes and whose frequency is such as to be in resonance with the motion of these particles. In other words, *the principle of synchronous acceleration is employed in a lineac.*

Construction and Working

As shown in Fig. 16.2, the lineac consists of a set of drift tubes of *increasing* lengths arranged linearly in a glass vacuum chamber. Alternate tubes are connected together, the odd-numbered tubes being joined to one terminal and the even-numbered ones to the other terminal of the high-frequency power supply *i.e.* an R.F. oscillator. Thus when tubes C_1, C_3 and C_5 are positive, the other tubes C_2, C_4 and C_6 are negative and reversal of potential takes place periodically according to the applied frequency. The positive ions produced by the source S travel along the axis of the tube and are accelerated on crossing the different gaps between the drift tubes. These ions are not accelerated while travelling *inside* the tubes because no electric field exists there.

To begin with, the positive ions move towards C_1 when it is negative. Then they travel through it with constant velocity. Length of C_1 is so selected in relation to the frequency of the oscillator that these ions arrive at the gap between C_1 and C_2 at the instant when C_2 has just become negative. This will happen only *when the time taken by ions to travel through C_1 is exactly equal to half the time-period of the oscillator*. The ions are, therefore, further accelerated across this gap and travel through C_2 at constant but increased velocity. The length of C_2 is such that they arrive at the C_2-C_3 gap at the instant when C_3 has just become negative so that they are again accelerated. Thus, at each gap, the ions are accelerated and are emitted from the final drift tube towards the target with extremely high velocities. It is obvious from the above discussion that the ions are kept in phase with the applied reversing high-frequency voltage in order that they gain additional energy

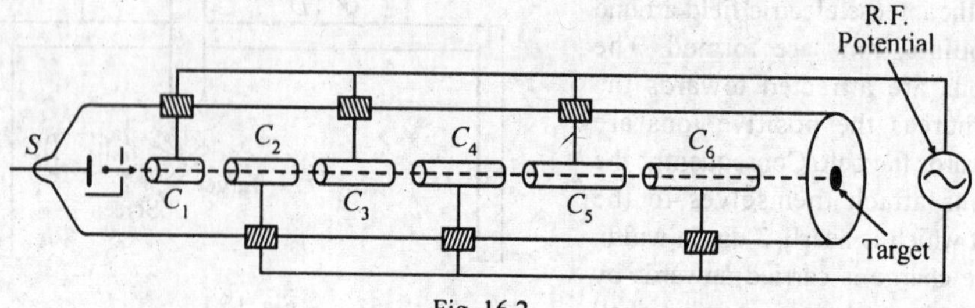

Fig. 16.2

every time they pass from one tube to another. If n is the number of tubes, V_m is the maximum voltage of the oscillator, then energy acquired by an ion of charge Q is given by

$$nQV_m = \frac{1}{2}mv_n^2 \qquad \text{...(i)}$$

where m is the mass of the ion and v_n its velocity while *entering* the nth tube.

Now, time required to travel through any tube is *equal to half the time-period of the applied a.c. voltage*, hence

$$t = \frac{T}{2} = \frac{1}{2f} \qquad \text{...(ii)}$$

If l_n is the length of the nth tube and v_n the velocity of the ion while travelling through it, then

$$t = \frac{l_n}{v_n}$$

$$\frac{l_n}{v_n} = \frac{1}{2f} \quad \text{or} \quad l_n = \frac{v_n}{2f} \qquad \text{...(iii)}$$

Substituting the value of v_n From Eq. (*i*) above, we get

$$l_n = \sqrt{\frac{nQV_m}{2mf^2}}$$

where f is the oscillator frequency.

With the help of oscillators having frequencies as high as 3000 MHz, it has been possible to design linear accelerators giving electron energies up to 40 GeV* and proton energies up to 70 MeV.

As compared to cyclotron (Art. 16.4), lineacs have the following advantages :—

(*i*) they are more economical when very high energies are required.

(*ii*) they produce a well-collimated beam of charged particles automatically whereas ejection of these particles is one of the main problems in circular machines.

However, the main disadvantage of lineac lies in the fact that an ion, instead of passing through the *same* gap over and over again through the *same* alternating field as in a cyclotron, has to pass through a number of alternating fields and a number of a.c. sources across many different gaps. This multiplicity of a.c. sources and fields requires many pieces of high-frequency apparatus which increases both the complications and the cost. Moreover, for a very long tube, it becomes extremely difficult to keep the alternating field and the ions in phase with each other.

16·4. Cyclotron

It is also known as Lawrence cyclotron (after the name of its inventor) or *fixed-frequency* cyclotron and employs the principle of synchronous acceleration in conjunction with *magnetic* field. Under the influence of this vertical magnetic field, the charged particles move round and round in circular orbits and are repeatedly accelerated across the *same gap* with the help of a horizontal *electric* field. This field need not be very strong because it is utilized over and over again.

Construction and Working. As shown in Fig. 16.3 (*a*), the heart of the cyclotron consists of two hollow flat semicircular (or cylindrical) metal boxes D_1 and D_2 (called 'dees' because of their shape). The two 'dees' are separated by a narrow parallel air-gap and are mounted inside a vacuum chamber between the poles of a powerful electromagnet. The two 'dees' are connected to a high-frequency oscillator which establishes an alternating electric field in the air-gap. Consequently D_1 and D_2 become alternately positive and negative at the same rate as the frequency of the oscillator.

*This is expected to be the output of the electron lineac recently built at Stanford.

When a trace of hydrogen is admitted into the evacuated chamber, the hot filament located at S between the 'dees' ionizes some of the hydrogen atoms thereby producing protons to be used as atomic bullets. These protons are accelerated by the electric field in the gap towards that 'dee' which happens to be *negative* at that instant. Suppose it is D_2. Due to the presence of an intense magnetic field, the protons travel along a circular path through D_2. Since there is no electric field *inside* the 'dees', protons travel with constant velocity along circles of constant radius while passing through them. If by the time the protons emerge from D_2, the polarity of the applied potential is reversed so that D_1 becomes negative and D_2 positive, the protons would be again accelerated while crossing the gap. Since their velocity would also be increased, they will now move through D_1 along a circular arc of *greater* radius as shown. If their time of travel through D_1 is equal to *half the time-period of the oscillator voltage*, the protons while coming out of D_1 will find the field reversed again and so would be once again accelerated while crossing the D_1-D_2 gap. In this way, as the potential of the 'dees' reverses periodically, the protons travel faster and faster, moving in ever-expanding circles until they reach the outer edge of the dees where they are deflected out through the window W by means of an auxiliary electric field and strike the target.

It would be realized that the fundamental principle on which the cyclotron works is that *the time required for a charged particle to make one complete turn within the 'dees' is the same for all the speeds and is equal to the time-period of the oscillator*. The faster a particle travels, the larger is the circle it must travel, thus keeping the time constant. This results in a stream of protons emerging from the window W.

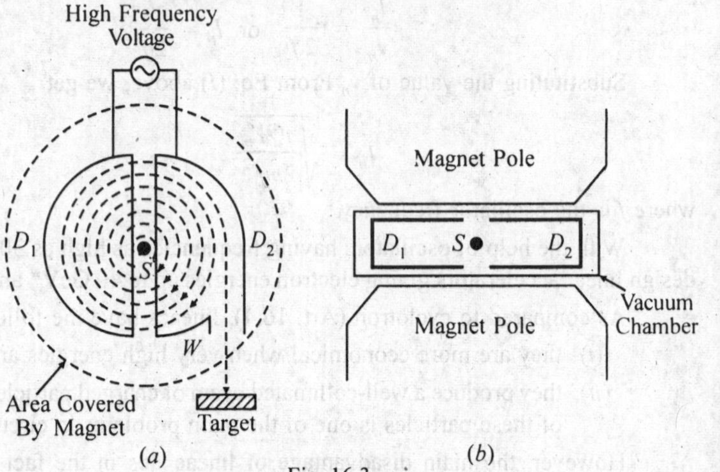

Fig. 16.3

Theory of cyclotron. When a particle of mass m and charge Q moves with a velocity of v in a magnetic field of flux density B, then force experienced by it is BQv (Art. 1.12). If it moves in a circle of radius R, then centripetal force acting on it is mv^2/R. Equating the two forces we get

$$mv^2/R = BQv \qquad \text{... }(i)$$

If V is the *equivalent* voltage through which the particle has been accelerated, then

$$QV = \frac{1}{2}mv^2 \qquad \text{... }(ii)$$

(*a*) From Eq. (*i*), we get

$$v = BQR/m$$

It shows that for a constant magnetic field B and for a fixed value of Q/m for a given particle, velocity is directly proportional to radius. It means that time of travel of a particle is the same whether it moves along a circle of larger radius or smaller radius.

(*b*) Eliminating v from Eqs (*i*) and (*ii*), we get the value of the *equivalent* potential as

$$V = \frac{1}{2}\frac{Q}{m} \cdot B^2 R^2$$

(*c*) For resonance between the applied a.c. voltage and the moving particle, the time taken by the particle to travel the circular arc within any of the 'dees' must be equal to *half* time-period of the applied oscillator voltage.

$$\therefore \quad \frac{2\pi R}{v} = \frac{T}{2}$$

$$\therefore \quad T = \frac{2\pi R}{v} = \frac{2\pi R}{BQR/m} = \frac{2\pi}{B} \cdot \frac{m}{Q} \qquad \text{... (iii)}$$

Hence, the orbital frequency (also called the cyclotron resonance frequency) is given by

$$f_0 = \frac{1}{T} = \frac{BQ}{2\pi m} = \frac{1}{\pi R} \cdot \sqrt{\frac{E}{2m}} \text{ Hz}$$

where E is the particle energy in joules.

For particles with given value of Q/m, the value of B for giving resonance with the applied a.c. voltage can be found from the above equation. *Higher the oscillator frequency, stronger is the magnetic field required.* It is customary to keep oscillator frequency fixed and to adjust the value of B till resonance is obtained.

(*d*) It is seen from Eq. (*iii*) above that time-period T is independent of v and R and is the same for all particles.

(*e*) The kinetic energy of the particle is given by

$$E = \frac{1}{2} mv^2 = \frac{1}{2} m \left(\frac{BRQ}{m} \right)^2$$

$$= \frac{1}{2} \cdot \frac{Q^2}{m} \cdot B^2 R^2 = QV$$

where V is equivalent accelerating voltage.

If R_∞ is the radius of the outermost orbit, then maximum energy of the particle at the time of leaving the cyclotron is

$$E_{max} = \frac{1}{2} \cdot \frac{B^2 Q^2 R_\infty{}^2}{m}$$

Now, 1 MeV = $1{\cdot}6 \times 10^{-13}$ J, hence

$$E_{max} = \frac{B^2 Q^2 R_m{}^2}{2m \times 1{\cdot}6 \times 10^{-13}}$$

$$= \frac{3{\cdot}12 \times 10^{12} B^2 Q^2 R_m{}^2}{m} \text{ MeV}$$

The maximum available particle energy from a cyclotron is limited by various factors such as (*i*) the maximum strength of the magnetic field which can be produced, (*ii*) the frequency and power limit of the oscillator and (*iii*) the fact that as particle approaches the velocity of light, the energy acquired on crossing the gap goes to increasing the effective mass rather than the velocity of the particle.

16·5. Synchrocyclotron or Frequency-modulated Cyclotron

The resonance frequency of a cyclotron is

$$f_0 = \frac{BQ}{2\pi m}$$

Due to relativistic increase* in mass with velocity, there is decrease in the orbital frequency of the particle. Hence, the particle gradually gets out of phase with the applied a.c. voltage whose frequency is held constant throughout.

This decrease in particle frequency may be compensated for in the following two ways :—

(*i*) by increasing magnetic flux density B in such a way that B/m remain constant. In this case, the frequency of the high frequency oscillator is kept fixed. Such a machine is called *synchrotron.*

(*ii*) by decreasing the oscillator frequency in step with the decrease in orbital frequency of the particle. In this case, value of B is held constant. Such a machine is called frequency-modulated (FM) cyclotron or synchrocyclotron.

In a synchrocyclotron, the gradual decrease in oscillator frequency is brought about by using a variable capacitor. Such machines have been used to gave α-particle energies up to 720 MeV.

Some of the basic differences between a cyclotron and a synchrotron are summarized below :

(*i*) a cyclotron uses two 'dees' while a synchrocyclotron uses only one.

(*ii*) in a cyclotron, the oscillator frequency is fixed whereas in a synchrotron it is variable.

(*iii*) a cyclotron provides a continuous supply of accelerated particles. In a synchrocyclotron, the particles starting out from the centre are subject to a frequency modulation as they approach the periphery and so come out in bursts of a few hundred per second with each burst lasting about 100 μs.

16·6. Betatron

It is used for speeding up electrons to extremely high energies with the help of an expanding magnetic field and was invented by D.W. Kerst at the University of Ilinois, U.S.A. This device differs from a cyclotron in at least two fundamental respects :

(*i*) firstly, the electrons are accelerated by an expanding magnetic field.

(*ii*) secondly, the circular orbit of the electron has a constant radius.

Construction

It consists of a highly-evacuated annular tube or toroidal chamber placed between the poles of an a.c. electromagnet excited by an alternating current usually of frequency 60 or 180 Hz (Fig. 16.4). The annular tube is often referred to as 'doughnut'. It lies in that part of the air-gap of the electromagnet *where flux density is one-half of that of the main field flux density.*

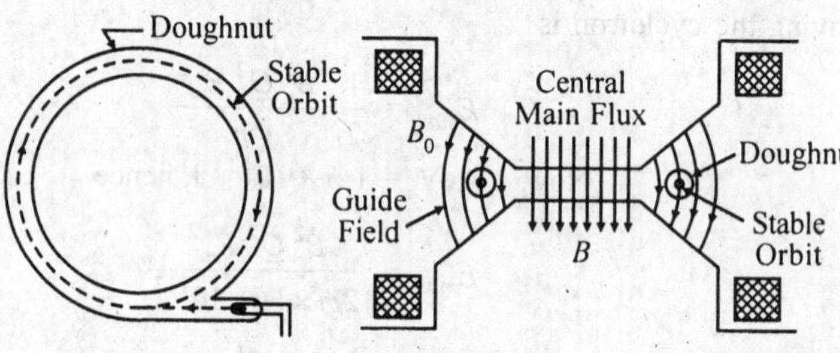

Fig. 16.4

The field traversing the 'doughnut' is known as *guide* or *control field* and its function is to *stabilize* the electrons in an orbit of constant radius in the 'doughnut'.

Bunches of high-velocity electrons from an electron gun are injected into the 'doughnut' at the instants when the main or central flux is rising in one direction *i.e.* at the beginning of each cycle of this flux (Fig. 16.5). This increasing magnetic flux gives rise to a voltage gradient (*i.e.* electric field) round the 'doughnut' which accelerates the orbiting electrons and so increase their energy.

*The mass of the particle at any velocity v is given by

$$m = \frac{m_0}{\sqrt{1-\beta^2}} \text{ where } \beta = \frac{v}{c}$$

If $v = 0{\cdot}8\ c$, then $\beta = 0{\cdot}8$ and $m = 1{\cdot}66\ m_0$.

Principle of Operation

An electron moving in a circular orbit may be looked upon as a ring-shaped conductor. When this conductor is linked by a changing flux, then as per Faraday's Laws of Electromagnetic Induction an e.m.f. is induced in it. The magnitude of the induced voltage per 'turn' is $d\Phi/dt$ volt. If R is the radius of the electron orbit within the 'doughnut' and E the tangential electric force acting on the electron, then

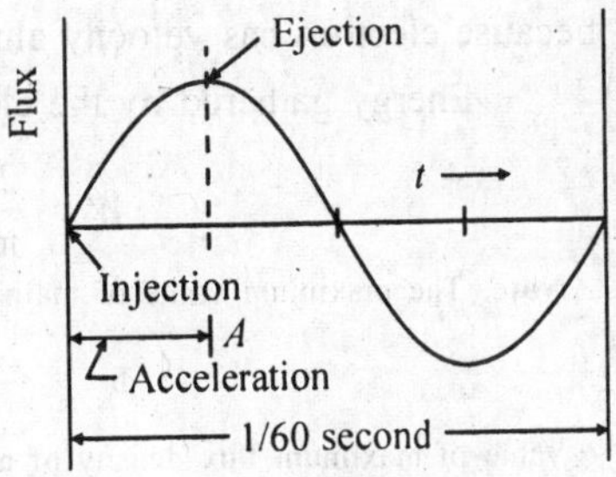

Fig. 16.5

$$E = \frac{\text{induced voltage}}{\text{orbit length}} = \frac{d\Phi / dt}{2\pi R}\ \text{V/m}$$

Under the influence of this electric field, the velocity and hence the energy of each electron increases as it goes round and round the orbit. Since the velocity of the electron increases, its radius will also tend to increase so that electron will spiral outwards. This unstable condition is prevented with the help of the guide field which is located at the 'doughnut' but perpendicular to it. It has been found that for *orbit stability, the flux density B_0 of the guide field should be half the flux density B of the main or central field.* In other words $B_0 = B/2$.

It should be noted that electron velocity increases only during the time the main flux rises from zero to its maximum value *i.e. during one-quarter cycle of the alternating field current.* During this time, the electron makes a number of revolutions* round the 'doughnut'. Since its velocity increases after each revolution, total increase in electron velocity during one quarter-cycle is very high.

Obviously, if the electron were to remain in the 'doughnut' during the second quarter-cycle (when the main flux is decreasing), it would be de-accelerated and would lose its energy. Hence, it has to be removed at the end of the first quarter-cycle itself. It is accomplished by applying a short voltage pulse to an auxiliary system of coils on the electromagnet poles at the end of the quarter-cycle. This causes the electron to spiral away from its orbit and hit a target which produces γ-rays or very penetrating X-rays.

No. of Orbits and Energy

Tangential force acting on the electron is

$$F = eE = \frac{e}{2\pi R}\cdot\frac{d\Phi}{dt}$$

Energy increase during one revolution is

$$W = 2\,\pi RF = \frac{e\,d\Phi}{dt}$$

If the equation of the main alternating flux is $\Phi = \Phi_m \sin \omega t$, then

$$W = e\cdot\frac{d}{dt}(\Phi_m \sin \omega t) = e\,\omega\,\Phi_m \cos \omega t$$

Average increase in energy per revolution is

$$W_{av} = \frac{2}{\pi}\times \text{maximum value} = \frac{2}{\pi}\cdot e\omega\Phi_m$$

Now, acceleration takes place in one quarter-cycle *i.e.* during $1/4f$ second or $\pi/2\omega$ second. If v_{av} is the electron velocity during this period, then distance travelled by it is $\pi \cdot v_{av}/2\omega$.

$$\therefore \qquad \text{No. of orbits} = \frac{\pi\cdot v_{av}}{2\omega}\cdot\frac{1}{2\pi R} = \frac{v_{av}}{4\omega R} \simeq \frac{c}{4\omega R}$$

*If the circumference of the 'doughnut' is about 3 m and electron has a velocity of 0.96 c, then the orbital frequency is $= v/2\pi R = 0{\cdot}96 \times 3 \times 10^8/3 = 96$ MHz. If the freqeuncy of the magnetic field is 60 Hz, then time taken for the first quarter-cycle = 1/240 second. Hence, number of orbits executed by electron per quarter = $96 \times 10^6/240 = 4 \times 10^5$.

because electron has velocity almost equal to the velocity of light.

Energy gathered by the electron during the quarter-cycle is

$$W = \frac{2}{\pi} \cdot \Phi_m = e\omega \times \frac{v_{av}}{4\omega R} = \frac{e\,v_{av} \cdot \Phi_m}{2\pi R} = \frac{e\,c\,\Phi_m}{2\pi R}$$

Note. The maximum value of main flux is

$$\Phi_m = \frac{2\pi r W}{ec}$$

Value of maximum flux density of main field is

$$B = \frac{\Phi_m}{\pi R^2} = \frac{2\pi R W}{e.c.\,\pi R^2} = \frac{2W}{ecR}$$

Maximum flux density of the guide field is

$$B_0 = \frac{B}{2} = \frac{2W}{2 \cdot ecR} = \frac{W}{ecR}$$

16·7. Electron Synchrotron

It is also known by the general name of 'bevatron' because it has been used to accelerate electrons up to energies of 6 GeV. In this machine, modulation is provided as explained for the synchrocyclotron. This is achieved by varying the flux density of the magnetic field used for accelerating the electron. This machine is otherwise similar to betatron except that pole pieces are annular and follow the outline of the doughnut as shown in Fig. 16.6. Some soft-iron flux bars placed in the central gap serve as the central core of the magnet to start up the machine as a normal betatron. Part of the 'doughnut' is covered with silver or copper to give a resonant cavity C which is connected to a vert high-frequency oscillator. When the oscillator is on, the electron is accelerated each time it passes through the resonator.

Primary electrons of about 100 keV energy are injected into the doughnut at the right time. They travel in circular paths and increase their energy as in a betatron. When the flux bars become magnetically saturated (at about 2 MeV), they can have no further effect and the betatron action ceases. Then resonant cavity action comes into operation. If oscillator frequency is so adjusted as to keep the electrons in phase, then these electrons receive increments of energy in each revolution. When they reach their maximum energy the oscillator is switched off. These high-energy electrons then strike the target which gives off short-wavelength X-rays.

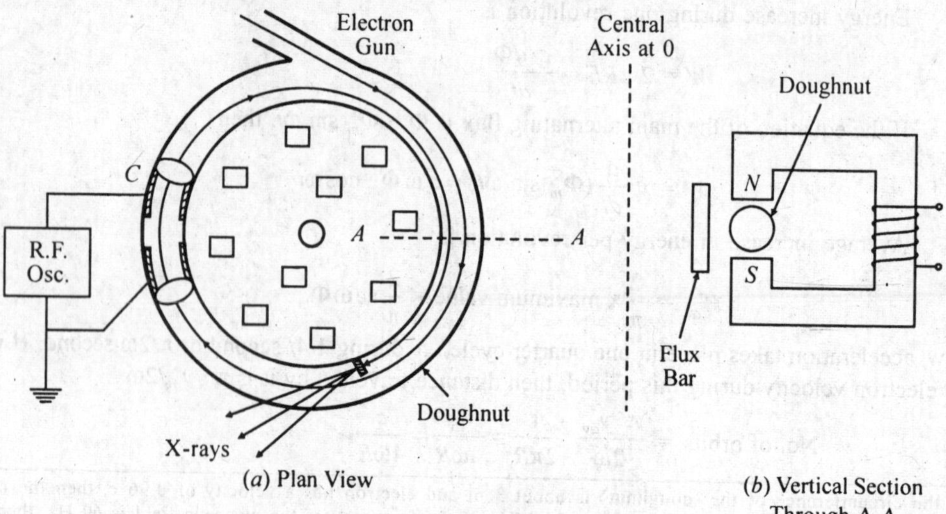

(a) Plan View (b) Vertical Section Through A-A

Fig. 16.6

16·8. Proton Synchrotron

In principle, it is similar to the electron synchrotron described above and gives protons up to 10 GeV energy. As shown in Fig. 16.7, ring-shaped magnet is used which is in four quadrants. Protons travel in the annular doughnut with constant radius. Protons are injected into the 'doughnut' from a linear accelerator or Van de Graaff machine and recovered by magnetic deflection as a pulsed beam after many revolutions. A high-frequency resonator cavity accelerator is used in one of the straight parts and its frequency is increased in step with the increase in the speed of protons. The field strength of the magnets is also increased to maintain the accelerated protons in a circular path of constant radius. When protons reach their maximum energy, the oscillator frequency is distorted so that the orbit expands or contracts and the protons strike the target.

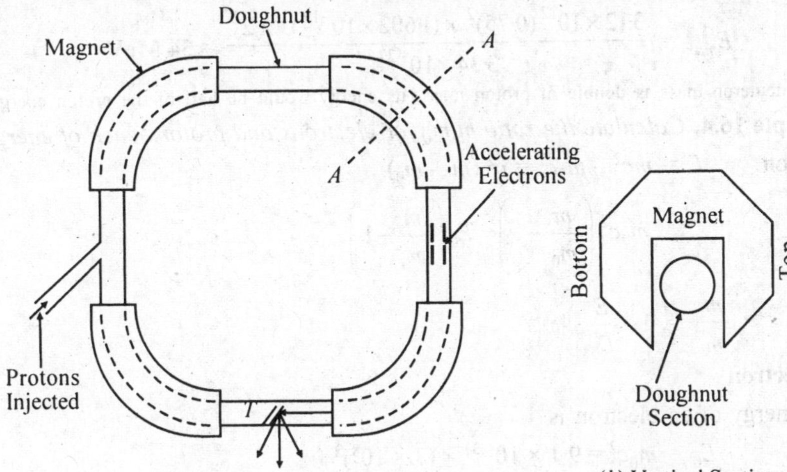

(a) Plan View (b) Vertical Section through A-A

Fig. 16.7

Example 16.1. *If the frequency of the oscillator potential applied to the dees of a cyclotron is 9 MHz, what must be the magnetic flux density B to accelerate α-particles? Take mass of an α-particle as 6·643 × 10^−27 kg.*

Solution. The orbital resonance frequency of the particle is given by

$$f_0 = \frac{BQ}{2\pi m} \text{ Hz}$$

Now for an α-particle, $m = 6{\cdot}643 \times 10^{-27}$ kg, $Q = 2 \times$ electron charge $= 2 \times 1{\cdot}602 \times 10^{-19}$ C $= 3{\cdot}204 \times 10^{-19}$ C. Also $f_0 = 9 \times 10^6$ Hz

$$\therefore \quad 9 \times 10^6 = \frac{B \times 3{\cdot}204 \times 10^{-19}}{2\pi \times 6{\cdot}643 \times 10^{-27}}$$

$$\therefore \quad B = \mathbf{1{\cdot}17\ Wb/m^2}$$

Example 16.2. *A cyclotron in which the flux density is 0·7 Wb/m² is employed to accelerate deuterons. What should be the frequency of the alternating potential to be applied between the dees, for the cyclotron resonance condition.*

Mass of deuteron = 3·34 × 10^−27 kg, charge Q = 1·6 × 10^−19 C

Solution. $f_0 = BQ/2\pi m$ Hz

Here, $B = 0{\cdot}7$ Wb/m², $Q = 1{\cdot}6 \times 10^{-19}$ C, $m = 3{\cdot}34 \times 10^{-27}$ kg

$$\therefore \quad f_0 = 0{\cdot}7 \times 1{\cdot}6 \times 10^{-19}/2\pi \times 3{\cdot}34 \times 10^{-27} = \mathbf{5{\cdot}3\ MHz}$$

Example 16.3. *A cyclotron with dees of radius 2 metres has a magnetic field of 0.75 Wb/m². Calculate the maximum energies to which (a) protons and (b) deuterons can be accelerated.*

For proton : $m = 1{\cdot}67 \times 10^{-27}$ kg, $Q = 1{\cdot}602 \times 10^{-19}$ C

For deuteron : $m = 3{\cdot}34 \times 10^{-27}$ kg, $Q = 1{\cdot}602 \times 10^{-19}$ C

Solution. (*a*) **Proton**

$$E_{max} = 3{\cdot}12 \times 10^{12}\, B^2Q^2R_m^{\ 2}/m$$

$$= \frac{3{\cdot}12\times10^{12}(0{\cdot}75)^2\times(1{\cdot}602\times10^{-19})^2\times2^2}{1{\cdot}67\times10^{-27}} = \mathbf{108\ MeV}$$

(*b*) **Deuteron**

$$E_{max} = \frac{3{\cdot}12\times10^{12}(0{\cdot}75)^2\times(1{\cdot}602\times10^{-19})^2\times2^2}{3{\cdot}34\times10^{-27}} = \mathbf{54\ MeV}$$

Note. Since deuteron mass is double of proton mass, its energy would be half of the proton energy.

Example 16.4. *Calculate the ratio m/m_0 for electrons and protons each of energy 1 MeV.*

Solution. $E = mc^2 - m_0c^2 = c^2(m - m_0)$

$$= m_0c^2\left(\frac{m}{m_0}-1\right) = E_0\left(\frac{m}{m_0}-1\right)$$

$$\therefore \quad \frac{m}{m_0} = 1+\frac{E}{E_0}$$

(*a*) **Electron**

Rest energy of an electron is

$$E_0 = m_0c^2 = 9{\cdot}1\times10^{-31}\times(3\times10^8)^2\ \text{J}$$

$$\therefore \quad E_0 = 9{\cdot}1\times10^{-31}\times9\times10^{16}/1{\cdot}6\times10^{-13} = 0{\cdot}51\ \text{MeV}$$

$$\frac{m}{m_0} = 1+\frac{1}{0{\cdot}51} = \mathbf{2{\cdot}96}$$

(*b*) **Proton**

Rest energy, $E_0 = 1{\cdot}67\times10^{-27}\times(3\times10^8)^2$ J

$$= \frac{1{\cdot}67\times10^{-27}\times9\times10^{16}}{1{\cdot}67\times10^{-13}} = 938\ \text{MeV}$$

$$\therefore \quad \frac{m}{m_0} = 1+\frac{1}{938} = \mathbf{1{\cdot}001067}$$

The above result shows that whereas electron must be treated relativistically, protons of this energy need not be.

Example 16.5. *In a certain betatron, the maximum magnetic field at the electron orbit is 0·5 Wb/m². The diameter of the stable orbit is 1·5 m. If the frequency of the alternating current through electromagnet coils is 59 Hz, calculate for the electrons*

(*i*) *average energy gained per revolution*

(*ii*) *final energy.*

Solution. It should be kept in mind that electron velocity in betatron is nearly equal to velocity of light *c*. Since the acceleration time is equal to the quarter time-period *i.e.* $1/4f$ or $\pi/2\omega$ second, total distance travelled by the electron is $= c \times \pi/2\omega$. If *R* is the radius of the stable orbit, the number of revolutions made is

$$N = \frac{c\pi/2\omega}{2\pi R} = \frac{e}{4\omega R}$$

$$= \frac{3\times10^6}{4\times(2\pi\times50)\times1{\cdot}5} = 3{\cdot}1\times10^5$$

Now, electrons have to be treated relativistically, hence the momentum of an electron = E/c.

$$\therefore \quad mv = E/c$$

Also $$\frac{mv^2}{R} = Bev$$

$$\therefore \quad mv = BeR \text{ or } E/c = BeR$$

$$\therefore \quad E = BeRc$$

(*ii*) $$E = \frac{0{\cdot}5\times1{\cdot}6\times10^{-19}\times0{\cdot}75\times3\times10^8}{1{\cdot}6\times10^{-13}} \text{ MeV}$$

$$= \mathbf{112{\cdot}5\ MeV}$$

(*i*) The average energy per revolution is

$$= 112{\cdot}5\times10^4/3{\cdot}1\times10^5 = \mathbf{363\ eV}$$

Example 16.6. *Calculate* (*i*) mass and (*ii*) *velocity of electrons with a kinetic energy of 0·51 MeV. If a beam of such electrons enters a uniform magnetic field, find the flux density required to cause them to move in a circular orbit of radius 15 cm. The rest mass of electron is 9·1 × 10^{-31} kg and charge 1·6 × 10^{-19} C.*

Solution.

(*i*) $$\frac{m}{m_0} = 1+\frac{E}{E_0}$$

Now, rest energy of an electron is 0·51 MeV.

$$\therefore \quad \frac{m}{m_0} = 1+\frac{0{\cdot}51}{0{\cdot}51} = 2$$

$$\therefore \quad m = 2m_0 = 2\times9{\cdot}1\times10^{-31}$$

$$= \mathbf{18{\cdot}2\times10^{-31}\ kg.}$$

(*ii*) The mass at any speed is given by

$$m = \frac{m_0}{\sqrt{1-\beta^2}} \qquad \text{where } \beta = v/c$$

$$\therefore \quad \sqrt{1-\beta^2} = m_0/m = 1/2$$

or $$1-\beta^2 = 1/4 \text{ or } \beta^2 = 3/4 \text{ or } \beta = 0{\cdot}866$$

$$\therefore \quad v = 0{\cdot}866\,c = 0{\cdot}866\times3\times10^8$$

$$= \mathbf{2{\cdot}6\times10^8\ m/s}$$

Now, $$\frac{mv^2}{R} = Bev$$

$$\therefore \quad B = mv/eR$$

$$= 9{\cdot}1\times10^{-31}\times2{\cdot}6\times10^8/1{\cdot}6\times10^{-19}\times0{\cdot}15$$

$$= \mathbf{0{\cdot}00985\ Wb/m^2.}$$

Example 16.7. *Determine the frequency of a generator feeding a cyclotron that accelerates deuterons up to 4 MeV when the maximum radius of the particle orbit is 75 cm. Mass of deuteron = 3·34 × 10⁻²⁷ kg.*

Solution. As shown in Art. 16.4,

$$f_0 = \frac{1}{\pi R} \cdot \sqrt{\frac{E}{2m}} \text{ Hz}$$

$$= \frac{1}{\pi \times 0{\cdot}75} \cdot \sqrt{\frac{4 \times 1{\cdot}6 \times 10^{-13}}{2 \times 3{\cdot}34 \times 10^{-27}}} = \mathbf{4{\cdot}15 \times 10^6\ Hz}$$

Example 16.8. *The magnetic flux within a stable orbit in a betatron changes at a constant rate of 15 Wb/s. What would be the energy of an electron which undergoes 10^6 revolutions?*

Solution. Increase of energy per revolution

$$= e \cdot \frac{d\Phi}{dt} \text{ joules} = \frac{d\Phi}{dt} \text{ electron-volt}$$

$$= 15 \text{ eV} = 15 \times 10^{-6} \text{ MeV}$$

Final energy of the electron after 10^6 revolutions = $15 \times 10^{-6} \times 10^6$ = **15 MeV**

Example 16.9. *Find the maximum energy and the corresponding wavelength of γ-rays produced by suddenly stopping in matter electrons which have been accelerated in a betatron under the following conditions :*

Period of acceleration = 4×10^{-4} s, orbit radius = 10 cm, the flux within the orbit increases from zero at a constant rate of 15 Wb/s.

Solution. It will be assumed that the electrons are moving with the velocity of light. Number of revolutions made in one second

$$= \frac{c}{2\pi R} = \frac{3 \times 10^8}{2\pi \times 0{\cdot}1}$$

No. of revolutions made in 4×10^{-4} second

$$= \frac{3 \times 10^8 \times 4 \times 10^{-4}}{0{\cdot}2\pi} = 1{\cdot}91 \times 10^5$$

Energy increase/revolution = $e\, d\Phi/dt$ joules = $d\Phi/dt$ eV = 15 eV

∴ Maximum energy = $1{\cdot}91 \times 10^5 \times 15 \times 10^{-6}$ = **2·865 MeV**

Now, $\gamma = \dfrac{h}{p}$ —Art. 9.2

Also, $p = E/c = 2{\cdot}865 \times 1{\cdot}6 \times 10^{-13}/3 \times 10^8$ kg-m/s

$h = 6{\cdot}62 \times 10^{-34}$ J-s

$$\therefore \quad \gamma = \frac{6{\cdot}62 \times 10^{-34} \times 3 \times 10^8}{2{\cdot}865 \times 1{\cdot}6 \times 10^{-13}} = \mathbf{4{\cdot}33 \times 10^{-13}\ m}$$

TUTORIAL PROBLEMS

1. A cyclotron in which flux density is 1·4 Wb/m² is employed to accelerate protons. What should be the frequency of the alternating field applied to the dees ? Mass of proton = $1{\cdot}67 \times 10^{-27}$ kg and charge = $1{\cdot}6 \times 10^{-19}$ C. **[42·7 MHz]**
2. Calculate the frequency of the alternating potential that must be applied to the cyclotron dees in which deuterons are accelerated, given that magnetic flux density is 3 Wb.m², mass of deuteron = $3{\cdot}32 \times 10^{-27}$ kg and charge = $1{\cdot}6 \times 10^{-19}$ C. **[23 MHz]**

3. Protons are accelerated in a cyclotron with dees of radius slightly greater than 32 cm and magnetic field of flux density 0·65 Wb/m^2. Calculate (*i*) the velocity of the protons, (*ii*) the energy in MeV and (*iii*) frequency of the alternating potential between the dees. Mass of proton = 1·66 × 10^{-27} kg.
[(*i*) 2 × 10^7 m/s; (*ii*) 2·07 MeV; (*iii*) 10 MHz]

4. Deuterons in a cyclotron describe a circle of 32 cm just before emerging from the dees, the frequency of the applied a.c. voltage being 10 MHz. Find (*i*) flux density of the magnetic field and (*ii*) the velocity of deuterons emerging out of the cyclotron. Mass of a deuteron = 3·32 × 10^{-27} kg and charge = 1·6 × 10^{-19} C. **[(*i*) 1·3 Wb/m^2; (*ii*) 2 × 10^7 m/s]**

5. In a 70-MeV betatron, the radius of the stable orbit is 28 cm. Calculate (*i*) the angular velocity of the electron, (*ii*) the frequency of the applied electric field and (*iii*) the maximum value of the magnetic flux density at the orbit for this energy. Assume electron velocity equal to velocity of light in vacuum.
Hint. See note of Art. 16.6. **[(*i*) 1·07 × 10^7 rad/s; (*ii*) 1·7 × 10^8 Hz; (*iii*) 0·833 Wb/m^2]**

6. A cyclotron with a radius of the dee as 50 cm is used to accelerate deuterons. The frequency of oscillator is 2 MHz. Calculate their energy as they emerge.
Mass of deuteron = 3·32 × 10^{-27} kg, charge = 1·67 × 10^{-19} C. **[0·41 MeV]**

HIGHLIGHTS

1. In a linear accelerator, the length of the *n*th drift tube is given by

$$l_n = \sqrt{\frac{nQV_m}{2mf^2}}$$

where *f* is the oscillator frequency.

2. In a cyclotron, following relations hold good

(*i*) $$\frac{1}{2}mv^2 = QV$$

where *V* is the *equivalent* voltage through which particle is accelerated.

(*ii*) $$mv^2/R = BQv$$

(*iii*) $$T = \frac{2\pi}{B}\cdot\frac{m}{Q}$$

(*iv*) $$f_0 = \frac{BQ}{2\pi m} = \frac{1}{\pi R}\cdot\sqrt{\frac{E}{2m}}$$ where E is the particle energy in joules.

(*v*) $$E = \frac{B^2Q^2R^2}{2m}$$

(*vi*) $$E_{max} = B^2Q^2R_m{}^2/2m = 3{\cdot}12 \times 10^{12}\, B^2Q^2R_m{}^2/m \text{ MeV.}$$

3. In a betatron, where electron moves with nearly the velocity of light, following relations hold good :

(*i*) Energy increase during one revolution = $e \cdot d\Phi/dt$

(*ii*) No. of orbits made by the electron during one quarter-cycle of time-period is

$$= \frac{c}{4\omega R}$$

(*iii*) Energy gathered by the electron during one quarter

$$= \frac{ec\Phi_m}{2\pi R}$$

17 ULTRASONICS

Ultrasonic Waves — Magnetostriction Effect — Magnetostriction Oscillator — Piezoelectric Effect — Piezoelectric Oscillator — Detection of Ultrasonic Waves — Applications of Ultrasonic Waves — Highlights — Can You Answer This ?

17·1. Ultrasonic Waves

These are the sound waves whose frequency lies above the audible limit of about 20 kHz*. Ultrasonics must be differentiated from supersonics which is the study of effects produced by objects that travel through a medium at a faster speed than the waves they generate. Supersonics is essentially confined to planes and missiles which fly through air at a speed greater than the speed of sound in air and hence give rise to shock waves.

Ultrasonics are usually produced by the application of (*i*) magnetostriction effect and (*ii*) piezoelectric effect.

17·2. Magnetostriction Effect

This phenomenon was discovered by Joule in 1847. It is found that when a rod of a ferromagnetic material is placed in a steady magnetic field parallel to its length, the rod undergoes a small change in length. If the steady field is replaced by an alternating magnetic field, the rod undergoes alternate contractions and extensions at a frequency twice the frequency of the applied magnetic field**. In other words, the rod is thrown into longitudinal vibrations thereby producing ultrasonic waves in the surrounding medium. The range of frequency depends on the dimensions of the magnetostrictive material and extends up to 3×10^5 Hz.

The longitudinal vibrations thus produced are exactly like those produced by a rod which is clamped at the mid-point but has both ends free. The frequency of vibration of such a rod is

$$f = \frac{1}{2l}\sqrt{\frac{E}{\rho}}$$

where E is Young's modulus, l the length of the rod and ρ its density.

17·3. Magnetostriction Oscillator

It generates ultrasonic waves and is based on the magnetostriction effect. It was first designed by G.W. Pierce and is schematically shown in Fig. 17.1. The iron rod is clamped in the middle and two coils L_1 and L_2 are wound round its two halves. Coil L_1 is included in the plate-cathode circuit and L_2 in the grid-cathode circuit of a triode oscillator valve and the two are inductively coupled. The rod gets magnetised by the plate current passing through coil L_1. Any change in the plate current results in the change of magnetisation of the rod which causes a change in its length. This change in length gives rise to change of flux linked with the grid coil L_2 thereby inducing an e.m.f. in it. This e.m.f. is amplified and passed on to the plate circuit so that oscillations are maintained. The frequency of oscillations is

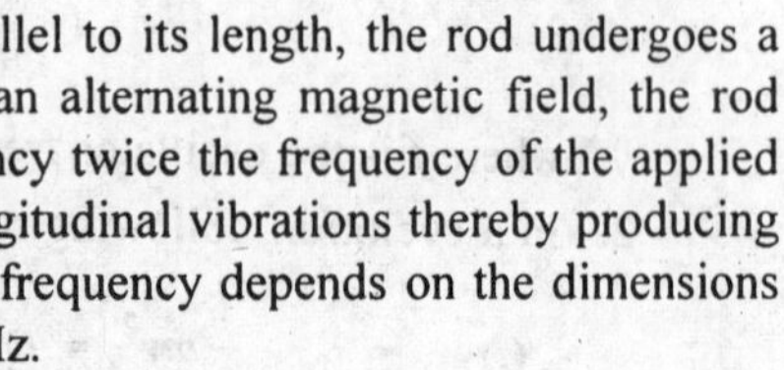

Fig. 17.1

*Those sound waves whose frequency lies below the audible limit are known as infrasonic waves. Their frequency is less than 20 Hz.

**It is due to this effect that the iron cores of the transformers start humming however rigidly they may be made.

controlled by the variable capacitor *C*. If this frequency matches the natural frequency of the longitudinal vibrations of the rod, then oscillations are vigorous because of the resonance effect. The variations in the deflection of the microammeter will give definite indication of the resonance.

17·4. Piezoelectric Effect

As early as 1880, the two brothers J and P Curie discovered that thin slices of certain crystals like quartz, tourmaline and Rochelle salt develop a potential difference across their opposite faces when subjected to distortion such as squeezing, twisting or bending. The magnitude of the p.d. developed is proportional to the extent of the deformation produced and its polarity reverses, when the direction of the deforming force is reversed. For example, if the voltage produced by the crystal is positive when twisted clockwise, then it would be negative when twisted anticlockwise. This phenomenon is known as *piezoelectric effect*.

Later on, the converse of the above phenomenon was also found to be true. It was discovered that if a p.d. is applied across the two faces of the crystal, a mechanical contraction or expansion is produced in the crystal. By changing the polarity of the applied voltage, a contraction changes into expansion and *vice versa*. Obviously, if an alternating voltage is applied across a slice of the crystal, it will alternately contract and expand and hence start vibrating. If the frequency of the applied alternating voltage happens to be equal to that of the crystal, the latter will be thrown into resonant vibrations and hence produce ultrasonic waves.

Natural quartz crystal has the shape of a hexagonal prism with a pyramid attached to each end. As shown in Fig. 17.2, the line joining the end points of these pyramids is called the optic axis or *Z*-axis. The three lines which pass through the opposite corners of the crystal constitute its three *X*-axes and are also known as *electrical* axes. Similarly, the three lines which are *perpendicular* to the sides of the hexagon form the three *Y*-axes of the crystal and are also known as *mechanical* axes. Thin plates of the quartz crystal cut perpendicular to *Z*-axes are known as *X*-cut plates whereas those cut perpendicular to *Y*-axes are known as *Y*-cut plates.

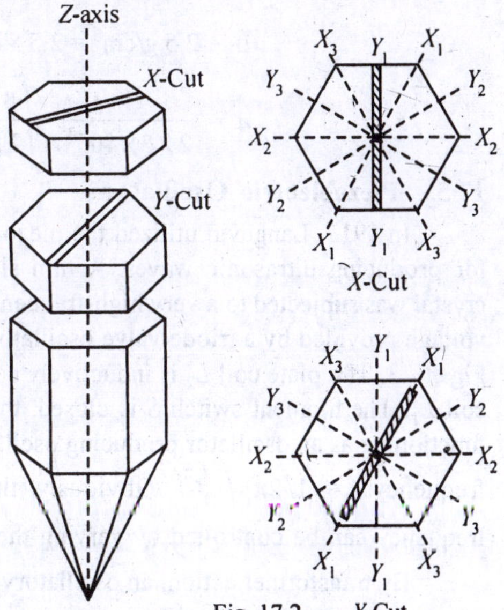

Fig. 17.2

Suppose there is an *X*-cut crystal plate of thickness *t* and length *l* (along optic axis). When an alternating voltage is applied across the faces of this plate along the electrical axis, then alternating stresses and strains are set up both in its thickness and length. If, as said above, the frequency of the alternating voltage happens to be equal to the natural frequency of vibration of the plate, then resonance will occur resulting in a large amplitude of vibration.

The frequency of the thickness vibration (*i.e.* those along *X*-axis) is

$$n = \frac{p}{2t} \cdot \sqrt{\frac{E}{\rho}}$$

Similarly, the frequency of the length vibrations is given by

$$n = \frac{p}{2l} \cdot \sqrt{\frac{E}{\rho}}$$

where $p = 1, 2, 3$.....etc. for fundamental, first overtone and second overtone respectively, E is Young's modulus along the appropriate direction and ρ the density of the crystal plate.

If, $E = 8 \times 10^{10}$ N/m², $\rho = 2{\cdot}654 \times 10^3$ kg/m³

$t = 2 \times 10^{-8}$ m $\quad p = 1$

then fundamental frequency of thickness vibrations is

$$n = \frac{1}{2 \times 2 \times 10^{-3}} \sqrt{\frac{8 \times 10^{10}}{2{\cdot}654 \times 10^3}} = 4{\cdot}34 \times 10^5 \text{ Hz}$$

Example 17.1. *Calculate the frequency of the fundamental note emitted by piezoelectric crystal. Use the following data :*

Vibrating length = 3 mm ; Young's modulus = 8 × 10¹⁰ N/m² and density of crystal = 2·5 g/cm³.

Solution. As seen from Art. 17.4, the fundamental frequency of length vibration is given by

$$n = \frac{p}{2l} \cdot \sqrt{\frac{E}{\rho}}$$

Here, $l = 3 \times 10^{-3}$ m ; $E = 8 \times 10^{10}$ N/m²

$\rho = 2{\cdot}5$ g/cm³ $= 2{\cdot}5 \times 10^3$ kg/m³

$$\therefore \quad n = \frac{1}{2 \times 3 \times 10^{-3}} \cdot \sqrt{\frac{8 \times 10^{10}}{2{\cdot}5 \times 10^3}} = \mathbf{0{\cdot}943 \times 10^4 \text{ Hz}}$$

17·5. Piezoelectric Oscillator

In 1917, Langevin utilized the piezoelectric effect for producing ultrasonic waves. A thin slice of quartz crystal was subjected to a very high-frequency alternating voltage provided by a triode valve oscillator as shown in Fig. 17.3. The plate coil L_2 is inductively coupled to grid coil L_1. The moment switch S is closed, the valve starts functioning as an oscillator producing oscillations having frequency $f = 1/2\pi\sqrt{L_1 C}$. Obviously, the oscillation frequency can be controlled by varying the capacitor C.

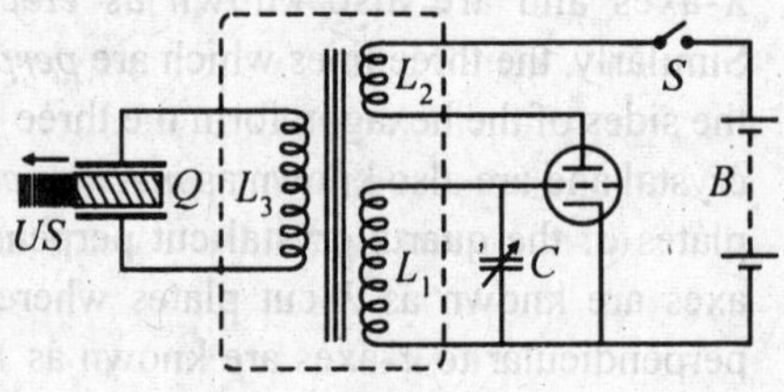

Fig. 17.3

By transformer action, an oscillatory e.m.f. is induced in coil L_3. The two ends of the coil L_3 are joined to two metallic plates which enclose a thin slice of quartz or tourmaline crystal. Capacitor C is varied till frequency of oscillation matches with the natural frequency of vibration of the crystal slice. The vibrating crystal slice produces the required ultrasonic waves in the surrounding air. Frequencies as high as 500 MHz have been produced by making a tourmaline crystal slice vibrate in higher mode of vibration.

17·6. Detection of Ultrasonic Waves

Since such waves cannot be detected by human ear, following indirect methods are used for the purpose :—

(*i*) Kundt's Tube Method

This method may be used provided the wavelength is not *very small*. As in the case of ordinary sound waves, in this case also, stationary supersonic waves are produced in air contained in a long glass tube supported horizontally. The lycopodium powder spread along the surface of the tube collects in small heaps at the nodes and is blown off at the antinodes. The calculation of wavelength is done in the same way as for sound waves.

(*ii*) **Acoustic Diffraction Method**

This method is based on the fact that ultrasonic waves which consist of alternating compressions and rarefactions change the density of the medium through which they pass. This results in the change of refractive index of the liquid which causes the diffraction of light.

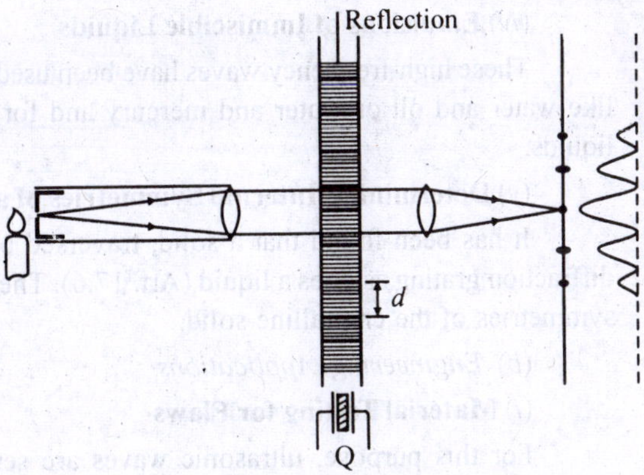

Fig. 17.4

As shown in Fig. 17.4 stationary ultrasonic waves are produced in a liquid contained in a glass tube. The liquid density and hence refractive index is maximum at nodal points and minimum at antinodal points. Consequently, nodal areas act as opacities and antinodal regions as transparencies and the whole liquid column acts as an *acoustic grating*. If a beam of monochromatic light is passed through this grating, a diffraction pattern, as shown in Fig. 17.4, is produced. It consists of a central maxima flanked by first-order and second-order maxima etc. If the distance between two consecutive nodal or antinodal regions in the liquid *i.e.* the grating element, is *d*, then

$$d \sin \theta = n\lambda$$

where θ is the angle for *n*th-order maxima, *n* the order of the maxima and λ the wavelength of light. Since other things are measurable or known, *d* can be found. Also $d = \lambda'/2$ or $\lambda' = 2d$ where λ' is the wavelength of the ultrasonic waves. The frequency *f* of the ultrasonics is known from the frequency of the oscillator. Hence, velocity of these waves can be found from the relation

$$v = f\lambda'$$

17·7. Applications of Ultrasonic Waves

Ultrasonic waves have found many industrial, scientific and medical uses.

(*a*) *Scientific Applications*

(*i*) **Ultrasonic Signalling**

Ultrasonic waves of 40 kHz frequency can be obtained in the form of a narrow beam which can travel many kilometres in water before being absorbed. Hence, they have been found very suitable for direction signalling in submarines and for talking from one vessel to another. The quartz vibrations are modulated by the speech frequencies in the same manner as in radio transmission.

(*ii*) **Depth Sounding and Submarine Detection**

Such applications use the echo sounding technique. For depth sounding, a beam of ultrasonic waves is directed towards the bottom of the sea from where it is reflected back to a suitable receiver. The time taken by the waves to go and come back is recorded accurately. If the velocity of ultrasonic waves in water is known, the sea depth can be calculated easily.

The same method is used for finding the distance and direction of a submarine or a shoal of fish in the sea. The whole system is known as *sonar* which stands for *sound navigation and ranging*.

(*iii*) **Photographic Emulsions**

Ultrasonic waves have been used to improve the homogeneity and stability of photographic emulsions. Their colour sensitivity has also been increased by thorough dispersion of the dye in the emulsion under the influence of ultrasonic waves.

(*iv*) **Emulsions of Immiscible Liquids**

These high-frequency waves have been used for forming stable emulsions of immiscible liquids like water and oil or water and mercury and for accelerating the crystallisation of supersaturated liquids.

(*v*) **Determining Internal Symmetries of a Solid**

It has been found that a solid, traversed by ultrasonic waves, acts like a three-dimensional diffraction grating as does a liquid (Art. 17.6). The diffraction patterns so obtained reveal the internal symmetries of the crystalline solid.

(*b*) *Engineering Applications*

(*i*) **Material Testing for Flaws**

For this purpose, ultrasonic waves are sent through the metal to be tested and either the duration of the emitted sound is noted against the standard specimen, or the resulting reflection or absorption of waves is measured. A noticeable reflection or absorption of the waves occurs at any blow hole or crack in the interior of the metal specimen. It is so because, in either case, there is change of medium and this results in the reflection of waves both from the blow hole and the back surface of the metal. This time interval between two reflections is recorded by an oscilloscope and is utilised for locating the position of the flaw in the metal. Cracks having thickness as small as 0·005 mm can be detected in this way. The detection of such flaws in metals prior to their failure while in use is of great practical importance.

(*ii*) **Soldering**

Metals have been soldered or brazed by subjecting them to ultrasonic vibrations.

(*iii*) **Drilling**

Ultrasonics have also been used for drilling and metal cutting purposes.

(*iv*) **Metallurgical Use**

Ultrasonics have been put to many metallurgical uses. Many metals such as iron-lead, aluminium-cadmium and zinc-lead etc. which cannot be alloyed in their liquid state can be brought together when subjected to high-intensity ultrasonic waves.

(*c*) *Medical and Biological Applications*

(*i*) **Biological Effect**

It has been found that smaller animals like frog and fish are either maimed or killed outright when exposed to ultrasonic waves because of the destruction of their blood vessels. In the case of micro-organisms like bacteria etc. these waves either promote or retard their agglutination and virulence.

(*ii*) **Disease Treatment**

Ultrasonic therapy has been used to treat diseases such as bursitis, abscesses and lumbago etc.

(*iii*) **Surgical Use**

Surgical use of ultrasonic waves includes the selective cutting of the tissue during an operation.

(*iv*) **Locating Abnormal Growths**

Cerebral ventricles are explored by ultrasonic waves for locating abnormal growths in the brain. For this purpose, the waves are passed through the brain and then picked up by means of separate transmitters and receivers. A special equipment is used to examine the different acoustic losses from different parts of the head both by reflection and absorption in it.

(*v*) **Dental Cutting**

Ultrasonic waves have been found very useful for dental cutting because (*i*) they make the cutting almost painless, (*ii*) they cut the hard material very easily and (*iii*) they do not require any mechanical device for cutting purpose.

(*vi*) **Cleaning and Washing**

Recently, ultrasonic waves have been used industrially for large-scale washing of clothes especially silken fabrics in hospitals, factories and textile industry. Other uses include cleaning of utensils, removing dust and soot from chimney gases, dispersing of fog at airports for enhanced visibility and removing water from paper during its manufacture.

HIGHLIGHTS

1. Ultrasonic waves are those whose frequency lies above the audible limit of about 20 kHz.
2. These waves are usually produced by utilizing either the magnetostriction effect or piezoelectric effect.
3. Such waves can be detected indirectly either by using Kundt's tube method or more generally, using acoustic diffraction method.
4. Ultrasonic waves find a wide variety of applications in scientific research, industry and medicine.
   (*i*) Scientific applications include direction signalling in submarines and depth sounding, improving homogeneity and stability of photographic emulsions.
   (*ii*) Engineering applications include detection of flaws in metals, soldering, drilling and formation of binary alloys in metallurgy.
   (*iii*) Medical uses include treatment and cure of special diseases, for surgical cutting, locating abnormal growths in the brain, dental cutting and cleaning and washing in hospitals and hotels etc.

Can You Answer This ?

1. Write a short note on ultrasonic testing method.
2. Write a note on ultrasonics and their uses in industry.
3. Write a note on ultrasonic testing.
4. Describe in detail any method of producing ultrasonics.
5. Write a short note on production of ultrasonics.
6. Enumerate the different methods of producing ultrasonic waves and describe any one method in detail.
7. Write a note on ultrasonics.
8. (*a*) What is Piezoelectric effect? How is this effect employed for production of ultrasonics?
   (*b*) Write briefly on four uses of ultrasonics in industry, science and medicine.
9. Write a short note on production and detection of ultrasonic waves.
10. (*a*) Explain what are ultrasonic waves. How can they be produced and detected?
    (*b*) Describe any two methods which employ ultrasonic waves in testing or research.
11. Write a short note on Piezoelectric effect.

APPENDIX A

DIFFRACTION GRATING

It is based on the phenomenon of diffraction of light and consists of a very large number of extremely narrow parallel slits separated by equal opaque spaces. The diffraction gratings are of two types — *reflection* type and *transmission* type.

The earliest transmission gratings consisted of fine wires placed very close side by side at regular intervals. Afterwards, Rowland produced *transmission* gratings by ruling fine lines extremely close together on the surface of a glass plate by using a fine sharp diamond point. In such gratings, the diamond scratches act as opaque spaces and the transparent portions act as slits. In this way, Rowland succeeded in producing very accurate gratings containing as many as 1000 lines/mm !

THEORY : The following figure represents the section of a grating whose slits are perpendicular to the plane of the paper. Only five slits have been shown although an actual grating contains about 1000 lines/mm of its width. Let the width of each slit be 'a' and distance between two slits (or width of opaque portion) be 'b'. The distance $(a + b)$ is known as '*grating element*' or '*grating space*'. Points like A and C, C and K etc. which are separated by a distance $(a + b)$ are known as *corresponding* points. Obviously, the total number of silts in the grating is $N = l/(a + b)$.

Let a parallel beam of light (or plane wavefront) of wavelength λ be incident on the grating. Since slits are narrow, light is diffracted by each one of them. The diffracted rays interfere with one another to produce the final diffraction pattern which consists of

(*i*) widely-spaced maxima (known as primary maxima such as M_0, M_1, M_2 etc.)

(*ii*) and a large number of faint subsidiary maxima and minima (such as those at points P, P etc.) in between the primary maxima.

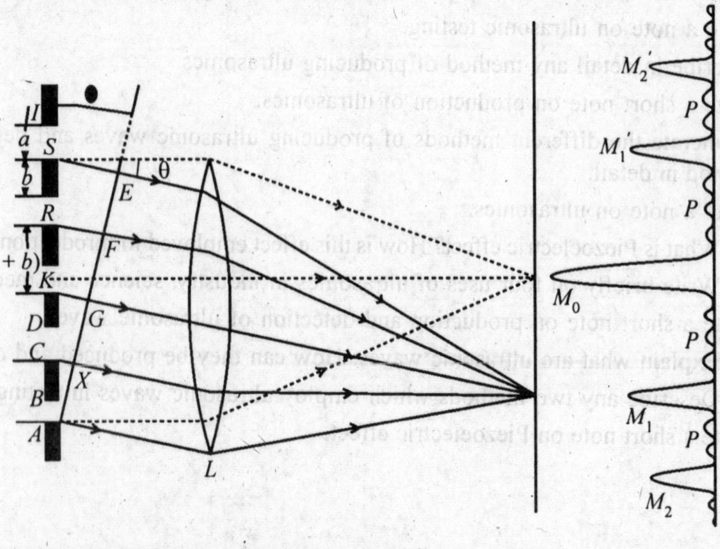

Let us consider the formation of primary maxima.

Most of the incident light proceeds straight and when collected by lens L produces a strong bright image at M_0. It is known as *central diffraction maxima.* Next consider the light diffracted in a direction making an angle θ with that of the incident beam. From point A, draw a normal on line SE. It is obvious that there is a path difference between the rays starting from various slits and reaching point M. Suppose that angle θ is such that $CX = \lambda$. In that case, $KG = 2\lambda$, $RF = 3\lambda$ and $SE = 4\lambda$. Obviously, waves from corresponding points (like A and C) would reinforce each other and so produce a first-order primary maxima at M_1.

Similarly, if angle θ is increased to such a value that $CX = 2\lambda$, then $KG = 4\lambda$, $RF = 6\lambda$ and $SE = 8\lambda$ and so on, again waves from corresponding points reinforce each other and produce second-order primary maxima at point M_2. In general, we can say that when CX equals $n\lambda$, we get nth order primary maxima.

Hence, condition for primary maxima is

$$CX = n\lambda$$

or

$$(a + b) \sin \theta_n = n\lambda$$

where θ_n is the angle of direction in which nth primary maxima is situated and $n = 0, 1, 2, 3, ...$ for zero-order, 1st-order, 2nd-order and 3rd-order maxima etc.

APPENDIX B

VALUES OF PHYSICAL CONSTANTS

Avogadro's number $N = 6.02 \times 10^{23}$ per g-mol or g-atom

$= 6.02 \times 10^{26}$ per kg-mol or kg-atom

Velocity of light in vacuum $c = 3 \times 10^8$ m/s

Electronic charge $e = 1.602 \times 10^{-19}$ C

Planck's constant $h = 6.625 \times 10^{-34}$ J-s

Rydberg constant $R_\infty = 1.097 \times 10^7 \text{ m}^{-1}$

Rest masses :

neutron $= 1.675 \times 10^{-27}$ kg

$= 1.0087$ amu

$= 939.6$ MeV

proton $= 1.67 \times 10^{-27}$ kg

$= 1.0073$ amu

$= 938.3$ MeV

electron $= 9.11 \times 13^{-31}$ kg

$= 5.487 \times 10^{-4}$ amu

$= 0.511$ MeV

Atomic mass unit (amu) $= \dfrac{1}{N} = 1.66 \times 10^{-27}$ kg

$= 931$ MeV

e/m ratio for electron $= 1.759 \times 10^{11}$ C/kg

APPENDIX C

TABLE OF ELEMENTS

Element	*Symbol*	*Atomic number* Z	*Atomic weight* A	*Element*	*Symbol*	*Atomic number* Z	*Atomic weight* A
Hydrogen	H	1	1.008	Germanium	Ge	32	72.60
Helium	He	2	4.033	Arsenic	As	33	74.91
Lithium	Li	3	6.940	Selenium	Se	34	78.96
Beryllium	Be	4	9.013	Bromine	Br	35	79.916
Boron	B	5	10.82	Krypton	Kr	36	83.80
Carbon	C	6	22.011	Rubidium	Rb	37	85.48
Nitrogen	N	7	14.008	Strontium	Sr	38	87.63
Oxygen	O	8	16.000	Yttrium	Y	39	88.92
Fluorine	F	9	19.00	Zirconium	Zr	40	91.22
Neon	Ne	10	20.183	Niobium	Nb	41	92.91
Sodium	Na	11	22.991	Molybdenum	Mo	42	95.95
Magnesium	Mg	12	24.32	Technetium	Te	43	99
Aluminium	Al	13	26.98	Ruthenium	Ru	44	101.1
Silicon	Si	14	28.09	Rhodium	Rh	45	102.91
Phosphorus	P	15	30.975	Palladium	Pd	46	106.7
Sulphur	S	16	32.066	Silver	Ag	47	107.88
Chlorine	Cl	17	36.467	Cadmium	Cd	48	112.41
Argon	Ar	18	39.944	Indium	In	49	114.20
Potassium	K	19	39.100	Tin	Sn	50	118.70
Calcium	Ca	20	40.08	Antimony	Sb	51	121.76
Scandium	Sc	21	44.96	Tellurium	Te	52	127.61
Titanium	Ti	22	47.90	Iodine	I	53	126.91
Vanadium	V	23	50.95	Xenon	Xe	54	131.3
Chromium	Cr	24	52.01	Caesium	Cs	55	132.91
Manganese	Mn	25	54.94	Barium	Ba	56	137.36
Iron	Fe	26	55.85	Lanthanum	La	57	138.92
Cobalt	Co	27	58.94	Cerium	Ce	58	140.13
Nickel	Ni	28	58.89	Praseodymium	Pr	59	140.92
Copper	Cu	29	63.64	Neodymium	Nd	60	144.27
Zinc	Zn	30	65.38	Promethium	Pm	61	145
Gallium	Ga	31	89.72	Samarium	Sm	62	150.43

Element	*Symbol*	*Atomic number* Z	*Atomic weight* A
Europium	Eu	63	152.0
Gadolinium	Gd	64	156.9
Terbium	Tb	65	158.39
Dysprosium	Dy	66	162.46
Holmium	Ho	67	164.94
Erbium	Er	68	167.2
Thulium	Tm	69	168.94
Ytterbium	Yb	70	173.04
Lutecium (Lutetium)	Lu	71	174.99
Hafnium	Hf	72	178.6
Tantalum	Ta	73	180.95
Tungsten (Wolfram)	W	74	183.92
Rhenium	Re	75	186.31
Osmium	Os	76	190.2

Element	*Symbol*	*Atomic number* Z	*Atomic weight* A
Iridium	Ir	77	192.2
Platinum	Pt	78	195.23
Gold	Au	79	197.0
Mercury	Hg	80	200.61
Thallium	Tl	81	204.39
Lead	Pb	82	207.21
Bismuth	Bi	83	209.00
Polonium	Po	84	210
Astatine	At	85	210
Radon	Rn	86	222
Francium	Fr	87	223
Radium	Ra	88	226.05
Actinium	Ac	89	227
Thorium	Th	90	232.05
Protactinium	Pa	91	231
Uranium	U	92	238.07

TRANSURANIC ELEMENTS

Element	*Symbol*	*Atomic number* Z	*Range of mass number*
Neptunium	Np	93	229 – 242
Plutonium	Pu	94	232 – 246
Americium	Am	95	236 – 248
Curium	Cm	96	236 – 246
Berkelium	Bk	97	240 – 250
Californium	Cf	98	242 – 253
Einsteinium	Es	99	244 – 254
Fermium	Em	100	246 – 255
Mendelevium	Md	101	250 – 256
Nobelium	No	102	253 – 255
Lawrencium	Lr	103	257
Kurchatovium (Rutherfordium)	Rf	104	261
Dubnium	Db	105	262

Element	*Symbol*	*Atomic number* Z	*Range of mass number*
Seaborgium	Sg	106	263
Bohrium	Bh	107	262
Hassium	Hs	108	265
Meitnerium	Mt	109	266
Darmstadtium	Ds	110	262
Roentgenium	Rg	111	272
Ununbium	Uub	112	277
Ununtrium	Uut	113	—
Ununquadium	Uuq	114	289
Ununpentium	Uup	115	—
Ununhexium	Uuh	116	—
Ununseptium	Uus	117	—
Ununoctium	Uuo	118	—

INDEX

Q

R

S